Handbook of Environmental Engineering

Volume 19

Series Editors

Lawrence K. Wang
PhD, Rutgers University, New Brunswick, NJ, USA
MS, University of Rhode Island, Kingston, RI, USA
MSCE, University of Missouri, Rolla, MO, USA
BSCE, National Cheng Kung University, Tainan, Taiwan, ROC

Mu-Hao Sung Wang
PhD, Rutgers University, New Brunswick, NJ, USA
MS, University of Rhode Island, Kingston, RI, USA
BSCE, National Cheng Kung University, Tainan, Taiwan, ROC

The past 30 years have seen the emergence of a growing desire worldwide to take positive actions to restore and protect the environment from the degrading effects of all forms of pollution: air, noise, solid waste, and water. The principle intention of the Handbook of Environmental Engineering (HEE) series is to help readers formulate answers to the fundamental questions facing pollution in the modern era, mainly, how serious is pollution and is the technology needed to abate it not only available, but feasible. Cutting-edge and highly practical, HEE offers educators, students, and engineers a strong grounding in the principles of Environmental Engineering, as well as providing effective methods for developing optimal abatement technologies at costs that are fully justified by the degree of abatement achieved. With an emphasis on using the Best Available Technologies, the authors of these volumes present the necessary engineering protocols derived from the fundamental principles of chemistry, physics, and mathematics, making these volumes a must have for environmental pollution researchers.

More information about this series at http://www.springer.com/series/7645

Lawrence K. Wang • Mu-Hao Sung Wang
Yung-Tse Hung • Nazih K. Shammas
Editors

Environmental and Natural Resources Engineering

Springer

Editors
Lawrence K. Wang
Rensselaer Polytechnic Institute
Troy, NY, USA
Lenox Institute of Water Technology
Newtonville, NY, USA

Mu-Hao Sung Wang
Lenox Institute of Water Technology
Newtonville, NY, USA
Krofta Engineering Corporation
Lenix, MA, USA

Yung-Tse Hung
Department of Civil and Environmental Engineering, Cleveland State University
Cleveland, OH, USA

Nazih K. Shammas
Lenox Institute of Water Technology
Newtonville, NY, USA
Krofta Engineering Corporation
Lenix, MA, USA

ISSN 2512-1359 ISSN 2512-1472 (electronic)
Handbook of Environmental Engineering
ISBN 978-3-030-54625-0 ISBN 978-3-030-54626-7 (eBook)
https://doi.org/10.1007/978-3-030-54626-7

This Springer imprint is published by the registered company Springer Nature Switzerland AG
The registered company address is: Gewerbestrasse 11, 6330 Cham, Switzerland

The Handbook of Environmental Engineering (HEE), Volume 19, Environmental and Natural Resources Engineering, *is published in memory of both Dr. Nazih K. Shammas (Left) and Dr. Donald B. Aulenbach (Right), who contributed significantly as the authors/ coeditors the HEE series and both were environmental humanitarian engineering professors at the Lenox Institute of Water Technology (LIWT), USA, for two decades. Dr. Aulenbach also taught at the Rensselaer Polytechnic Institute (RPI), USA. Their former colleagues of HEE, LIWT, and RPI and their friends have jointly contributed to this HEE*

Volume 19 to salute and forever remember these two outstanding scholars.

An Alternative Poem of "Thousand Winds"
For: Environmental Scholars gone with Winds like beautiful autumn leaves
By: Lawrence K. Wang and Mu-Hao Sung Wang

Do not stand at my grave and weep
I am not there, I do not sleep
I am in a thousand winds that blow
I am the fresh air up there
I am the fertile soil on land
I am the swimming fish in water

When you find me in the autumn's hush
I am in the graceful rush
Of beautiful leaves in circling current
I am the scholar with passion
I am in the homes with love
I am on the Internet
I am the leaves that fly

I am in each lovely thing

Do not stand at my grave and cry
I am not there, I do not die

Note: Original "Thousand Winds" author was Mary Elizabeth Frve.

Preface

The past 75 years have seen the emergence of a growing desire worldwide that positive actions be taken to restore and protect the environment from the degrading effects of all forms of pollution – air, water, soil, thermal, radioactive, and noise. Since pollution is a direct or indirect consequence of waste, the seemingly idealistic demand for "zero discharge" can be construed as an unrealistic demand for zero waste. However, as long as waste continues to exist, we can only attempt to abate the subsequent pollution by converting it to a less noxious form. Three major questions usually arise when a particular type of pollution has been identified: (1) How serious are the environmental pollution and water resources crisis? (2) Is the technology to abate them available? and (3) Do the costs of abatement justify the degree of abatement achieved for environmental protection and natural resources conservation? This book is one of the volumes of the Handbook of Environmental Engineering series. The principal intention of this series is to help readers formulate answers to the above three questions.

The traditional approach of applying tried-and-true solutions to specific environmental and water resources problems has been a major contributing factor to the success of environmental engineering and has accounted in large measure for the establishment of a "methodology of pollution control." However, the realization of the ever-increasing complexity and interrelated nature of current environmental problems renders it imperative that intelligent planning of pollution abatement systems be undertaken. Prerequisite to such planning is an understanding of the performance, potential, and limitations of the various methods of environmental protection available for environmental scientists and engineers. In this series of handbooks, we will review at a tutorial level a broad spectrum of engineering systems (natural environment, processes, operations, and methods) currently being utilized, or of potential utility, for pollution abatement, environmental protection, and natural resources conservation. We believe that the unified interdisciplinary approach presented in these handbooks is a logical step in the evolution of environmental engineering.

Treatment of the various engineering systems presented will show how an engineering formulation of the subject flows naturally from the fundamental principles and theories of chemistry, microbiology, physics, and mathematics. This emphasis on fundamental science recognizes that engineering practice has in recent years become more firmly based on scientific principles rather than on its earlier dependency on empirical accumulation of facts. It is not intended, though, to neglect empiricism where such data lead quickly to the most economic design; certain engineering systems are not readily amenable to fundamental scientific analysis, and in these instances we have resorted to less science in favor of more art and empiricism.

Since an environmental natural resources engineer must understand science within the context of applications, we first present the development of the scientific basis of a particular subject, followed by exposition of the pertinent design concepts and operations, and detailed explanations of their applications to natural resources conservation or environmental protection. Throughout the series, methods of mathematical modeling, system analysis, practical design, and calculation are illustrated by numerical examples. These examples clearly demonstrate how organized, analytical reasoning leads to the most direct and clear solutions. Wherever possible, pertinent cost data or models have been provided.

Our treatment of environmental natural resources engineering is offered in the belief that the trained engineer should more firmly understand fundamental principles, be more aware of the similarities and/or differences among many of the engineering systems, and exhibit greater flexibility and originality in the definition and innovative solution of environmental system problems. In short, the environmental and natural resources engineers should by conviction and practice be more readily adaptable to change and progress.

Coverage of the unusually broad field of environmental natural resources engineering has demanded an expertise that could only be provided through multiple authorships. Each author (or group of authors) was permitted to employ, within reasonable limits, the customary personal style in organizing and presenting a particular subject area; consequently, it has been difficult to treat all subject materials in a homogeneous manner. Moreover, owing to limitations of space, some of the authors' favored topics could not be treated in great detail, and many less important topics had to be merely mentioned or commented on briefly. All authors have provided an excellent list of references at the end of each chapter for the benefit of the interested readers. As each chapter is meant to be self-contained, some mild repetition among the various texts was unavoidable. In each case, all omissions or repetitions are the responsibility of the editors and not the individual authors. With the current trend toward metrication, the question of using a consistent system of units has been a problem. Wherever possible, the authors have used the British system (fps) along with the metric equivalent (mks, cgs, or SIU) or vice versa. The editors sincerely hope that this redundancy of units' usage will prove to be useful rather than being disruptive to the readers.

The goals of the Handbook of Environmental Engineering series are: (1) to cover entire environmental fields, including air and noise pollution control, solid waste processing and resource recovery, physicochemical treatment processes, biological treatment processes, biotechnology, biosolids management, flotation technology, membrane technology, desalination technology, water resources, natural control processes, radioactive waste disposal, hazardous waste management, and thermal pollution control; and (2) to employ a multimedia approach to environmental conservation and protection since air, water, soil, and energy are all interrelated.

This book (Volume 19) and its two sister books (Volumes 17 and 20) of the Handbook of Environmental Engineering series have been designed to serve as natural resources engineering reference books as well as supplemental textbooks. We hope and expect they will prove of equal high value to advanced undergraduate and graduate students, to designers of natural resources systems, and to scientists and researchers. The editors welcome comments from readers in all of these categories. It is our hope that the three natural resources engineering books will not only provide information on natural resources engineering, but will also serve as a basis for advanced study or specialized investigation of the theory and analysis of various natural resources systems.

This book, *Environmental and Natural Resources Engineering, Volume 19*, covers the topics on understanding, conservation, and protection of precious natural resources – bees; waste reclamation for reuse; biological processes for water resources protection and water reuse; removal of endocrine disruptors for environmental protection; cooling and reuse of thermal discharges; basic hydrology, water resources, and DAF boat plant for lake restoration; cadmium detoxification by sintering with ceramic matrices; treatment of vegetable oil refining wastes; environmental engineering education; environmental control of pests and vectors; new book reviews; and glossary of environmental and natural resources engineering.

This book's first sister book, *Natural Resources and Control Processes, Volume 17*, covers the topics on management of agricultural livestock wastes for water resources protection; application of natural processes for environmental protection; proper deep well waste disposal; treating and managing industrial dye wastes; health effects and control of toxic lead in the environment; municipal and industrial wastewater treatment using plastic trickling filters for BOD and nutrient removal; chloride removal for recycling fly ash from municipal solid waste incinerator; recent evaluation of early radioactive disposal and management practice; recent trends in the evaluation of cementitious material in radioactive waste disposal; extensive monitoring system of sediment transport for reservoir sediment management; and land and energy resources engineering glossary.

This book's second sister book, *Integrated Natural Resources Management, Volume 20,* covers the topics on effect of global warming, climate change on glacier and salmons; community-based latrine development with Engineers Without Borders – USA; surface water quality and analysis; water quality control of tidal rivers and estuaries; treatment of electrical and electronic component manufacturing wastes; geographic information systems and remote sensing applications in environmental and water resources; investigation and management of water losses from wet infrastructure; lake restoration and acidic water control; biohydrogen production through mixed culture dark anaerobic fermentation of industrial waste; agricultural waste-derived absorbents for decontamination of heavy metals; removal of heavy metal ions using magnetic materials; and biohydrogen production from lignocelluosic biomass by extremely halotolerant bacterial communities.

The editors are pleased to acknowledge the encouragement and support received from Mr. Aaron Schiller, Executive Editor of the Springer Nature Switzerland, and his colleagues during the conceptual stages of this endeavor. We wish to thank the contributing authors for their time and effort, and for having patiently borne our reviews and numerous queries and comments. We are very grateful to our respective families for their patience and understanding during some rather trying times.

Newtonville, NY, USA — Lawrence K. Wang
Newtonville, NY, USA — Mu-Hao Sung Wang
Cleveland, OH, USA — Yung-Tse Hung
Pasadena, CA, USA — Nazih K. Shammas

Contents

1 **Understanding, Conservation, and Protection of Precious Natural Resources: Bees** 1
Cynthia Li

2 **Water Reclamation and Reuse for Environmental Conservation** 53
Nazih K. Shammas, Lawrence K. Wang, and Mu-Hao Sung Wang

3 **Biological Processes for Water Resource Protection and Water Recovery** 73
Lawrence K. Wang, Mu-Hao Sung Wang, and Nazih K. Shammas

4 **Removal of Endocrine Disruptors for Environmental Protection** 169
Nazih K. Shammas, Lawrence K. Wang, and Mu-Hao Sung Wang

5 **Cooling and Reuse of Thermal Discharges** 195
Yung-Tse Hung, James Eldridge, Jerry R. Taricska, Kathleen Hung Li, and William W. Shuster

6 **Basic Hydrology, Water Resources, and DAF Boat Plant for Lake Restoration** 229
Donald B. Aulenbach, Lawrence K. Wang, Mu-Hao Sung Wang, and Nazih K. Shammas

7 **Advances in Cadmium Detoxification/Stabilization by Sintering with Ceramic Matrices** 299
Minhua Su, Kaimin Shih, and Diyun Chen

8 **Advances in Treatment of Vegetable Oil Refining Wastes** 325
Devrim Kaya and Yung-Tse Hung

9 Environmental Engineering Education in the Modern Era 377
Henry R. Bungay III

10 Environmental Control of Pests and Vectors 387
Lenore S. Clesceri and Erika J. Clesceri

11 Book Reviews: Water Environment 411
Lawrence K. Wang and Mu-Hao Sung Wang

12 Glossary of Natural Resources and Environmental Pollution Control 421
Mu-Hao Sung Wang and Lawrence K. Wang

Index 495

About the Editors

Lawrence K. Wang has more than 28 years of experience in facility design, environmental sustainability, natural resources, resources recovery, global pollution control, construction, plant operation, and management. He has expertise in water supply, air pollution control, solid waste disposal, water resources, waste treatment, and hazardous waste management. He is a retired acting President/director/VP of the Lenox Institute of Water Technology, Krofta Engineering Corporation, and Zorex Corporation, respectively, and was an Assistant Professor/Associate Professor/Professor of Rensselaer Polytechnic Institute, Stevens Institute of Technology, and the University of Illinois, respectively, in the USA. He has represented the US government to serve the United Nations Industrial Development Organization (UNIDO) as a Senior Advisor in Vienna, Austria. Dr. Wang is the author of over 700 papers and 45 books, and is credited with 24 US patents and 5 foreign patents. He received his BSCE degree from National Cheng-Kung University, Taiwan, ROC; his two MS degrees from the University of Missouri (Missouri University of Science and Technology) and the University of Rhode Island, USA; and his PhD degree from Rutgers University, USA. Currently he is the chief series editor of the *Advances in Industrial and Hazardous Wastes Treatment* series (CRC Press of Taylor & Francis Group), the *Handbook of Environmental Engineering* series (Springer), the *Evolutionary Progress in Science, Technology, Engineering, Arts, and Mathematics* series (Lenox Institute Press), and the *Environmental Science, Technology, Engineering and Mathematics* series (Lenox Institute Press). He is also the coeditor of the *Handbook of Environment and Waste Management* series (World Scientific) and the coauthor of the *Water and Wastewater Engineering* series (John Wiley).

Mu-Hao Sung Wang has been an engineer, editor, and professor serving private firms, governments, and universities in the USA and the Republic of China (ROC) for over 25 years. She is a licensed professional engineer and a diplomate of the American Academy of Environmental Engineers. Her publications have been in the areas of water quality, modeling, environmental sustainability, waste management, NPDES, flotation, and analytical methods. Dr. Wang is the author of over

50 publications and an inventor of 14 US and foreign patents. She received her BSCE degree from National Cheng Kung University, Taiwan, ROC; her MSCE degree from the University of Rhode Island, USA; and her PhD degree from Rutgers University, USA. She was an Adjunct Associate Professor at Stevens Institute of Technology, NJ, USA; an Adjunct Professor at National Cheng Kung University, Taiwan, ROC, and a professor at the Lenox Institute of Water Technology, MA, USA. She is the co-series editor of the *Handbook of Environmental Engineering* series (Springer), and the coeditor of the *Environmental Science, Technology, Engineering and Mathematics* series (STEM) (Lenox Institute Press). She is a member of AWWA, NEWWA, WEF, and OCEESA.

Yung-Tse Hung has been a professor of civil engineering at Cleveland State University since 1981. He is a fellow of the American Society of Civil Engineers and has taught at 16 universities in 8 countries. His research interests and publications have been involved with biological processes and industrial waste treatment. Dr. Hung is credited with over 470 publications and presentations on water and wastewater treatment. He received his BSCE and MSCE from National Cheng-Kung University, Taiwan, and his PhD from the University of Texas at Austin, USA. He is the editor of *International Journal of Environment and Waste Management, International Journal of Environmental Engineering,* and *International Journal of Environmental Engineering Science.*

Nazih K. Shammas was an environmental consultant and professor for over 45 years. He was an ex-dean/director of the Lenox Institute of Water Technology and an advisor to Krofta Engineering Corporation, USA. Dr. Shammas was the author of over 250 publications and 16 books in the field of environmental engineering. He had experience in environmental planning, curriculum development, teaching, scholarly research; and had expertise in water quality control, wastewater reclamation and reuse, physicochemical and biological processes, and water and wastewater systems. He received his BE degree from the American University of Beirut, Lebanon; his MS degree from the University of North Carolina at Chapel Hill, and his PhD degree from the University of Michigan, USA. He was the coeditor of the *Advances in Industrial and Hazardous Wastes Treatment* series (CRC Press of Taylor & Francis Group), the *Handbook of Environmental Engineering* series (Springer), the *Environmental Science, Technology, Engineering and Mathematics* series (Lenox Institute Press), and the *Handbook of Environment and Waste Management* series (World Scientific). In addition, Dr. Shammas was the coauthor of the *Water and Wastewater Engineering* series (John Wiley).

Contributors

Donald B. Aulenbach PhD, PE, DEE Rensselaer Polytechnic Institute, Troy, NY, USA
Lenox Institute of Water Technology, Lenox, MA and Newtonville, NY, USA

Henry R. Bungay III PhD H. P. Isermann Department of Chemical and Biological Engineering, Rensselaer Polytechnic Institute, Troy, NY, USA

Diyun Chen PhD School of Environmental Science and Engineering, Guangzhou University, Guangzhou, China

Erika J. Clesceri PhD Bureau for Democracy, Conflict and Humanitarian Assistance, US Agency for International Development, Washington, DC, USA

Lenore S. Clesceri PhD Department of Biology, Rensselaer Polytechnic Institute, Troy, NY, USA

James Eldridge Sales Department, Lantec Product, Agoura Hills, CA, USA

Yung-Tse Hung PhD, PE, DEE, F-ASCE Department of Civil and Environmental Engineering, Cleveland State University, Cleveland, OH, USA

Devrim Kaya PhD Department of Civil and Environmental Engineering, University of Maryland, College Park, MD, USA

Cynthia Li MS Lenox Institute of Water Technology, Lenox, MA and Newtonville, NY, USA

Kathleen Hung Li MS NEC Business Network Solutions, Inc., Irving, TX, USA

Nazih K. Shammas PhD Lenox Institute of Water Technology, Lenox, MA and Newtonville, NY, USA
Krofta Engineering Corporation, Lenox, MA, USA

Kaimin Shih PhD Department of Civil Engineering, The University of Hong Kong, Hong Kong, China

William W. Shuster PhD Department of Chemical and Environmental Engineering, Rensselaer Polytechnic Institute, Troy, NY, USA

Minhua Su PhD School of Environmental Science and Engineering, Guangzhou University, Guangzhou, China

Jerry R. Taricska PhD, PE Environmental Engineering Department, Hole Montes, Inc., Naples, FL, USA

Lawrence K. Wang PhD, PE, DEE Lenox Institute of Water Technology, Lenox, MA and Newtonville, NY, USA
Rensselaer Polytechnic Institute, Troy, NY, USA
Krofta Engineering Corporation, Lenox, MA, USA

Mu-Hao Sung Wang PhD, PE, DEE Lenox Institute of Water Technology, Lenox, MA and Newtonville, NY, USA
Krofta Engineering Corporation, Lenox, MA, USA

Chapter 1
Understanding, Conservation, and Protection of Precious Natural Resources: Bees

Cynthia Li

Contents

1 Introduction 3
2 Pollinators and Pollination 5
2.1 Wild Bees 6
2.2 Honey Bees 6
3 Pollinator Contributions to Agriculture 15
4 Pollinator Decline: Bees 17
4.1 History, Statistics 17
4.2 Causes 18
4.3 Effects of Pollinator Decline (Bee Loss) 28
5 Conservation and Protection of Bees 31
5.1 Beekeepers 31
5.2 Farmer: Habitat Diversity 35
5.3 Farmer: Agrochemical Practices 39
5.4 Government: Legislation 40
6 Conclusions 42
Glossary 44
References 46

Abstract This chapter concerns the role of pollinators, especially bees, in relation to our agriculture and economy. Besides being a critical participant in the reproductive process of most flowering plants, the bees' pollination services increase flowering crop yields, improve nutritional quality, and contribute significantly to the global economy. However, beekeepers that manage the domesticated European honey bee had noticed that huge numbers of their bees were dying in recent years, and apiculture is closely wedded to agriculture. Honey bees suffer from a variety of maladies attributed to pesticides, climate change, parasites, and disease, all of which contribute to increased bee mortality and catastrophic hive death such as Colony Collapse Disorder (CCD). A number of these issues appear to affect wild pollinators

C. Li (✉)
Lenox Institute of Water Technology, Newtonville, NY, USA
e-mail: Cynthia.Li@mnsu.edu

L. K. Wang, M. -H. S. Wang, Y. -T. Hung, N. K. Shammas (eds.),
Environmental and Natural Resources Engineering, Handbook of Environmental Engineering 19, https://doi.org/10.1007/978-3-030-54626-7_1

as well. In order to conserve and protect our pollinator resources, this chapter describes a number of major stressors that lead to bee mortality and offers suggestions to improve bee and ultimately general pollinator survival. Recommended actions for beekeepers include adoption of honey bee strains that are bred for specific climates and survival traits as well as maintaining diverse forage around the apiary. Farmers are recommended to maintain areas of wild plants for diverse foraging sources, where a healthy bee diet also aids in crop pollination. In order to protect food availability, control prices, and support agricultural sustainability, governments can review ongoing research concerning known bee and pollinator stressors and implement legislation and guidelines to control human-created stressors such as pesticides while encouraging pollinator-friendly land management.

Keywords Pollinators · Pollination · Wild bees · Honey bees · Agriculture · Crops · Bee conservation and protection

Acronyms

ABPV	Acute Bee Paralysis Virus
AIV	Apis Iridescent Virus
ARS	Agricultural Research Service
CCD	Colony Collapse Disorder
CRP	European Union Conservation Reserve Program
DWV	Deformed Wing Virus
EU€	European Union Euro
FAO	Food and Agriculture Organization of the United Nations
IAPV	Israeli Acute Paralysis Virus
IPCC	Intergovernmental Panel on Climate Change
KBV	Kashmir Bee Virus
NASS	National Agricultural Statistics Service
NRCS	National Resources Conservation Service
NWRS	National Wildlife Refuge System
OPMP	USDA's Office of Pest Management Policy
US EPA	United States Environmental Protection Agency
US FWS	United States Fish and Wildlife Service
US$	United States Dollar
USDA	United States Department of Agriculture
VSH	Varroa-Sensitive Hygiene

1 Introduction

Pollinators are animals that allow 75% of flowering plants to reproduce; these plants include most fruits, vegetables, and nuts, where 87 of the leading 115 food crops are dependent on pollination services, where pollinators contribute an estimated one-third of global food production. Pollinators also play a key role in our agricultural economy, where they contributed more than $24 billion to the United States' economy in 2014, and honey bee pollination services specifically contributed more than $15 billion.

There are thousands of bee species. Most wild (native) bees in America are solitary. Honey bees, our most commonly managed pollinator, were imported to the Americas from Europe during colonial times in the 1600s. These are called European or Western honey bees; in this chapter, we will refer to them as European honey bees.

Honey bees are social insects; each colony has one queen, several drones, and thousands of worker bees living in a wax comb hive that they built themselves. In this hive, they store food and rear their brood (eggs, larvae, and pupae). Work responsibilities are caste based. The queen bee is responsible for reproduction and maintaining colony population; a drone's sole responsibility is to mate with a queen; worker bees do all other duties in maintaining the colony. Worker bee responsibilities include rearing the brood, caring for the queen, maintaining the wax comb structure, defending the colony, and foraging. Plant pollination occurs when worker bees are foraging on their flowers. The nectar and pollen gained from foraging are prepared and stored in the hive as the primary food supplies, where the nectar is converted to honey. When flowers begin to bloom again in the spring, the queen begins to lay eggs as worker bees leave the hive to forage and pollinate plants along the way, continuing the cycle. How do honey bees execute their colony responsibilities in a systematic way? They communicate with each other using pheromones and dances. Pheromones are chemical substances secreted from the sender's glands and received by the recipient's antennae or other body parts. All colony members from queen to even brood members can produce pheromones. Besides pheromones, honey bees also perform different dances to convey information or request services. The most famous dance is the "waggle dance," which conveys the direction and distance of a food resource.

For a very long time, the majority of honey bees in the Americas had been European honey bees. However, a scientist in Brazil imported pure African honey bees in 1956 with the intention of breeding more tropically adapted honey-producing bees. Just 1 year later, 26 African queen bees escaped into the wild. Since then, those African queens have hybridized with feral European honey bee populations, becoming Africanized honey bees (AHBs). Six decades later, Africanized honey bees have spread from South America through Central America and to North America. By 2009, Africanized honey bees had been confirmed in ten southern US states.

Africanized and European honey bees are cousins; they are of same species though different strains, and they share the basic characteristics. The Africanized honey bees' aggressive defensive behavior earned them the "killer bee" nickname. They are also more difficult to manage due to frequent swarming and high absconding rates. Beyond those negatives, however, Africanized bees exhibit more positive differences from their European cousins, such as resistant to *Varroa* mite, using more propolis, and adapting well to tropical zones. Since the introduction of Africanized honey bees in Mexico, honey production had dropped, which might have been attributed to beekeepers dropping out of the profession. Mexico's honey production has since recovered, and now, they prefer Africanized honey bees.

In the United States, managed honey bee colonies have steadily declined in the past 60 years from six million in 1947 to only 2.5 million in 2014. Starting in 2006, commercial beekeepers in the United States reported an average of 30% per winter losses as opposed to the historical rates of 10–15% colony losses. In the United States, where honey bees contribute billions of dollars in added value to the agriculture, these bee losses must be taken seriously. For many years, the commercial managed honey bee losses were due to Colony Collapse Disorder (CCD), where a specific set of symptoms led to catastrophic death of a colony. The trend was alarming, but according to the newest reports from the US Department of Agriculture, CCD was not among the major cause of winter losses in the last couple of years.

Many stressors are correlated with CCD, such as pesticides (insecticides, herbicides, fungicides), parasites, disease, and poor nutrition. Many scientists believe that CCD is caused by a combination of the above factors, where the compounding variables make it much harder to study.

There is no doubt about there is economic impact due to honey bee population decline, but how much of an impact? We know that humans will not all die in 4 years if bees disappear from the surface, because some of our most productive crops are wind-pollinated or self-pollinated. However, our food will contain less variety, cost more to produce and to purchase, and have less nutritional quality.

Beekeepers should follow best management practices to care for their bees: monitor colonies for any pest infestation and apply treatments at the best time, rotate treatment methods to prevent pests from developing of resistance, requeen if necessary, and use resistant-bred queens when possible. Since selective breeding programs are available, beekeepers may consider purchasing resistant-bred queens or even breeding themselves. Some of the potential breeding possibilities include breeding for *Varroa*-sensitive hygiene (VSH) trait, for the tropical hardiness and mite resistance of Africanized honey bees, and for cold weather and mite resistance of the Russian honey bee.

Everyone from individuals to governments should diligently contribute to protect and conserve the honey bees and other pollinators. The general public can grow bee-friendly flowering plants in their backyards and window sills; farmers can increase the habitat diversity in farmlands by planting native and nonnative, flower-rich herbaceous plants in appropriate areas, which will in turn help nearby crop fields; beekeepers can also improve the bee yards by creating or allowing nectar-rich wild flower areas on the premises, planting a variety of flowers that can provide nectar and pollen across honey bees' foraging seasons. Government departments can apply pollinator-friendly best practices on government-controlled lands and encourage its citizens to cooperate through incentives.

Besides increasing habitat diversity for bees, everyone who must use pesticides should follow the pesticide usage best practices. Pesticides have been shown to cause notable problems for honey bees and other pollinators even if the effects are not outright lethal; research is still ongoing. Best practices include but are not limited to the following: use approved least-toxic pesticides, follow product instructions, start with the lowest recommended dose, select appropriate timing including when bees are not active, manage pesticides drift through weather conditions, application method, equipment settings, and spray formulation. At the same time, governments are reviewing research and may control pesticide use to protect honey bees and all pollinators in general.

2 Pollinators and Pollination

Pollinators are animals that facilitate the reproductive cycle of over 75% of flowering plants, including most fruits, and vegetables. These animals include small mammals such as bats, birds, and many types of insects including the most well-known and accessible honey bee.

The reproductive cycle of flowering plants includes pollination, which is the transfer of pollen grains (male haploid gametophytes) to the female gametophytes, which contain the ovules. The pollen grains produce male gametes, which in turn fertilize the ovules; these fertilized ovules develop into plant seeds. The ovary in which seeds are contained may develop into fruit. The seeds and/or fruit are then disseminated by various methods to fertile land in which they may grow as new plants. These plants then supply significant resources to other organisms including humans: food, building materials, fibers, and so on.

To incorporate pollinators into their life cycle, these plants produce flowers that contain pollen and nectar, which are both sources of food for bees and other pollinating insects. Besides providing food, serving as shelter and nesting spots for their various pollinators, plants also produce resins. Resin is an important compound that honey bees specifically retrieve to produce propolis, which is used to fortify the hive, improve hygiene by preventing parasites and inhibiting fungi and bacteria, and mummify carcasses they cannot remove from the hive [1].

As the pollinator travels and gathers food and materials from these flowering plants, pollen grains stick to its body and are carried from flower to flower, fertilizing other plants of the same species along the way. Thus, pollinators and flowering plants have tightly integrated and mutually beneficial relationships [2].

Pollination, particularly that provided by insects, is essential to agriculture, which not only provides food to humans but also fibers, fuel, and feed to our domesticated animals that further provide sources for meat.

While there is a very wide variety of animal pollinators contributing to the ecosystem, most animal pollinators are insects, and within insects, most are bees. Bees consist of thousands of species but are broadly split between two basic categories for the purposes of study in agriculture and other industries of interest to humans: wild (native) bees and honey bees.

2.1 Wild Bees

Wild bees, also referred to as native bees for a given locale, are those bees not managed by humans, where management involves breeding and generally tending to the bees' basic needs such as providing shelter and supplemental food. Wild bees may be either social or solitary.

More than 90% of the world's 20,000 plus bee species are in fact solitary, which makes both management and research very challenging compared to the highly social nature of honey bees that live in structured colonies. A solitary bee is simply one that lives alone, where each nest is built and occupied by a single female with no caste or role hierarchy [3].

Bumblebees are social bees like honey bees, but their nest population numbers in the low hundreds, not the tens of thousands honey bees in a colony. Bumblebees are not as widely adopted into commercial apiculture (beekeeping), because they produce only small amounts of honey-like substance sufficient for their own consumption and are difficult and expensive to raise. Certain species are reared for commercial pollination in greenhouses, but most species of bumblebee are wild [4].

Roughly 4000 species of wild bee are native to the United States, and they are still credited with the majority of pollination activities with respect to native plants. For example, the imported honey bee either cannot or poorly pollinates tomato, eggplant, pumpkins, cherries, blueberries, and cranberries [5].

However, consistent and accurate studies for wild native bees, most of which are solitary, can be difficult to achieve. The accessibility of managed honeybee colonies means more data on their pollinator role and environmental effects upon them, particularly in close relation to agricultural production.

2.2 Honey Bees

Farmers manage only 11 of over 20,000 bee species worldwide, with the European honey bee (*Apis mellifera*) being the most commonly managed bee species [6].

Since honey bees are the primary subject of most pollinator-crop studies due to the close relationship between the beekeeping and agriculture industries, a basic understanding of their characteristics is necessary. Subspecies and hybrids exhibit variations in behavior and morphology, but all honey bees share the following general characteristics.

2.2.1 Honey Bee Colony Social Structure

The honey bee colony's social structure is particularly important when considering stressors that are transferred from individual to hive, such as infectious diseases and parasites that target concentrated populations of bees.

Honey bees are highly sociable insects. A honey bee colony consists of three types of adult bees: one queen, several drones, many workers, and brood (eggs, larvae, and pupae). All of them live in a hive constructed of wax comb containing many hexagonal cells; for managed hives, the comb is suspended within a frame, which can be plastic or wood. Within the cells, the bees store pollen and honey, and rear the brood (Table 1.1).

There is only one active queen bee per colony—a fertile female bee larger than other bees and whose main responsibility is to reproduce by laying eggs to increase and maintain colony size. The queen bee mates only once with several drones and remains fertile for her lifespan, which can be several years. There are different mechanisms to replace a queen if it dies, such as feeding a larva royal jelly and allowing it to develop into a virgin queen, or adopting a queen exiting another colony that already has a queen. Under managed conditions, beekeepers (apiarists) will commonly purchase a fertilized queen bee from a reputable source.

Drone bees are fertile male bees and number approximately 300–800 per colony; stingless and lacking the physiology to forage or build hive comb, the drone's sole purpose is to mate with a queen, and dies shortly after doing so.

Worker bees are the smallest and most numerous colony members, numbering approximately 20,000–80,000 bees per colony and constituting over 98% of its population. They are sterile, female, and are responsible for numerous duties that are split between two main worker roles: house bee or foraging field bee. House bees are responsible for hive upkeep including cleaning, brood feeding (as "nurse" worker bees), caring for the queen, building hive comb and food storage, and defense. Field bees are responsible for foraging, which is the retrieval of various resources (pollen, nectar, propolis, water). House bees are typically young worker

Table 1.1 Queen, drone, and worker bees [7]

	Queen	Drone	Worker
Sex	Female	Male	Female
Number	Typically one per colony	~300 to 800	~20,000 to 80,000 (about 98% of the colony population)
Fertility	Fertile	Fertile	Sterile
Stinger	Yes; only for fighting other queens	No	Yes; for colony defense. Dies after issuing a sting that pulls out the stinger apparatus
Body size	Generally larger than drone and worker when mature	Larger than worker	Smallest
Anatomy	No pollen-collecting apparatus	No pollen-collecting apparatus	Has pollen-collecting apparatus (pollen basket)
Responsibility	Mating and laying eggs to produce all the colony's offspring	Mate with a virgin queen from another colony; dies after successful mating	Hive upkeep, defense, foraging

bees, while field foragers are more mature worker bees. All worker bees have a sting and usually die after using it.

Flower nectar converted into honey is the bees' primary source of carbohydrates for energy. Pollen or "bee bread" is their primary source of protein, fatty acids, minerals, and vitamins, which are crucial for overall hive health and young bee development [8].

Besides collecting pollen, nectar, water, and resin while foraging, worker bees also produce "royal jelly" to raise a queen, secrete enzymes to ripen honey, and produce wax for hive comb construction. One foraging worker bee makes a dozen or more trips in a day, visiting several thousand flowers in a range of 2–5 miles away from the hive. They typically limit themselves to one plant species per trip, which aids in the pollination process. A foraging bee can return to its colony bearing a pollen load weighing nearly 35% of its own body weight. The pollen gathered by worker bee "pollen baskets" located on their legs is also a food supplement used by humans and some other animals [7, 9].

2.2.2 Honey Bee Reproduction

Individual

Worker bees build hexagonal comb cells using wax produced from their wax glands, and these cells serve as food storage for honey and pollen as well as incubation for the bee brood. The "brood" consists of eggs, larvae, and pupae, which are the bees' early developmental stages.

Sometime after her mating flight and being fertilized by multiple drones, the queen bee lays one egg in each cell; in the spring and early summer, she may lay up to 1500 eggs per day. Fertilized eggs become female worker bees (sterile), while unfertilized eggs become male drones (fertile).

In 3 days, the egg hatches into a larva. Inside its beeswax cell, the larva is fed by the "nurse" worker bees, and after a few days, the worker bees seal the cell with a wax "cap." After being capped, the larva enters the pupa or transformation stage, where its grub-like form changes into an adult bee. The developed adult bee emerges from the cell 7.5–14.5 days after the capping, depending on its caste [8].

Swarming

Swarming, essentially colony-level reproduction, is the process of forming a new colony by leaving the parent colony and nesting in a new location. When environmental variables are favorable and food sources plentiful, a colony's population will increase to a certain point that causes crowding within the hive. At this point, the bees will begin to raise a new queen. As the new queen reaches maturity, the original queen flies away with a swarm consisting of at least half the existing worker bees and some of the drones, while the new reared queen stays with the parent colony. When the new queen matures, she will take mating flights and begin to lay eggs to increase the bee population as most had left with the swarm.

Swarms look intimidating but are usually harmless and rarely sting, because the bees do not yet have brood, honey, or hive to defend. The exiting swarm will land somewhere nearby to rest, and the worker bees will cluster around the queen to protect her while some scout bees leave to search for another suitable home. Once a home location is found, the swarm will create a hive and continue their activities in foraging and raising bees. Under ideal conditions, European honey bee colonies swarm approximately once or twice a year [10].

2.2.3 Honey Bee Communication

Pheromones

The honey bee has two primary modes of communicating with other members of its colony: pheromones and dances, particularly the "waggle" dance. Pheromones are chemicals secreted from honey bee glands that affect behavior in other bees and thus affect all aspects of colony life. The waggle dance is a physical activity done to convey spatial information of a resource location.

Pheromones are transmitted by direct contact as liquid or vapor and received by bees' antennae and other body parts to trigger behavioral or physiological responses, usually in the same species [11]. Pheromones are secreted by all colony members from queen, drone, worker bee, to brood (particularly the larvae), and allow communication among all castes. They are involved in all aspects of colony life including reproduction (queen mating and swarming), brood development, defense, foraging, building hive comb, and other activities [12].

Some glands and pheromones are caste-specific. There are two types of pheromone effects: releaser, which temporarily affects the recipient's behavior, and primer, which has long-term physiological effects. The queen signal and brood pheromones are of the primer type, while most worker bee pheromones are considered the releaser type.

Queen bee pheromone (queen mandibular pheromone) has different effects on different bee castes. For example, it serves as a sexual attractant for drones during the mating flight while also serving to attract the queen retinue, which is a group of worker bees whose role is to attend to the queen. The pheromone is responsible for inhibiting additional queen rearing, suppressing fertility in worker bees, keeping a swarming bee cluster together, stimulating worker bee activity around the hive, and generally calms a swarm.

Drone pheromones are concentrated upon sexual activities, such as attracting other flying drones to aggregate at sites suitable for mating with a virgin queen [13].

Brood pheromones secreted by larval salivary glands and serve to regulate brood development and care by chemically communicating needs to nurse worker bees. Along with queen pheromones, brood pheromones suppress worker ovarian development, since fertile worker bees do not work as much as sterile. Finally, the level of brood pheromones can modulate behavioral maturation in worker bees: a low level decreases the adult bee foraging age, promoting a larger proportion of worker bees to become foragers; a high level of brood pheromone increases worker foraging age, stimulating young adult workers to remain as nurse bees for an extended period of time [14].

Worker honey bee pheromones cover a wide range of functions. Depending on the age and role differentiation of the worker bee, glands develop at different times or produce different pheromones depending on the situation. For example, wax glands develop sooner and are more active in young worker bees for building comb, while alarm pheromone production rises in older guard and forager bees. The glands responsible for producing pheromone that suppresses worker bee fertility also produce alarm pheromone when the worker bee becomes a forager [15].

Honey Bee Dances

All of the known species and subspecies of honey bee exhibit dancing behavior, but the form and details of execution vary among species [16].

The following describes *Apis mellifera*, as it is in the most widely managed honey bee species. There are several known honey bee dances, including the waggle dance, round dance, tremble dance, and grooming dance. Each dance conveys information, but the use of said information may be context-dependent [17].

The waggle and round dances are used by forager bees to communicate foraging locations. The round dance is performed by a returning scout or forager when it wishes to communicate a food source less than about 35 yards away from the hive; the bee will move in circles alternately to the left and right, without indicating specific direction, but it will share the scent of the food source [18].

When a resource is more than 35 yards away, the honey bee will perform the waggle dance instead; this resource may be food, water, or even a new nest site. This is a more complex dance that communicates direction relative to the hive, distance, and odor of the target resource if applicable [19, 20].

The waggle dance is usually performed on the vertical surface of the hive comb. The dancing bee first moves forward for a certain distance along the vertical line of the comb, and this vertical line serves as the reference base for the direction. Then the bee moves back to the starting point and begins her dance pattern. Each dance involves running through a figure-eight pattern with one straight phase and two half circle return phases; the dance move pattern is then repeated. On the straight stretch phase, the bee “waggles.” The angle between the straight stretch and the vertical line is the angle of the target sources relative to the sun. If experiencing a long delay and the sun has moved, the dancing bee adjusts the angle accordingly. The distance to the target resource is expressed by the duration of the straight waggle phase. One second is approximately for 1 km of distance. The time taken to complete the waggle phase is directly related to the distance, and the speed of the waggle is inversely related to the actual distance [19].

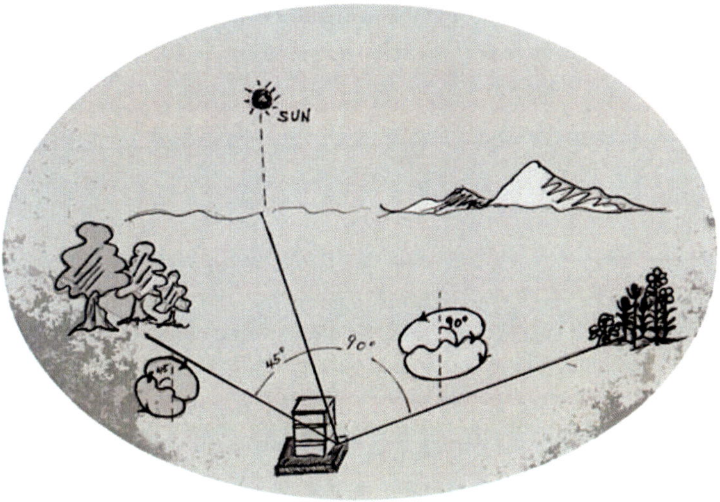

Fig. 1.1 Honey bee waggle dance [19]

The speed or the frequency of the repeating dance pattern performed is an indication of the quality of the resource; the higher the speed of the dance means the better the quality of the resource. When dance repetition is low or reduced, it indicates potential risk. If a dead bee is found at a potential foraging site, for example, returning foragers and scouts perform fewer waggle dances relative to that location [21, 22] (Fig. 1.1).

2.2.4 European Honey Bees

The honey bee as we know it was not native to the Americas; they were, in fact, imported from Europe during colonial times in the 1600s. *Apis mellifera* was instead native to Europe, Asia, and Africa, and evolved into different subspecies by geography throughout Africa and Eurasia. These honey bees are referred to as European or western honey bees [23]. The majority of available pollination information is derived from European honey bee data for this reason.

Most crops grown in the United States are not native to the Americas either. Both the crops and honey bees evolved in other parts of the world and were brought to United States by European settlers. Today, more than 90 commercially produced crops in the United States rely on bee pollination [8].

Honey bee population numbers and temporal activity patterns are easily assessed by visual inspection. Physically managed hives can produce important information on hive conditions and activity, the timing of flowering and nectar access, and details about the interaction between bees and the environment [24].

It is important to note at this point that "feral" honey bees are not considered the same as wild bees, which are native to the locale. Rather, feral honey bees are those that escape apiaries and establish colonies away from human management, but are still an alien species.

2.2.5 Africanized Honey Bee

All subspecies of *Apis mellifera* can interbreed or hybridize. The Africanized honey bee is a hybrid of the European honey bee and African bee, which are two separate strains of the same *Apis mellifera* species. Africanized bees acquired the appellation "killer bee" due to their highly aggressive defense behavior, which can often result in serious injury and death for humans and other animals [25].

Pure African honey bees, *Apis mellifera scutellata,* were imported into São Paulo, Brazil, in 1956. The intention was to breed the African honey bees, which were better adapted to a tropical environment, with European honey bees in order to improve country's honey production. Just 1 year later in 1957, 26 African queen bees were accidentally released or escaped from the managed facility into the wild. The cross-bred descendants of these African and European honey bees quickly established a large feral population of Africanized honey bees. This event unintentionally demonstrated the Africanized honey bee's superior adaptation to the tropical environment. In just over four decades, these wild AHB migrated to most of the tropical and subtropical area of America, from South to Central then to North America.

The first permanent colonies arrived in City of Hidalgo, Texas, from Mexico was in 1990; the first documented AHB case in the State of Florida near Tampa was 2001; by 2003, AHB had spread to Tampa and throughout of the Florida state. By July 2009, AHBs have been confirmed in ten southern United States: Texas, New Mexico, Arizona, California, Nevada, Utah, Louisiana, Oklahoma, Arkansas, and Florida [26] (Fig. 1.2).

Africanized Honey Bee Characteristics Summary

Beyond the very aggressive hive defense behavior, Africanized bees exhibit additional differences from their European cousins (Table 1.2).

Heightened Defensiveness

The Africanized honey bee is notorious for highly aggressive defense behavior. However, they do not randomly seek out victims to attack. They are very sensitive to alarm pheromone and produce much more of it than European honey bees [31].

They ardently guard and protect their hives within a wide range of at least 30 m. Africanized bees also have a low threshold for disturbance; direct hive disturbance is

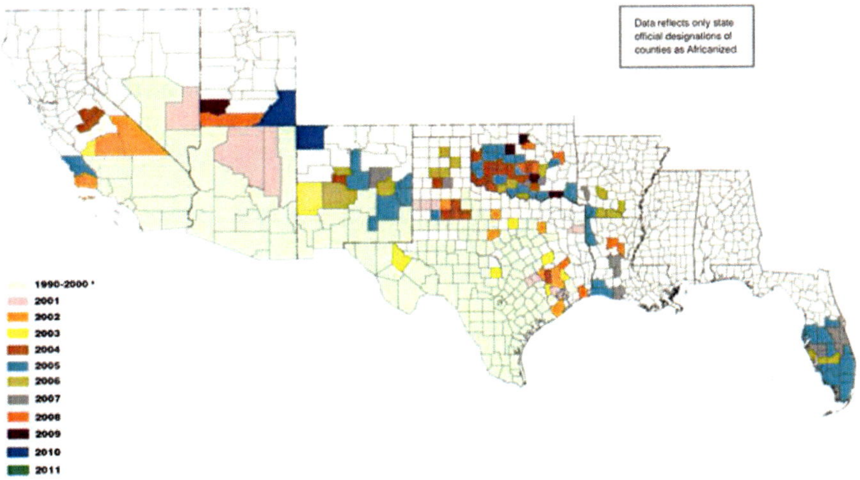

Fig. 1.2 Spread of Africanized honey bees in the southern United States to 2011. (Source: USDA [27])

Table 1.2 Differences between European and Africanized honey bees [26, 28–30]

	European	Africanized
Defensive behavior	May send 10–20 guard bees up to 20 feet away from hive. 10–15 minutes to calm down. Calmed down by smoke. Disturbance may result in 10–20 stings.	May send hundreds of guard bees up to 120 feet away from hive. Hours to calm down or until sunset. Disturbance may result in 100–1000 stings.
Climate adaptation	Temperate zones.	Tropical zones.
Swarm behavior	Swarm 1–2 times per year, only if crowded or to make new nest. Large swarms requiring larger volume sites.	Can swarm 10+ times per year. Smaller swarms requiring smaller nest sites.
Abscond (nest abandonment)	Rare.	Often, relocating in times of stress (shortage of resources, infected nest, etc.).
Nest characteristics	Larger space (~38 liters).	Smaller cavities (~3–19 liters).
Survival strategy	Expend energy in producing and storing honey and pollen to prepare for resource shortage (e.g., overwintering).	Use energy and resources in producing large numbers of progeny and generating many reproductive swarms.
Varroa susceptibility	Susceptible to *Varroa* mites, which shuts down the bees immune system.	Resistant to *Varroa* mites.
Age-based worker behavior	Start foraging at older ages and harvest less pollen and more concentrated nectar.	Start foraging at young ages and harvest resources with low concentrations of sucrose that include water, pollen, and low concentration of nectar.
Physical differences (usually only distinguishable by tool)	Slightly larger.	Slightly smaller.

not necessary to initiate an attack. Merely noise or vibration from equipment can induce a quick, vicious response that persists for a farther distance than European bees. Their venom is no more potent than European honey bees, but they attack in much greater numbers, in the hundreds, where European honey bees may attack the same intruder with no more than 10–20 bees [32].

Swarming

Africanized honey bees swarm frequently and unpredictably, which can be as often as every 6 weeks with more than one separate swarm each time. A European bee hive will typically swarm once a year. Africanized bee swarms are also smaller than European bee swarms [27].

Abscond

During times of food resource shortage or repeated nest disturbance, Africanized bees have a high tendency of abandoning the entire nest including the brood, food stores, and hive. This is known as "absconding." The entire colony composed of adult bees relocates elsewhere. Absconding is something European honey bees rarely do, which also makes their management that much easier.

Nesting Locations

European honey bees prefer nesting in larger spaces such as tree hollows, chimneys, and so forth. Africanized honey bees are less selective about their nesting locations, and their choices rarely overlap with European honey bee hive location preferences. In general, Africanized honey bees will occupy much smaller spaces, because their swarm sizes are smaller: hollow tree limbs, man-made structures like cement blocks, junk piles, meter boxes. Even holes in the ground can become their home nests. This is one reason why humans encounter Africanized bees more frequently.

Survival Strategy

European honey bees devote their efforts to producing and storing honey in order to survive through the winter (overwintering); Africanized honey bees, on the other hand, endeavor to generate many swarms and produce larger number of descendants. The European bees' strategy makes them suited for temperate environments, while the Africanized bees' strategy allows them to quickly replace lost colonies either due to harsh environmental conditions or predators. In response to such stressors, Africanized honey bees are more likely to abscond, while European honey bees seldom do. This explains why Africanized bee colonies store proportionately less food than that of European honey bees [26].

3 Pollinator Contributions to Agriculture

Honey is the most well-known direct product of honey bees, but these insects produce altogether six hive products that humans collect and use: honey, beeswax, propolis, pollen, royal jelly, and venom.

According to the US Department of Agriculture's (USDA) National Agriculture Statistics Service, US honey production in 2013 was 67 million kilograms, valued over US$315 million. Beeswax is the second most important hive product from an economic point of view. It is commonly used for making candles, as an ingredient in artists' materials, in leather and wood polishes, and in cosmetics; drug companies use beeswax as binding agent, time-release feature, and drug carrier. The United States is a major producer of raw beeswax and supplier of refined beeswax.

However, the greatest contribution of honey bees to agriculture and human health is indirect, through their services as crop pollinators. Globally, 87 of the leading 115 food crops are dependent at least to some extent on animal pollinators, contributing an estimated 35% of global food production [33].

When considering the scale of our dependency upon pollinators, primarily bees, crop studies have investigated both quantitative and qualitative effects. Quantitative studies look at amounts of seed and/or fruit produced via pollination, and qualitative studies look at the quality of the final fruit, particularly from a commercial perspective. Both quantitative and qualitative assessments of pollinator participation gain greater meaning when the results are estimated as economic values.

In quantitative studies, it is important to differentiate the plants' reliance upon pollination. Some plants are entirely dependent on external pollination services to produce seed and fruit, while some are partially dependent but can still produce over 90% of their yield without pollination help. Yet other crops achieve pollination by wind, gravity, or water [34].

The value of directly pollinator-dependent crops in the United States reached an estimated US$15.12 billion in 2009; examples of such crops that require or benefit from insect pollination include apples, almonds, blueberries, and the gourd family (squash, melons, etc.). Of that 2009 crop value, US$11.68 billion was attributed to honey bees and US$3.44 billion attributed to other pollinators. The honey industry from the managed honey bee accounted for over US$300 million in the US economy [8, 35]. Bee pollinators contribute at least EU€22 billion per annum to the European agriculture industry as well [36].

Some crops like almonds are almost exclusively pollinated by honey bees. For example, California's almond industry requires 60% of all US honey bee pollination services, about 1.4 million hives, to produce 80% of the worldwide almond crop, which is worth an estimated US$4.8 billion per annum [37].

Animal-assisted pollination enhances both yield and quality of different crops to different extents. Yield is a measurement more economically relevant to farmers and is the amount of crop harvested per unit area as opposed to merely total product.

Rapeseed or oilseed rape, known in the United States primarily by its canola cultivars, is the third-largest source of vegetable oil in the world. The United States is also responsible for nearly a third of the world's strawberry production. In 2012, canola crop in the United States was valued at US$644.8 million, while strawberry production was valued at over $2 billion [38–40].

In a 2014 report, both rapeseed and strawberry crop yields increased about 20% while field bean increased by 40% and buckwheat by 71%, thanks to pollinators. Despite these average numbers illustrating significant impact by pollinating insects, region and variety can change a crop's animal-pollination dependency, where differences can vary up to 30% in rapeseed varieties [34].

For rapeseed, strawberry, field bean, and buckwheat crops, an increase in wild insect visitation was shown to improve fruit and seed set (formation) by twice as much as an equivalent increase in managed honey bee visitation. However, more research needs to be done to tease apart the differences in wild bee and honey bee effectiveness [41].

The nutritional value of plants directly affects human health. The animal-pollinated plants above that are responsible for providing oils for our use are also primary sources of fat-soluble vitamins. Ninety-eight percent of vitamin C, a water-soluble vitamin, is provided by animal-pollinated citrus and other fruits and vegetables. Over 50% of available folate, a B vitamin, is found in crop plants including beans and green leafy vegetables, where animal pollination is responsible for over 7% direct yield increase. Of the plants that supply more than 70% of available vitamin A and nearly all of our carotenoids, pollinators are credited for at least 40% direct crop yield increase.

Besides vitamins, actively pollinated plants also contribute significant minerals to the human diet. For example, about 60% of both calcium and fluoride come from different crops that include heavily pollinator-dependent plants, such as fruits and nuts. We derive an estimated 29% of iron from crops that see a 6% increase in yield due to animal pollination [42].

Quality is represented as cosmetically marketable product as well as nutritional value. For buckwheat, strawberry, and rapeseed crops, insect pollination has been shown to be directly responsible for overall fruit and seed quality.

Rapeseed plants produced more oil in their seeds when properly pollinated, which is critical considering the vast economic value and uses in animal feed, human-consumed vegetable oil, and biofuel. In strawberries, the size of a strawberry's red, edible portion (receptacle) was found to be directly related to the number of pollination-fertilized achenes, which are the seed-like formations on the receptacle [34].

Bee-pollinated strawberries were found to be heavier, physically superior with fewer malformations, and were categorized in higher commercial grades than fruits fertilized by wind- or self-pollination. The bee-pollinated strawberries also had improved shelf life with reduced sugar-acid ratios and increased firmness, where the longer shelf life translated to reduced fruit loss by over 11%. This longer shelf life and market value has a direct effect on trade revenue, representing reduced waste [43].

In a 2012 experiment, the cape gooseberry plant responded to bee pollination with increased fruit size, successful seed set, and germination rates compared to manual- and self-pollination [44].

While experiments for animal-pollinated crops tend to be highly specific to cultivar, region, and other variables, they are important data points and support the conclusion that pollinator-dependent crops, including those partially dependent, show significant improvement in quantity and quality of yields when properly pollinated.

4 Pollinator Decline: Bees

4.1 History, Statistics

A significant decline in pollinator populations, particularly bees, is currently being investigated, with the majority of studies centered in the United States and Europe. While wild bees are known to be important for pollination activities, lack of data as to the full extent of their involvement means most of the extant information concerning pollinator decline necessarily comes from human-managed honey bees at this time.

Cumulatively, European countries have also seen significant losses in a similar time frame, and 2013–2014 overwintering mortality rate was not as severe as 2012–2013, leading to the suggestion that climate and weather patterns across regions may have contributed to differences [45].

For the 2014–2015 winter season, the international COLOSS honey bee research association in Switzerland collected data from 31 countries and reported a range of loss data ranging from 5% in Norway to 25% in Austria. The proportion of colonies lost to either overwintering or post-winter queen problems was estimated to be 17.4%, or twice that of the previous winter [46].

In the United States, managed honey bee colonies have steadily declined in the past 60 years from six million in 1947 to only 2.5 million in 2014. Starting in 2006, commercial beekeepers in the United States reported a significant increase in overwinter losses, averaging 30% per winter as opposed to earlier historical rates of 10–15% colony losses. The 2013–2014 overwintering loss was over 23%, less than the over 30% losses of the previous season but still notably higher than earlier historical data. Hives had an estimated value of US$200 each as of 2014; the cumulative loss has been estimated at ten million hives thus far. To compensate, beekeepers must quickly rebuild colonies at great expense, which translates to increasing costs of commercial pollination. For example, the cost of renting honey bee hives for almond pollination in the United States rose from US$50 per hive in 2003 to US$150–$175 in 2009 [37]. These increased costs are passed on to the agricultural producer and finally transferred to the consumer.

The above statistics are primarily derived from overwintering records supplied by beekeepers via survey. However, recent trends indicate that honey bees face significant stressors throughout the year and not just winter, at least in the United States.

For the 2014–2015 season beginning and ending in April, the United States Department of Agriculture reported a 42.1% loss of managed honey bee colonies. Winter colony loss records, considered from October 2014 through April 2015, posted a small improvement over previous years, but 2014 summer losses exceeded winter for the first time at 27.4%. The 2014–2015 US losses exceeded the previous period's 34.2%.

Noncommercial beekeepers were more prone to mite infestations due to less aggressive mite control methods than commercial beekeepers, but commercial beekeepers saw more losses during the summer, indicating other factors were and are still at play [47].

4.2 Causes

The notable decrease in honey bee colonies and probable wild bee populations is believed to be caused by numerous factors acting in conjunction: climate change, exposure to certain pesticides or herbicides, loss of natural foraging sources leading to inadequate diets, parasitic mite infestations and ensuing diseases, among other things [24, 37].

The interactions among variables correlating with pollinator decline can be complex, with some of the posited variables themselves under contention. Some pesticides affect pollinators synergistically instead of additively. Many elements from chemical exposure to difficult weather and poor food sources can suppress immune response, which in turn promotes parasites and disease. Regional differences can also alter each variable's effect or produce a unique combination of stressors specific to a geographical area [48].

Seasonal mortality plays a significant role in overall honey bee population health, particularly in northern climates across North America and Europe, where cold winters and lack of flowering plants make foraging impossible. This is called the overwintering period. By late summer or early autumn, honey bees will have accumulated significant honey and pollen stores in their hives that serve as their only food source over winter. Overwintering is thus the period of winter survival and a key setting in which the many variables interact to increase stress and mortality rate on bees [49].

However, given the latest data (2014–2015) on summer colony losses in the United States, overwintering is only one period of expected bee stress [47]. The fact that data for annual including summer losses were only added recently in the major US Department of Agriculture honey bee reports means there is still much to learn about how bee stressors interact to produce mortality across the seasons.

4.2.1 Colony Collapse Disorder: Honey Bee

At the forefront of the overall pollinator decline concern was the phenomenon of Colony Collapse Disorder (CCD); though CCD is visible only due to the beekeeping industry and specific to the managed Western honey bee, the catastrophically sudden nature of hive mortality was enough to turn attention to pollinator health in 2006.

According to the available research, a combination of variables contributes to CCD, investigated primarily in the United States and Europe. CCD is a complex phenomenon that affects managed honey bee colonies (*Apis mellifera*); the primary symptom is a rapid loss of adult worker bees, and since these bees are responsible for the majority of hive tasks, the death of the entire colony typically follows.

CCD has a defined set of symptoms and is just one form of honey bee colony loss. Between the years of 2007 through 2014, 30–60% of winter colony losses in the United States conformed to CCD signs and symptoms, which include [50]:

1. Rapid loss of adult worker bees. Sudden decline varies between days and a few weeks.
2. Remaining worker bees are young.
3. The queen is still present.
4. There are no dead bees inside or around the hive.
5. At the final stage, there are no adult bees left in the hive.
6. Presence of capped brood in an abandoned colony.
7. Ample food stores in the hive (honey and pollen).
8. Delayed invasion of pests such as small hive beetles and wax moths.
9. Kleptoparasitism from neighboring honey bee colonies (also known as "robbing").
10. Colonies suffering from CCD are more likely to neighbor each other.

A complex combination of variables correlates with CCD:

1. Pesticides: insecticides such as neonicotinoids used in crops, miticides used in beekeeping, some herbicides and fungicides.
2. Parasites and disease:
   - *Varroa* mite infestation that causes direct damage by sucking bee hemolymph and indirect damage by promoting infection by virus, fungus, or bacteria
   - Viral infections include Acute bee paralysis virus (ABPV), Israeli acute paralysis virus (IAPV) and Kashmir bee virus (KBV), Deformed wing virus (DWV), Apis Iridescent virus (AIV)
   - Bacterial infection
3. Poor nutrition: pollen and nectar deficiencies from a variety of causes such as lack of flowers, habitat loss, etc.

Pesticides are broadly defined as any substance used to destroy or prevent various pests; in the context of agriculture, such chemicals include herbicides to kill weeds, fungicides to kill fungi, and insecticides to kill insects. These substances are most commonly used for crop protection, with global market revenues projected to be over US$50 billion in 2019 [51].

Other traditional pests and diseases such as foulbrood, chalkbrood, and *Nosema* fungal infection are not believed to be directly responsible for causing CCD, because they did not have a history of inducing CCD-like symptoms. However, the complexity of CCD due to a combination of stressors does not rule out the possibility that any previously singular malady exacerbates a hive situation that can lead to CCD [52].

A preliminary 2014–2015 honey bee loss report provided in collaboration between the USDA and Apiary Inspectors of America indicates that CCD has not been reported as the most troubling mortality cause to Western commercial honey bees in recent years, but its individual components may still be significant issues leading to current annual mortality rates that are still higher than acceptable levels [47]. This annual report relies on self-reported beekeeper survey data (Table 1.3; Fig. 1.3).

4.2.2 Insecticides

Neonicotinoid insecticides are effective insecticides chemically similar to nicotine and operate as selective agonists of nicotinic acetylcholine receptors, which are excitatory neurotransmitter receptors in insects. It is an extremely common class of pesticide used on 95% of United States corn and canola crops and the majority of numerous other crops such as cotton and a wide variety of popular fruits and vegetables. Besides direct crop application, it is applied to seeds for planting seasons. Neonicotinoids represent one-quarter to one-third of the global insecticide market and is thus a highly profitable product [54, 55].

Assertions of neonicotinoid pesticides' negative effects upon bees have been a source of contention since CCD came to light, with studies pointing to insecticide correlations.

There is evidence that sublethal doses of neonicotinoid pesticide—doses that do not cause direct mortality—have the potential to cause a variety of problems in foraging honey bees, including memory and learning dysfunction and interference with navigation. Because honey bees are social creatures that communicate direction and distance to food sources by way of the spatial waggle dance, these sublethal effects can prevent foragers from finding food sources or returning to their hives. Either risk is highly threatening to colony health, as the ongoing loss of adult bees stresses the entire social structure and leads to colony collapse.

A study published in 2012 described an experiment with pesticide-exposed foraging honey bee populations, using thiamethoxam at commonly encountered yet nonlethal doses. Scientists found substantial mortality due to homing failure, where exposed foragers were far less likely to return to the colony than control bees: about 10% to over 30% of exposed honey bees would fail to return when foraging on pesticide-treated crops. The probability that a bee forager would die due to homing

Table 1.3 The rate and leading causes of honey bee mortality

Winter (from October to April)	Total mortality rate	Acceptable mortality rate	Reported leading causes of mortality
2006/2007	32.0%		
2007/2008	36.0%		
2008/2009	29.0%		26% reported having at least one CCD loss.
2009/2010	33.8%	14.4%	28% reported having at least one CCD loss. Only 5% of beekeepers attributed CCD as the major cause for their losses. Starvation, weather, weak colonies in the fall, poor queens were listed as other causes.
2010/2011	30.0%	13.0%	31% reported having at least one CCD loss.
2011/2012	21.9%	13.6%	Poor wintering conditions, CCD, pesticides, weak condition in the fall, queen failure.
2012/2013	30.5%	14%	Queen failure, pesticides, starvation, weak colony in the fall.
2013/2014	23.2%	18.9%	Queen failure, poor winter conditions, *Varroa* mites.
2014/2015	23.1%	18.7%	Mites (noncommercial).
Nine-year average	28.7%		

Source: Bee Informed Partnership [53]

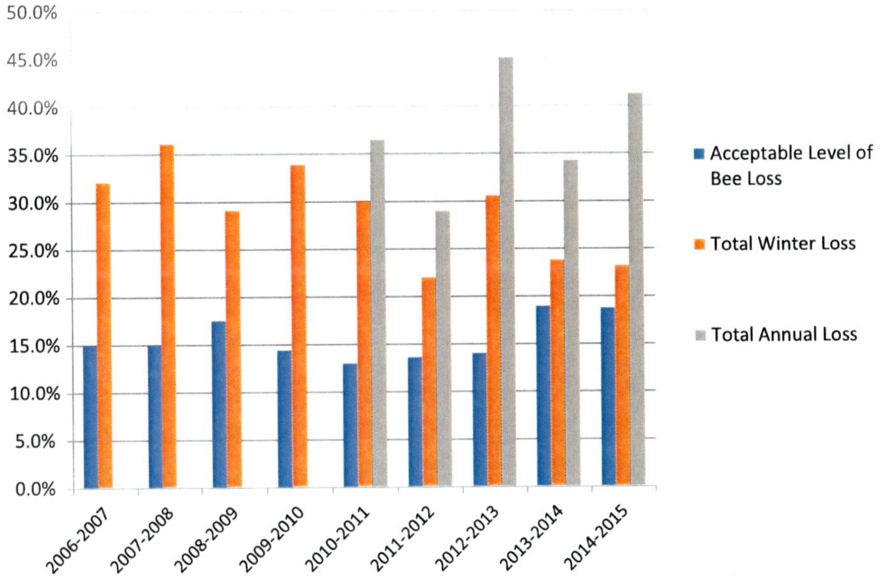

Fig. 1.3 Total US managed honey bee colony loss estimates (preliminary data for the 2014–2015 report). (Source: Bee Informed Partnership [53])

failure after exposure to treated crops was estimated to be twice that of natural death [56].

Outside Europe and North America, a study conducted in Japan drew a more direct connection between neonicotinoid pesticides and CCD. Three field experiments were conducted between July 2010 and August 2012 using approximately 10,000 adult bees across ten hives. The scientists observed effects of two neonicotinoids, which were primarily used in rice fields, upon foraging bees. They found that foraging bees were killed instantly wherever the insecticides were sprayed. The immediate death of foraging worker bees stressed the colonies and forced a role change from house worker bees to foraging bees in an effort to find sufficient food. However, the lack of house bees led to an imbalance in the workings of the colony; house bees are adult worker bees responsible for cleaning the hive, feeding the brood, caring for the queen, building, and so on. This social structure imbalance stressed the queen, which led to a decrease in egg-laying and eventually, due to constant mortality of adult foraging bees from the pesticide, collapse of the colony.

The Japan study found that even if foraging bees escaped instant death due to a sublethal dose of pesticide, the surviving forager bees returned to the hive with contaminated nectar and pollen. Hive members including the queen ingested and accumulated enough of the pesticide that CCD would still eventually be triggered [7, 57].

There have been numerous studies concerning varying effects of insecticides, particularly neonicotinoids, upon bee pollinator health and other ecosystem effects, though not all conclusions can be extrapolated to all situations. There are still knowledge gaps, but significant information now exists that indicates much greater caution should be exercised or alternatives sought [55].

4.2.3 Fungicides

According to the American Phytopathological Society in 2004, fungi are the number one cause of crop loss worldwide. Fungicides thus have an important place in agriculture, functioning to control crop diseases, which affect not only plant development but can be fatal to humans; increase plant productivity and reducing blemishes that affect market value; and improve the plants' storage life and quality [58].

The fungicides involved in preventing postharvest losses due to fungus-induced spoilage are primarily for the immediate benefit of human and animal consumption, but those chemicals applied on crops before harvest can have sublethal effects upon the crop pollinators.

A 2013 study that analyzed pollen from bees responsible for pollinating certain major fruit crops (apples, watermelons, pumpkins, cucumbers, blueberries or cranberries) found that fungicides were the most frequently found chemical substances, and the most common fungicide among the samples was chlorothalonil, which is widely used on apples and other crops. Bees that consumed the chlorothalonil-tainted pollen were found to be three times more susceptible to *Nosema* infection, a microspore fungal parasite; *Nosema* is known to be a major disease by beekeepers, where infected colonies fail to thrive shortly after overwintering [59].

To further knowledge of fungicide exposure effects upon other pollinators besides honey bees, a controlled experiment was conducted upon a native bumblebee species. Colonies were exposed over a month to flowers treated with field-relevant levels of a fungicide and then quantitatively assessed by number and biomass of the three bee growth stages (larvae, pupae, adult). Exposed bumblebee colonies were found to produce fewer adult workers bees and lower bee biomass, including smaller and lighter queens, than control colonies. The number of worker bumblebees was reduced from an average 43 bees in control cages to a mere 12 in fungicide-treated cages.

While these effects were sublethal, reduced individual and colony health has direct bearing on ability to resist variable environmental conditions and stressors [60].

Fungicides have been typically believed to be safe for bees, but sublethal effects observed in recent studies should be taken into consideration for future agricultural control efforts, otherwise the long-term result may be trading fungicide advantages for pollinator decline that can still lead to reduced crop production.

4.2.4 Herbicides

Herbicides are pesticides that target plants perceived to be weeds, or those plants that compete against crops or preferred ornamentals in the same land space.

The herbicide market is projected to be worth nearly US$30 billion by 2019, with glyphosate occupying the largest market segment [61]. Glyphosate is a broad-spectrum herbicide used for weed control, which some may remember as the original retail product Roundup. As of 2007, glyphosate was the most common herbicide used in the US agriculture market and the second-most used herbicide in the home garden market [62].

An experiment to study sublethal effects of this herbicide upon honey bees was conducted using field-realistic doses of glyphosate, exposing the bees chronically or acutely to the herbicide. In the groups chronically exposed to concentrations within the range of recommended doses, a reduced sensitivity to sucrose and reduced learning performance were found. Acute glyphosate exposure decreased learning and short-term memory retention when compared against control groups. These results implied that glyphosate at common concentrations found in standard spraying can reduce sensitivity to nectar reward and impair associative learning in honey bees. However, no effect on foraging-related behavior was found, which implies that successful forager bees could become a source of constant inflow of nectar with glyphosate traces that could then be distributed and stored in the colony, causing long-term negative consequences on colony performance [63].

Another sublethal experiment revealed that honey bees that had been fed with a glyphosate solution of a common concentration took a longer time to return home and took more indirect flight paths than control bees or bees treated with lower glyphosate concentrations. Furthermore, after a second release at a release site, the proportion of direct homeward flights performed increased in control bees but not in treated bees, where one would expect the bees to remember the local topography due to experience. This result suggested that exposure to glyphosate doses commonly found in agricultural settings impairs the cognitive capacities needed to successfully return to the hive, such as navigation and memory. Long-term negative effects on pollinator colony health are likely as forager bees are lost and house bees are forced to begin foraging while still young, creating a social structure imbalance [64].

4.2.5 Habitat Loss

Habitat loss, which is the reduction of natural and seminatural home environment, is believed to negatively affect pollinators by loss of diverse floral resources from their diet. This is likely more of an issue for wild bees, which are major yet largely invisible participants in crop pollination, than for managed honey bees. Managed beehives are rented out to farmers and thus can be physically moved between floral sources, but wild bees do not have such a luxury.

For honey bees, deficiencies in quantity and quality of pollen and nectar lead to decreased colony population. A pollen deficiency causes low colony populations and further reduces the number of foraging bees responsible for food gathering.

Modern industrial agriculture systems that use an annual cropping schedule and simple rotations across large swaths of land have resulted in monocultures, particularly wind- or self-pollinated crops, with greatly reduced plant diversity. Monocultures of bee-pollinated crops may not only have a short duration of massive flower blooming but also insufficient nectar and pollen variety for honey bee needs. Then, once the mass bloom is over, there is little else for bees to forage upon across the large monocropped landscape, which can prove challenging to their overwintering preparation.

Besides nectar as a basic food source, pollen quality and diversity have been shown to improve bee immune system health. Both are so important in the bee diet that beekeepers must often provide supplements to prevent nutritional deficiency and potential colony failure. However, pollen supplements and sugar syrup do not provide the same nutritional quality as natural pollen and nectar [65].

In comparing the seasonal contribution of rapeseed and sunflower mass-flowering crops versus other floral resources in the bee diet, scientists found that bees collected nectar mainly from the crops, but pollen came from a diverse range of plant species in nearby seminatural habitats or from weed plants among crops. Weed species actually constituted the bulk of the honey bee diet between the mass flowering crop periods [66].

One of the first studies to specifically address wild bee participation helped illustrate the importance of diverse habitat alongside crop lands. A study set in France investigated effects of forest edges upon pollinator activity in rapeseed fields. Scientists caught bees within the crop fields at measured distances from forest edges and found that, out of several thousand identified bees representing multiple taxa, both the number of bees and species diversity declined with increased distance from a forest edge. The bee taxa included those that nested in particular wilderness habitats. The data indicates that nearby natural and seminatural habitats such as forest edges may serve as "reservoirs" of pollinators and would directly benefit agriculture by providing shelter and mating sites appropriate for specific bee species [67].

To further the implication that uncultivated plants play a critical role in both bee health and crop success, several other past studies have shown that pollination services increase in crops near forest patches or other seminatural areas rather than in uniform crops surrounded by yet other crops [68].

It thus stands to reason that as pollinator activity increases with the presence of diverse wild habitat near croplands, loss of such habitat will reduce pollination potential at the same time. Moreover, if managed honey bees are imported to mitigate wild pollinator loss, the increased operating costs will be passed on to consumers as well.

4.2.6 Parasites

A number of pests and parasites target honey bee colonies and pose significant risks to individual bees as well as overall colony health due to the bees' close proximity to each other in the hive. Not all pests are associated with CCD, though their presence could precipitate additional issues such as infection that may lead to CCD symptoms.

Honey bee pests and parasites include *Varroa* and tracheal mites, *Nosema* fungal microsporidian infestation, small hive beetle, and wax moths; risks vary by region and climate.

The *Varroa destructor* mite is the most dangerous to honey bees and is associated with CCD; it is a parasitic mite about 1–2 mm long. It was first identified in the 1960s in Japan and the former USSR. It made its way across the world and was introduced (identified) in many European countries in the decades following, reaching the United States in 1987, with the latest identification in Hawaii in 2007 [69].

Parasitic mite infection begins when the female mite enters a honey bee brood cell. As soon as the cell is capped (sealed with food), the *Varroa* mite lays eggs on the bee larva, which typically hatch into several females and one male; they hatch at about the same time as the young bee develops, and the mites thus leave the cell with the host bee. Infestation occurs when the young developed bee emerges from the capped cell, allowing the *Varroa* mites to spread to other bees and larvae. The mite preferentially infests drone cells. The adults suck the "blood" (hemolymph) of adult honey bees, leaving wounds and weakening the adult bees such that their immune systems become compromised. These weakened, mite-infested bees become prone to a variety of viral, bacterial, and fungal infections [70].

The primary lethality of *Varroa* mite infestation occurs during the overwintering period, particularly in longer winters where honey bees are under environmental stress and surviving within their hives on food stored in the previous spring through fall seasons. Close proximity in hives promotes infestation, and because the bees are weakened by the mites, they are also less likely to fend off other infectious diseases. This compounded scenario of stressors is one of the primary variables believed to be responsible for CCD [71].

Chemical treatments, such as fluvalinate, have proven to be mostly successful, but a small percent of surviving mites after treatment become resistant and eventually immune. Treatments are still being developed. A number of other countermeasures may be used, whether chemical, physical, or behavioral, but the mite threat continues.

4.2.7 Disease

Honey bee colonies are also prone to a number of viruses and bacteria that thrive in the close quarters of the hive. There are numerous viruses, some of which are associated closely with *Varroa* mite infestation, and some bacterial infection that can produce significant rapid losses [72]. Any of these diseases can be associated with CCD as a compounding variable.

1. Bacterial
   - American foulbrood
   - European foulbrood
2. Fungal
   - Chalkbrood
   - Stonebrood
3. Viral
   - Cripaviridae
   - Dicistroviridae
   - Cloudy wing virus
   - Sacbrood virus
   - Iflaviridae
   - Iridoviridae
   - Secoviridae
   - Lake Sinai virus

While honey bee colonies most obviously suffer the consequences of infestation and infection due to the physical proximity of their members, new research suggests that managed bee services introduce parasites and disease to local wild populations when high concentrations of such services are imported to new areas. These alien stressors upon the local pollinator populations are associated with both declines and extinctions, a problem which must be addressed with some urgency because merely compensating with more managed bees is not a sustainable answer [73].

4.2.8 Climate Change

Climate change is a very broad, long-term, and complex process that affects global ecologies in a chain reaction across organismal relationships. Due to its long temporal nature, climate change in relation to the highly seasonal life cycle of bees can be challenging to study. The most symptomatic effect of climate change in the pollinator-plant relationship is the influence upon plant flowering time, which in turn can affect the pollinators' primary food sources.

An early observation of potential climate change effects upon pollinator activities, particularly the honey bee, came from a NASA oceanographer, Wayne Esaias, who observed changes in his hobbyist beehives; his background and access to large data sets allowed him to piece together the effects of climate change upon local flora and thus his honey bees. Esaias began noticing changes in his hives in the early 2000s, where the hive weight in preparation for overwintering showed symptoms that local Maryland flowering was occurring earlier in the year, an observation in-line with long-term trends from flowering records dating back to the 1970s.

The co-dependency of pollinators and plants can be interrupted by climate changes when their respective life cycles in response to the environment shift in different ways. The relationship between plants and their pollinators are at risk for desynchronizing when environmental cues are different and when there is not sufficient overlap of generalist pollinator populations. For example, some insects' larval stage is timed to mature when local flowers first bloom and nectar begins to flow. The pollinator for one plant may begin foraging when air temperature rises to a specific point, but that plant may begin to produce flowers and nectar flow when the snow melts [74].

While some of the earliest observations of climate change effects upon pollinators came from managed honey bees due to necessary human interaction and thus accessibility of data, wild pollinators have not been entirely ignored. However, because the habitats and life cycles of typically nonsocial wild species can vary substantially as opposed to the standardized rearing of the European honey bee, research on wild bees must be taken with understanding of native wild plants and the local ecology instead of standardized agricultural crops.

A Colorado, US study indicated that local environmental conditions are the primary cues for insect emergence, particularly between wild plants that have evolved alongside a number of wild pollinators. Furthermore, seasonal cycles of plants, bees, and wasps in the area are regulated similarly by temperature, but plants are more likely than insects to advance their cycle in response to spring warming. Thus, flowering before the emergence of local pollinators is a real possibility according to the 2011 Colorado study. Despite that risk, none of the native pollinators were specialized to a single plant species, which helps mitigate more catastrophic cycle shifts in specific plant species if the pollinator were completely dependent. However, even generalist bees may be affected as both larval development and survival can vary by available pollen diet.

The local wild bee pollinators were shown to have evolved specific coping mechanisms for a variable local environment on a year-by-year basis, but a prolonged succession of poor seasonal conditions could prove to be too much for localized mechanisms. The Colorado area saw warmer and drier years in the past few decades, which correlated with midsummer floral scarcity, while some plant species have suffered a series of poor flowering years due to spring frost damage. This combination of poor plant conditions over a longer period of time due to climate change may prove to be a much broader threat to the pollinator-plant relationship than just their life cycle asynchrony [75].

The Food and Agriculture Organization of the United Nations (FAO) took a broader approach and surveyed available research in 2011. They found a small number of studies that investigated how increased temperatures might instigate temporal mismatching between plants and pollinators. For example, a 2005 study centered on the Iberian Peninsula researched seasonal life cycle responses of plants and pollinators in relation to increased temperatures, finding a potential desynchronization between them. A 2009 study found that both *Apis mellifera* honey bees and *Pieris rapae* butterflies advanced their activity timing in response to warmer temperatures while their main floral resources did not, producing a temporal mismatch between pollinator and plant. A Japanese 2004 study found that early-flowering plants would flower earlier in a warm spring, but bumblebee queen emergence was not affected by spring temperatures. A 2007 study on the effects of increasing temperature on a plant-pollinator network showed that seasonal cycle shifts reduced floral resources for 17% to 50% of pollinator species. These studies indicate that temperature-based temporal mismatches between pollinator and plant do occur, but vary by species and region [24].

Out of the probable stressors upon pollinator health, climate change itself and its complex ecological effects are still nebulous, but extant research suggests it should not be ignored as a variable that can affect agriculture and our own food supply in the long term.

4.3 Effects of Pollinator Decline (Bee Loss)

Though variations of the inaccurately attributed quote "If the bee disappears from the surface of the earth, man would have no more than four years to live" have appeared in mass media when news about the declining bee population periodically resurface, the actual projected effects of pollinator loss are more protracted and complex [76].

There are numerous studies of specific crop improvements thanks to pollinator activity, but calculations for estimated impact of loss upon agriculture can differ substantially depending on the methods and data used. It is important to note that most of the studies focused upon attempted analysis of pollinator decline effects must extrapolate to some extent; much like climate change science, such information is intended for preventative use before negative impact becomes irreversible. Reports agree that both ecosystems and agriculture will be negatively impacted as pollinator populations fall. Three basic viewpoints from which bee loss impact can be analyzed and extrapolated are as follows:

1. Direct crop yield reduction
2. Economic impacts
3. Human health impacts

4.3.1 Direct Crop Yield Reduction

One approach to calculating the impact of pollinator loss is by measuring crop output and land use in relation to what we know about pollinator dependency and yield improvements. Just like prior yield improvement studies, however, it is important to remember that different crops in different regions rely on pollinators to different extents.

In a 2009 study attempted to estimate global crop losses should animal pollination fail. Using extensive 1961–2006 data from the Food and Agricultural Organization of the United Nations, pollination dependency was calculated for 87 important crops. Though about 75% of flowering crops benefit from pollinators to varying degrees, only 10% depend fully on pollinators for seed and fruit production; these dependent crops account for an estimated 2% of global agricultural production by 2009. If those crops with the highest pollinator dependency failed due to lack of pollination, global crop production would decrease by an estimated 8%.

However, the 8% crop impact is merely an estimate of a direct reduction and does not represent economic or nutritional effects, so the given calculation is an abstraction. As it turns out, indirect negative impacts can be significant. If the yield of a land unit is reduced, then more land would be turned over to agriculture, resulting in yet more pollinator habitat loss and other related ecological threats. The estimated increase in cultivated land area necessary to compensate for the production deficit was several times higher than normal, especially in the developing world, which already supports two-thirds of global agricultural land [33].

A 2013 study centered in the European Union approached the problem with a different method and reached a different conclusion. Instead of calculating pollinator dependency of individual crops, scientists assessed "relative pollination potential," which is the relative potential or capacity of an ecosystem to support crop pollination. When insects begin foraging, their host ecosystems or "land cover" such as forest edges have the potential to increase yields of adjacent pollinator-sensitive crops; this potential represents the wild pollinator's economic value and can be measured as yield contribution or the cost savings when compared to replacing with managed honey bees. Using existing foraging data of wild bees with short flight distances, pollination potential was linked to regional European crop production statistics. Representing data limited to those pollinating insects with short foraging distances, the crop production deficit was estimated to be 2.5%. However, as more data of species with increased flight range were added to the map, pollination potential would increase up to 25%.

That is, under this method and using aggregate data of different pollinator species, the yield reduction of partially and fully pollination-dependent crops without insect pollination was estimated to be at least 25%, also taking into account habitat diversity. These estimated results were specific to Europe [77].

There is thus some disparity among results from studies that attempt to estimate direct yield reduction, due to different methods, crops, as well as regions. However, a decline of some sort is noted in all cases.

4.3.2 Economic Impacts

Besides a direct yield reduction estimation, the economic impact to the agricultural and apiculture industries would be significant though variable depending on the cost-benefit analysis.

The total economic value of crop pollination, considering different levels of pollination dependency, was estimated to represent 9.5% of the world agricultural product value in 2005 [78].

In a study that employed geographically-bounded information on global yields of 60 crops that either benefitted or depended upon pollinators, scientists investigated economic vulnerability in relation to pollinator decline as the portion of agricultural economy depending upon pollination benefits. The general pollination dependency of agricultural economy appeared stable from 1993 to 2009, but producer prices for pollination-dependent crops increased in the same time period [79].

Some economic studies can only provide specific examples due to cultivar and regional specificity, but they are nonetheless illustrative of negative impact. A 1997 Canadian study estimated that pollination-improved apples provided returns of about 5–6% to the farmer compared to orchards without honey bees. With pollination services costing about 1% of production expenses, the improved yield represented a 700% return to the farmer. A loss of pollination services would introduce a notable reduction of economic value for the grower in this case [80].

More generally, a number of variables point to pollinator shortage as an actual, recorded impact upon food economics. The demand for pollination services has been increasing with the expansion of pollinator-dependent crops by acreage, at the same time the supply of managed honey bees in some regions is declining; the rate of growth of pollination service demand outstrips the managed bee supply rate of growth. The cost of managed honey bee colony rentals in regions of the United States has increased dramatically. For example, Pacific Northwest hive rentals increased from US$19.25 in 1992 to nearly US$90 by 2009; if pricing followed the US inflation rate, hive rental cost should have been US$25.73 in 2009. Colony Collapse Disorder losses by 2007 forced a loosening of trade restrictions of imported honey bee queens in order to sustain the vast almond industry; compensating for lost locally managed colonies thus introduced additional costs to rebuild hive populations and potentially new pathogens [81].

Taking into account economic factors such as calculated costs and crop conversion to monetary units, it is therefore possible to better understand the impact of pollinator decline in relation to economic trade as opposed to just a direct yield reduction.

4.3.3 Human Health Impacts

Significant pollinator decline would also affect the nutritional quality of agriculture produce. Pollinated crops are responsible for supplying the majority of available fats, vitamins A, C, and E, and minerals such as calcium, fluoride, and iron to the human diet. Crops that are partially or fully dependent upon pollinators account for more than 90% of our dietary vitamin C, for example [82].

A 2015 modeling analysis estimated that assuming a complete loss of pollinators, over 70 million people in low-income nations would become vitamin A deficient; 173 million people would experience folate deficiency. A globally estimated 22.9% of fruit supplies, 16.3% of vegetables, and 22.1% of nuts and seeds could be lost without pollination, with significant variation by country. Annual global human deaths due to malnutrition could increase by 1.42 million, while a 50% loss of pollination services is estimated to potentially cause 700,000 additional annual deaths [83].

Yield increases associated with pollinator activity in these crops are not insignificant, so a decline in crop yields means lower produce availability, nutritional quality, and increasing prices of varying urgency around the world. Even considering the possibility for human diet supplementation and fortification, many governments do not regulate their food industries to the same extent as more affluent countries. Moreover, the fact that many dietary supplements are actually derived from pollinator-dependent plants adds a catch-22 to the threat of pollinated-crop declines.

5 Conservation and Protection of Bees

5.1 *Beekeepers*

Recommendations for bee management will vary from region to region due to differences in climates and local fauna and flora. These recommendations may come from a variety of sources such as commercial apiary experts to government resources and apply to anyone seeking to manage their own bees, including hobbyists.

Best practices include general recommendations not only for direct bee monitoring and treatments but also managing the surrounding areas where bees are kept [84, 85].

1. Create or allow nectar-rich wild flower areas on the premises.
2. Plant a variety of flowers that can provide nectar and pollen across the length of the honeybee's foraging seasons, including early-flowering plants and plants for late autumn.
3. Plant and manage hedges of mixed flowering species that are good for bee foraging (the list will differ based on location, but examples include hawthorn, hazel, crab apple, willow).

4. Leave wild plants and weeds until they stop flowering before removing them.
5. Avoid cutting back nectar plants that flower late autumn into winter, or plants that continue to flower while still producing berries.
6. Mow lawns containing white clover only after the flower heads have turned brown.

Management of honeybee health and pests may vary among geographical regions, and the specifics will vary among bee strains as well. American beekeepers in the state of Florida, for example, must deal with local black bears that target and destroy honey bee hives [86]. Some bee management recommendations include the following [87]:

1. Use only products approved in your country.
2. Read and follow all instructions on disease or mite control product labels before application.
3. Monitor colonies for levels of mite infestation.
4. Apply treatments before infestations or infections become damaging. Autumn feed should be provided before temperatures are too low for bees to safely break cluster.
5. Rotate *Varroa* treatments to prevent development of resistance. For example, use a synthetic mite strip in spring and formic acid treatment in autumn.
6. Avoid using temperature-dependent treatments such as formic acid or thymol above the recommended thresholds.
7. Treat all colonies that require treatment at the same time.
8. Follow recommended treatment withdrawal times; do not use treatments when honey supers (portion of a managed beehive used to collect surplus honey) are attached, unless otherwise specified by treatment instructions.
9. Use oxalic acid only as a follow-up treatment in late autumn after a primary early autumn treatment, unless *Varroa* mite levels are low.
10. Use resistant-bred queens when possible.

Along with the above best management practices, beekeepers may also participate in breeding or acquire honey bees that have been specially bred through hybridization or trait selection. This approach introduces pest and pathogen resistance into European honey bee stock and allows beekeepers to reduce or altogether eliminate chemical treatments in their apiaries. Additionally, hybridization or trait breeding can increase colony survival rates when the selectively bred bee is more adapted to a given environment.

5.1.1 Varroa-Sensitive Hygiene (VSH) Genetic Trait

Varroa-sensitive hygiene (VSH) is a genetic trait that compels the honey bee to remove mite-infested pupae from the capped brood, which is the *Varroa* mite's preferred target. The capped brood are young, developing bees that are sealed inside cells of the comb with a protective layer of wax. Because the brood is hidden below the wax seals, the mites are sometimes difficult for bees to detect. While honey bees are naturally hygienic and often remove diseased brood from their nests, VSH is a specific form of nest cleaning aggressively focused on removing Varroa-infested pupae. The VSH honey bees will chew and cut through the cell cap and discard the infected brood plus mites from the broodnest.

After testing and hybridization, mite population growth was shown to be significantly lower in VSH and hybrid colonies than those without the trait. Hybrid colonies had half the VSH genes compared to pure VSH bees but still retained significant *Varroa* mite resistance [88].

Since the mid-2000s, USDA scientists in Baton Rouge, Louisiana, developed honey bees with high expression of this trait, and VSH bees are now commercially available to beekeepers [89].

5.1.2 Africanized Bees

The Africanized honey bee is particularly invasive in the Americas and has settled across numerous southern states of the United States, so it is important to understand their advantages and disadvantages in agriculture as they continue to hybridize with European honey bee stocks.

To most people, Africanized honey bees are perceived as a threat due to its fierce defensive behavior. In reality, while deaths attributed to "killer bee" attacks are violent and frightening, they number far fewer than other causes and, in North America, pose greater danger to the beekeeping industry and agriculture than to the general public.

European honey bees were selected and imported by beekeepers for their predictable and manageable traits such as gentleness, limited swarming behavior, and high honey production. On the other hand, Africanized honey bees often behave unpredictably in comparison. Once Africanized honey bees enter European colonies, mating between them results in more hybrid bees having African genes and expressing African traits over European ones. The behavior of the entire colony may suddenly shift to highly defensive and short-tempered. Even though the defensiveness of Africanized honey bees varies from colony to colony, just a few colonies within an apiary expressing excessive aggression could cause a beekeeper to abandon apiculture. The Africanized bees' higher tendency to abscond also makes them less desirable in managed beekeeping [90].

Due to higher swarming and absconding rates of Africanized honey bees, honey production per colony does decline. However, since there are many more, smaller Africanized colonies participating in pollination and honey-producing activities, the overall apiary yield may actually increase. When Africanized bees first arrived in Mexico, the country's honey exports initially dropped by more than 50%, which might be attributed to beekeepers abandoning their trade due to the bees' unmanageable aggression. Mexican honey production had recovered to 75% of historical levels by 1986 [91].

Over two decades since the first arrival of Africanized Honey Bees in the United States in the 1990s, there were no reports of significant effects upon the honey production and investment behavior of beekeepers in the United States [92].

In terms of pollination activity, competition for nectar and pollen resources among Africanized honey bees, wild bees, and European bees is expected. However, Africanized honey bees have shown good work ethic by working earlier, longer, and even under undesirable conditions of a desert or rainforest. At least in Mexico, it is believed that these hybridized bees helped the coffee plant to flourish [91].

Despite negative traits that make management difficult, Africanized honey bees have significant advantages over their European counterparts as well. Africanized bees have shown higher resistance to parasites such as *Varroa* mites and certain diseases, reducing or negating the need for chemical treatments such as antibiotics [93].

Africanized honey bees produce and use more propolis than European honey bees. Honey bees collect saps and resins from trees and mix with self-produced wax to make propolis, which has various antibiotic properties. Bees use propolis to seal cracks, close potential entries to the hive, to embalm dead intruders too large to easily remove, and, most importantly, to prevent diseases and parasites from invading the hive [32, 94]. In fact, Africanized honey bee lines bred for high propolis production were found to have superior hygienic behavior as well as increased honey and pollen storage [95].

Beekeepers in areas affected by significant Africanized bee populations such as Brazil and Mexico have since adapted their management practices and now prefer Africanized honey bees due to their greater hardiness and adaptation to those tropical and subtropical climates, as opposed to the European honey bee [96].

Thus, it is possible to selectively breed Africanized honey bee colonies for preferred traits, such as increased propolis usage, mite resistance, honey production, and so forth. This does not mean that the increased aggressiveness must be accepted in managing hybridized stock, however. In fact, selective breeding of Africanized honey bees has been shown to reduce aggressiveness after only a few generations [97, 98].

Feral honey bee populations cannot be controlled this way, of course, but for the purposes of apiculture and its relation with agriculture, selective breeding of Africanized stock in appropriate climate zones may prove to be beneficial for a variety of reasons.

5.1.3 Russian Bees

The Russian honey bee is actually derived from European honey bees that originated in the Primorsky Krai region of Russia. This bee strain was imported into the United States in 1997 in response to parasitic *Varroa* mite infestations during overwintering periods that would precipitate colony collapse.

The USDA bee lab in Baton Rouge compared Russian colonies to domestically raised European bee lines and found that Russian bees are resistant to three of the major honey bee stressors: *Varroa* mites, tracheal mites, and cold temperatures [99]. Specifically, Russian honey bees have been found to be about one-third to one-half more resistant to mite infection than the European honey bee, and mite infestation is a major variable in Colony Collapse Disorder [100].

They exhibit much higher hygienic behavior in opposition to parasitic mites. Experts reported that Russian bees are more aggressive in grooming themselves and each other, biting and removing mites and infected brood from the hive. Russian honey bees are not actually immune to *Varroa* mites but instead have developed a resistance via behavior adaptation after 150 years of exposure in Russia.

Durable overwintering is one of the Russian honey bee's strongest traits, a product of the strain's originating cold, northern climate. After five winters in northeast Iowa, entomologists observed that Russian bees are less likely than European bees to lose members to cold weather [99].

Russian queens shut down earlier, and the worker bee population drops to winter levels sooner than their European counterparts. In the overwintering months, Russian bees are frugal in their consumption of stored food compared to other managed bee strains, and their hives have plenty of leftover winter food stores by the time spring begins [101].

Importing or hybridizing with Russian honey bees has the primary advantage of developing mite resistance in locally managed honey bees, while the bees' honey production abilities were shown to be sufficient. Moreover, the bees' temperament is reported to be rather docile. A number of beekeepers have shifted to Russian or Russian-hybridized honey bees for their pollination and honey-producing capacities. Initial reports of honey production were conservative, where quantities were below European counterparts, but current estimates place their honey-producing capacity as in line with or more than European honey bees.

In one example, a US breeder in the late 1990s reported that after a harsh winter, 1200–1400 of his European bee stock perished, while only two out of 2000 Russian-bred colonies did not survive. The breeder also reported average honey yields of approximately 60–68 kg per hive, well above the usual 38 kilograms [71].

The US Department of Agriculture's research agency is continuing their Russian bee breeding program, where this strain of honey bee has a particular advantage versus European honey bees in more northern climates.

5.2 *Farmer: Habitat Diversity*

There are a number of actions that farmers can take to improve foraging conditions for honey bees and wild pollinators. It is important to note that habitat conservation to enhance nectar and pollen forage for wild bees would benefit managed honey bees, but the opposite may not be true [65].

Those in the agricultural industry can help mitigate effects of bloom de-synchronization, monoculture limitations, and poor diet by local planting practices. By improving landscape conditions for both wild and managed pollinators, farmers stand to improve their own crop yields and quality to remain competitive in the agricultural economy.

In order to sustain bee populations, support honey bee health, and assist beekeeping activities, diverse native and nonnative, flower-rich herbaceous plants should be maintained in fallow areas, field margins, and buffer strips. Landscapes with ample flowers containing good-quality pollen and nectars are vital for enriching pollinator health. This in turn also helps nearby crop fields [65].

By conserving and creating diverse habitats on farmland, farmers can help address bloom de-synchronization due to climate change or nectar and pollen shortage due to monoculture. Providing diverse supplemental forage resources to both wild bees and managed honey bees would both increase survival rates and improve the bees' capability to pollinate.

5.2.1 Cultivated Arable Farmland

Arable farmland is land that can be cultivated for agricultural use. Cover crops are annual or perennial plants temporarily sown to protect noncultivated land from erosion, prevent weed spread, disrupt crop pest life cycles by replacing pest host plants, and improve soil quality by adding nitrogen and organic matter to the soil, or reducing nitrogen runoff in water [102].

Annual Cropping Systems

In annual cropping systems, a single annual crop rotation has resulted in mostly monocultures of very low plant diversity across large expanses of land. To supplement honey bee diets and help wild pollinators as well, the following conservation activities are recommended:

11. Plant flowering plants in noncropped areas to supplement forage resources between the bloom times of cultivated crops.
12. Maintain large strips (6–12 m in width) of native or nonnative melliferous plants between crop fields as "bee pastures."
13. Mitigate herbicide usage within cropping systems to support bee foraging through the increase of flowering weeds near the crop areas.

To aid pollinators, use plants with short, rapid flowering life cycles as cover crops in the spring before the primary crop is sown: this will provide early forage for honey bees and other pollinators. Similarly, these plants can follow the primary crop in rotation to provide late season forage sources. Examples of plants suitable for this purpose are annual clovers and mustards. Using longer-lived or even perennial cover crop plants are preferred for a cover crop rotation between two *Gramineae* crops such as barley and corn [103].

Perennial Cropping Systems

Perennial cropping systems are those that do not require replanting on an annual basis, thereby reducing necessary inputs, erosion losses, and water pollution, as well as increase biological carbon sequestration [104].

Typical conservation efforts in perennial cropping systems consist of planting flowering hedgerows, strips of vegetative plants surrounding field margins, flowering meadows, and designated bee pastures. These practices serve multiple purposes including providing supplemental foraging for bees while crops are not in bloom, reducing erosion, serving as buffers against pesticide drift, and providing food for other wildlife [105].

However, the presence of these flowers has a potential risk of exposing bees to insecticide when crop is treated outside of crop's bloom period. Pesticide mitigation strategies should be incorporated with the supplemental nectar and pollen plants in perennial crops [65].

5.2.2 Non-cropped Lands

Noncropped lands are defined as natural and seminatural habitats that are field margins (strips bordering crop fields), hedgerows (linear scrub along field boundaries), ponds, woodlands, ditches, and fallow farm fields. These lands may or may not be arable, are less disturbed than cultivated lands, and they can better maintain overall biodiversity. The flowering plants in noncropped farmland can help restore and increase habitat for both honey bees and wild pollinators [106].

There are two basic methods to enhance floral diversity in noncropped areas: Allow native wildflowers to regenerate naturally, or sow a variety of annual and perennial flowering plants.

Fallow Areas

Fallow areas are defined as farmland plowed and harrowed but left unsown for some time to allow restoration of fertility as part of a crop rotation or to avoid surplus production. The original purpose of fallow farmland was the latter, where farmers were compensated by government incentive payments for removing part of their land from production, but now, fallow field objectives have expanded to include conservation of biodiversity [107].

For example, under the EU Conservation Reserve Program (CRP), participating farmers have been paid an annual rental fee for the conversion of highly erodible farmland to a land planted with a mixture of native grass and wildflowers [108].

As of 2010, the national limit of fallow set-aside programs in the United States was capped to 15.8 million hectares (about 39 million acres), and it was capped at 2.8 million hectares in the European Union [65].

Field Margins

Field margins are narrow strips of land adjacent to crop fields, roads and railways, hedgerows, and forest edges. Individually, these areas may seem negligible, but cumulatively, they amount to significant unexploited acreage. Field margins can serve a variety of purposes including river bank stabilization, erosion prevention, and buffer zones for pesticide drift. Many USDA Natural Resources Conservation Service programs actively manage field margins for erosion control and have adapted management standards to incorporate forage for pollinators as an additional benefit [105].

Plant Selection for Noncropped Land

The value of noncropped farmland to honey bees and other pollinators depends very much upon the plant species available within this land. Two strategies are recommended to improve nutritional resources for honey bees: sow plants that provide good pollen and nectar, and conserve native plants [65].

For example, legumes (*Fabaceae*) are among plants most visited by bees for pollen and nectar. Some legumes that are particularly attractive to honey bees are those having a long and multiannual flowering periods [109]. Some farmers may be concerned by legumes' slow growth due to lack of flowers the first year of planting, in which case they can plant a mixture of annual, biannual, and perennial plants including wildflowers to ensure early and extended flowering for both honey bees and wild pollinators [65].

Farm Practices for Noncropped Land

Besides actively encouraging plant diversity, land management in noncropped areas is important as well. Among recommended management practices are the limiting of chemical and mechanical destruction of noncrop floral resources and preventing pesticide drift to bee foraging areas.

To reduce mechanical destruction of floral resources, the following practices are recommended [103, 110]:

1. Avoid mowing high-nectar plants during flowering to reduce foraging honey bee mortality rates.
2. Delay the first cutting of vegetation in the spring to allow more flowering plants and thus more pollinator visits.
3. Cut grass at the end of summer or in autumn to conserve grasslands.

5.3 *Farmer: Agrochemical Practices*

Besides habitat conservation and agricultural land-use practices, both farmers and the general public can alter the type, amount, and timing of pesticide application to help reduce or eliminate risk to bees. Ideally, pesticides are best avoided whenever possible, but if they must be used, these strategies are recommended.

5.3.1 Pesticide Usage

Use the least toxic pesticides when possible and follow label instructions, starting with the minimum manufacture recommended dose. Granular form applications are usually the safest method around bees. Prior to spraying with insecticides, remove flowering weeds, which may attract bees; if ground-cover plants are in bloom, mow before spraying.

Stay away from bee hive areas. Notify beekeepers to move nearby bee colonies prior to treatment, so that the hives can be relocated or covered. Place buffers around important pollinator foraging, nesting, and overwintering areas [111].

Both Pyrethrin, and spinosad are commonly used pesticides in organic farming. These broad-spectrum insecticides will kill pests and beneficial species indiscriminately. However, other approved organic products, such as horticultural oils and insecticidal soaps, are safer to use if not applied where pollinators are present [112].

5.3.2 Timing of Application

Fungicides are recommended to be applied at the end of bloom. If this is not possible, then application several days before honey bees are brought to the orchard is recommended [113].

Try to avoid applying pesticide to blooming plants or on plants where bees are active. Applying at night or at lower temperature is one way to reduce the chance of bee exposure, since they typically are not active in those situations (though this does not avoid other potential pollinators).

5.3.3 Managing Pesticide Drift

While applying pesticides, the chemicals can move or drift beyond the target and cause harm to bees and other wildlife more than a mile away. Some contributing factors are weather conditions, application method, equipment settings, and spray formulation. To minimize wind-related drift, try to spray when wind speed is lowest, often during early morning or evening. Avoid mid-day spraying especially during low humidity and high temperatures; as the ground heats, the spray droplets evaporate into smaller particle size, which may remain aloft for a longer time and travel farther. The best weather conditions to spray chemicals with minimal drift is when temperatures are moderate, air is slightly humid and with a very mild steady wind or no wind.

The application methods and equipment settings when spraying pesticides are also important factors of chemical drift. Since small droplets will most likely drift farther, aerial applications and mist blowers should be avoided when possible. When using standard boom sprayer equipment, set the nozzle as low as possible and operate the sprayer at the lowest effective pressure; always calibrate the sprayers to make sure that no excess amount of pesticide will be applied [111, 112].

5.4 Government: Legislation

5.4.1 General

Different governments have reacted differently to pollinator decline threats based on beekeeper and agriculture industry input and available research. Governments may issue regulations or laws for the purposes of supporting pollinators, while departments and organizations under government oversight may issue their own guidelines. Not all governments recognize threats to honey bees or wild pollinators to the same extent. Moreover, regulations are constantly changing due to political influences, lobbyist pressure, and new findings. Legislation and regulations will often address pollinators in general as opposed to only managed honey bees.

For example, the European Union has restricted or banned the use of neonicotinoid insecticides and Fipronil in 2013, but the regulations are under review. The United Kingdom recently suspended a neonicotinoid ban, allowing use on 5% of England's rapeseed crop up to 30,000 hectares. The province of Ontario, Canada, began restricting neonicotinoid-treated seeds in July 1, 2015, under the Pesticides Act, where the number of acres using treated seeds will be reduced by 80% by 2017 [114–117].

In the United States, the US Environmental Protection Agency (US EPA) is empowered to implement legislation, share knowledge with stakeholders, issue guidelines, and other activities relevant to pollinators in relation to environmental health. The Saving America's Pollinators Act of 2013 (H.R.2692) requires the US EPA to suspend neonicotinoid insecticide registration from that point forward, until the agency has a full review of scientific evidence plus field studies to demonstrate that these chemicals have no unreasonable negative effect on animal pollinators [118].

In 2014, the National Wildlife Refuge (NWR) System under the United States Fish and Wildlife Service began to phase out the use of genetically modified crops and neonicotinoid pesticides in agricultural programs, where a third of crops grown across less than 50 system refuges were left for wildlife consumption. The use of genetically modified crops was deemed unnecessary for the scope of wildlife management, while neonicotinoid pesticides typically used as seed treatments were determined to be linked to bee population declines and a substantive risk to local ecosystems. The complete phasing-out of neonicotinoid pesticides across the NWR System is aimed for early 2016 [119].

Following President Obama's June 2014 memorandum that established the Pollinator Health Task Force between the US EPA and USDA, a research action plan was produced to serve as a roadmap for pollinator health research. The Pollinator Research Action Plan was released as part of the May 2015 National Strategy to Promote the Health of Honey Bees and Other Pollinators, which outlines an approach to reduce stressors on pollinator health. This strategy includes the following targeted outcomes:

1. Reduce honey bee colony winter losses to a sustainable level of no more than 15% in 10 years (by 2025).
2. Increase the Eastern wintering population of the monarch butterfly to 225 million butterflies in 5 years (by 2020).
3. Restore or enhance seven million acres of forage land for pollinators in 5 years (by 2020).

In addition, a comprehensive "Pollinator-Friendly Best Management Practices for Federal Lands" draft was produced to help facilitate the stated Strategy goals. The document lists six common habitat types located on US federal lands; some of these habitat management recommendations for increasing pollinator populations and diversity are as follows [111]:

5.4.2 Forests

Create canopy openings to allow understory pollinator-friendly plants to grow (determine specific species for tree thinning and shrub control, treat invasive species of infected tree stands, manage shrubs and mid-story trees based on a variety of factors, determine if seeding native plants is necessary to supplement foraging sources).

5.4.3 Roadsides

Manage roadside vegetation to provide for pollinator needs (plant native flowering herbaceous plants to provide food for pollinators' limited foraging ranges, choose exposed roadsides with access to sunlight, determine areas for tree removal to allow more sunlight).

5.4.4 Arid and Semiarid Western Shrub Lands

Restore and rehabilitate western rangelands to increase ground cover and food diversity for native pollinators (sow native flowering plants in arid and semiarid shrub lands recently scorched by fire, use annual and short-lived perennial native forb species to provide blooms in the first year, seed forbs with different flowering times, cover seeds with weed-free hays to retain moisture, consult local experts).

5.4.5 Grasslands

Open grasslands including parks, roadways, railways, wild lands, recreational areas can be managed to become quality pollinator habitats in addition to their intended uses (take regional approach, restore native prairie, convert nonnative grasslands to native prairies and meadows, increase native flowering species, reduce mowing or time after blooming, remove undesirable plant species, plant a variety of flowering species to provide floral resources across a longer period of time).

5.4.6 Riparian Areas

Manage areas around riverbanks, wet meadows, springs, fens, etc., to provide diverse plants for foraging (maintain hydraulic function within watersheds to aid water-dependent native plants, reduce wetland drainage that reduce riparian plant cover, seed native plants, maintain vegetative structure for pollinator nesting, control invasive plants, develop grazing policies if needed, minimize broad-spectrum pesticides).

5.4.7 Wildlife Openings

Artificial gaps in dense forest can stimulate more natural heterogeneous habitats and forage for pollinators (remove woody plants for more sun, replace nonnative species with native, use mixed wildflower seed to provide floral resources for a longer period of time, select plants that can serve as butterfly larval hosts, plant pollinator-friendly trees along field margins for long-term floral resources, maintain wildlife openings with mowing and control of woody species).

6 Conclusions

The honey bee is essentially our ambassador for pollinators. Given the importance of pollinators to not only our environment but a significant proportion of our agriculture and economy, it behooves us to actively support both managed honey bee and wild populations. Failure to address pollinator decline within our abilities will harm apiculture sustainability, reduce crop production, lower the quality of our plant food sources, and increase operating costs that cascade down to everyone who buys food to put on the table.

Our knowledge has been expanding to include information about wild pollinators as public awareness first began with managed honey bees, but the necessary research areas are still quite vast with many unknowns. Many studies have since acknowledged the importance of wild pollinators to agriculture, but challenges in measuring their contributions and declines persist; as new species of bee are still being identified, how many have also gone extinct? Merely importing more managed honey bees to offset other pollinator losses is not the ideal answer, as doing so merely increases operating costs and causes more problems for the remaining local populations.

Much like climate change, which is a long-term and very complex event covering multiple ecological systems, different parties must actively participate and cooperate to properly address pollinator decline.

Both hobbyist and commercial beekeepers should investigate many avenues to maintain hardy managed populations, which would lower operating costs and help reduce disease transmission to local wild pollinators. In light of available research, a few specific recommendations are as follows:

1. Beekeepers in northern regions with temperate and continental climates may adopt honey bee breeds such as the Russian bee, which are adapted to harsh winters and resistant to mites.
2. Beekeepers in southern regions of the Americas where Africanized honey bees have invaded may participate in breeding programs to select for the Africanized bee's positive traits while making it more manageable.
3. Introduce additional diversity to honey bee stocks such as bees carrying the *Varroa*-sensitive hygiene trait.

Farmers can assist long-term pollinator health by creating and maintaining diverse habitat on their lands and reducing pesticide use whenever possible. Even backyard gardens can be converted to pollinator-friendly micro-habitats.

1. Add diverse floral plants to appropriate areas around crops and other unused lands such as ditches and forest edges.
2. Favor native plants.
3. Use pesticides sparingly or not at all.

The pesticide industry has little incentive to protect pollinators as their revenues are not directly affected by pollinator decline, but government legislation will impact their interests as more research is accumulated and regulations created [120]. In the United States, the comprehensive 2015 "Pollinator-Friendly Best Management Practices for Federal Lands" document can be adopted at all levels of government including state and city. This and related federal pollinator documents are located at the US Department of Agriculture, Forest Service and the US Environmental Protection Agency website [120–123]. Private landowners can also adopt these general recommendations where applicable.

Two possible avenues to apply pollinator-friendly best management practices include incentives and state mandate:

1. A minimum acreage of public land in each state may be set aside under specific pollinator-friendly practices including a complete chemical ban; this would aid wild bees and other pollinators specific to each state to repopulate and sustain healthy ecosystems. Such protected pollinator areas could also serve as reservoirs for nearby crop pollination when necessary. The National Wildlife Refuge is already on track to ban neonicotinoid pesticide use from its crop lands by 2016, but this approach may be further expanded.
2. A tax incentive may encourage private landowners and Homeowner Associations common in the United States to adopt pollinator-friendly practices on their residential properties. Though private lands do not individually provide much area for pollinators, numerous small oases would provide increased forage diversity and also help migratory pollinators such as certain butterflies.

Finally, consumers who buy food products to feed ourselves should become aware of how pollinator decline and its challenges would affect us, such as increased costs and lowered nutritional value and availability. Public awareness correlates with political awareness and increased dialogue, a necessary ingredient for the creation and implementation of national strategies to address the complex problems our pollinators face.

Glossary

Absconding swarm: An entire colony of bees abandons the hive, often because of disease, wax moth, excessive heat or water, lack of resources, or other reasons.

Annual plant: A plant that completes its life cycle from germination to the production of seed within 1 year and then dies.

Apiary: Colonies, hives, and other equipment assembled in one location for bee-keeping operations; also known as a bee yard.

Apiculture: Beekeeping; the keeping of bees, particularly on a commercial scale.

Breeding: In the context of managing domesticated animals, to cause an animal to produce offspring in a controlled and organized manner.

Brood: Immature bees that have not yet emerged from their cells; brood of different ages can be in the form of eggs, larvae, or pupae.

Colony: All the worker bees, drones, queen, and developing brood living together in one hive or other dwelling.

Drone: The male honey bee.

Fallow: (Of farmland) plowed and harrowed but left unsown for a period in order to restore its fertility as part of a crop rotation or to avoid surplus production.

Field margins: Generally, the least productive areas of a field and just a 1-metre grass strip between the outer edge of the hedge and the crop edge can benefit wildlife in many ways.

Forage: (Of a person or animal) search widely for food or provisions.

Fungicide: A chemical that destroys fungus.

Herbicide: A substance that is toxic to plants and is used to destroy unwanted vegetation.
Hive: The structure used by bees for a home.
Honey crop: The amount of honey produced per bee colony per season.
Insecticide: A substance used for killing insects.
Melliferous: Yielding or having to do with honey.
Miticide: A substance that kill mites.
Nectar flow: A time when nectar is plentiful and bees produce and store surplus honey; also called honey flow.
Parasite: An organism that lives in or on another organism (its host) and benefits by deriving nutrients at the host's expense.
Perennial crops: Crops developed to reduce inputs necessary to produce food by greatly reducing the need to replant crops from year to year.
Pesticide: A substance used for destroying insects or other organisms harmful to cultivated plants or to animals.
Pheromones: Chemical substances secreted from bee glands and used as a means of communication; honey bees secrete many different pheromones.
Pollination: The transfer of pollen from the anthers of a flower to the stigma of the same flower or of another flower; pollination is a prerequisite for fertilization, and fertilization allows the flower to develop seeds.
Pollinator: The biotic agent (vector) that moves pollen from the male anthers of a flower to the female stigma of a flower to accomplish fertilization or "syngamy" of the female gametes in the ovule of the flower by the male gametes from the pollen grain.
Primary swarm: The first swarm to leave the parent colony, usually with the old queen (see secondary swarm).
Propolis: Sap or resinous materials collected from trees or plants by bees and used to strengthen the comb and to seal cracks; also called bee glue.
Queen: A female bee with a fully developed reproductive organ, and larger and longer than a worker bee.
Riparian: Of, relating to, or situated on the banks of a river.
Robbing: Stealing of nectar or honey by bees from other colonies, which occurs more often during a nectar dearth.
Super: A part of commercial or other managed beehive that is used to collect surplus honey. Normally, it is placed over or above the brood chamber.
Surplus honey: Honey removed from the hive, which exceeds what needed by bees for their own use.
Swarm: A large number of worker bees, drones, and usually the old queen leave the parent colony to establish a new colony.
Swarming: The natural process of honey bee reproduction at colony level.
Worker bee: A female bee with undeveloped reproductive organ; the majority of the honey bees in a colony are worker bees, and they do all the work in the colony except laying fertile eggs.

References

1. Simone-Finstrom M, Spivak M (2010) Propolis and bee health: the natural history and significance of resin use by honey bees. Apidologie 41(3):295–311. https://doi.org/10.1051/apido/2010016
2. U.S. Fish and Wildlife Service (2015) Pollinators, USA. http://www.fws.gov/pollinators/. Updated November 2015
3. International Bee Research Association., IBRA.org.uk. Solitary bees. http://www.ibrabeeworldproject.com/solitary-bees/. Retrieved Nov 2015
4. Hood WM. Bumble bees as pollinators. Department of Entomology, Soils, and Plant Sciences, Clemson University, SC, USA. http://www.clemson.edu/extension/beekeepers/factsheets/bumble_bees_as_pollinators.html. Retrieved Nov 2015
5. Moisset B, Buchmann S (2011) Bee basics: an introduction to our native bees. USDA Forest Service, Pollinator Partnership, Washington, D.C., San Francisco, USA
6. Parker FD, Batra SWT, Tepedino VJ (1987) New pollinators for our crops. Agric Zool Rev 2:279–304
7. Adjare S (1990) Chapter 2 Colony life and social organization. In: Beekeeping in Africa, FAO Agricultural Services Bulletin, 68/6. ISBN 92-5-102794-3. http://www.fao.org/3/t0104e/T0104E00.htm. Retrieved Oct 2015
8. U.S. Food Drug Administration (2015) Helping agriculture's helpful honey bees, USA. http://www.fda.gov/AnimalVeterinary/ResourcesforYou/AnimalHealthLiteracy/ucm309134.htm
9. Mid-Atlantic Apiculture Research and Extension Consortium. The colony and its organization, USA. https://agdev.anr.udel.edu/maarec/honey-bee-biology/the-colony-and-its-organization/. Retrieved Nov 2015
10. University of Arkansas, Division of Agriculture. Beekeeping in Arkansas: what is a swarm? USA. http://uaex.edu/farm-ranch/special-programs/beekeeping/swarms.aspx. Retrieved Nov 2015
11. Free JB (1987) Pheromones of social bees. Chapman and Hall, London, UK
12. Trhlin M, Rajchard J (2011) Chemical communication in the honey bee (Apis mellifera L.): a review. Vet Med, Czech 56(6):265–273
13. Bortolotti L, Costa C (2014) Chapter 5 - Chemical communication in the Honey Bee Society. In: Neurobiology of chemical communication. http://www.ncbi.nlm.nih.gov/books/NBK200983/
14. Sagili RR, Pankiw T, Metz BN (2011) Division of labor associated with brood rearing in the honey bee: how does it translate to colony fitness? PLoS One 6(2):e16785. http://www.ncbi.nlm.nih.gov/pmc/articles/PMC3035648/
15. Katzav-Gozansky T, Soroker V, Hefetz A (2002) Honey bees Dufour's gland: idiosyncrasy of a new queen signal. Apidologie 33:525–537
16. Dyer FC, Seeley TD (1991) Dance dialects and foraging range in three Asian honey bee species. Behav Ecol Sociobiol 28:227–233. https://doi.org/10.1007/bf00175094
17. Sherman G, Visscher PK (2002) Honeybee colonies achieve fitness through dancing. Nature 419:920–922. Bibcode:2002Natur.419..920S. https://doi.org/10.1038/nature01127
18. NOVA Online (1999) Dances with bees. http://www.pbs.org/wgbh/nova/bees/dancesroun.html. Retrieved Nov 2015
19. Riley J, Greggers U, Smith A, Reynolds D, Menzel R (2005) The flight paths of honeybees recruited by the waggle dance. Nature 435(7039):205–207. https://doi.org/10.1038/nature03526. Bibcode:2005Natur.435.205R. PMID 15889092
20. Seeley TD, Visscher PK, Passino KM (2006) Group decision making in honey bee swarms. Am Sci 94:220–229. https://doi.org/10.1511/2006.3.220
21. Walker M (2009) Honeybees warn of risky flowers. BBC Earth News. http://news.bbc.co.uk/earth/hi/earth_news/newsid_8176000/8176878.stm
22. Abbott KR, Dukas R (2009) Honeybees consider flower danger in their waggle dance. Anim Behav 78(3):633. https://doi.org/10.1016/j.anbehav.2009.05.029

23. Whitfield CW, Behura SK, Berlocher SH, Clark AG, Johnston JS, Sheppard WS, Smith DR, Suarez AV, Weaver D, Tsutsui ND (2006) *Thrice out of Africa: ancient and recent expansions of the honey bee, Apis mellifera* (PDF). Science 314(5799):642–645. https://doi.org/10.1126/science.1132772
24. Kjøhl M, Nielsen A, Stenseth NC (2011) Potential effects of climate change on crop pollination. Pollination Services for Sustainable Agriculture. Food and Agriculture Organization of the United Nations, Rome. isbn:978-92-5-106878-6
25. Smithsonian. Killer bees. Information Sheet Number 45. http://www.si.edu/encyclopedia_si/nmnh/buginfo/killbee.htm
26. Hall HG, Zettel-Nalen C, Ellis JD. African honey bee: what you need to know. University of Florida, Entomology and Nemotology Department. Fact Sheet ENY-114 (MG113) http://edis.ifas.ufl.edu/mg113
27. United States Department of Agriculture, Agricultural Research Service (2011) Africanized honey bees. http://www.ars.usda.gov/Research/docs.htm?docid=11059&page=1
28. O'Malley MK, Ellis JD, Zettel Nalen CM (2008, Revised 2015) Differences between European and African honey bees. University of Florida, Institute of Food and Agricultural Sciences Extension. #ENY-147. http://edis.ifas.ufl.edu/in784
29. U.S. Department of Agriculture, Agriculture Research Service (2007) More buzz about a bad bee Africanized honey bees in the news again. AgResearch Magazine. http://agresearchmag.ars.usda.gov/2007/feb/bees/
30. Pankiw T (2003) Directional change in a suite of foraging behaviors in tropical and temperate evolved honey bees (Apis mellifera L). Behav Ecol Sociobiol 54(54):458–464. https://doi.org/10.1007/s00265-003-0640-1. JSTOR 25063290
31. Winston ML (1988) The impact of a tropical-evolved honey bee in temperate climates of North America. In: Needham GR et al (eds) Africanized honey bees and bee mites. Halsted Press, New York, USA, pp 136–137
32. Ellis J, Ellis A (2008) African honey bee, Africanized honey bee, killer bee. Institute of Food and Agricultural Sciences, University of Florida, USA. Publication:EENY-429 http://entnemdept.ufl.edu/creatures/misc/bees/ahb.htm
33. Aizen MA, Garibaldi LA, Cunningham SA, Klein AM (2009) How much does agriculture depend on pollinators? Lessons from long-term trends in crop production. Ann Bot 103:1579–1588. https://doi.org/10.1093/aob/mcp076, www.aob.oxfordjournals.org
34. Bartomeus I et al (2014) Contribution of insect pollinators to crop yield and quality varies with agricultural intensification. PeerJ 2:e328. https://doi.org/10.7717/peerj.328
35. Calderone NW (2012) Insect pollinated crops, insect pollinators and US agriculture: trend analysis of aggregate data for the period 1992–2009. PLoS One 7(5):e37235. https://doi.org/10.1371/journal.pone.0037235
36. European Commission (2015) Honey bees http://ec.europa.eu/food/animals/live_animals/bees/index_en.htm
37. U.S. White House (2014) Fact sheet: the economic challenge posed by declining pollinator populations. https://www.whitehouse.gov/the-press-office/2014/06/20/fact-sheet-economic-challenge-posed-declining-pollinator-populations
38. U.S. Department of Agriculture, Foreign Agricultural Service (2015) Oilseeds: world markets and trade. http://apps.fas.usda.gov/psdonline/circulars/oilseeds.pdf
39. Boland M, Brester G (2003, updated 2013) Canola profile. Iowa State University, Agricultural Marketing Resource Center. http://www.agmrc.org/commodities__products/grains__oilseeds/canola-profile/
40. Iowa State University, Agricultural Marketing Resource Center (2014) Commodity strawberries. http://www.agmrc.org/commodities__products/fruits/strawberries/commodity_strawberries.cfm
41. Garibaldi LA et al (2013) Wild pollinators enhance fruit set of crops regardless of honey bee abundance. Science 339(6127):1608–1611. https://doi.org/10.1126/science.1230200

42. Eilers EJ, Kremen C, Greenleaf SS, Garber AK, Klein A-M (2011) Contribution of pollinator-mediated crops to nutrients in the human food supply. PLoS One 6(6):e21363. https://doi.org/10.1371/journal.pone.0021363
43. Klatt BK, Holzschuh A, Westphal C, Clough Y, Smit I, Pawelzik E, Tscharntke T (2014) Bee pollination improves crop quality, shelf life and commercial value. Proc R Soc Lond B 281:20132440. https://doi.org/10.1098/rspb.2013.2440
44. Chautá-Mellizoa A et al (2012) Effects of natural and artificial pollination on fruit and offspring quality. Basic Appl Ecol 13:524–532
45. European Commission (2015) Study on honey bee colony mortality. http://ec.europa.eu/food/animals/live_animals/bees/study_on_mortality/index_en.htm
46. COLOSS, Institute of Bee Health, University of Bern (2015) Losses of honey bee colonies over the 2014/15 winter: Preliminary results from an international study. http://www.coloss.org/announcements/losses-of-honey-bee-colonies-over-the-2014-15-winter-preliminary-results-from-an-international-study
47. Kaplan K (2015) Bee survey: lower winter losses, higher summer losses, increased total annual losses. United States Department of Agriculture, Agricultural Research Service. http://www.ars.usda.gov/is/pr/2015/150513.htm
48. Goulson D, Nicholls E, Botías C, Rotheray EL (2015) Bee declines driven by combined stress from parasites, pesticides, and lack of flowers. Science 347(6229):1255957. https://doi.org/10.1126/science.1255957
49. Ellis JD, Hammons K (2013) Overwintering honey bee colonies in Northern Climates. University of Florida, Institute of Food and Agricultural Sciences Extension. #ENY-161. https://edis.ifas.ufl.edu/in1006
50. vanEngelsdorp D, Evans JD, Saegerman C, Mullin C, Haubruge E, Nguyen BK et al (2009) Colony collapse disorder: a descriptive study. PLoS One 4(8):e6481. https://doi.org/10.1371/journal.pone.0006481
51. https://en.wikipedia.org/wiki/Pesticide. Retrieved Nov 2015
52. Ellis J (2007, revised 2013) Colony Collapse Disorder (CCD) in honey bees. University of Florida, Institute of Food and Agricultural Sciences Extension, ENY-150 (IN720). https://edis.ifas.ufl.edu/in720
53. Bee Informed Partnership. https://beeinformed.org/results-categories/winter-loss/. Retrieved Nov 2015
54. Grossman, E. (2013) Yale Environment360. Declining bee populations pose a threat to global agriculture. Yale School of Forestry & Environmental Studies. http://e360.yale.edu/feature/declining_bee_populations_pose_a_threat_to_global_agriculture/2645/
55. Pisa L et al (2014) Effects of neonicotinoids and fipronil on non-target invertebrates. Environ Sci Pollut Res. https://doi.org/10.1007/s11356-014-3471-x
56. Henry M, Béguin M, Requier F, Rollin O, Odoux JF, Aupinel P, Aptel J, Tchamitchian S, Decourtye A (2012) A common pesticide decreases foraging success and survival in honey bees. Science 336(6079):348–350
57. Yamada T (2012) Influence of dinotefuran and clothianidin on honey bee colony during long-term field experiment in apiary. IUCN Tokyo Forum on Systemic Pesticides, September 2012, Tokyo. Unpublished conference paper. Japan for Sustainability, 2015. PDF
58. McGrath MT (2004) What are fungicides. Plant Health Instr. https://doi.org/10.1094/PHI-I-2004-0825-01
59. Pettis JS, Lichtenberg EM, Andree M, Stitzinger J, Rose R et al (2013) Crop pollination exposes honey bees to pesticides which alters their susceptibility to the gut pathogen Nosema ceranae. PLoS One 8(7):e70182. https://doi.org/10.1371/journal.pone.0070182
60. Bernauer OM, Gaines-Day HR, Steffan SA (2015) Colonies of bumble bees (Bombus impatiens) produce fewer workers, less bee biomass, and have smaller mother queens following fungicide exposure. Insects 6:478–488. https://doi.org/10.3390/insects6020478
61. MarketsandMarkets.com (2014) Herbicides market by type (glyphosate, atrazine, acetochlor, 2, 4-D), by crop type (cereals & grains, oilseeds & pulses, fruits & vegetables), by mode of

action (selective & non-selective) & geography - trends and forecasts to 2019. http://www.marketsandmarkets.com/Market-Reports/herbicides-357.html. Retrieved Oct 2015
62. https://en.wikipedia.org/wiki/Glyphosate. Retrieved Nov 2015
63. Herbert LT, Vázquez DE, Arenas A, Farina WM (2014) Effects of field-realistic doses of glyphosate on honey bee appetitive behavior. J Exp Biol 217(Pt 19):3457–3464. https://doi.org/10.1242/jeb.109520. Epub 2014 Jul 25. http://www.ncbi.nlm.nih.gov/pubmed/25063858
64. Balbuena MS, Tison L, Hahn M-L, Greggers U, Menzel R, Farina WM (2015) Effects of sub-lethal doses of glyphosate on honey bee navigation. J Exp Biol. https://doi.org/10.1242/dev.117291
65. Decourtye A, Mader E, Desneux N (2010) Landscape enhancement of floral resources for honey bees in agro-ecosystems. Apidologie 41(3):264–277. https://doi.org/10.1051/apido/2010024
66. Requier F et al (2015) Honey bee diet in intensive farmland habitats reveals an unexpectedly high flower richness and a major role of weeds. Ecol Appl 25(4):881–890
67. Bailey S, Requier F, Nusillard B, Roberts SPM, Potts SG, Bouget C (2014) Distance from forest edge affects bee pollinators in oilseed rape fields. Ecol Evol 4(4):370–380
68. Carvalheiro LG, Seymour CL, Veldtman R, Nicolson SW (2010) Pollination services decline with distance from natural habitat even in biodiversity-rich areas. J Appl Ecol 47:810–820
69. U.S. Department of Agriculture (2009) Luring Varroa mites to their doom. Agricultural Research Magazine, July 2009. http://agresearchmag.ars.usda.gov/2009/jul/mites
70. https://en.wikipedia.org/wiki/Varroa_mite. Retrieved Oct 2015
71. U.S. Department of Agriculture (2001) Russian honey bee earning its stripes. United States Department of Agriculture, Agriculture Research Magazine. AgResearch Magazine, October 2001. http://agresearchmag.ars.usda.gov/2001/oct/bee/
72. https://en.wikipedia.org/wiki/List_of_diseases_of_the_honey_bee
73. Graystock P, Blaneb EJ, McFredericka QS, Goulsonc D, Hughesc WOH (2016) Do managed bees drive parasite spread and emergence in wild bees? Int J Parasitol Parasites Wildl. https://doi.org/10.1016/j.ijppaw.2015.10.001. Retrieved Nov 2015
74. Lindsey R (2007) Buzzing about climate change. National Aeronautics and Space Administration, Earth Observatory. http://earthobservatory.nasa.gov/Features/Bees/bees.php
75. Forrest JRK, Thomson JD (2011) An examination of synchrony between insect emergence and flowering in Rocky Mountain meadows. Ecol Monogr 81(3):469–491
76. Quote Investigator (2013) If the bee disappeared off the face of the earth, man would only have four years left to live. http://quoteinvestigator.com/2013/08/27/einstein-bees/
77. Zulian G, Maes J, Paracchini ML (2013) Linking land cover data and crop yields for mapping and assessment of pollination services in Europe. Land 2:472–492. http://www.mdpi.com/2073-445X/2/3/472
78. Gallai N, Salles J-M, Settele J, Vaissière BE (2009) Economic valuation of the vulnerability of world agriculture confronted with pollinator decline. Ecol Econ 68(3):P810–P821. https://doi.org/10.1016/j.ecolecon.2008.06.014
79. Lautenbach S, Seppelt R, Liebscher J, Dormann CF (2012) Spatial and temporal trends of global pollination benefit. PLoS One 7(4):e35954. https://doi.org/10.1371/journal.pone.0035954
80. Kevan PG, Phillips TP (2001) The economic impacts of pollinator declines: an approach to assessing the consequences. Conserv Ecol 5(1):8. http://www.consecol.org/vol5/iss1/art8/
81. Bauer DM, Wing IS (2010) Economic consequences of pollinator declines: a synthesis. Agric Resour Econ Rev 39:368–383
82. Eilers EJ, Kremen C, Smith Greenleaf S, Garber AK, Klein A-M (2011) Contribution of pollinator-mediated crops to nutrients in the human food supply. PLoS One 6(6):e21363. https://doi.org/10.1371/journal.pone.0021363
83. Smith MR, Singh GM, Mozaffarian D, Myers SS (2015) Effects of decreases of animal pollinators on human nutrition and global health: a modelling analysis. Lancet. https://doi.org/10.1016/S0140-6736(15)61085-6

84. U.S. Department of Agriculture, Natural Resources Conservation Service (2015) Conservation work for honey bees. http://www.nrcs.usda.gov/wps/portal/nrcs/detail/national/plantsanimals/pollinate/?cid=stelprdb1263263
85. International Bee Research Association. The BeeWorld project: honeybees. http://www.ibrabeeworldproject.com/honeybees/. Retrieved Nov 2015
86. Sanford MT, Ellis JD (2012) Florida bears and beekeeping. University of Florida, Institute of Food and Agricultural Sciences, ENY-105. https://edis.ifas.ufl.edu/aa133
87. Ontario Ministry of Agriculture, Food and Rural Affairs (2014) Ontario treatment recommendations for honey bee disease and mite control. http://www.omafra.gov.on.ca/english/food/inspection/bees/2014-treatment.htm
88. Flores A (2009) Honey bees selected by ARS Toss Out Varroa Mites. United States Department of Agriculture, Agricultural Research Service. http://www.ars.usda.gov/is/pr/2009/090910.htm
89. Danka RG, Harris JW, Villa JD (2011) Expression of Varroa Sensitive Hygiene (VSH) in commercial VSH honey bees (Hymenoptera: Apidae). J Econ Entomol 104(3):745–749. https://doi.org/10.1603/EC10401
90. Delaplane KS (2010) Africanized honey bees (B 1290). University of Georgia Cooperative Extension. http://extension.uga.edu/publications/detail.cfm?number=B1290. Retrieved Oct 2015
91. Camazine S (1986) Differential reproduction of the mite, Varroa jacobsoni (Mesostigmata: Varroidae), on Africanized and European Honey Bees (Hymenoptera: Apidae). Ann Entomol Soc Am 79(5):801–803. https://doi.org/10.1093/aesa/79.5.801
92. Livanis G, Moss CB (2010) The effect of Africanized honey bees on honey production in the United States: an informational approach. Ecol Econ 69(4):P895–P904. https://doi.org/10.1016/j.ecolecon.2009.11.013
93. Medina-Flores CA, Guzmán-Novoa E, Hamiduzzaman MM, Aréchiga-Flores CF, López-Carlos MA (2014) Africanized honey bees (Apis mellifera) have low infestation levels of the mite Varroa destructor in different ecological regions in Mexico. Genet Mol Res 13 (3):7282–7293
94. Walker M (2009) Honeybees sterilise their hives. BBC Earth News, July 2009. http://news.bbc.co.uk/earth/hi/earth_news/newsid_8152000/8152574.stm
95. Nicodemo D et al (2013) Honey bee lines selected for high propolis production also have superior hygienic behavior and increased honey and pollen stores. Genet Mol Res 12 (4):6931–6938. Available at: http://hdl.handle.net/11449/113315
96. Dejong D (1984) Africanized bees now preferred by Brazilian beekeepers. Am Bee J 124:116–118
97. Guzmán-Novoa E, Page RE Jr (1993) Backcrossing Africanized honey bee queens to European drones reduces colony defensive behavior. Ann Entomol Soc Am 86(3):352–355. https://doi.org/10.1093/aesa/86.3.352
98. Guzmán-Novoa E, Page RE Jr (1999) Selective breeding of honey bees (Hymenoptera: Apidae) in Africanized areas. J Econ Entomol 92(3):521–525. https://doi.org/10.1093/jee/92.3.521
99. Peabody E (2006) The latest buzz on Russian bees. United States Department of Agriculture, Agricultural Research Service. http://www.ars.usda.gov/is/pr/2006/060809.htm
100. Raloff J (1998) Russian queen bee-little mites' impact. Sci News 154(6):84
101. Tubbs H, Harper C, Bigalk M, Bernard SJ, Delatte GT, Sylvester HA, Rinderer TE (2003) Commercial management of ARS Russian honey bees. Am Bee J 143:819–820
102. Nearing MA et al (2005) Modelling response of soil erosion and runoff to changes in precipitation and cover. Catena 61:131–154
103. Kleijn D et al (2006) Mixed biodiversity benefits of agri-environment schemes in five European countries. Ecol Lett 9(3):243–254. http://www.ncbi.nlm.nih.gov/pubmed/16958888

104. Wahlquist A (2010) Perennial crops a win for food security. The Australian, 30 June 2010. http://www.theaustralian.com.au/higher-education/perennial-crops-a-win-for-food-security/story-e6frgcjx-1225885860575
105. U.S. Department of Agriculture, Natural Resources Conservation Service (2008) Using farm bill programs for pollinator conservation. Technical Note No. 78. http://www.xerces.org/guidelines/using-farm-bill-programs-for-pollinator-conservation/
106. Corbet SA (1995) Insects, plants and succession: advantages of long-term set-aside. Agric Ecosyst Environ 53(3):201–217. https://doi.org/10.1016/0167-8809(94)00581-X
107. Firbank LG, Telfer MG, Eversham BC, Arnold HR (1994) The use of species-decline statistics to help target conservation policy for set-aside arable land. J Environ Manag 42:415–422
108. European Commisson (2006) Halting the loss of biodiversity by 2010 and beyond: sustaining ecosystem services for human well-being. http://ec.europa.eu/environment/nature/biodiversity/comm2006/
109. Lagerhöf J, Wallin H (1993) The abundance of arthropods along two field margins with different types of vegetation composition: an experimental study. Agric Ecosyst Environ 43 (2):141–154
110. Fluri P, Frick R (2002) Honey bee losses during mowing of flowering fields. Bee World 83:109–118. http://www.conservationevidence.com/individual-study/1819
111. U.S. Environmental Protection Agency (2015) Pollinator-friendly best management practices for federal lands. 11 May 2015. http://www.fs.fed.us/wildflowers/pollinators/BMPs/
112. The Xerces Society for Invertebrate Conservation. Managing pesticides to protect bees. Portland. http://www.xerces.org/pollinator-conservation/agriculture/managing-pesticides-to-protect-bees/. Retrieved Nov 2015
113. Alarcón R, DeGrandi-Hoffman G, Hayden C (2009) Fungicides can reduce, hinder pollination potential of honey bees. http://westernfarmpress.com/fungicides-can-reduce-hinder-pollination-potential-honey-bees
114. European Commission (2013) Bee health: EU takes additional measures on pesticides to better protect Europe's bees. Brussels. http://europa.eu/rapid/press-release_IP-13-708_en.htm
115. Oregon State University (2009) Fipronil general fact sheet. National Pesticides Information Center, OR, US. http://npic.orst.edu/factsheets/fipronil.html#wildlife
116. European Commission (2013) Bees & pesticides: commission to proceed with plan to better protect bees. Brussels. http://europa.eu/rapid/press-release_IP-13-379_en.htm
117. Ontario Ministry of Agriculture, Food and Rural Affairs (Updated November 2015) Neonicotinoid regulations. Ontario, Canada. http://www.ontario.ca/page/neonicotinoid-regulations
118. U.S. Congress (2013) Saving America's Pollinators Act of 2013. Congress Bill H.R.2692. 113th (2013–2014). https://www.congress.gov/bill/113th-congress/house-bill/2692/text
119. U.S. National Wildlife Refuge Association (2014) U.S. fish and wildlife service bans GMOs and neonicotinoid insecticides. August. http://refugeassociation.org/2014/08/u-s-fish-and-wildlife-service-bans-gmos-and-neonicotinoid-insecticides/
120. McNeil MEA (2015) The national strategy to promote the health of honey bees and other pollinators: what is it – smoke and mirrors or bountiful possibility? 20 Nov 2015. http://www.beeculture.com/the-national-strategy-to-promote-the-health-of-honey-bees-and-other-pollinators/
121. US EPA (2016) Pollinator risk assessment guidance. US Environmental Protection Agency, Washington, DC, USA. https://19january2017snapshot.epa.gov/pollinator-protection/pollinator-riskassessment-guidance_.html. 16 Aug 2016
122. US EPA (2018) How we assess risks to pollinators. US Environmental Protection Agency, Washington, DC, USA. https://www.epa.gov/pollinator-protection/how-we-assess-risks-pollinators. 8 June 2018
123. US EPA (2019) Pollinator risk assessment guidance. US Environmental Protection Agency, Washington, DC, USA. https://www.google.com/search?client=firefox-b-1-d&q=us+epa+pollinator+risk+assessment+guidance. 3 May 2019

Chapter 2
Water Reclamation and Reuse for Environmental Conservation

Nazih K. Shammas, Lawrence K. Wang, and Mu-Hao Sung Wang

Contents

1 Introduction ... 54
2 Flotation/Filtration Cell ... 55
3 Two-Stage Flotation System ... 57
4 Treatment of Primary Municipal Effluent ... 59
5 Treatment of Raw Municipal Wastewater ... 60
6 Treatment of Activated Sludge Effluent ... 62
7 Treatment of RBC Effluent ... 63
8 Treatment of Trickling Filter Effluent ... 66
9 Treatment of Lagoons Effluent ... 67
10 Conclusion ... 68
Glossary ... 68
References ... 69

Abstract The strategy for long-term planning and management of water resources is more and more being based on the renovation and utilization of wastewater for use in agricultural and landscape irrigation as well as in industrial production. Conventional wastewater treatment, even when it is economically feasible, is costly because of biosolids handling and tertiary sedimentation tanks. A newly developed flotation/filtration cell is an advanced water clarification package plant, using a combination of chemical flocculation, dissolved air flotation (DAF), and rapid granular filtration in one unit. The average processing time from start to finish is less than 15 min. This innovation replaces a conventional process requiring five separate tanks with a single, compact, and cheaper unit. The aim of this chapter is to discuss six applications that illustrate the versatility and effectiveness of the flotation/filtration cell, to demonstrate the performance of the compact unit in secondary and tertiary treatment of various wastewater effluents, and to illustrate the system's usefulness for wastewater reclamation, recycling, and reuse.

N. K. Shammas · L. K. Wang (✉) · M.-H. S. Wang
Lenox Institute of Water Technology, Newtonville, NY, USA

Krofta Engineering Corporation, Lenox, MA, USA

L. K. Wang, M. -H. S. Wang, Y. -T. Hung, N. K. Shammas (eds.),
Environmental and Natural Resources Engineering, Handbook of Environmental Engineering 19, https://doi.org/10.1007/978-3-030-54626-7_2

Keywords Flotation-filtration cell · Flotation cell · Package treatment plant · Wastewater renovation · Recycling · Reuse · Dissolved air flotation · DAF · Flotation-filtration · DAFF · Two-stage DAF · Primary flotation clarification · Secondary flotation clarification

Acronym

BOD	Biochemical oxygen demand
COD	Chemical oxygen demand
DAF	Dissolved air flotation
DAFF	Dissolved air flotation and filtration
F-Cell	Dissolved air flotation cell or DAF
FF-Cell	Dissolved air flotation and filtration cell or DAFF
MGD	Million gallons per day
PAC	Polyaluminum chloride
RBC	Rotating biological contactor
STP	Sewage treatment plant
TSS	Total suspended solids

1 Introduction

In many parts of the world, the limited availability of both ground and fresh surface waters makes it imperative to conserve water and to utilize every drop of available wastewater for reuse in beneficial purposes. The strategy for long-term planning and management of water resources is more and more being based on the renovation and utilization of wastewater for use in agricultural and landscape irrigation as well as in industrial production.

Water quality standards or guidelines have been initiated in many states and regions in order to protect public health, prevent nuisance conditions, and preclude damage to crops, soils, and groundwater. Risk-based wastewater reclamation criteria often require full tertiary treatment, especially in applications, which have high potential exposure such as in using the reclaimed water for unrestricted irrigation [1–5]. Conventional full tertiary treatment consists of a train of multiple processes, which include rapid mixing, flocculation, sedimentation, and granular filtration in addition to disinfection [2, 3, 6]. Such processing, even when it is economically feasible, is costly because of biosolids handling and tertiary sedimentation tanks [6, 7]. Thus, research has been directed toward developing an innovative alternative capable of producing a comparatively highly clarified effluent [8–14].

A newly developed flotation/filtration cell (FF-Cell) is an advanced water clarification package plant, using a combination of chemical flocculation [15], dissolved air flotation (DAF) [8], and rapid granular filtration [6] in one unit. The average processing time from start to finish is less than 15 min. The unique compact and efficient design is made possible by the use of the space (water head) above the filter for flotation, a space or water head, which, in any case, is necessary for filtration. Two more features were used to reduce space requirements [10, 16, 17]:

1. A static hydraulic flocculator was built into the central portion of the tank to combine a third process in the single tank.
2. A segmented continuous backwash filter was used to eliminate requirements for large tanks for clearwell and backwash storage. Therefore, the end result replaces a conventional process requiring five separate tanks with a single, compact, and cheaper unit.

The aim of this chapter is to document six applications to illustrate the versatility and effectiveness of the flotation/filtration cell, to demonstrate the performance of the compact unit in secondary and tertiary treatment of various wastewater effluents, and to illustrate the system's usefulness for wastewater reclamation, recycling, and reuse. The five applications are as follows [18–21]:

1. Treatment of primary municipal wastewater effluent at Hoboken, NJ., USA
2. Treatment of raw municipal wastewater at Lee, MA, USA
3. Tertiary treatment of activated sludge effluent at Oak Meadows STP, Licking County, OH, USA
4. Tertiary treatment of RBC Effluent at Jimmy Peak, MA, USA
5. Tertiary treatment of trickling filter effluent at Norwalk, OH, USA
6. Tertiary treatment of lagoons effluent at Arpin, WI, USA

2 Flotation/Filtration Cell

The flotation/filtration pilot plant used in these applications is shown in Fig. 2.1 and its details are illustrated in Fig. 2.2. The flotation-filtration tank is 1.5 m in diameter and has a design nominal capacity of 150 L/min of wastewater flow [18].

The influent flow is mixed with flocculant and coagulant chemicals, and gently flocculated in the central zone of the tank. The backwash recycled from the filter is mixed with the inflowing water at the flocculator inlet. This eliminates the need for disposal of the backwash separately, and also, in some cases, provides a "seed" of solids for better floc formation. The floc size required for removal of the solids is smaller than that required for settling. This reduces the space requirement for flocculation [10, 22].

When the flocculated solids reach the upper part of the flocculation zone, they are mixed with the recycle flow, which contains millions of microscopic (20–100 micron diameter) air bubbles. The air bubbles are generated by injecting air into recirculated clarified water under pressure (60 psi), followed by rapid decompression under high shear conditions. The amount of recirculated water used varies depending on the amount and type of solids to be removed, but it is generally 15–30% of the incoming flow. The air bubbles attach to the flocculated solids, or are entrapped in the floc to produce air-solid agglomerate that rapidly rises to the surface of the tank.

Fig. 2.1 Flotation/filtration cell (FF-Cell)

The accumulated float (thickened biosolids) at solids content of 2–3% is removed by the spiral scoop and discharged to the biosolids handling system via the central sludge well [8, 16, 17].

As the biosolids rise to the top, the clarified water flows downward through the filter bed. The bottom of the tank is composed of multisections of sand filter segments. Each of the 27 segments is individually isolated and backwashed, while the remaining parts of the filter are on line. The filter media consist of 0.28 m deep high-grade silica sand. The effective size and uniformity coefficient for the sand are 0.35 mm and 1.55, respectively. The backwashing is uniformly extended over the complete filtering time. This enables the use of a smaller backwash pump, and minimizes or even eliminates any overloading on the unit that may result from the recycling of decanted backwash water [23].

The backwash hood, pump, and motor are mounted on a carriage that rotates on the upper rim of the main tank. The filter segments are set up for backwashing at a predetermined time interval that can be adjusted depending on head loss and the accumulation of solids. The backwash water containing the solids captured by the sand is recirculated back to the flocculator. The clearwell is located immediately below the sand bed. The clear water is utilized directly for backwash as needed.

The filter assures the removal of any solids not removed in the flocculation and flotation stages, the flocculation is enhanced by the recirculated backwash solids, and the filtration is protected by the removal of solids in the flotation stage. The combination of the three operations in a single tank reduces the head loss and turbulence between stages, thus increasing the efficiency of removal of the fragile flocculated solids. No storage is required for the clarified water or the backwash water. Discharge of backwash water is eliminated, and the only discharge is the float, thickened biosolids suitable for handling, thus possibly eliminating the necessity for a biosolids thickener [24–26].

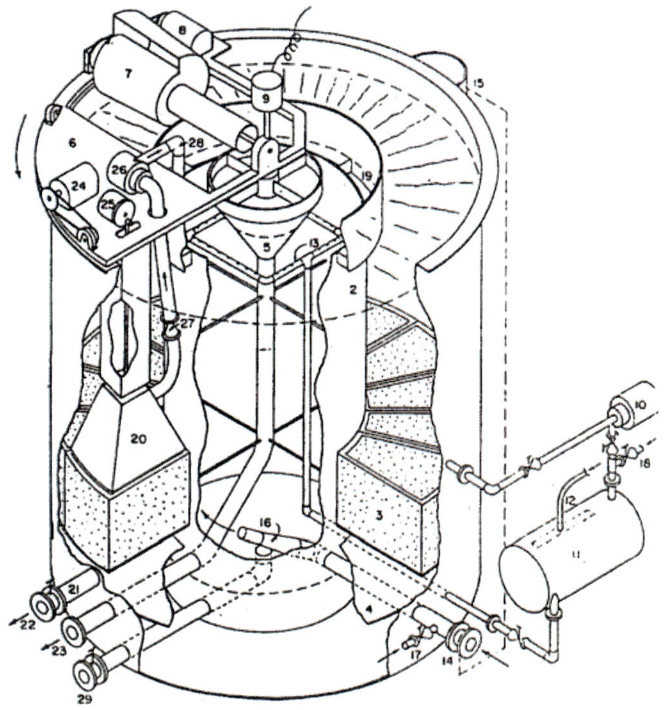

1. Outside tank
2. Inside flocculation tank
3. Sandbed assembly with screen
4. Tank bottom
5. Sludge collection funnel
6. Moveable carriage assembly
7. Spiral scoop
8. Scoop variable speed drive
9. Electrical rotary contact
10. Pressure pump
11. Air dissolving tube
12. Compressed air addition point
13. Aerated water distribution pipes
14. Raw water inlet regulating valve
15. Tank level control sensor
16. Raw water inlet jet nozzle
17. Coagulant addition point
18. Polyelectrolyte addition point
19. Deflector ring into flotation tank
20. Backwash hood assembly
21. Clarified water pipeline
22. Clarified water flow regulating valve
23. Floated sludge discharge pipe
24. Main carriage drive
25. Motor to lift backwash hood assembly
26. Backwash suction pump
27. Check valve (backflow preventor)
28. Dirty backwash water recycle pipe
29. Drain line

Fig. 2.2 Details of the flotation/filtration cell (FF-Cell) [18]

3 Two-Stage Flotation System

The flow diagram for the two-stage flotation pilot-plant is shown in Fig. 2.3. The system, mounted on a mobile trailer, consists of a combination of a primary dissolved air flotation clarifier, F-Cell [7], and a secondary flotation/filtration clarifier, FF-Cell [18].

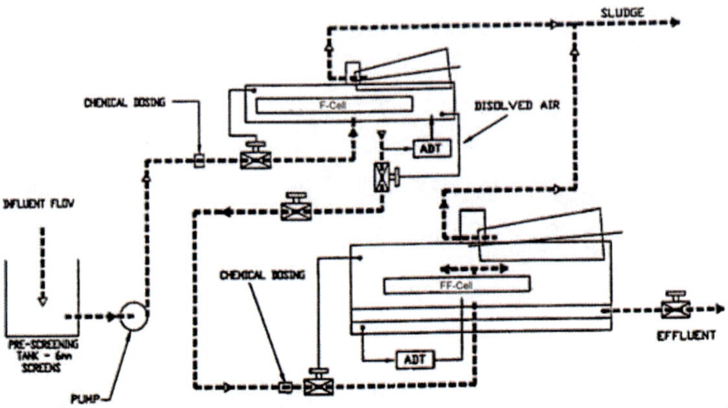

Fig. 2.3 Flow diagram of two-stage flotation system [20]

The flotation tank is 1.2 m (4 ft) in diameter with 0.41–0.46 m (16–18 in) effective water depth. The system is complete with feed pump, pressure pump, compressor, air dissolving tube, collection tank, and chemical feed equipment. The flotation tank is equipped with a spiral scoop for collection and removal of floated biosolids (Fig. 2.4).

The inlet, outlet, and biosolids removal mechanisms are contained in the central rotating section. This section and the spiral scoop rotate around the tank at a speed synchronized with the flow. The system is operated in the recycle flow pressurization mode, whereby a portion of clarified effluent (30–40%) is continuously recycled from the collection tank to the air dissolving tube under 400 kPa (60 psi) pressure. After pressure release, the aerated water is mixed with the influent flow just before the inlet to the distribution duct that moves with the same velocity, but in opposite direction to the incoming flow, thus creating a quiescent state in the flotation chamber. Fine bubbles generated in this manner attach to the suspended particles and float them to the surface. The spiral scoop takes up the floated biosolids, pouring them into the stationary center section where they are discharged by gravity. Clarified water is removed by extraction pipes, which are attached to the moving center section and discharged into the collection tank [8, 16, 17].

Wiper blades attached to the moving distribution duct scrape the bottom and the sides of the tank and discharge settled biosolids into the built-in sump, for periodic purging. The variable speed gear motor drives the rotating elements and scoop. Electrical current for the gear motor feeds from a rotary contact mounted on the central shaft.

The second stage FF-Cell is identical to the cell described previously in Sect. 2. The main characteristics of the two cells are listed in Table 2.1.

For further information on the subject of flotation the reader is referred to the literature [27–32]

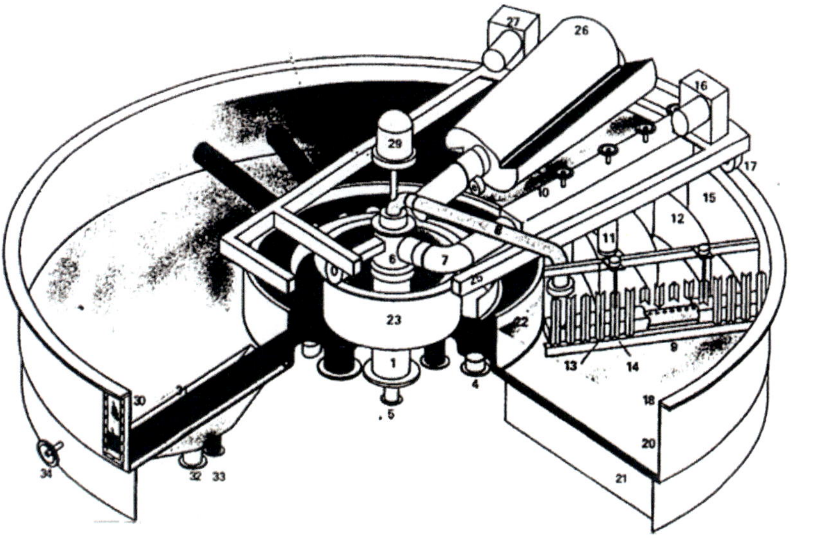

1) RAW WATER INLET
2) CLARIFIED WATER OUTLET
3) FLOATED SLUDGE OUTLET
4) CLARIFIED WATER RECYCLE OUTLET
5) PRESSURIZED WATER INLET
6) ROTARY JOINT
7) RUBBER PIPE CONNECTION
8) PRESSURIZED WATER PIPING
9) PRESSURIZED WATER DISTRIBUTION HEADER
10) RAW WATER DISTRIBUTION HEADER
11) DISTRIBUTION HEADER OUTLET PIPES
12) FLOW CONTROL CHANNELS
13) TURBULENCE REDUCTION BAFFLES
14) ADJUSTABLE HEIGHT BAFFLE ATTACHMENT
15) FLOW CONTROL CHANNEL OUTER WALL
16) ROTATING CARRIAGE GEARMOTOR DRIVE
17) CARRIAGE DRIVE WHEEL
18) WHEEL SUPPORT RIM
20) TANK WALL
21) TANK FLOOR SUPPORT STRUCTURE
22) ROTATING CLARIFIED WATER CONTAINMENT WALL
23) SLUDGE WELL
24) LEVEL CONTROL OVERFLOW WEIR
25) ROTATING CARRIAGE STRUCTURE
26) REVOLVING SPIRAL SCOOP
27) SPIRAL SCOOP GEARMOTOR DRIVE
28) CLARIFIED WATER EXTRACTION PIPES
29) ELECTRICAL SLIP RING
30) TANK WINDOW
31) SEDIMENT REMOVAL SUMP
32) FINAL DRAIN
33) SEDIMENT PURGE OUTLET
34) LEVEL CONTROL ADJUSTMENT HANDWHEEL

Fig. 2.4 Details of the flotation cell (F-Cell)

4 Treatment of Primary Municipal Effluent

Hoboken is a city located on the west bank of the Hudson River, between Lincoln Tunnel and Holland Tunnel in New Jersey, USA. It is an old city with an approximate population of 45,000. The primary treatment at the existing 68,000 m^3/d (18 MGD) Hoboken Wastewater Treatment Plant consists of screens and sedimentation clarifiers [18]. The effluent was continuously treated by the FF-Cell pilot plant to check its suitability for upgrading the quality of the plant effluent. At chemical dosages of 20 mg/L alum and 2 mg/L of nonionic polymer, the primary effluent was successfully treated, as is demonstrated in Table 2.2, by lowering the total suspended

Table 2.1 Characteristics of pilot plant units [20]

Parameter	First stage F-Cell	Second stage FF-Cell
Nominal capacity, L/min	150	150
Cell diameter, m	1.2	1.5
Cell depth, m	0.6	0.9
Air feed, L/min	2.4	2.4
Air pressure, bar	6	6.5
Sludge scoop speed, Rev./min	2	2
Backwash rate, L/min	Not applicable	38
Backwash time, s	Not applicable	45

Table 2.2 Treatment of primary municipal wastewater effluent by FF-Cell at Hoboken, NJ, USA [18]

Parameter	Primary effluent	FF-cell[a] effluent	Removal[b] %
Turbidity, NTU	45	3.3	93
Phosphate-P, mg/L	5.6	0.04	99
BOD, mg/L	103	12	88
COD, mg/L	260	80	69
TSS, mg/L	57	2	97
Coliforms, No/100mL	240,000	5,000	98

[a]FF-Cell = Flotation/filtration cell
[b]Chemicals addition: alum (as Al_2O_3) 20 mg/L and polymer 2 mg/L

solids and BOD by 97% and 88%, respectively. The Hoboken primary effluent contents (BOD = 103 mg/L and TSS = 57 mg/L) were reduced to 12 mg/L, and 2 mg/L, respectively. Turbidity and phosphates were also significantly removed; turbidity was lowered from 45 to 3.3 NTU and phosphate-P was brought down from 5.6 to 0.04 mg/L. It is important to note that the FF-Cell was able to reduce the total coliforms from 240,000 per 100 mL to 5000 per 100 mL without the use of any disinfectant.

Other applications of flotation for the treatment of primary wastewater effluents have been reported by Schneider et al. [33] and Mennell et al. [34].

5 Treatment of Raw Municipal Wastewater

The pilot plant results of operating the two-stage system for a period of 6 months at the sewage treatment plant (STP) in Lee, MA, USA. [19] are presented in Table 2.3. All relevant parameters were measured on the raw municipal wastewater effluent water after preliminary flotation in the F-Cell, and final effluent after the secondary flotation-filtration unit (FF-Cell).

Table 2.3 Treatment of raw municipal wastewater[1] by two-stage F-Cell[2] and FF-Cell[3] [19]

Parameter	Raw Wastewater		F-Cell effluent		FF-Cell effluent		Total removal (%)
	Range	Average	Range	Average	Range	Average	
TSS (mg/L)	98–1191	303	15–240	73	ND-23	4	99
COD (mg/L)	251–1325	571	142–475	290	74–342	145	75
BOD (mg/L)	108–494	242	68–222	122	30–127	59	75
Turb. (NTU)	44–147	76	20–87	44	0.6–55	12	84
P (mg/L)	1.2–12.8	56	1-1-5.8	4.1	0.04–3.3	0.98	82
TKN (mg/L)	25–65.2	39	23–61	32	8.4–50	27	30
NH_3-N (mg/L)	20–31	23	17–26	21	10–23	19	19
NO_3-N (mg/L)	0.7–16.9	2.2	0.1–10.6	1.5	ND-8.6	0.74	66
COD/BOD	2.3–2.7	2.4	2.1–2.1	2.4	2.4–2.7	2.4	–

1. Temperature = 15–20 °C, pH_{Raw} = 6.9–8.1, $pH_{F\text{-}Cell}$ = 6.8–7.9, $pH_{FF\text{-}Cell}$ = 6.2–7.8
2. F-Cell = Flotation cell
3. FF-Cell = Flotation/filtration cell

Satisfactory results were obtained in a broad range of inlet organic loadings. Chemical oxygen demand (COD) values, which varied from 251 to 1325 mg/L, were reduced by 75%, and biochemical oxygen demand (BOD) values, which ranged between 108 and 494 mg/L, were also reduced by 75%. The total suspended solids' (TSSs) cumulative removal of 99% was very close to the theoretical maximum with an average effluent TSS of 4 mg/L. The overall corresponding turbidity removal was 84%. Also obtained were the following nutrient reductions: phosphorous 82%, total nitrogen 30%, and ammonia-nitrogen 19% [19].

All above-mentioned reductions have been attained over a wide range of flow rates (76–170 L/min) and varying recycle ratios (5–63%). Also using various combinations and concentrations of coagulants and flocculants, it was possible to produce biosolids with a solids content of 3%, making it easier and less costly to manage and dispose. The hydraulic loadings were up to 235 $L/m^2/min$ on the primary F-Cell unit and 155 $L/m^2/min$ on the secondary FF-Cell unit. These hydraulic loading rates are 7- to 11-fold greater in comparison to the 20 $L/m^2/min$ loading of a conventional sedimentation system.

In addition to proving the effectiveness of the two-stage system in treating raw municipal wastewater, it was possible to optimize the coagulant and flocculant chemicals and doses to the following types and their concentrations [21]:

1. *Primary clarification* (F-Cell)

   Ferric sulfate: 75 mg/L
   Magnifloc 496 C: 4.4 mg/L

2. *Secondary clarification* (FF-Cell)

   Magnifloc 496 C: 3.3 mg/L

Enhanced results were obtained for total phosphorous and nitrogen removals as well as for the reduction of various forms of nitrogen. Changing from ferric sulfate to polyaluminum chloride (PAC) in the F-Cell caused a significant reduction in total phosphorous from 7.3 to 0.04 mg/L. This could be explained as an enhanced co- and postprecipitation of phosphorous when PAC was used instead of ferric sulfate. Another application of flotation for treatment of raw wastewater effluents has been reported by Bratby [35]

6 Treatment of Activated Sludge Effluent

An FF-Cell pilot study was performed at the Oak Meadows Sewage Treatment Plant in Licking County, OH, USA [18]. The objective of this study was to demonstrate the FF-Cell's performance in the treatment of their secondary clarifier effluent following an aeration basin. The FF-Cell was run at various chemical dosages using up to 5.6 mg/L alum and 0.5 mg/L polymer. All results reported in Table 2.4 were based on daily composite samples. The FF-Cell was proven to be

Table 2.4 Tertiary treatment of activated sludge effluent by FF-Cell at Oak Meadows STP, Licking County, OH, USA [18]

	Chemical Addition		Suspended Solids		Turbidity	
Period[a]			mg/L		NTU	
	Chemical	mg/L	In	Out	In	Out
1	Alum (as Al_2O_3)	3.6	23	1.0	2.6	0.95
2	Alum (as Al_2O_3)	4.3	39	2.6	4.3	0.50
	Polymer: 1849A	0.4				
3	Alum (as Al_2O_3)	4.9	5.9	1.0	4.2	0.29
	Polymer: 1849A	0.4				
4	Alum (as Al_2O_3)	5.3	24	2.6	14	0.46
	Polymer:					
	Percol LT-25	0.3				
5	Alum (as Al_2O_3)	5.6	–	–	3.7	0.34
	Polymer:					
	Percol LT-25	0.5				
	Average		23	1.8	5.8	0.50
	Removal %		92		91	

[a]All tests were done an daily composite samples

effective in the reduction of both suspended solids (92% removal) and turbidity (91% removal). The secondary treatment plant effluent had a total suspended solids content in the range of 6–39 mg/L, which was reduced to an average value of below 2 mg/L. In a similar performance, the treatment plant effluent had turbidities in the range of 2.6–14 NTU, while the FF-Cell was able to attain an average effluent turbidity of 0.5 NTU. The BOD values were consistently at or below 1.0 mg/L. Another application of flotation to tertiary treatment has been reported by Kiuru [36].

7 Treatment of RBC Effluent

An FF-Cell unit was installed at the Jimmy Peak Wastewater Treatment Plant in Hancock, Massachusetts, USA. (Fig. 2.5). The 400 m^3/d resort area domestic wastewater flows through an aerated equalization tank, two rotating biological contactors (RBCs) [37, 38], one circular secondary sedimentation clarifier (3.6 m diameter), one rectangular tertiary sand filter (3.1 m^2 area), and finally two ultraviolet disinfection units [39] before being discharged to a leaching field. The FF-Cell was fed from the RBC's effluent just before the flow enters the secondary sedimentation clarifier (see Fig. 2.5).

Table 2.5 documents the chemical dosages applied to the FF-Cell influent, as well as the performance data for both the FF-Cell and the conventional combination of secondary clarifier and sand filter as they were run in parallel. The results demonstrate that with optimization of chemical dosages, it is possible to produce effluents that can satisfy the most strict standards or guidelines with BOD & TSS

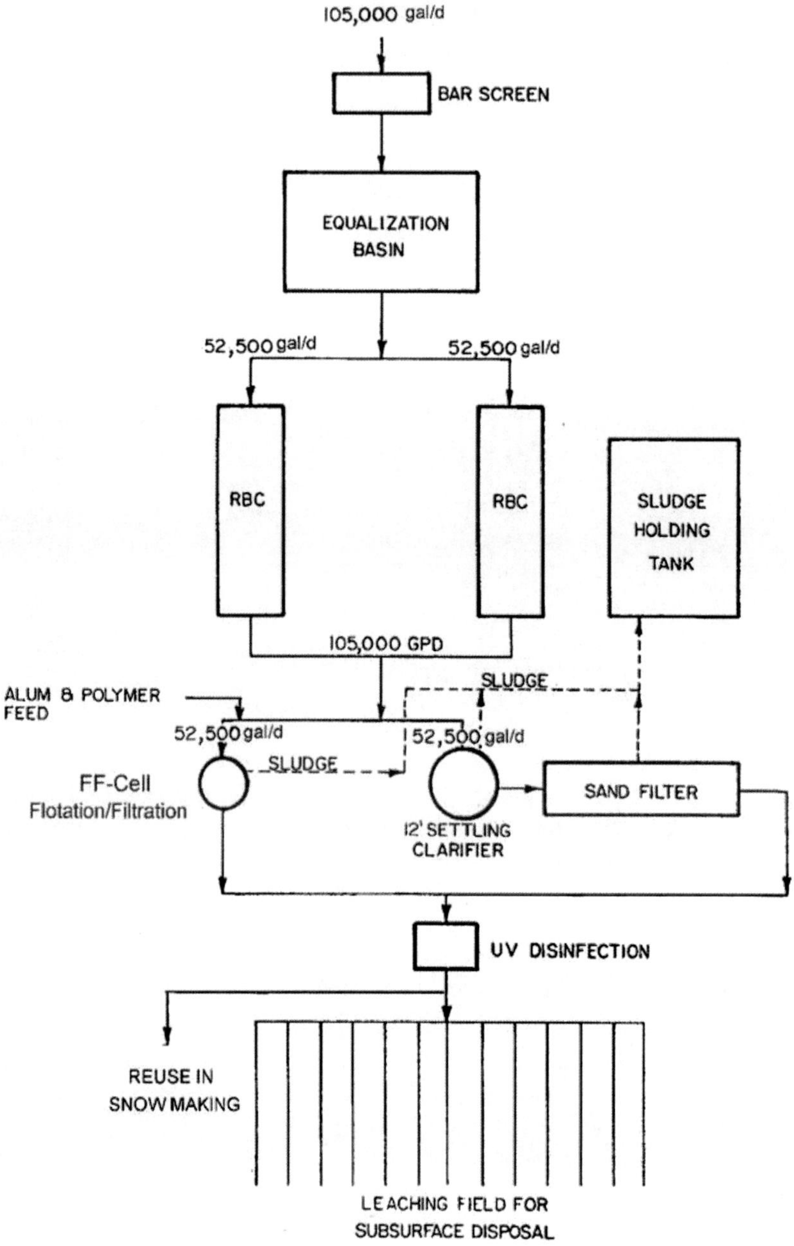

Fig. 2.5 FF-Cell for secondary clarification and tertiary filtration at Jiminy Peak Resort, MA, USA

requirements of less than 5 mg/L [18]. Another important conclusion can be drawn from the parallel performance of the FF-Cell unit and the conventional secondary sedimentation plus filtration: the innovative FF-Cell is superior to conventional

Table 2.5 Tertiary treatment of RBC effluent by FF-Cell at Jiminy Peak, MA, USA [18]

Period	Chemicals and dosages mg/L	BOD. mg/L RBC Effluent	SC+SF	FF-Cell	COD. mg/L RBC Effluent	SC+SF	FF-Cell	TSS. mg/L RBC Effluent	SC+SF	FF-Cell	Total coliforms No./100 mL × 1000 RBC Effluent	SC+SF	FF-Cell
1	A = 16. C = 1.5	9.0	4.2	2.1	40	20	25	19	ND	4	21	8	9
2	A = 20.5. D = 1.3	38	24	15	63	39	15	7	1	4	251	146	171
3	A = 60. D = 1.3	70	28	25	216	88	70	116	39	10	TNTC	280	65
4	A = 90.2. C = 0.03	102	27	26	105	75	63	89	18	10	TNTC	430	112
5	A = 43.1. C = 0.004	174	23	11	350	77	81	312	15	30	1630	1.5	1
6	A = 10.5. D = 1.2	150	8	4.3	150	19	13	131	21	12	600	ND	ND
7	B = 103.8. E = 0.73	118	–	1.6	192	–	5	179	–	15	37.6	–	2.8
	Minimum	9.0	4.2	1.6	40	19	5	7	ND	4	21	ND	ND
Range	Maximum	174	28	26	350	88	81	312	39	30	TNTC	430	171
Average		94	19	12	159	53	39	122	16	12	–	144	52

A: Alum as Al_2O_3 ; B: Ferric chloride; C: Nalco 2PD-462; D: Nalco 7533; E: Nalco 7766
RBC rotating biological contactors, *SC* secondary clarifier, *SF* sand filter, *FF-Cell* flotation/filtration, *ND* not detectable, *TNTC* too numberous to count

Table 2.6 Tertiary treatment of trickling filter effluent by FF-Cell at Norwalk, OH, USA [18]

	TSS, mg/L		BOD	mg/L	Phosphates-P, mg/L	
Period	In	Out	In	Out	In	Out
1	104	11	54	5	8.8	0.8
2	136	6	51	3	9.2	0.4
3	122	12	–	–	9.6	1.1
4	106	1	45	2	–	0.3
5	72	4	43	3	6.7	0.3
6[a]	28	9	20	4	7.0	0.7
7[a]	132	7	32	4	4.5	0.2
8[a]	103	8	39	4	7.8	0.6
Range:						
min.	28	1	20	2	4.5	0.2
max.	136	12	54	5	9.6	1.1
Average	100	7.2	41	3.6	7.7	0.6
Removal, %	93		91		92	

[a]Optimized chemistry: Alum = 120 mg/L as alum and anionic polymer Nalco 7769 = 0.25 mg/L
All tests were done on daily composite samples

tertiary treatment, not only in removal of BOD, COD, TSS, and coliforms, but also in land space requirement and consequently in capital cost required to be invested for reusing the treated effluent in snow making machines.

8 Treatment of Trickling Filter Effluent

The effectiveness of the FF-Cell in treating the secondary trickling filter [40] effluent at the Norwalk Wastewater Treatment Plant in Ohio, USA was investigated in this part of the applications. Norwalk is located in northern Ohio, approximately 50 miles southwest of Cleveland and supports a population of 14,500 people. The plant treats combined domestic sewage and food processing waste [18].

Secondary trickling filter effluent was pumped to the FF-Cell with a submersible sump pump at 164 m^3/d. Treatment chemicals were added directly to the feed line using masterflex variable speed peristaltic dosing pumps. Alum was added approximately 12 m ahead of the FF-Cell to allow for thorough mixing. Anionic polymer (Nalco 7769) was added in line just prior to the FF-Cell inlet compartment. TSS, BOD, and phosphate-P tests were all performed on daily composite samples collected from the influent and effluent of the FF-Cell. Experimental results are summarized in Table 2.6. It can be seen that on the average, the FF-Cell met the tertiary effluent standards (TSS = 10 mg/L, BOD = 10 mg/L and $P = 1$ mg/L) with removals above 90%. Considering the final three testing periods (# 6, 7 and 8 in Table 2.6) when chemical dosages, 120 mg/L of alum and 0.25 mg/L of polymer, were optimized, the FF-Cell effluent met the above-mentioned effluent standards all

Table 2.7 Tertiary treatment of lagoon effluent by FF-Cell at Arpin, WI, USA [18]

	TSS, mg/L		BOD, mg/L	
Period	In	Out	In	Out
1	26	14	20	2
2	26	8	26	9
3	28	5	44	21
4	28	2	41	14
5	28	4	30	13
6	28	2	17	14
7	28	5	17	11
8	40	11	20	12
Range	26–40	2–14	17–44	2–21
Average	29	6	27	12
Removal, %	79		56	

Optimized chemical dosage: Alum = 10 mg/L (as Al_2O_3); Anionic polymer Hercofloc 1018 = 1 mg/L

the time. In addition, the average biosolids consistency was 3.2% solids, which negates the need for a biosolids thickener.

9 Treatment of Lagoons Effluent

Arpin Wastewater Treatment Plant in Arpin, WI, USA is a system of aerated lagoons [41] treating a combination of dairy processing wastewater and domestic sewage with a high proportion (over 75%) of the flow coming from the dairy. The lagoons effluent contained a lot of colloidal substances and had high color (green) and algae count [18].

A sump pump was used to feed the pilot plant from the third lagoon. Alum and polymer were added to the feed line at dosages of 6–40 mg/L of alum and 0.5–1 mg/L of anionic polymer. The FF-Cell proved itself capable of treating the lagoons effluent and producing a clarified effluent below 20 mg/L in BOD and TSS. The average total suspended solids and BOD values in the effluent were 6 mg/L and 12 mg/L, respectively (see Table 2.7). The optimized chemical dosage was 10 mg/L for alum and 1 mg/L for this anionic polymer. At all times during the study, the high consistency green floated biosolids layer was removed by the spiral scoop. Consistencies of over 2% solids were obtained when the FF-Cell's biosolids scoop was operated intermittently allowing a thick biosolids layer to build up. Existing lagoons experiencing operational problems can be improved or upgraded for effluent reuse by the addition of an FF-Cell in series for tertiary treatment.

For further detailed information on wastewater reclamation and reuse, the reader is referred to the literature especially references [42–48].

10 Conclusion

The technical feasibility of the innovative flotation-filtration FF-Cell system for producing a high-quality effluent has been successfully demonstrated by continuous pilot plant operation. An existing secondary biological wastewater treatment plant can be easily upgraded by the addition of an FF-Cell to produce an effluent having a water quality compatible with water reuse requirements. The capital cost of such a system is low because of its short detention time and unique compact design.

Glossary [49–51]

DAF Dissolved air flotation process.

DAFF Dissolved air flotation and filtration.

Dissolved air flotation (DAF) One of dissolved gas flotation (DGF) processes when air is used for generation of gas bubbles. See dissolved gas flotation (DGF).

Dissolved gas flotation (DGF) It is a process involving pressurization of gas at 25–95 psig for dissolving gas into water, and subsequent release of pressure (to 1 atm) under laminar flow hydraulic conditions for generating extremely fine gas bubbles (20–80 microns), which become attached to the impurities to be removed and rise to the water surface together. The impurities or pollutants to be removed are on the water surface are called float or scum, which scooped off by sludge collection means. The clarified water is discharged from the flotation clarifier's bottom. The gas flow rate is about 1% of influent liquid flow rate. The attachment of gas bubbles to the impurities can be a result of physical entrapment, electrochemical attraction, surface adsorption, and/or gas stripping. The specific gravity of the bubble-impurity agglomerate is less than 1, resulting in buoyancy or nonselective flotation (i.e. Save-All).

F-Cell Dissolved air flotation cell or DAF.

FF-Cell Dissolved air flotation and filtration cell or DAFF.

Flotation cell A dissolved air flotation cell or unit process.

Flotation-filtration cell A package treatment plant, which consists of both dissolved air flotation and filtration unit processes.

Package treatment plant A prefabricated water or wastewater treatment plant consisting of two or more unit processes.

Primary flotation clarification A dissolved air flotation clarification process, which is used for primary clarification of wastewater, with a main purpose of total suspended solids removal.

Secondary flotation clarification A dissolved air flotation clarification process, which is used for secondary clarification of wastewater, with a main purpose of either: (a) removal of biological sludge from a biological oxidation process, or nitrification process or (b) removal of chemical sludge from a secondary physicochemical process mainly for removal of dissolved organics, heavy metals, etc.

Tertiary flotation clarification A dissolved air flotation clarification process, which is used for advanced or tertiary treatment, with a main purpose of nutrient removal or wastewater renovation.

Two-stage DAF A series of two consecutive dissolved air flotation treatment using different chemicals or operational conditions.

Two-stage treatment A series of two consecutive water or wastewater treatment using different chemicals, operational conditions, or even different unit processes.

Wastewater renovation Treatment of wastewater to a degree that the wastewater becomes a useful water resource.

References

1. Shammas NK (1987) Wastewater management and reuse in housing projects. In: Water reuse symposium IV, implementing water reuse, AWWA Research Foundation, Denver, CO, USA, pp 1363–1378, 2–7 Aug 1987
2. Metcalf and Eddy, Asano T, Burton SL, Leverenz HL, Tsuchihashi P, Tchobanoglous G (2007) Water reuse. McGraw Hill, New York, USA
3. Crook J (1991) Regulatory issues associated with reuse practices throughout the world. In: AWWA 1991 annual conference proceedings, resources, engineering and operations of the New Decade, Philadelphia, USA, 23–27 June, pp 225–239
4. WPCF (1989) Water reuse. Manual of Practice SM3, Water Pollution Control Federation, Alexandria, VA, USA
5. Shammas NK, El-Rehaili A (1986) Tertiary filtration of wastewater for use in irrigation. In: Symposium on the effect of water quality on the Human Health and Agriculture in the G.C.C. *States*, Al-Khobar
6. Wang LK, Hung YT, Shammas NK (eds) (2005) Physicochemical treatment processes. The Humana Press, Inc., Totowa, NJ, USA
7. Shammas NK, DeWitt N (1992) Flotation: a viable alternative to sedimentation in wastewater treatment. In: Water environmental federation 65th annual conference, New Orleans, USA, 20–24 Sept, pp 223–232
8. Wang LK, Fahey EM, Wu Z (2005) Dissolved air flotation. In: Wang LK, Hung YT, Shammas NK (eds) Physicochemical treatment processes. The Humana Press, Inc., Totowa, NJ, USA
9. Krofta M, Wang LK (1984) Development of innovative flotation-filtration systems for water treatment, Part A: first full-scale Sandfloat Plant in U. S. In: Proceedings of water reuse symposium III, 26–31 Aug, San Diego, USA, pp 1226–1237
10. Guss D (1991) Improving water quality with flotation/filtration. In: 1991 Nonwoven conference, TAPPI Proc., Technology Park/Atlanta, USA, pp 323–326
11. Wang LK, Shammas NK, Selke WA, Aulenbach DB (2010) Flotation technology. The Humana Press, Inc., Totowa, NJ, USA, p 680
12. Krofta M, Wang LK (1984) Tertiary treatment of secondary effluent by dissolved air flotation and filtration. In: Civil engineering for practicing and design engineers, vol 3, pp 253–272. (NTIS-PB83-171165)
13. Krofta M, Wang LK (1986) Wastewater treatment by a biological-physicochemical two-stage process system. In: Proceedings of the 41st industrial waste conference, IN, USA, 13–15 May 1986
14. Ross CC, Valentine GE Jr. (2015) Coupled biological/dissolved air flotation processes for treatment of food and dairy processed wastewater. Environmental Treatment Systems, Inc. http://www.etsenvironmental.com/white-papers/coupled-biological

15. Shammas NK (2005) Coagulation and flocculation. In: Wang LK, Hung YT, Shammas NK (eds) Physicochemical treatment processes. The Humana Press, Inc., Totowa, NJ, USA
16. Krofta M, Wang LK (2000) Flotation engineering, Technical Manual No. Lenox/1-06-2000/368, Lenox Institute of Water Technology, Lenox, MA, USA
17. Krofta M, Wang LK (1999) Flotation and related adsorptive bubble separation processes, Technical Manual No. Lenox 7-25-1999/348, 4th ed. Lenox Institute of Water Technology, Lenox, MA, USA
18. Shammas NK, Krofta M (1994) A compact flotation-filtration treatment unit for wastewater reuse. In: Proc. Water Reuse Symposium, AWWA, Dallas, USA, pp 97–109, 27 Feb–2 Mar 1994
19. Krofta M, Miskovic D, Shammas NK, Burgess D, Lampman L (1994) An innovative multiple stage flotation-filtration low cost municipal wastewater treatment system. In: IAWQ 17th Biennial international conference, Budapest, Hungary, 24–30 July 1994
20. Shammas NK (1997) Physicochemically-enhanced pollutants separation in wastewater treatment. In: Proceedings of international conference: Rehabilitation and development of civil engineering infrastructure systems - upgrading of water and wastewater treatment facilities, organized by The American University of Beirut and University of Michigan, Beirut, Lebanon, 9–11 June 1997
21. Krofta M, Miskovic D, Shammas NK, Burgess D (1994) Pilot-scale applications of a primary-secondary flotation system on three municipal wastewaters. In: Specialist conference on flotation processes in water and sludge treatment, Orlando, USA, 26–28 Apr 1994
22. Krofta M, Guss D, Wang LK (1988) Development of low-cost flotation technology and systems for wastewater treatment. In: Proceedings of 42nd industrial waste conference, Lewis Publishers, Chelsea, PA, USA, 185
23. Wang LK (1991) Water and wastewater treatment using advanced dissolved air flotation. In: The 1991 annual conference of the Korea Society of water pollution research and control, Seoul, Korea, 22–23 Feb 1991
24. Wang LK, Shammas NK, Hung YT (eds) (2007) Biosolids treatment processes. The Humana Press, Inc., Totowa, NJ, USA
25. Wang LK, Shammas NK, Selke WA, Aulenbach DB (2007) Flotation thickening. In: Wang LK, Shammas NK, Hung YT (eds) Biosolids treatment processes. The Humana Press, Inc., Totowa
26. Cizinska S, Matejo V, Wase C, Kiasson Y, Krejci J, Daihainmar G (1992) Thickening of waste activated sludge by biological flotation. Water Res 26:139
27. Environmental Treatment Systems, Inc. (2015) Dissolved air flotation. http://www.etsenvironmental.com/products-and-services/dissolved-air-flotation
28. Dissolved Air Flotation Corporation (2015) What is a dissolved air flotation clarifier? http://www.dafcorp.com/aboutdaf. Kaukauna
29. Aerofloat 100 DAF (2013) Aerofloat DAF solution for treatment of production wastewater. www.aerofloat.com.au. 28 Mar 2013
30. Al-Zoubi H, Al-Thyabat S, Al-Khatib L (2009) A hybrid flotation-membrane process for wastewater treatment: an overview. Desalin Water Treat 7(1–3):60–70
31. Ksenofontov BS, Ivanov MV (2013) A novel multistage kinetic modeling of flotation for wastewater treatment. Water Sci Technol 68:807–812, IWA Publishing
32. Hydro Flow Technologies (2015) Dissolved air flotation theory and operation. http://www.dissolved-air-flotation.hydroflotech.com/Engineering%20Data/Dissolved%20Air%20Flotation%20Theory%20of%20Operation.htm
33. Schneider IAH, Neto VM, Soares A, Rech RL, Rubio J (1995) Primary treatment of soybean protein bearing effluent by dissolved air flotation and by sedimentation. Water Res IAWQ 29(1):69–75
34. Mennell M, Merrill DT, Jorden RM (1974) Treatment of primary effluent by lime precipitation and dissolved air flotation. J Water Pollut Control Fed 46:2471
35. Bratby JR (1982) Treatment of raw wastewater overflows by dissolved air flotation. J Water Pollut Control Fed 54:1558

36. Kiuru HJ (1990) Tertiary wastewater treatment with flotation filters. Wat Sci Tech 22:139
37. Wang LK, Pereira NC, Hung YT, Shammas NK (eds) (2009) Biological treatment processes. The Humana Press, Inc., Totowa, NJ, USA
38. Wang LK, Shammas NK, Hung YT (eds) (2009) Advanced biological treatment processes. The Humana Press, Inc., Totowa, NJ, USA
39. Wang LK, Hung YT, Shammas NK (eds) (2006) Advanced physicochemical treatment. The Humana Press, Inc., Totowa, NJ, USA
40. Wang LK, Zucheng W, Shammas NK (2009) Trickling filters. In: Wang LK, Pereira NC, Hung YT, Shammas NK (eds) Biological treatment processes. The Humana Press, Inc., Totowa, NJ, USA
41. Shammas NK, Wang LK, Zucheng W (2009) Waste stabilization ponds and lagoons. In: Wang LK, Pereira NC, Hung YT, Shammas NK (eds) Biological treatment processes. The Humana Press, Inc., Totowa, NJ, USA
42. Hung YT, Wang LK, Shammas NK (eds) (2012) Handbook of environment and waste management: air and water pollution control. World Scientific, Singapore, p. 1227
43. Hung YT, Wang LK, Shammas NK (eds) (2014) Handbook of environment and waste management: land and groundwater pollution control. World Scientific, Singapore, p. 1091
44. Lincoln Wastewater System (2015) Wastewater reclamation and reuse. Blue Ribbon Presentation, Department of Public Works and Utilities. http://lincoln.ne.gov/city/pworks/waste/index.htm
45. Jimenez B, Asano T (2008) Water reuse: an international survey of current practice, issues, and needs, Scientific and Technical Report No. 20. International Water Association Publishing, London, UK
46. Asano T, Bahri A (2015) Global challenges to wastewater reclamation and reuse. www.worldwaterweek.org/documents
47. Water Action Hub (2015) Effluent management / wastewater reclamation / reuse. Website: http://wateractionhub.org/action_areas/view/18
48. U.S. EPA (2015) Water recycling and reuse: the environmental benefits, USA. http://www.epa.gov/region9/water/recycling
49. Wang LK, Wang MHS, Hung YT, Shammas NK (eds) (2020) Environmental and resources engineering. Springer, New York, USA
50. Wang LK, Wang MHS, Shammas NK, Aulenbach DB (eds) (2020) Environmental flotation engineering. Springer, New York, USA
51. Wang LK, Wang MHS, Hung YT, Shammas NK (eds) (2020) Integrated natural resources management. Springer, New York, USA

Chapter 3
Biological Processes for Water Resource Protection and Water Recovery

Lawrence K. Wang, Mu-Hao Sung Wang, and Nazih K. Shammas

Contents

1 Concepts and Physical Behavior 77
1.1 Definition of Process 78
1.2 Principles of Biological Oxidation 79
1.3 Energy Flow 84
1.4 Synthesis and Respiration 86
2 System Variables and Control 87
2.1 Kinetics of Sludge Growth and Substrate Removal 87
2.2 Process Variables, Interactions, and Their Significance in Process Operation and Performance 92
2.3 Aeration Requirements 95
2.4 Temperature Effect 97
3 System Modifications and Design Criteria 99
3.1 Conventional Activated Sludge Process 99
3.2 Step Aeration Process 100
3.3 Complete Mix Process 101
3.4 Extended Aeration Process 101
3.5 Contact Stabilization Process 102
3.6 Kraus Process 104
3.7 Design Criteria 104
3.8 Other Processes 105
3.9 Nitrification and Denitrification for Nitrogen Removal 109
3.10 Membrane Bioreactor 109
4 Trickling Filter 110
4.1 Process Description of Attached-Growth Systems 110
4.2 Historical Development and Applicability of Attached-Growth Systems 112
4.3 Microbiology and Ecology 115
5 Theories and Mechanisms 116
5.1 Transfer of Oxygen in Slime Layer and Liquid Film 116
5.2 Transfer of Substrate in Liquid Film and Slime Layer 117
6 Types of Trickling Filters 119
6.1 General Description 119

L. K. Wang (✉) · M.-H. S. Wang · N. K. Shammas
Lenox Institute of Water Technology, Newtonville, MA, USA

Krofta Engineering Corporation, Lenox, MA, USA

L. K. Wang, M. -H. S. Wang, Y. -T. Hung, N. K. Shammas (eds.),
Environmental and Natural Resources Engineering, Handbook of Environmental
Engineering 19, https://doi.org/10.1007/978-3-030-54626-7_3

6.2 Low-Rate, High-Rate, and Super-Rate Filters 120
6.3 Single- and Multi-stage Trickling Filter Plants 124
7 Performance Models and Design Procedures 125
7.1 National Research Council Models 126
7.2 Velz Model 127
7.3 Upper Mississippi River-Great Lakes Board Model 128
7.4 Howland Models 128
7.5 Eckenfelder Models 129
7.6 Galler and Gotaas Model 130
7.7 Biofilm Model 131
7.8 US Army Design Formulas 132
7.9 US Environmental Protection Agency Model 133
8 Design and Construction Considerations 133
9 Recent Advances in Biological Waste Treatment Processes 136
Glossary 139
References 160

Abstract Current practice in the secondary treatment of wastewater for pollution control calls for the use of biological oxidation to remove organic substances. When it comes to selecting the method of biological oxidation, a pollution control engineer has at his or her disposal a variety of treatment processes, among which activated sludge and trickling filters are the most popular. The function and limitations of the activated sludge processes are reviewed. The principles of biological oxidation and of the energy flow concept are described, and the relationship of synthesis and respiration is discussed in relation to the importance of activated sludge process control.

A contact bed, contact aerator, trickling filter, rotating disks, or other attached-growth systems consist of a bed of coarse contact media such as granite, limestone, clinkers, wood slats, plastic tubes, corrugated plastic sections, hard coal, or other material over which wastewater is distributed or contacted. Wastewater flows over the contact media on which a biological slime layer develops. Dissolved organic pollutants in the wastewater are transported into the slime layer, where biological oxidation takes place. Organic pollutants are removed by the biological slime film, which consists of various microorganisms. In the outer portions of the film, organic pollutants are degraded by aerobic and facultative bacteria. Transfer of oxygen in slime layer and liquid film, transfer of substrate in liquid film and slime layer, types of trickling filters, performance models, and design procedures are discussed. This book chapter also introduces recent advances in biological wastewater treatment processes involving the use of dissolved air flotation for primary flotation clarification, secondary flotation clarification, tertiary flotation clarification, flotation sequencing batch reactor, and flotation sludge thickening.

Keywords Biological treatment · Biological oxidation · Activated sludge · Trickling filters · Zoogleal slime film · Synthesis · Respiration · Attached-growth system · Models · Design · Dissolved air flotation · Primary flotation clarification · Secondary flotation clarification · Tertiary flotation clarification · Flotation sludge thickening · Flotation sequencing batch reactor

Nomenclature

a	Media-specific surface area per unit volume, L^2/L^3
A	Surface area of the trickling filter, L^2
A_c	Biofilm area, L^2
b	Endogenous respiration or decay coefficient, $1/t$
C	Constant
C_o	DO concentration of the floc surface, M/L^3
C_i	DO concentration in the innermost cell of floc, M/L^3
D	Dilution rate, reciprocal of hydraulic retention time, $1/t$
D	Diameter of a sphere or cylinder or the thickness of a plane or parallel floc, L
D	Depth of the trickling filter, L
D_c	Diffusion coefficient within the biofilm, L^2/T
D_w	Diffusion coefficient of the chemical species through water, L^2/T
E	Fractional efficiency of BOD removal
E	Electrode potential, millivolts
Eo	Electrode potential when [oxidant] = [reductant], millivolts
E_a	Activation energy, cal/mol
E_n	Energy consumption, kwh/yr
E_T	Efficiency at temperature, T °C
E_1	Percent efficiency of BOD removal of single-stage trickling filter, calculated by the NRC model, %
E_2	Percent efficiency of BOD removal of second-stage trickling filter, calculated by the NRC model, %
E_{20}	Efficiency at 20 °C, %
F	Faraday, 23,060 cal/eV
F	Recirculation factor
F_5	Sludge production factor, M solids/M BOD_5
F/M	Food-to-microbial mass ratio, mass of substrate applied for unit biomass per day, $1/t$
H	Discharge head, L
J	Flux of substrate into the biofilm, $M/L^2/T$
J_o	Surface flux of the chemical species, M/T
K	Maximum utilization rate of the rate-limiting substrate, M/T/M
Ka	Adsorption removal constant
km	Maximum rate of specific substrate utilization, $1/t$
K	A constant
k_1	Reaction rate at temperature T_1
k_2	Reaction rate at temperature T_2
K_{20}	Cell growth rate or substrate utilization rate at 20 °C
K_s	Michaelis-Menten constant, or half-velocity coefficient, numerically equal to the substrate concentration when $U = ½\ k$. M/L^3
K_T	Cell growth rate or substrate utilization rate at temperature T
k_T	Reaction rate at temperature T

k'	A constant
k''	A constant
K_{ace}	Coefficient used in the model developed by the US Army Corps of Engineers
K_e	Rate of BOD removal, base e
K_{nrc}	Coefficient used in the National Research Council Model for trickling filter design = 0.0085
K_p	Performance measurement parameter
K_{10}	Rate of BOD removal, base 10
L	Applied removable BOD, M/L^3
L_a	Applied BOD after dilution by recirculation, M/L^3
L_D	BOD remaining in the effluent at depth 0. M/L^3
L_e	BOD of effluent, M/L^3
L_o	BOD of influent, M/L^3
M	A constant
N	A constant
n	Number of electron moles transferred per mole of substrate utilized for energy or station number downstream of each subaerator
Q	Hydraulic loading, $L^3/L^2/T$
Q	Flow rate, L^3/T
Q	Influent wastewater flow rate, L^3/T
Q_{ave}	Average daily flow, L^3/T
Q_{pd}	Peak daily flow, L^3/T
Q_{pm}	Peak monthly flow, L^3/T
Q_r	Recirculation flow rate, L^3/T
Q_r	Recirculated flow rate, L^3/T
q_w	Wetting rate (surface application rate), $L^3/T/L^2$
Q_w	Wasted sludge flow rate, L^3/T
R	Recirculated flow ratio, Qr/Q
S	Substrate concentration, M/L^3
S	Substrate concentration, M/L^3
S_{ao}	Initial concentration of organics removable by adsorption M/L^3
S_a	Concentration of remaining organics removable by adsorption, M/L^3
S_1	Effluent substrate concentration of the plug flow process, M/L^3
S_c	Substrate concentration within the biofilm cellular matrix, M/L^3
S_e	Effluent substrate concentration, M/L^3
S_i	Interior substrate concentration, M/L^3
S_m	Substrate mass, M
S_0	Bulk liquid substrate concentration, M/L^3
S_o	Initial substrate concentration, M/L^3
S_s	Substrate concentration at the biofilm surface, M/L^3
t	Time, t
T	Temperature, °F or °C
U	Specific substrate utilization rate, change of soluble substrate concentration per unit time per unit microbial mass, $1/t$

V	Volume, L^3
V_S	Volume of final clarifier, L^3
W'	BOD loading to a second-stage filter, M/T
X	Microbial concentration, M/L^3
$-X$	Average microbial concentration in the plug flow process, M/L^3
X_c	Sludge concentration in the final clarifier effluent, M/L^3
X_c	Bacterial concentration within the biofilm, M/L^3
X_r	Return sludge concentration, M/L^3
X_v	Volatile biological solid concentration, M/L^3
X_θ	Sludge production rate for a complete mix model with no sludge recycle, M/L^3
$X_{\theta c}$	Sludge production rate for a complete mix model with sludge recycle, M/L^3
X_v	MLVSS, mixed liquor volatile suspended solids, mg/L
Y	Growth yield coefficient, mass microbial growth per unit mass substrate utilized
Y	Depth of a stagnant liquid layer outside the slime-liquid interface, L
Y_c	Depth of a biofilm within the cellular matrix, L
Y_e	Effective depth of biofilm at which $S_c = S_i$, L
Z	Direction z
α	Specific consumption of oxygen, M/L^3t
ΔG	Gibbs free energy, kcal/electron mole
$\mu.$	Net specific growth rate, change of microbial concentration per unit time per unit microbial concentration, $1/t$
P	A coefficient
P_s	Sludge produced, M/T
θ	Temperature coefficient
θ	Hydraulic retention time, t
θ_c	Sludge retention time, sludge age, or mean cell residence time, t
θ_T	Temperature coefficient, a constant

1 Concepts and Physical Behavior

Current practice in the secondary treatment of wastewater for pollution control calls for the use of biological oxidation to remove organic substances. When it comes to selecting the method of biological oxidation, the pollution control engineer has at his or her disposal a variety of treatment processes, among which activated sludge and trickling filters are currently the most popular. In this section, the function and limitations of the activated sludge processes are reviewed. The principles of biological oxidation and of the energy flow concept are described, and the relationship of synthesis and respiration is discussed in relation to the importance of activated sludge process control. Other biological treatment processes will be dealt with in later sections.

1.1 Definition of Process

Activated sludge consists of biological flocs that are matrices of microorganisms, nonliving organic matter, and inorganic materials. The microorganisms include bacteria, fungi, protozoa, and higher forms of animals such as rotifers, insect larvae, and worms. An activated sludge process can be defined as a system in which biological flocs are continuously circulated to come into contact and to oxidize the organic substances in the presence of oxygen. The fact that an "active" mass of biological forms is maintained in the system for continuous and successful biological oxidation explains why the process is designated "activated sludge" treatment.

The objectives of activated sludge treatment are twofold: (1) to obtain the maximum possible removal of organic substances with the shortest possible time and (2) to produce flocculant biological flocs having a good settling characteristic. Both are essential in controlling the secondary effluent quality. From the economic point of view, it is also desirable to meet both objectives since small aeration tank (s) and final clarifier(s) can be used. The two objectives, however, are not compatible. Biological flocs that are very efficient in removing organic substances at a rapid rate are flocs that normally settle poorly and vice versa. The tradeoff is manifested in the performance of various activated sludge processes. Design engineers and plant operators should be fully aware of the incompatibility of these two objectives for proper design and operation of a plant so that certain specific treatments can be accommodated and optimization of treatment performance can be planned intelligently.

It is important to recognize the capability as well as the limitations of activated sludge processes. In the aeration tank, biodegradable organics are converted to inorganics. A complete oxidation of organics can be expressed as:

$$\text{Organics}\ (C, H, O, N, P, S) + O_2 \rightarrow CO_2 + H_2O + NO^-{}_3 + PO_4{}^{3-} + SO_4{}^{2-} + H^+$$

The equation above assumes an infinite period of aeration time, and plentiful microorganisms needed to carry out the complete oxidation, including nitrosomonas and nitrobacters, are present. Economic constraints do not allow sufficient time for complete oxidation even for an extended aeration process. Nor is it feasible to maintain a steady population of the less competitive organisms (e.g., nitrosomonas and nitrobacters) in the aeration tank with the present operational scheme. Most noticeable is the lack of nitrification in the process, and a significant amount of ammonia nitrogen exists in the effluent as result. A typical activated sludge treatment process may yield the following:

Ammonia nitrogen	12 mg/L as N
Phosphate	10 mg/L as $PO_4{}^{3-}$
Nitrate	0.1 mg/L as N
Nitrite	0.01 mg/L as N
BOD	20 mg/L

Suspended solids 30 mg/L

Other than high residues of nitrogen and phosphates in the effluent, treatment plant operators are content with a BOD of 20 mg/L and suspended solids of 30 mg/L or thereabouts. The July 1977 effluent guidelines for publicly owned secondary treatment plants issued by the Environmental Protection Agency under the authority of the 1972 amendments to the Federal Water Pollution Control Act [1] specify the maximum monthly average effluent BOD and suspended solid requirements to be 30 and 45 mg/L, respectively. One should not overlook also the presence of refractory organics in the effluent that could eventually exert an oxygen demand on the receiving water. Tertiary treatment processes will be needed to polish the effluent in order to achieve the goal of zero pollution discharge.

1.2 Principles of Biological Oxidation

In the biological oxidation of wastewater, both synthesis and oxidation occur. Many groups of activated sludge microorganisms take part in carrying out the process. The important groups are described in Table 3.1.

In an activated sludge system that is properly operated, biological flocs are produced that incorporate most of the important groups of microorganisms. Metazoa and the fast moving protozoa are not part of the biological flocs because they can break away from them. Nevertheless, metazoa and protozoa constantly graze on biological flocs and are consequently found together in a highly stabilized wastewater. Although the ecosystem of an activated sludge process is complex, some general principles of biological oxidation can be applied [2–10].

1.2.1 Physical Adsorption

It has been observed that a fast initial removal of organics usually occurs when the wastewater is contacted with activated sludge in an aeration tank. This initial removal can be accomplished in a few minutes. The removal rate depends upon the wastewater characteristics and the volatile solid concentration of the activated sludge. The removal is interpreted as an adsorption phenomenon and was reported as early as 1939 by Ruchhoft [2].

The initial adsorption removes primarily discrete and colloidal particles. The adsorbed organic matter is subsequently oxidized or used in the synthesis of cellular components. It is important to recognize that, once adsorbed onto the biological flocs, the removal of organics from the wastewater is complete since the flocs are to be separated in the final clarifier. However, initial adsorption has little or no practical effect on dissolved organics. This explains why a contact stabilization process can take full advantage of initial adsorption only when it is applied to the treatment of wastewater containing large amounts of colloids and easily adsorbed solids [11, 12].

Table 3.1 Important groups of activated sludge microorganisms

Group	General characteristics	Identified organisms
Bacteria		
1. *Zooglea*-forming	A growth from of various species, nonspore forming, motile, and capsulated rods, nonitrifying, etc. each may rapidly oxidize carbohydrates and produce ammonia from gelatin, casein and peptone, produce well-organized flocs	*B. subtilis*: aerobic spore former, g+, nonnitrifying *Alcaligenes*: g $^-$ rods, proteolytic *Chromobacterium*: g $^-$ rods, proteolytic (*Flavobacterium*) *Pseudomonas*: g $^-$ rods, prefer carbohydrates *Achromobacterium*: g $^-$ rods *Spirillum*: spiral or curved rods, not flocculating
2. Filamentous	Form loose flocs having poor settling characteristics	*Sphuerotilus natans*: (*Leptothrix*, *Cladothrix*) *Bacillus mycoides*: *Crenothrix*, *Beggiatoa*
3. Others	Specialized or controversial functional importance	Nitrifying bacteria: *Nitrosomonas*, *Nitrobacter* Intestinal group: *B. coli*, *B. aerogenes*. Some proteolytic effect and little action in carbohydrates
Fungi	Rarely abundant except under abnormal circumstances. Filamentous	*Geotrichoides paludosus*, *Pullularia pullulans*, *Phoma* *Oospora (Geotrichum) Sporotrichum* *Fusarium*: non-filamentous. Good floc forming
Protozoa	Feed on bacteria, but also utilize soluble organics at high concentrations Control bacteria population to effect better organic removal	Ciliate: free swimming (paramecium), stalked (*Vorticella opercularia*) Sutoria: early ciliated, free-swimming stage and an adult stalked stage: feed on bacteria and protozoa Flagellata (Mastigophora): Flagellated, only found in poorly aerated environment
Metazoa	Higher forms of life often found in activated sludge; at times they may become so abundant as to be considered a factor in the ecological system. Feed on bacteria, protozoa, and algae	Rotifer: multicellular, rotating motion of two sets of cilia at one end. Aerobic Nematode: aerobic worms can metabolize solid organics not readily degraded by other organisms

Source: US EPA

1.2.2 Substrate Intake

With few exceptions, most organic compounds of high molecular weight (macromolecules such as polysaccharides and proteins) cannot penetrate into cells for use in cellular metabolism. Specific exoenzymes are excreted by the cells to digest the macromolecules adsorbed on the floc surface or in the wastewater. The digestive enzymes are hydrolases that catalyze the hydrolytic decomposition of their substrate.

Examples of hydrolases are alpha-amylase (which breaks down glucose polymers), cellulase (for cellulose), proteolytic exoenzyme (breaks down peptide bonds in protein molecules), and lipase (which hydrolyzes fats and other esters).

Without digestive enzyme, only low molecular weight substances gain entrance into cells. There are indications that hydrophilic groups containing OH, COOH, NH_2, sulfonate, etc. with 12 carbons or under can pass through cell membranes, whereas hydrophobic groups with more than 8 carbons cannot diffuse into cells. Nonhydrolyzed, hydrophobic compounds gain entrance in a different manner. According to Mickinney [3], these hydrophobic compounds are attracted to the lipid fraction of the cytoplasmic membrane where they are soluble. By penetrating into the lipid fraction, the remainder of the molecule can be brought into the cell.

Not all diffusible substance can penetrate into cells. The cell can absorb and retain certain substances selectively while excluding or excreting others. Although all cells have this same general property, different organisms differ markedly from each other in their ability to accept certain specific organic nutrients from the wastewater for use in their metabolism. The selective nature of a cytoplasmic membrane is caused in part by its lipoprotein composition. There are specific "combining sites" in the membrane that effect the selective transport of particular compounds and ions into the cell. The possession of specific penetration or transport mechanisms plays an important role in the substrate specificity of many bacteria. Stainer [4] cites an example in which many bacteria are unable to oxidize citrates as organic nutrients simply because these compounds do not enter the cell. The same bacteria, however, possess all the enzymes necessary for citrate oxidation and produce citric acid constantly as an intermediate metabolite.

1.2.3 Intracellular Enzymatic Actions

Inside the cell, chemical transformations take place with the help of intracellular enzymes. The CoA portion of the enzymes react with the carboxyl group of short chain acids, amino acids or hydroxy acids, to form a CoA-acid complex. A number of reactions will follow. Most often, β-oxidation takes place in which enzymes remove H_2 and add H_2O to the organic molecules. The resulting acetyl-CoA will then enter the tricarboxylic acid cycle for terminal oxidation. Acetate metabolism is very common with microorganisms, and it is the key intermediate for energy and synthesis according to Mickinney [5]. The tricarboxylic acid cycle, which is well accepted as the terminal oxidation scheme for acetate, is presented in Fig. 3.1. Six CO_2 molecules are produced in three different steps to show the fate of the organic carbon in glucose.

With short chain alcohols, aldehydes, ketones, and amines, the reactions are directed toward conversion of the hydrophilic group to a carboxyl group so that reaction with CoA is possible. Chemical changes of organic substances in wastewater by microorganisms are summarized in Fig. 3.2.

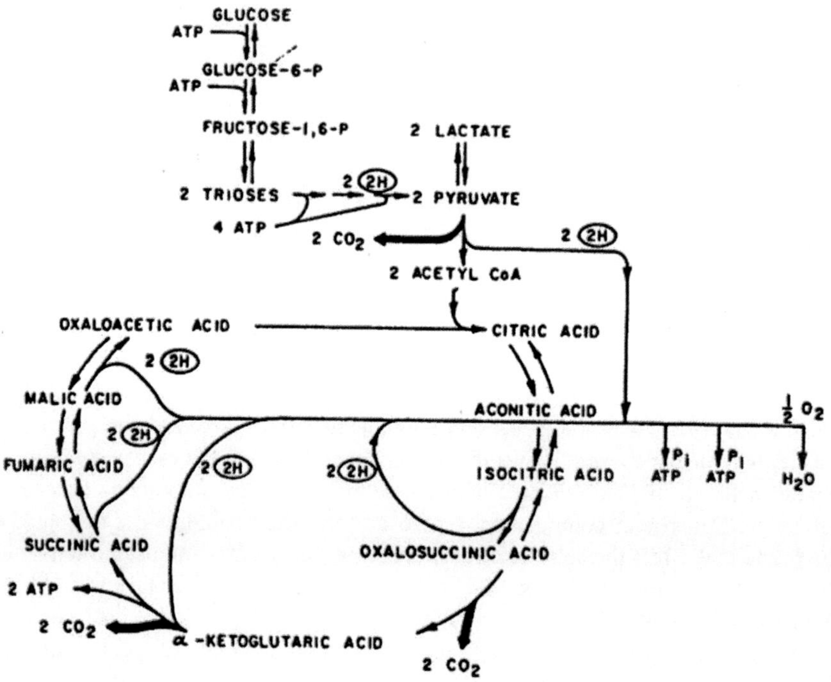

Fig. 3.1 Oxidation of glucose through acetate metabolism and tricarboxylic acid cycle. (Source: US EPA)

1.2.4 Hydrogen Transfer

The oxidation of organic compounds requires the removal of hydrogen. Activated sludge treatment, being an aerobic process, utilizes oxygen as the final hydrogen acceptor, as shown in Fig. 3.1. Before this final step is taken, however, hydrogen removal is brought about by the coenzyme and cytochrome systems. The sequence of enzymatic reactions that mediate between the oxidation of a substrate and the reduction of oxygen is called the electron transport system.

For example, the oxidation of lactic acid by lactic dehydrogenase requires continuous regeneration (reoxidation) of NADH. NADH cannot be oxidized directly by molecular oxygen. Its regeneration requires a second coenzyme FAD:

$$\mathrm{NADH} + \mathrm{H}^+ + \mathrm{FAD} \rightarrow \mathrm{NAD}^+ + \mathrm{FADH}_2$$

where FAD is flavine adenine dinucleotide and NAD is nicotinamide adenine dinucleotide. FAD becomes reduced in the process and is regenerated by the reduction of a third catalyst, the third by a fourth, and so on until finally a catalyst

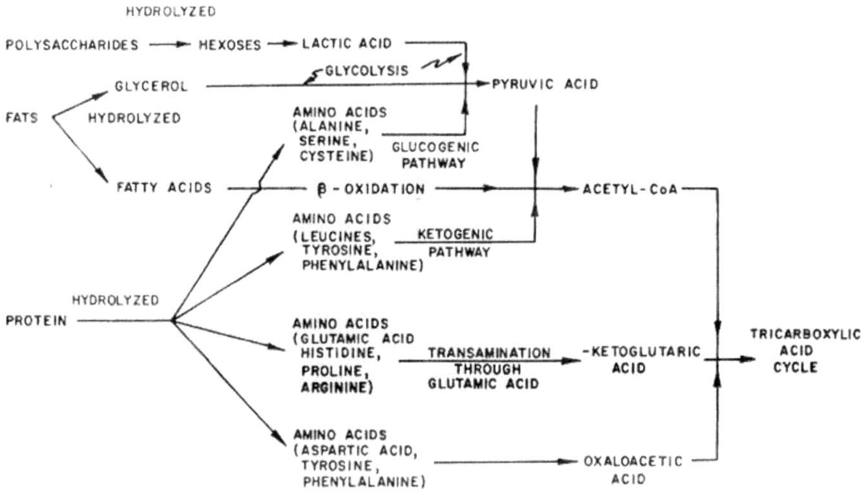

Fig. 3.2 General scheme of chemical changes by wastewater microorganisms. (Source: US EPA)

that can react with oxygen becomes reduced. The general scheme of an electron transport system can be illustrated as follows:

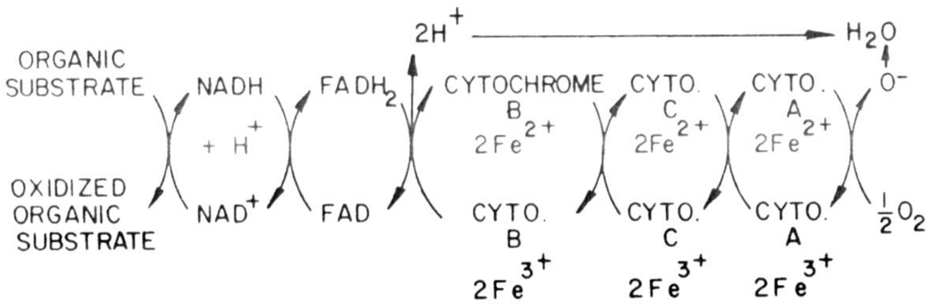

The cytochrome complement associated with respiration in eukaryotic organisms is uniform, whereas many different cytochrome systems exist in bacteria. It should be noted that in the regeneration of FAD, there is a net liberation of hydrogen ions, since only electrons are transferred to the cytochrome systems:

$$FADH_2 + 2\,Fe^{3+} \rightarrow FAD + 2H^+ + 2Fe^{2+}$$

$$4Fe^{2+} + O_2 \rightarrow 4Fe^{3+} + 2O^{2-}$$

The hydrogen ions are consumed in the subsequent reduction of oxygen to form water as a final product.

1.3 Energy Flow

Oxidation of organic substrates yields large amounts of energy. The electron transport systems provide for a smooth release of this energy. Some energy can be readily utilized by the microorganisms, and the remaining can be converted into the phosphate bond energy of ATP for storage. The amount of energy available from oxidation of substrates depends on the nature of the substrate and on the metabolic pathways used by an organism. The potential energy available from the complete oxidation of glucose with molecular oxygen, according to McCarty [6], is 28.7 kcal/electron equivalent or 688,000) cal/mol of glucose. Conversion of glucose to alcohol and CO_2 has a potential energy of only 58,000 calories. The alcoholic fermentation of a molecule of glucose results in the net generation of two molecules of ATP. Assuming the bond energy of ATP is 7000 cal, it can be calculated that approximately 25% of the potential energy can be tapped for use by the microorganisms. In the case of complete oxidation, one molecule of glucose yields 38 molecules of ATP or approximately 266,000 cal. The efficiency, at 40% of the potential energy, is high compared to mechanical systems.

The potential energy (Gibbs free energy change, ΔG) for a substrate oxidation reaction can be calculated from published data on free energy for half reactions. The following is a list of substrates commonly found in wastewater and their potential energy with complete oxidation in the presence of molecular oxygen:

Substrate	ΔG kcal/electron mole	Substrate	ΔG kcal/electron mole
Glucose	−28.7	Glutamate	−26.3
Fructose	−28.7	Butanol	−25.8
Lactose	−28.7	Benzoate	−25.6
Sucrose	−28.7	Butyrate	−25.5
Glycine	−27.1	Propionate	−25.3
Alanine	−26.3	Acetate	−25.3

In an activated sludge process, the substrates in the wastewater are assimilated by heterotrophic bacteria. Part of the energy is spent for supporting various metabolic activities. The bacterial population supports the growth of protozoa with a further energy loss as a result of the predator activities. A similar loss in energy takes place at the higher level (metazoa as predators). The utilizable energy in the process is therefore gradually diminishing. Given a long aeration time in the process, more energy is consumed and a higher effluent quality is obtained as a result.

The chemical transformation brought about by biological treatment processes involves oxidation-reduction reactions. The general equation for oxidation-reduction potential is given as:

$$E = E_0 + (RT/n\mathrm{F}) \quad \ln\,[\text{oxidant/reductant}]$$

where E_0 is the potential when [oxidant] = [reductant]; R is the universal gas constant, 1.99 cal/mol/deg; T is temperature in degrees Kelvin; n is number of electron moles transferred per mole of substrate utilized for energy; and F is Faraday, 23,060 cal/eV.

The electron activity defined as pE = −log [e -] is conceptually related to free energy and redox relationships. pE is a convenient measurement of the oxidizing intensity of a system at equilibrium and is related to the redox potential and Gibbs free energy according to Stumin [7]:

$$\mathrm{p}E = E(\mathrm{F}/2.3RT) = -\Delta G/(2.3\ nRT)$$

or

$$\mathrm{p}E = E/0.059 = -\Delta G/1,362n \quad \text{at } 25^{\circ}\mathrm{C}$$

It can be seen that pE, as an intensity factor, is conceptually similar to pH. It represents the electron free energy level per mole of electrons. When the substrates in the wastewater are oxidized to release energy, aerobic system is characterized by a high pE value, indicating high energy availability, whereas an anaerobic system is characterized by a low pE value indicating low energy availability.

For systems involving the coupling of electron transfer with hydrogen ion transfer, the general equation for oxidation-reduction potential becomes:

$$E = E_0 - 0.03 \quad \log\,[\text{oxidants/reductants}] - 0.03 \quad (\mathrm{pH})$$

The above equation describes the redox potential for a thermodynamically equilibrated system. The activated sludge process is a steady-state system, but is thermodynamically not at equilibrium. Nevertheless, the pE concept and the general redox potential are useful in a qualitative manner to describe a biological treatment system. It is, however, suggested that the measured potential should be referred to as the "electrode potential, E_c" rather than "redox potential" since the two have slightly different meanings.

A typical curve showing the relationship between E_c and substrate concentration in wastewater is shown in Fig. 3.3. The curve shows that the positive E_c value increases with the degree of treatment. A well-stabilized wastewater effluent therefore should have a high positive E_c value. Dirasian [8] reported potential values ranging from +100 to +550 mV for aeration tanks of numerous activated sludge processes. The wide range of values should not surprise anyone since the measured potential depends on wastewater characteristics, temperature, pH, DO, efficiency of treatment, location where the reading is taken, and type of reference electrode used. If one measures the electrode potential over a lengthy period of time in a treatment plant at a specific location, the range of values can be narrowed down, and the potential should reflect closely the degree of treatment.

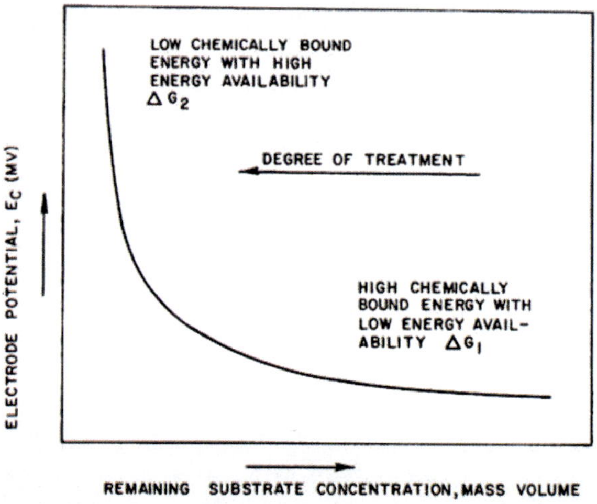

Fig. 3.3 Relationship between E_C and substrate concentration in wastewater. (Source: US EPA)

Measurement of E_c for monitoring the efficiency of treatment has many advantages over other monitoring techniques, (1) low cost, (2) real-time monitoring, and (3) no training required. In conjunction with DO and pH measurements, E_c monitoring can provide valuable information to aid the operation of activated sludge processes.

1.4 Synthesis and Respiration

In the process of biological oxidation of organic substrate, synthesis of cellular protoplasm takes place. The same enzymes that attack the substrate entering the cell also oxidize the cellular protoplasm continuously. Thus, this latter action, often called endogenous respiration or self-metabolism, occurs simultaneously with synthesis. When there is plenty of organic substrate available, synthesis exceeds endogenous loss, resulting in a net growth of cells. Conversely, when substrate is depleted, endogenous loss exceeds synthesis, resulting in net loss of cell mass in the system. The relative amount of synthesis and endogenous respiration depend mainly on the time of aeration. This is demonstrated in Fig. 3.4. Increasing the aeration time will result in more endogenous mass loss. The effect of sludge age on the distribution of synthesis and respiration is more pronounced. Different activated sludge processes employ different aeration times and sludge ages. Consequently varying amounts of biological solids are produced. Factors affecting biological growth in the system also include the nature of substrate; the substrate in biomass ratio (F/M ratio), which is related to sludge age; substrate concentration; and temperature. The growth of biological solids and its engineering control are discussed in later sections of this chapter.

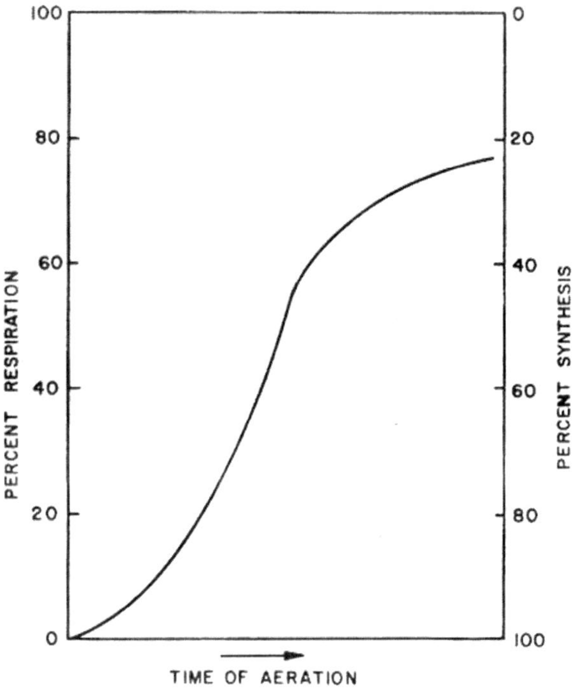

Fig. 3.4 Distribution of synthesis and respiration at varying aeration time. (Source: US EPA)

2 System Variables and Control

The activated sludge process involves a highly complex biological system. In order to design and operate the treatment system properly, one needs to understand the system variables as well as the control alternatives that are available. Engineers have in the past identified the major system variables that significantly affect the performance of an activated sludge process. Among these variables are hydraulic retention time, organic loading, mean cell residence time (sludge retention time), aeration, mixing, and enzyme regulation. It is possible to control these variables, partially if not completely, through process design and operation. Discussions on these system variables, their relationships, and their significance in process control and performance are presented in this section.

2.1 Kinetics of Sludge Growth and Substrate Removal

The success of an activated sludge process in producing a high-quality effluent depends on the continuous growth of biological flocs having a good settling characteristic. The growth of these biological flocs is accompanied by organic substrate removal, where the rate of microbial growth and the rate of substrate utilization are interrelated. If one assumes that the Michaelis-Menten enzymatic kinetics can be applied to the substrate utilization by microorganisms in the process [13, 14], then

$$U = \frac{dS/dt}{X} = \frac{k_m S}{K_s + S} \tag{3.1}$$

in which U = specific substrate utilization rate, change of soluble substrate concentration per unit time per unit microbial mass; S = substrate concentration in mass per unit volume; X = microbial concentration in mass per unit volume; k_m = maximum rate of specific substrate utilization; and K_s = Michaelis-Menten constant, or half-velocity coefficient, which numerically equals the substrate concentration when $U = ½\ k_m$ in mass per unit volume.

Biological growth is the result of the coupled synthesis-endogenous respiration reactions described in the previous section. The net result can be expressed as:

$$\mu = \frac{(dX/dt)}{X} = YU - b \tag{3.2}$$

in which μ = net specific growth rate, change of microbial concentration per unit time per unit microbial concentration, time-1; Y = growth yield coefficient, mass microbial growth per unit mass of substrate utilized; and b = endogenous or decay coefficient, time-1.

Considering a simple biological reactor with complete mixing and no sludge return, a microbial mass balance equation can be written for the reactor:

$$V\ (dx/dt) = V(YUX - bX) - QX \tag{3.3}$$

in which V reactor volume and Q = wastewater flow rate through the reactor in volume per unit time.

At steady state, i.e., (dX/dt) = 0, Eq. (3.3) yields:

$$D = 1/\theta = YU - b = \mu \tag{3.4}$$

which establishes the relationship between dilution rate (reciprocal of hydraulic retention time, θ) and the net rate of specific growth as well as the specific substrate utilization rate. It should be noted that $D = \mu$ at the steady state, since a constant microbial concentration X can be maintained in the reactor only when the net specific growth is continuously washed out. Engineers recognize the fact that in order to operate this system properly, the hydraulic flow should be controlled such that the dilution rate is smaller or equal to the net specific growth rate ($D \leq \mu$). When $D > \mu$, the microbial concentration decreases because of the high washout rate, and system failure occurs. However, this problem can be minimized in a system with sludge return.

2.1.1 Complete Mix and No Recycle Model

By combining Eqs. (3.1) and (3.4), the following equation is obtained:

$$S = \frac{K_s(1 + b\theta)}{\theta(Yk_m - b) - 1} \tag{3.5}$$

in which S = effluent substrate concentration at steady state. The effect of hydraulic retention time on the system performance can be illustrated by plotting S vs θ in Fig. 3.5 which is similar to a diagram presented by Herbert [9].

From a substrate mass balance on the biological reactor, one can derive the following:

$$X = \frac{Y(S_0 - S)}{1 + b\theta} \tag{3.6}$$

in which X = effluent microbial concentration at steady state and S_0 = initial substrate concentration in wastewater. Assuming negligible amount of biological sludge leaving the final clarifier and knowing the rate of wastewater flow, the net sludge production rate can be easily calculated.

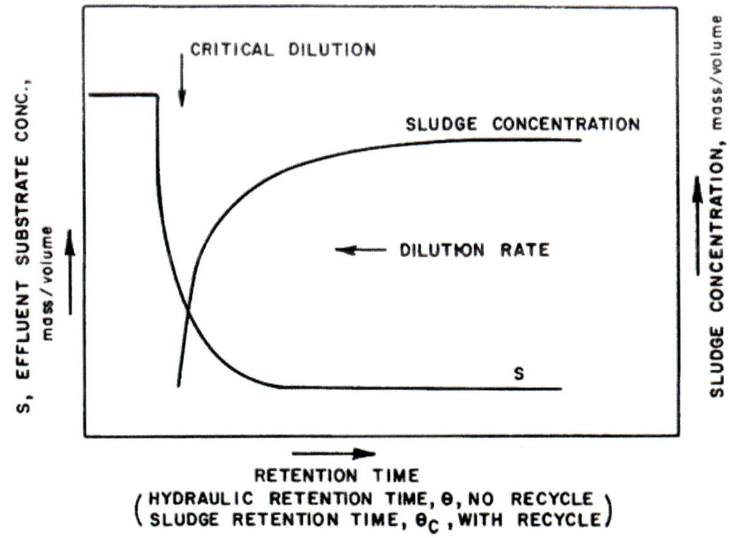

Fig. 3.5 Effect of hydraulic retention time on the performance of an activated sludge process (complete mix model). (Source: US EPA)

2.1.2 Complete Mix and Sludge Recycle Model

With biological sludge recycled from the final clarifier, the mean cell residence time or sludge retention lime is longer than the hydraulic retention time. The sludge retention time is calculated as θ_c in the following:

$$\theta_c = \frac{VX}{Q_w X_r + (Q - Q_w) X_e} \tag{3.7}$$

in which Q_w = wasted sludge flow rate, volume per unit time; X_r = return sludge concentration, mass per unit volume; and X_e = sludge concentration in the treatment effluent from the final clarifier. By writing the mass balance equation for sludge in the entire system and assuming both X_e and X_o are in negligible amounts (X_0 = sludge concentration in the primary effluent), one can develop the following:

$$\frac{1}{\theta_c} = \mu = YU - b \tag{3.8}$$

Following the same procedure in the development of working equations for the no recycle model, one derives the following:

$$S = \frac{Ks(1 + b\theta_c)}{\theta_c(Yk_m - b) - 1}$$

$$X = \frac{\theta_c}{\theta} \frac{Y(S_0 - S)}{(1 + b\theta_c)} \tag{3.9 and 3.10}$$

One readily recognizes the similar nature of Eqs. (3.5) and (3.9). Equation (3.5) expresses the effect of the hydraulic retention time on system performance for a complete mix no recycle process as is shown in Fig. 3.5. It is important to know from Eq. (3.9) that the performance of a complete mix with recycle system does not depend on hydraulic retention time. For a specific wastewater, a biological culture, and a particular set of environmental conditions, all coefficients K_s, b, Y, and k_m become constant. It is apparent from Eq. (3.9) that the system performance is a function of θ_c. Thus it is possible to regulate θ_c to achieve good treatment efficiency without increasing the hydraulic retention time. This is basically the advantage of a recycle system over a no recycle system.

2.1.3 Plug Flow and Sludge Recycle Model

The plug flow model does not provide longitudinal mixing for adjacent elements of wastewater. The increasing microbial concentration and a concurrent decreasing substrate concentration along the axis of flow make the development of a kinetic model difficult. Lawrence [10] has developed a simplified model in which a constant microbial concentration, X, in the reactor is assumed. This is believed to be valid as long as $(\theta_c/\theta) > 5$. Substituting the term X in Eq. (3.1) with the average microbial concentration $\overline{X}$, integrating the equation over the hydraulic retention time of the wastewater in the aeration tank and simplifying, one obtains the following equation:

$$\frac{1}{\theta_c} = \frac{Yk_m(S_0 - S_1)}{(S_0 - S_1) + eK_s} - b \tag{3.11a}$$

in which $e = (1 + R)\ \ln[(RS_1 + S_0)/(1 + R)S_1]$ and S_1 = effluent substrate concentration of the plug flow process. When R, the recirculated flow ratio Q_r/Q, approaches zero, $e = \ln(S_0/S_1)$; therefore

$$\frac{1}{\theta_c} = \frac{Yk_m(S_0 - S_1)}{(S_0 - S_1) + K_s \ln(S_0/S_1)} - b \tag{3.11b}$$

The equation applies as long as the volumetric recycle ratio (ratio of return sludge flow to influent wastewater flow) is less than unity. Equation (3.11b) shows that θ_c is a function of the influent as well as the effluent wastewater concentration, which is a unique characteristic of a plug flow process.

With a given set of values for the coefficients K_s, b, Y, and k_m, one can calculate, from the equations given above, the sludge retention time θ_c required to produce a predetermined effluent substrate concentration. For the treatment of wastewater to obtain an effluent BOD of 20 mg/L or below, a shorter sludge retention time is required for the plug flow process. In other words, for a given θ_c value, a plug flow process can obtain a lower effluent substrate concentration than that of a complete mix process.

2.1.4 Sludge Growth

Previously, it has been shown that the amount of cell synthesis depends on the length of time the cells are exposed to aeration. This phenomenon is expressed by Eq. (3.4) for a no sludge recycle process or Eq. (3.8) for a sludge recycle process. Both equations state that a shorter retention time results in a higher specific growth rate and therefore more sludge growth. Since the effluent microbial concentration for a no sludge recycle process is found to be $X = Y(S_0 - S)/(1 + b\theta)$ in Eq. (3.6), the daily microbial sludge production can be calculated as:

$$X_\theta = \frac{VY(S_o - S)}{\theta(1 - b\theta)} \tag{3.12}$$

in which X_θ = daily microbial sludge production with no sludge recycle, mass/d when θ is expressed in d. A negligible amount of sludge loss with the secondary clarifier effluent is assumed in the calculation.

For a plug flow or complete mix process with sludge recycle, the daily sludge production is the amount of sludge wasted from the system, $Q_w X_r$. Neglecting the sludge in the secondary clarifier effluent, and combining with Eqs. (3.9 and 3.10), one finds:

$$X_{\theta c} = Q_w X_r = \frac{VX}{\theta_C} = \frac{VY(S_o - S)}{\theta(1 + b\theta_C)} \tag{3.13}$$

The similarities between Eqs. (3.12) and (3.13) are apparent. For a plug flow system, the effluent substrate concentration S_1 should replace the steady-state substrate concentration S in the above equation.

2.2 *Process Variables, Interactions, and Their Significance in Process Operation and Performance*

The important process variables and their interactions have been presented mathematically. They are further delineated here to help engineers to see better their interactions. The significance of these variables in process operation and performance of the treatment system will be discussed in this section.

The single most important variable in activated sludge process is sludge retention time. The term sludge retention time is synonymous with sludge age and mean cell residence time. Equation (3.8) in the form of $1/\theta_c = \mu = YU - b$ relates sludge retention time to specific growth of biomass and specific substrate utilization rate. Figure 3.6 depicts their relationship. It is apparent from Fig. 3.6 that within the limit of the capability of its biosynthesis, a young activated sludge (short sludge retention time) will grow faster and utilize the soluble substrate at a faster rate. This is desirable from the standpoint of minimizing the aeration time and aeration tank volume. Unfortunately the effluent substrate concentration increases with increasing specific substance utilization rate. Considering the rate of substrate removal as proportional to the existing biomass and substrate concentrations, $dS/dt = kXS$ where k is a proportionality constant, and writing a material balance around the aeration tank at the steady state,

$$0 = Q(S_O - S) - VXkS$$

solving for S,

$$S = k\frac{S_O - S}{XT} = kU \tag{3.14}$$

The relationship expressed by Eq. (3.14) is illustrated by Fig. 3.7.

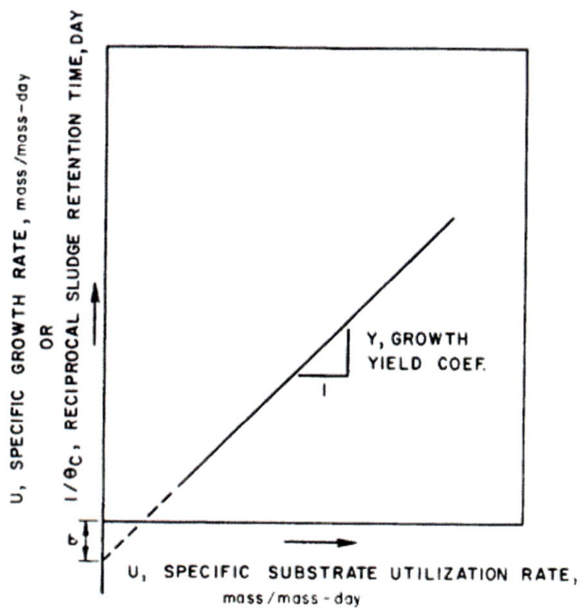

Fig. 3.6 Relationship between sludge retention, specific growth rate, and substrate utilization rate. (Source: US EPA)

In order to meet the requirement of a low effluent substrate concentration, therefore, a lower specific substrate utilization rate (or a longer sludge retention time) should be used. The terms food-to-biomass ratio (F/M) and specific substrate utilization rate have been used interchangeably by engineers. The two have slightly different meaning in definition in that F/M is the mass of substrate applied for unit biomass per day, whereas U is the mass of substrate actually utilized by unit biomass per day. The value of U approximates F/M only when nearly all of the substrate applied to the system is utilized, a situation found in the secondary treatment of organic waste in which a very high treatment efficiency is to be expected. However, it is not correct to assume such relationship when the system is not achieving high treatment efficiency.

At this point, it is necessary to reiterate the fact that all equations presented in the previous section and relationships between process variables aforementioned apply mainly to soluble substrates. With a significant portion of the substrate in the solid form of organics, the system may behave differently in its treatment performance. This fact is manifested in the contact stabilization operation in which the kinetics of growth and substrate removal are completely different from that of other modifications of activated sludge processes.

The effect of sludge retention time θ_c on effluent substrate concentration can be found in Eq. (3.9), as is illustrated in Fig. 3.8.

The aforementioned kinetics also assumes a near-complete removal of biomass from the final clarifier. In other words, the settling characteristics of the activated sludge are completely overlooked. Sludge retention time has significant effects on the sludge settling characteristics. A short sludge retention time and the resulting

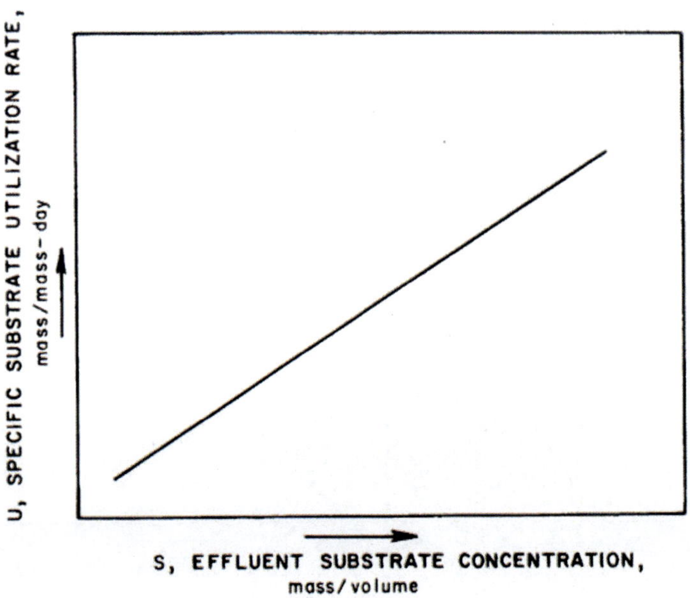

Fig. 3.7 Relationship between specific substrate utilization rate and effluent substrate concentration. (Source: US EPA)

high substrate utilization rate will force the operation in exponential growth phase of the biomass. The biomass forms flocs poorly and does not settle well in the final clarifier resulting in low treatment efficiency. Conversely, with a prolonged sludge retention time and its resulting low specific substrate utilization rate, the sludge is subjected to an extended period of endogenous respiration and becomes inactive. As a result, pinpoint flocs will develop and settle poorly in the clarifier. In either case, the treated effluent contains a significant amount of BOD in the form of biological solids. Thus the treatment efficiency is low, as is illustrated in Fig. 3.8. To produce the good settling flocs essential for a successful treatment, engineers select a θ_c value that gives neither a high nor low substrate utilization rate in the process operation. Suggested θ_c values for various activated sludge processes are presented later in this chapter.

Sludge retention time is normally controlled by wasting a portion of the settled sludge before returning to the aeration tank. To extend the sludge retention time, one needs to waste the biological sludge less often. This operation in effect generates less sludge from the treatment plant. This relationship between θ_c and sludge production is expressed in Eq. (3.13) and is illustrated in Fig. 3.9. It shows a significant drop of sludge production when the sludge retention time is very short. This is the result of poor sludge settling associated with a small θ_c value. The sludge leaves the system with the treated effluent in significant amounts and therefore reduces the amount that needs to be wasted in order to maintain a steady biomass in the aeration tank.

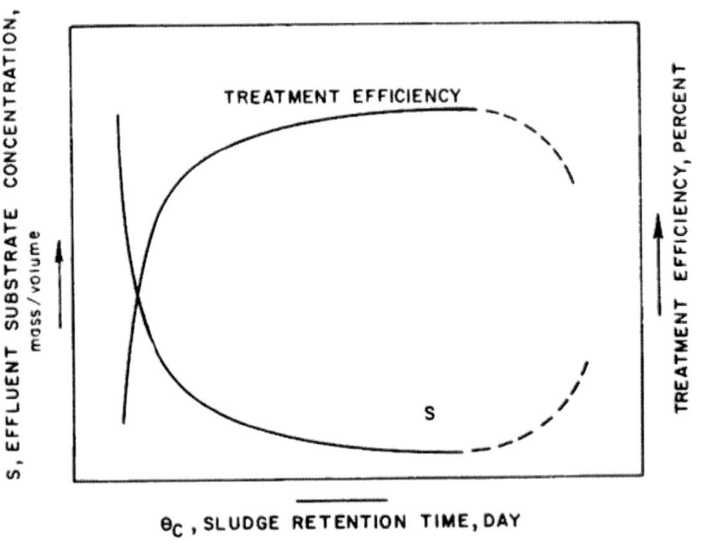

Fig. 3.8 Relationship between effluent substrate utilization and sludge retention time. (Source: US EPA)

Recognizing the important process variables and their interactions in the operation, it is evident to engineers that they cannot obtain a high-quality effluent and at the same time achieve fast substrate removal producing good settling flocs and a minimal amount of biological sludge. One can optimize the performance and cost of the treatment for an existing process or can select among various modifications of the activated sludge process to suit special needs. The various process modifications will be discussed in Sect. 3.

2.3 Aeration Requirements

Air is supplied to the aeration tank to satisfy the biochemical oxygen demand in the process of organic oxidation. In addition, diffused air is required for turbulent mixing to keep the biological sludge in suspension and in intimate contact with the substrate. This is particularly true for diffused aeration, although mechanical aeration provides good mixing without relying on the diffused air in the wastewater. It is also believed that turbulent mixing by diffused air facilitates mass transfer of oxygen into the biological flocs and transfer of CO_2 and other waste products out of the flocs.

In the activated sludge process, the oxygen requirement consists of the amount of oxygen needed for both synthesis and respiration. Consequently one needs to know the ultimate BOD (BOD_L) of the wastewater, which can be obtained from BOD_5

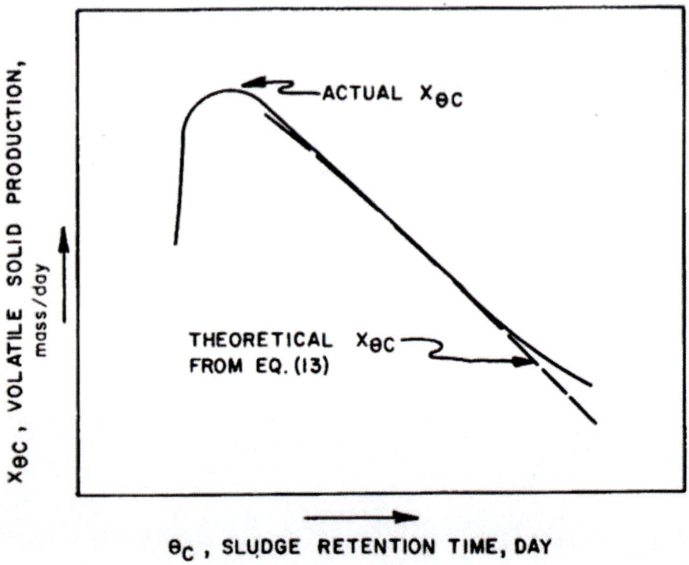

Fig. 3.9 Relationship between sludge production and sludge retention time. (Source: US EPA)

using an appropriate conversion factor, The respiration oxygen demand is 1.42 g O_2/g VSS based on the empirical equation [15, 16]:

$$C_5H_7NO_2 + 5O_2 \rightarrow 5CO_2 + 2H_2O + NH_3$$

Since part of the VSS produced is wasted in the process operation for the control of sludge retention time, the respiration oxygen demand is reduced by an amount prepositional to the amount of wasted sludge. The theoretical oxygen requirement for an activated sludge process is therefore:

$$\begin{aligned} &\text{Theoretical } O_2 \text{ requirement} = (\text{BOD}_\text{L} \text{ of wastewater utilized/d}) \\ &\quad -1.42(\text{VSS wasted/d}) \end{aligned} \tag{3.15}$$

in which all terms are expressed in mass per day.

In practice, air is supplied to the aeration tank liquid to maintain a minimum dissolved oxygen concentration of 1–2 mg/L. The objective is to maintain a dissolved oxygen gradient across the liquid-floc interface to ensure an effective oxygen transfer into the biological flocs. The critical O_2 tension for the biological floc is believed to be in the neighborhood of 0.1 mg/L DO. Based on the geometry of floc particles, the passive transport of O_2 molecules through aggregates of living cells by diffusion has been mathematically resolved by Wuhrmann [17] as follows:

$$\text{Spherical floc} \quad d^2 = (C_0 - C_i)\ (24D/\alpha) \tag{3.16a}$$

$$\text{Cylindrical floc} \quad d^2 = (C_0 - C_i)\ (16D/\alpha) \tag{3.16b}$$

$$\text{Plane floc} \quad d^2 = (C_0 - C_i)\ (8D/\alpha) \tag{3.17}$$

$$\text{Parallel floc (biological film)} \quad d^2 = (C_0 - C_i)\ (2D/\alpha) \tag{3.18}$$

in which d = diameter of a sphere or cylinder or the thickness of a plane or parallel floc, cm; C_o = DO concentration at the surface of the floc (same as the DO concentration in the liquid medium), g/cm^3; C_i = DO concentration in the innermost cell of the floc, g/cm^3; B = diffusion coefficient, 5×10^{-6} cm^2/s at 15 °C; and α = specific consumption of O_2 molecules by the cells in the floc, g/cm^3/s and at 15 °C; $\alpha_{min} = 1.0 \times 10^{-4}$ mg O_2/cm^3/s, whereas $\alpha_{max} = 5 \times 10^{-3}$ mg O_2/cm^3/s.

With C_i = 0.1 mg/L. and a minimum C_o = 1–2 mg/L maintained throughout the aeration rank, biological flocs with maximum diameter of 340–490 μm can be adequately supplied with oxygen without interference of normal respiration. Since the average floc size in activated sludge process is approximately 200 μm or below, a DO concentration of 1–2 mg/L in the aeration tank is adequate.

Since the oxygen transfer efficiency in wastewater medium is very low, much more air is supplied to the aeration tank than is present in the liquid to satisfy the oxygen demand and to maintain the minimum DO of 1–2 mg/L. The transfer efficiency varies depending on the nature of the wastewater and the geometry of the aeration tank. Eckenfelder [18] reported transfer efficiency ranging from 6.4% to 24.8%. Most diffused aeration devices have transfer efficiency between 8% and 12%. For mechanical aeration devices, the average transfer is between 1 and 1.5 kg O_2/hp/h (2.2–3.3 lb O_2/hp/h). Detailed discussion on aeration and its design is presented in Chap. 4, Submerged Aeration.

Actual air requirements range from 3.75 to 15 m^3/m^3 of wastewater (0.5–2.0 ft^3/gal) depending on the strength of the wastewater. The volume of air required per unit mass of BOD removal therefore is a better basis for design. A range of 500–900 ft^3/lb of BOD removal (31–56 m^3/kg of BOD removal) has been recommended for the design of aeration in activated sludge treatment plants [19, 20]. In addition, the Ten State Standards recommend a minimum air flow of approximately 3 cfm/ft of aeration tank length (17 m^3/h/m) for maintaining adequate mixing velocities in order to avoid deposition of solids.

2.4 Temperature Effect

The rates of all chemical reactions are affected by temperature. It is generally accepted that the rates of all biological reactions vary with temperature according to the van't Hoff-Arrhenius relationship. The integrated form of the van't Hoff-Arrhenius equation is:

$$\ln \frac{k_2}{k_1} = \frac{E_a}{R}\left(\frac{1}{T_2} - \frac{1}{T_1}\right) \tag{3.19}$$

in which k_2 and k_1 are reaction rates at temperature T_2 and T_1, respectively, E_a = activation energy in cal/mol, and R is the universal gas constant, 1.99 cal/mol (°K). Most often a modified Arrhenius equation is used to predict the temperature effect [15, 21, 22]:

$$K_T = K_{20}\theta_T^{(T-20)} \tag{3.20}$$

in which K_{20} is the cell growth rate or substrate utilization rate at 20 °C and K_T is the corresponding rate at some temperature, T; θ_T is constant called be temperature coefficient. Values of θ_T for activated sludge process are generally in the range of 1.0–1.03.

Equation (3.20) is obviously oversimplified. The θ_T value is by no means a constant for all activated sludge processes. It varies even for a given activated sludge process depending on many factors. If one examines Eqs. (3.19) and (3.20), it can be seen that θ_T is a function of the activation energy. Inorganic chemical reactions generally have activation energies that are changed only slightly by temperature variations, and their Arrhenius plots are essentially linear. The effects of temperature on activation energy for biochemical reactions are far more complex. Activation energy may change in a system with a heterogeneous microbial population because of (1) a shift of predominant species as temperature changes and (2) a change in nutrient substrates being utilized as the population shifts with changes in temperature [23]. Other investigators have found that θ_T depends on substrate concentration, chemical nature of substrate, food-to-microorganism ratio, number of test temperatures used, method of chemical analysis, and the procedure for evaluation of the rate constant [24–27]. In addition, the aeration process, as an energy-controlled kinetic rate process with its rate depending upon temperature, has been found that the activation energy and therefore the θ_T value vary linearly with temperature [23].

It can be seen that the exponential form of temperature correction relationship, as is presented by the modified Arrhenius equation, is limited in its applicability because θ_T is not a constant, but varies with temperature and other factors. Over a moderate and narrow temperature range, the θ_T value does not change significantly. Wastewater temperature in activated sludge processes varies much less than in trickling filters and lagoons. A temperature variation of no more than ±7 °C from 20 °C is expected for activated sludge processes in most treatment plants. Provided that this is true, Eq. (3.20) can be used for temperature correction of treatment performance. It is suggested that a small θ_T value (θ_T = 1.0) be used when the food-to-microorganism ratio is very low (0.2/d) and a high θ_T value (θ_T = 1.03) be used when the F/M ratio is high (0.6/d or above).

3 System Modifications and Design Criteria

In the past, the conventional activated sludge process was used exclusively because engineers were handling relatively small-volume and low-strength wastewaters. Later in dealing with wastewater of greater volume and strength and with industrial wastewater in particular, which has a complex chemical nature, engineers were forced to operate the activated sludge process differently as well as to modify the process design in order to suit special needs. This led to the development of many process modifications that have become standardized today [28].

In this section, both the conventional activated sludge process and its modifications are discussed. The characteristics, removal efficiency, and application and design parameters are listed for comparison [29].

3.1 Conventional Activated Sludge Process

The process employs a long rectangular aeration tank. Biological sludge is collected from the final clarifier. Part of the sludge is wasted from the sludge return line, and the recycled portion, together with the wastewater influent (overflow from primary clarifier), enters the aeration tank at the head end. The mixed liquor is aerated for approximately 4–8 h during which time synthesis and respiration take place with the organic matter being oxidized. The mixed liquor has a MLSS (mixed liquor suspended solids) of 1500–3000 mg/L. The sludge settles in the final clarifier until there are 8000–10,000 mg/L, when it is withdrawn off for partial return. The return rate is approximately 15–50% of the influent flow rate (Fig. 3.10a).

Hydraulically, the conventional process is operated in a quasi-plug flow mode using a long and deep rectangular tank. Some longitudinal dispersion and some short circuiting are expected, and a true plug flow regime is not obtained. A true plug flow system is a more efficient process, as was discussed earlier. It is, however, more susceptible to shock loads if the aeration tank is divided into a series of complete mix reactors; Jenkins [30] has shown that improvement in treatment performance can be obtained without a major loss in the ability of the system to handle shock loads. Toerber [31] reports that in response to a severe shock load, the plug flow operation in long rectangular tanks has a removal efficiency only 10% smaller than does the complete mix operation on a BOD basis. A conventional process in reality comes close to a series of complete mix reactors, and consequently its treatment performance is one of the best among all activated sludge processes in use. Nevertheless, the conventional process is designed for and applicable to low-strength wastewater only, for example, domestic wastewater. The requirement of a larger aeration tank, and consequently a larger land area, further inhibits its use.

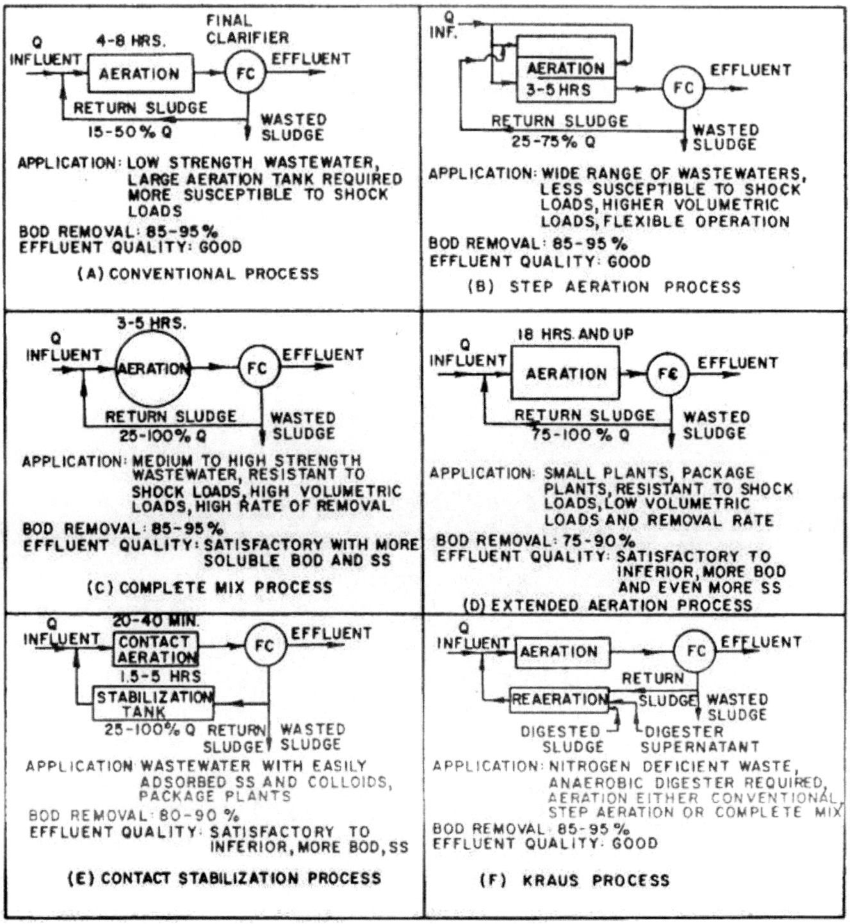

Fig. 3.10 Flow diagrams and applications of major activated sludge processes. (Source: US EPA)

3.2 Step Aeration Process

The step aeration process represents a significant improvement of the conventional process with very little physical modification. Return sludge enters the head end of the tank with a portion of the wastewater influent. The piping is so arranged that an increment of wastewater be discharged into the aeration tank at subsequent steps. In doing this, the waste load is more uniformly spread over the length of the aeration tank, resulting in better utilization of the oxygen supplied. More importantly, the load is more uniformly spread so that the system is much less susceptible to shock load. The biological sludge is maintained highly active throughout the tank so that more organics are removed in a shorter contact time. Consequently, a high waste load (volumetric loading, kg BOD/m^3 or lb BOD/1000 ft^3 of aeration tank volume) is possible, resulting in a significant reduction in unit size compared to the conventional process.

Either a long rectangular aeration tank or a rectangular tank subdivided into four or more parallel channels can be used. In either case, the first portion of the aeration tank can be used for reaeration of the return activated sludge alone. This adds to the flexibility of operation since recondition of return sludge by sludge aeration sometimes is essential for maintaining top treatment performance. The arrangement of subdividing the aeration tank into multiple parallel channels (Fig. 3.10b) further adds flexibility to the process operation. In small cities and communities where fluctuations of wastewater flow and strength are expected, a portion of the aeration tank can be shut off to accommodate lower flow. This operation is very important if uniform treatment performance is to be maintained. Flexibility in operation provided is therefore an important feature of the step aeration process.

3.3 Complete Mix Process

The complete mix process is most useful for the treatment of wastewater with fluctuating organic strength because it is least susceptible to shock loads among all activated sludge processes. For this reason, the process has found increasing popularity among those particularly concerned with the treatment of industrial wastewater with moderate to high strength. A mechanically stirred reactor test simulates a complete mix condition (Fig. 3.10c). When diffused aeration in a rectangular tank is used, the mixture of wastewater influent and the return sludge should enter the aeration tank at several points in a central location with the effluent going into channels on the sides of the aeration tank.

The process can be better described mathematically than other activated sludge processes (see Sect. 2). In general, both volumetric loading and F/M ratio are higher than that used in conventional and step aeration processes. A high MLVSS concentration is maintained to accommodate the strong organic wastewater that leads to the higher sludge recycle ratio required, even though the sludge retention time is comparable to those of the step aeration and conventional processes.

It should be emphasized here that although comparable treatment efficiencies are obtained for the complete mix process, the conventional process, and the step aeration process, the higher F/M ratio employed in the operation of a complete mix process tends to leave more soluble in the effluent. Consequently an effluent with a lower quality is expected.

3.4 Extended Aeration Process

The extended aeration process is operated hydraulically, rather more like a complete mix than any other activated sludge process (Fig. 3.10d). This is made possible by providing a very long hydraulic retention time, allowing the mixing of incoming fluids with those in the rest in the aeration tank with little or no chance of short circuiting. Wasting of sludge is done only periodically to avoid building up inert solids as well as to avoid excessive carryover of solids in the effluent.

The extended aeration time and the near-complete return of all solids impose an endogenous growth condition on the system. Volumetric loading is very low, and the sludge normally has an inferior settling characteristic. One can expect on the average a BOD removal efficiency of only 85% [32–34].

Since a long aeration time is employed, the process is generally applicable only to small treatment plants of less than 3785 m^3/d (1.0 MGD) capacity. Prefabricated package plants commercially available for housing subdivisions, small communities, institutions, schools, military bases, etc. generally use the expended aeration process. No primary sedimentation is provided and no exercise of return sludge control is attempted. The objective is to simplify the process in construction and in operation. Sludge production is very small because of prolonged endogenous oxidation, which minimizes the problem of sludge treatment and disposal. Needless to say, the oversimplified approach does not guarantee good treatment efficiency. In fact, more often than not, the process yields an effluent of lower quality. Most state water pollution control agencies are reluctant to approve the installation of such treatment plants, at least not as secondary treatment facilities.

3.5 Contact Stabilization Process

When activated sludge is contacted with wastewater in the aeration tank, there is an initial removal of organic matter that is often considered an adsorption phenomenon; this initial removal is rapid. The contact stabilization process takes advantage of the rapid adsorption characteristics of the activated sludge. The wastewater influent is allowed 20–40 min contact time with the return sludge in the aeration tank. During this period, much colloidal and suspended organic matter and some dissolved organics are adsorbed by the sludge flocs to some extent. Although very little oxidation occurs in this brief period, the adsorbed organics are nonetheless removed as the sludge is separated from the treated effluent in the final clarifier. The settled sludge is then allowed 1.5–5 h of aeration in a sludge stabilization tank in which the adsorbed organics are synthesized and oxidized. After the stabilization step, the sludge is partially returned to the incoming wastewater in the aeration tank (Fig. 3.10e).

The advantage of the process is evident in that the shorter aeration time reduces the aeration tank volume considerably. Although the settled sludge needs aeration as well, the volume is small, whereas in the other processes, the entire volume of the mixed liquor is aerated to achieve sludge stabilization. Aeration tank volume reduction by 50% over the conventional and step aeration processes is feasible. Many existing conventional plants are too small to accommodate the increasing volume and strength of the wastewater that they are converted to contact stabilization plants out of necessity. There are new treatment facilities, particularly package plants, designed and built as contact stabilization plants. The limit of the process application, for some reason, is not properly recognized by many engineers. The process should not be applied for the treatment of wastewaters whose organics occur mostly in the dissolved form. For municipal wastewater or even for domestic wastewater, laboratory tests should be performed to determine (1) the removable fraction of the organic matters and (2) the contact time and sludge concentration required to affect specified treatment efficiency.

Rich [35] assumes that adsorption removal per unit concentration of activated sludge is related to the concentration of the removable portion of the organics:

$$-\frac{dS_{\mathrm{a}}}{dX} = k_{\mathrm{a}} S_{\mathrm{a}} \tag{3.21a}$$

or

$$\frac{S_{\mathrm{a}}}{S_{\mathrm{ao}}} = -K_{\mathrm{a}} X \tag{3.21b}$$

in which S_{a} = concentration of remaining organics that can be removed by adsorption, mass per volume; S_{ao} = initial concentration of organics that can be removed by adsorption; X = initial concentration of activated sludge; and K_{a} = adsorption removal constant, which is equal to 0.434 k_{a} in Eq. (3.21a).

The following procedure can be followed for a laboratory development of design criteria using Eq. (3.21b):

1. Use four or more batch reactors, each with a different amount of acclimated activated sludge.
2. Allow a contact period of 20 min with the wastewater in each reactor.
3. Analyze the initial solid concentration and the initial and final BODs of the wastewaters in each reactor.
4. Plot figures similar to Fig. 3.11a, b to establish the removable fraction and removal constant at various solid concentrations.

The procedure may be repeated with different contact periods (30 min, 40 min, etc.) to see whether a larger fraction of organics could be removed and whether such operation is economically justifiable.

From Fig. 3.11, suppose a certain sludge concentration X is selected for operation of the adsorption process with a $(S_{\mathrm{a}}/S_{\mathrm{ao}})$ value of 0.1; then the total BOD removal of the process is 90% – (90%)(10%) = 81%. If a total removal of 90% is desirable, it can be seen that 100% of the organics removable by adsorption is required, which is highly unlikely for any wastewater. The limitation of the process application is therefore self-evident.

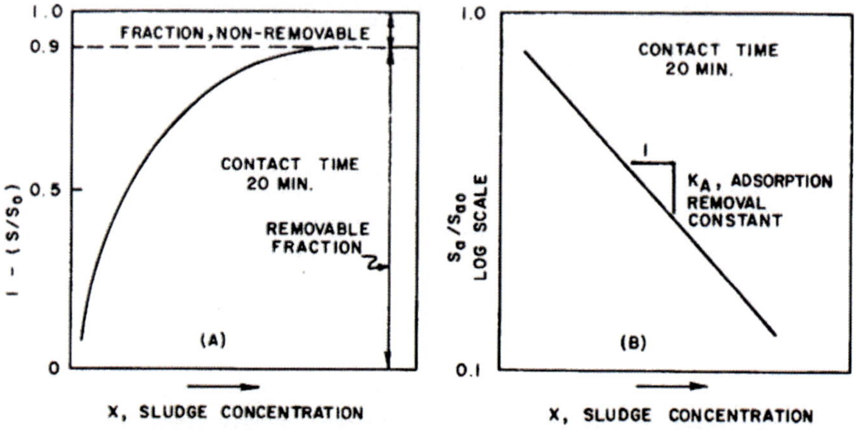

Fig. 3.11 Relationship between sludge concentration and removable BOD by adsorption. (Source: US EPA)

3.6 Kraus Process

When nitrogen deficiency occurs in biological waste treatment and exogenous supply of nitrogen is expensive, the Kraus process can be useful in that an internal or endogenous supply of nitrogen is used to maintain the growth of active biomass. As shown in Fig. 3.10f, some digested sludge and digester supernatant, together with a portion of the settled sludge from the final clarifier, are aerated in a reaeration tank. The released ammonia nitrogen from the sludge and the supernatant is converted to nitrate. When the content from the reaeration tank is introduced into the aeration tank with the wastewater, nitrate serves as the supplemental nitrogen source for synthesis. There is an improvement of the settleability of the MLSS also because of the presence of the inert fraction of the aerated, digested sludge. Additions of nitrogen in the form of ammonia, nitrate, or urea can be added to the reaeration tank for better control of nitrogen supply if needed.

3.7 Design Criteria

From past experience in the activated sludge process operation, design parameters have been established for the various processes. Table 3.2 lists the significant design parameters for various activated sludge processes [36]. Design parameters for modified aeration and high-rate aeration from this source are omitted here for reasons to be given later.

For comparison purposes, Tables 3.3 and 3.4 are presented to show the current design criteria for activated sludge processes in New York and Illinois States,

Table 3.2 Design parameters for major activated sludge processes

Process modification	Parameter F/M ratio kg BOD/kg ML VSS/d	Volumetric lb BOD/10^3 ft^3/d	Loading kg BOD/m^3/d	MLSS mg/L	Sludge retention time d	Hydraulic retention time h
Conventional	0.2–0.4	20–40	0.32–0.64	1500–3000	5–15	4–8
Step aeration	0.2–0.4	50–60	0.64–0.96	2000–3500	5–15	3–5
Complete mix	0.2–0.6	50–120	0.80–1.90	3000–6000	5–15	3–5
Extend aeration	0.05–0.15	10–25	0.16–0.40	3000–6000	20–30	18–36
Contact stabilization	0.2–0.6	60–75	0.96–1.20	1000–3000	5–15	0.5–10
Kraus process	0.3–0.8	40–100	0.64–1.60	2000–3000	5–15	4–8
Pure oxygen system	0.25–1.0	100–250	1.60–4.0	6000–8000	8–20	1–3

Source: US EPA

respectively. It can be seen that whereas the design parameters of the two states tend to be slightly conservative, they are in general in conformity with those suggested in Table 3.2.

3.8 Other Processes

3.8.1 Tapered Aeration

Tapered aeration is only a slight modification of the conventional process in that diffusers are spaced close together at the head end of the aeration tank so that the higher oxygen demand is met by the more abundant oxygen supply. The spacing of diffusers increases toward the tank outlet as the oxygen demand is lowered rapidly. It is believed that the process can eliminate oversupply of oxygen at the outlet end and thereby inhibit the growth of nitrifying organism that further tax the oxygen supply. The belief that less air is required in tapered aeration with consequent lower operating costs is unfounded because the reduction of air used at the outlet end is canceled out by the increase in air used at the head end. However, a better BOD removal performance can be expected in a tapered aeration process because of its more effective use of the air.

3.8.2 Modified Aeration

This process uses shorter aeration times, down to 1.5–3 h. A low MLSS concentration is maintained at approximately 500 mg/L, resulting in a high F/M ratio. The

Table 3.3 Design criteria for activated sludge processes for New York State –aeration tank capacities and permissible loadings

Process	Plant design flow MGD	Aeration period h (based on design flow)	Plant design 1b BOD_5/d	Aerator loading lb. BOD_5/ 1000 ft^3/d	MLSS[a], lb/BOD_5, lb/d
Conventional	To 0.5	7.5	To 1000	30	2/1–4/1
	0.5–1.5	7.5–6.0	1000–3000	30–40	2/1–4/1
	1.5 and up	6.0	3000 and up	40	2/1–4/1
Modified or "high rate"	All	2.5 and up	2000 and up	100	1/1 (or less)
Step aeration	0.5–1.5	7.5–5.0	1000–3000	30–50	2/1–5/1
	1.5 and up	5.0	3000 and up	50	2/1–5/1
Contact stabilization	To 0.5	3.0 (in contact zone)[b]	To 1000	30	2/1–5/1
	0.5–1.5	3.0 to 2.0 (in contact zone)[b]	1000–3000	30–50	2/1–5/1
	1.5 and up	1.5 to 2.0 (in contact zone) [b]	3000 and up	50	2/1–5/1
Extended aeration	All	24	All	12.5	As low as 10/1 to as high as 20/1

[a]MLSS/BOD_5: normally recommended values at the ratio of mixed liquor suspended solids under aeration (lb) to BOD loading (lb/d)
[b]Contact zone: 30–35% of total aeration capacity. The reaeration zone comprises the balance of the aeration capacity

purpose is to operate the system at the exponential growth phase of the activated sludge for fast removal of BOD. The short aeration time allows only 60–75% BOD removals. Furthermore, the sludge in the exponential growth phase does not flocculate properly. The net result of the poor settling characteristics of the sludge is a high effluent suspended solid concentration. Because of this inferior performance, the process is primarily suited to intermediate treatment and should not be designed as a secondary treatment unit. In fact the modified aeration process was originally planned as a modification of the old activated sludge treatment plants to accept wastewater with the flow rate and strength above their design capacities. The effluent quality is far below the secondary effluent standards specified by US EPA today.

Table 3.4 Design criteria for activated sludge processes for the State of Illinois – tank capacity and permissible loadings

Process type	BOD and SS % Removal	Design Flow MGD	BOD loading lb/1000 ft^3/d	Detention time h	MLSS/ BOD lb/lb/d	Detention time h	Surface Settling rate gpd/ft^2	Weir overflow rate gpd/ft
Extended aeration	70	All	To 20	24	10/1 to 20/1	(To 0.05 MGD) 4.0	300	5000
						(0.05–0.15 MGD) 3.6	300	5000
						(0.15 MGD up) 3.0	600	5000
Contact stabilization	90	To 0.5	30	3.0 mix, 6.0 reaer.	2/1 to 5/1	3.6	500	5000
		0.5–1.5	30–50	2.0–3.0 mix, 4.0–6.0 reaer.		3.0	600	10,000
		1.5 up	50	1.5–2.0 mix, 3.0–4.0 reaer.		2.5	700	10,000
Step aeration	90	0.5–1.5	30–50	7.5–5.0	2/1 to 5/1	(To 0.05 MGD) 3.0	600	5000
	90	1.5 and u	50	5.0		(0.5–1.5 MGD) 2.5	700	10,000
		p				(1.5 MGD up) 2.0	800	10,000
Complete mix	90	All	35	6.0	2/1	2.0	500	5000
Conventional	90	To 0.5	30	7.5	2/1 to 4/1	3.0	600	5000
		0.5–1.5	30–40	7.5–6.0		2.5	700	10,000
		1.5 and up	40	6.0		2.0	800	10,000

Source: US EPA

3.8.3 High-Rate Aeration

Similar to the modified aeration process, a high-rate aeration process uses short aeration times, 0.5–2 h, to take advantage of the fast BOD removal at the exponential growth phase of biological sludge. A high MLSS concentration (4000–8000 mg/L) is maintained to increase the sludge retention time in the system. Most often the system is operated with a very high volumetric loading (up to 500 lb/1000 ft^3) and a very high substrate utilization rate that poor settling of sludge in the final clarifier occurs. Even though a BOD removal of 75–85% can be obtained, the effluent still contains a significant amount of soluble and solid BOD that the effluent quality is no better or inferior to that of the modified aeration.

3.8.4 Step Feed

In all standard process modifications that use plug flow mode operation, the return sludge enters the head end of the aeration tank. Balmér et al. [37] developed a process in which the pre-aerated return sludge is returned to the aeration tank in equal portions at various locations. In the standard step aeration process, part of the wastewater entering the outlet end of the aeration tank will have a short retention time compared to the part entering the outlet end of the aeration tank. The BOD removal will not be uniform, and a lower efficiency should be expected at the outlet end. With step feeding of sludge, more biomass exists at the outlet end to ensure efficient BOD removal. Compared with conventional and step aeration processes, the step feed process maintains a higher MLSS concentration with a better adsorption property and thus can receive a higher BOD loading.

Balmér's experience showed good-quality effluent at high *F/M* or volumetric loading. An 85% BOD removal was obtained at 4 kg BOD/m^3/d (250 lb BOD/1000 ft^3/d) loading and 66–79% removal at 9.6 kg BOD/m^3/d or 600 lb BOD/1000) ft^3–/d loading. The process is able receive ten times as much volumetric loading as an extended aeration process with equivalent treatment performance.

3.8.5 Oxidation Ditch

Using a shallow channel with a circular path, the wastewater is introduced into the ditch continuously or intermittently for a long hydraulic retention time. The screened wastewater is aerated by an aeration rotor and circulates at about 30–60 cm/s (1–2 ft/s). The process is operated in a similar fashion to an extended aeration process or an aerated lagoon with equivalent performance [38].

3.9 *Nitrification and Denitrification for Nitrogen Removal*

Due to the stricter effluent standards, in particular for nutrients, many existing wastewater treatment plants need to be upgraded. Upgrading of plants with conventional processes may require the extension of tank volumes and the construction of new reactors. It is necessary to develop innovative processes, which satisfy the following requirements: high removal efficiencies, upgrading with minimal interference with the existing facilities, low investments, and simple technologies with low operating costs [39].

Conventional activated sludge treatments for nitrogen removal include several phases with different oxygen concentrations or may employ only one reactor in which alternating aerobic and anoxic phases are achieved in time or space. In such processes, the oxygen concentration must be controlled in order to obtain the simultaneous (in time and in space, i.e., with constant conditions and in the same reactor) activity of denitrifying and nitrifying bacteria. It has been shown, in fact, that denitrification reaction takes place also under aerobic conditions and that nitrification is possible at low levels of dissolved oxygen. As far as denitrification is concerned, "aerobic denitrification" and "microzone denitrification" can be defined [40]. In the former, microorganisms utilize nitrate and oxygen simultaneously as terminal electron acceptors; and in the latter, denitrification occurs in the anoxic microzones with a biological floc. This last phenomenon is supposed to occur due to the particular aeration conditions at very low dissolved oxygen concentration.

Low oxygen concentrations can be maintained by direct DO control or indirect DO control, like oxidation-reduction potential (ORP). Since nitrification and denitrification have become major aims in wastewater treatment, attempts have been made to use ORP as a control parameter, cheap but reliable control strategies, for wastewater treatment plants [41]. Most ORP controllers based on curve interpretation also have implemented time-control strategies as an emergency control whenever characteristic points cannot be found by controller [42].

Other studies for the nitrification-denitrification process include nitrification performance in an integrated fixed activated sludge process [43] and iterative design of a nitrate controller using an external carbon source in an activated sludge process [44]. In an activated sludge process, the nitrogen removal efficiency may be improved by adding an external carbon source. Automatic control of the nitrate level can be achieved by regulating external carbon flow [45–47].

3.10 *Membrane Bioreactor*

In recent years, new membranes specially developed for use in wastewater treatment have made membrane bioreactor application become a promising alternative to the well-known aerobic processes. Ultra- or micro-filtration membrane possesses advantage in possible replacing conventional sedimentation for the separation of the treated water from the sludge [48]. The use of submerged membranes has reduced the power consumption of membrane bioreactors and hence increased their potential for the application of membranes in wastewater treatment. Moreover, ultra- or micro-filtration membranes with a pore size of 0.2 μm or less retain not only bacteria but also viruses [49]. The complete retention of sludge allows operation at much higher biomass concentrations. The higher the concentration, the lower the F/M ratio becomes, with the effect that the microorganisms utilize a growing portion of the carbon content of the feed for maintenance.

Urbain et al. [50] and Rosenberger et al. [51] demonstrated the good performance of aerobic treatment of municipal wastewater using a membrane bioreactor. Rosenberger et al. [51] pilot plant comprised an anoxic zone to enable denitrification. The hydraulic retention time (HRT) varied between 10.4 and 15.6 h.; accordingly, the volumetric loading rate was between 1.1 and 1.7 kg COD/m^3/d. The mixed liquor suspended solid concentration gradually increased to 18–20 g MLSS/L. The F/M ratio varied according to the operation conditions but decreased to a value of 0.07 kg COD/kg MLSS/d. Treatment performance was very stable and on a high level. The COD was reduced by 95%. Nitrification was complete and up to 82% of the total nitrogen could be denitrified [52–58].

4 Trickling Filter

4.1 Process Description of Attached-Growth Systems

A contact bed, contact aerator, trickling filter, rotating disks, or other attached-growth systems consist of a bed of coarse contact media such as crushed trap rock, granite, limestone, clinkers, wood slats, plastic tubes, corrugated plastic sections, hard coal, or other material over which wastewater is distributed or contacted [14, 36, 58–64]. Wastewater flows over the contact media on which a biological slime layer (i.e., zoogleal slime) develops. Dissolved organic pollutants in the wastewater are transported into the slime layer, where biological oxidation takes place. Organic pollutants are removed by the biological slime film, which consists of various microorganisms, as shown in Fig. 3.12. In the outer portions of the film, organic pollutants ($C_aH_bO_cN_dP_eS_f$) are degraded by aerobic and facultative bacteria under aerobic conditions according to a biochemical reaction approximately expressed by Eq. (3.22).

$$
\begin{aligned}
&4\,C_aH_bO_cN_dP_eS_f + (4a + b - 2c - 3d + 5e + 6f)\,O_2 \rightarrow \\
&\quad 4a\,CO_2 + (2b - 6d - 6e - 4f)\,H_2O + 4d\,NH_3 + 4e\,PO_4^{3-} \\
&\quad +4f\,SO_4^{2-} + (12e + 8f)\,H^+
\end{aligned}
\tag{3.22}
$$

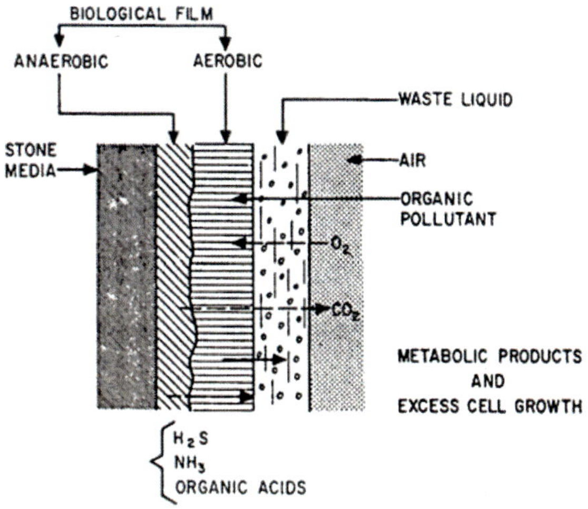

Fig. 3.12 Schematic diagram of trickling filter process. (Sources: US EPA)

in which the end product NH_3 can be further oxidized to NO_2^- and/or NO_3^-. The microorganisms in the biological slime film grow following Eq. (3.23) and/or Eq. (3.24) [16, 65, 66]:

$$NH_4^+ + 4\ CO_2 + HCO_3^- + H_2O \rightarrow C_5H_7O_2N + 5\ O_2 \quad (3.23)$$

$$C_{10}H_{19}O_3N + 1.5\ NH_3 + 2.5\ CO_2 \rightarrow 2.5\ C_5H_7O_2N + 3\ H_2O \quad (3.24)$$

in which $C_5H_7O_2N$ is the empirical formula of bacterial cell and $C_{10}H_{19}O_3N$ is the empirical composition of domestic wastewater [67]. When the thickness of the biological film increases to a degree that the diffused oxygen is almost totally consumed before it can reach the full depth of the film, an anaerobic (or anoxic) film is established near the surface of the contact media as shown in Fig. 3.12. Equation (3.25) describes the anaerobic biological reaction performed by facultative and anaerobic bacteria assuming all intermediate organic acids are degraded [68].

$$\begin{aligned} &8\ C_aH_bO_cN_dS_e + (8a - 2b - 4c + 6d + 4e)\ H_2O \rightarrow \\ &(4a - b + 2c + 3d + 2e)\ CO_2 + 8d\ NH_3 + (4a + b - 2c - 3d - 2e)\ CH_4 \\ &\quad +8e\ H_2S \end{aligned} \quad (3.25)$$

where $C_aH_bO_cN_dS_e$ is the general formula of the dissolved organic substrate penetrating the full depth of the biological film.

Since the biological slime film further increases in thickness to a degree that both dissolved organic substrate and oxygen are almost consumed before they can reach the anoxic zone, no external organic source is available for cell carbon. The facultative and anaerobic bacteria near the media face enter into a phase of "endogenous nitrate respiration" [16, 65]:

$$C_5H_7O_2N + 4\ NaNO_3 + H_2O \rightarrow 4\ NaHCO_3 + NH_4HCO_3 + 2\ N_2 \quad (3.26)$$

in which $NaNO_3$ represents the nitrate source penetrating the full depth of the biological film. If NO_3^- is not available, the bacteria will simply decay under anaerobic or anoxic conditions:

$$2\ C_5H_7O_2N + 6\ H_2O \rightarrow 5\ CO_2 + 2\ NH_3 + 5\ CH_4 \quad (3.27)$$

In either case, the microorganisms lose their ability to cling to the media surface. The wastewater then washes the old biological slime layer off the contact media, and a new slime layer will start to grow.

Sometimes the diffused oxygen can reach the microorganisms near the media face, but there is no external carbonaceous source available for cell assimilation because of either low organic loading or thick slime layer. The microorganisms near the media face will enter into an "endogenous oxygen respiration" phase [16, 65, 69, 70]:

$$\begin{aligned} C_5H_7O_2N + 5O_2 &\rightarrow 4CO_2 + NH_4HCO_3 + H_2O \\ \text{or} &\rightarrow 5\ CO_2 + NH_3 + 2H_2O \end{aligned} \quad (3.28)$$

Similarly, the microorganisms will lose their ability to cling to the media surface, and the old slime layer will be washed off the media by the shear force of the wastewater flow.

The phenomenon of losing the biological slime layer in a contact bed or a similar microbial slime system is called "sloughing" and is a function of the organic and hydraulic loading on the process bed. Therefore, although nonsettleable solids and soluble organic pollutants are removed by an adsorption-oxidation phenomenon occurring at the biofilm-wastewater interface, the effluent from the process bed usually contains solids that are generated by periodic or continuous sloughing of the biological slime film from the process bed medium. The effluent is then subjected to a further solid-liquid separation process such as clarification. Clarification can be accomplished by either conventional sedimentation or innovative dissolved air flotation [71].

4.2 *Historical Development and Applicability of Attached-Growth Systems*

Various contact beds and trickling filters employed for the transfer of dissolved organic matter and fine suspended solids from settled wastewater to contact surfaces have been developed [72–97]. One of the artificial wastewater treatment processes used in the eighteenth century was the double-contact bed with dosing and draining siphons. The first installation of trickling filter with distribution by spray nozzles was reported to be at Lawrence Experiment Station in Massachusetts in 1891 [74].

Figure 3.13 shows an early type of trickling filter with fixed nozzles and automatic dosing tank. Figure 3.14 shows a trickling filter with rotary distributor.

Biological filtration was sufficiently popular that in 1940 about 60% of all wastewater treatment plants, which provided secondary treatment in the United States, utilized trickling filters. It was in 1946 [80] that the National Research Council first proposed a mathematical formulation for the design of trickling filters. Although the number of conventional trickling filter plants has been increasing each year, the percentage has been declining since 1957.

The development of plastic packings as high-rate trickling filter media has been one of the most significant advances in the field of biological wastewater treatment. The plastic medium was initially developed by the Imperial Chemical Industries, Ltd., in England during the early 1960s [81]. Today on a worldwide basis, there are a large number full-scale biological filtration plants using plastic filter media and operating on wastes of widely differing character [69, 81, 82, 98–100].

Aerated-contact beds (also called submerged-contact aerators) were developed for more efficient continuous operation in the early 1900s [75]. When a submerged-contact aerator is continuously operated, air is usually provided in sufficient volume to keep the waters and slime surfaces aerobic and with sufficient intensity to tear away aging slime and solid accumulations for subsequent solid-water separation in a secondary settler.

Rotating disk systems have been used for biological wastewater treatment in Europe since 1960 [76]. Research and development work has been conducted on this process in the United States since 1965 [76–78]. In 1971 Torpey [79] received a patent on the invention of a new method and apparatus for substantially upgrading

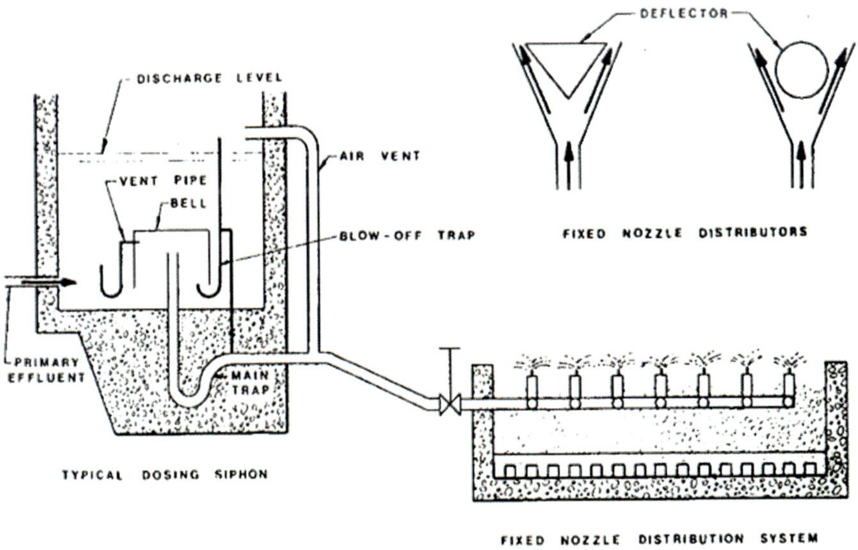

Fig. 3.13 Fixed nozzle distribution system. (Source: US EPA)

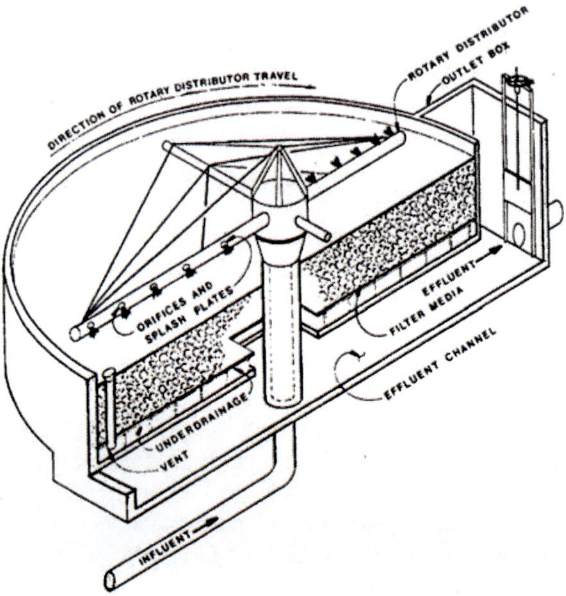

Fig. 3.14 Typical trickling filter in cross section. (Source US EPA)

the operating efficiency of a biological rotating disk system. Torpey's invention utilizes forcibly rotating bodies that are partially submerged in the wastewater to provide surface area on which biological slimes develop for the purpose of removing pollutants from the wastewater.

The application of contact beds, trickling filters, or rotating contact systems to wastewater treatment has been found profitable in areas where:

(a) Wastewater treatment personnel may be limited
(b) Small flows exist
(c) A plant effluent of from 20 to 30 mg/L of BOD_5 is acceptable
(d) Land area requirements dictate height to be increased to achieve the designed bio-oxidation capacity
(e) Intermittent discharges of toxic or inhibitory waste stream create shock conditions
(f) Partial treatment or a specific treatment may be required on an industrial waste stream

A new process, activated biofilters (ABF), which combines both attached and suspended-growth biological treatment systems, has been developed by Neptune Microfloc. The process entails the use of a redwood-medium trickling filter followed by aeration. The return sludge from the clarifier is recycled to the biocell along with the biocell underflow. Although the redwood medium is recommended by the manufacturer, plastic media may provide equally effective treatment [101].

Another development in the field of attached-growth wastewater treatment processes is the anaerobic filter [102]. It is primarily a column filled with various types of filter media used for the anaerobic carbonaceous oxidation of organic matter in wastewater. The wastewater flows upward through the column, contacting the media on which anaerobic bacteria grow and are retained. Because the bacteria are retained on the media and not significantly washed off in the effluent, mean cell residence times on the order of 100 d can be obtained. Anaerobic filtering appears to be a viable process for the pretreatment of high-temperature and high-strength industrial wastes. It achieves high carbonaceous removal at relatively low operating cost and low sludge production compared to aerobic biological systems. Besides, it also produces methane as a useable end product according to Eq. (3.25).

The emphasis of this chapter is on the principles of well-established aerobic attached-growth systems and the design of trickling filters. The design of rotating biological contactors, anaerobic filters, and activated biofilters is presented in detail elsewhere.

4.3 *Microbiology and Ecology*

A conventional microbial slime system is designed to contain an air-renewable surface to which a wastewater stream containing organic substrate and minerals is applied. Microorganisms can metabolize the substrate in wastewater with net production of energy for growth or sufficient energy to maintain the existing population. Although the system is classed as an aerobic treatment device, it is not truly aerobic, but rather must be considered as facultative because the great majority of microorganisms in the system are facultative bacteria, such as *Pseudomonas*, *Alcaligenes*, *Flavobacterium*, *Micrococcus*, and so on, which live aerobically as long as dissolved oxygen is present and anaerobically when the oxygen is almost depleted. Aerobic bacteria, such as *Bacillus*, are found in the upper, aerobic slime surfaces [92]. When an anoxic or anaerobic zone is formed in a thick slime layer, the obligate anaerobe, such as *Desulfovibrio*, and some other sulfur-reducing bacteria have been isolated from the slime-medium interface [93]. Under such circumstances the microbial slime system can develop odors, and conceivably sloughing can occur as the result of gas generation in the interior slimes.

Fungi are aerobic microorganisms living in the aerobic zone of the slime and also decomposing the organic substrate in the wastewater. The contribution of fungi is significant only under low pH conditions or with unusual industrial effluents because the fungi cannot compete successfully with the bacteria for their food under normal environmental conditions.

Algae growing on the surface of a microbial slime system, such as trickling filter, are usually an inconsequential element of the microorganism's population, limited to illuminated surfaces. Algal microorganisms on a slime system are clearly tolerant of organic substances and high levels of carbon dioxide. Although algae add oxygen to the wastewater, they have been charged with responsibility for bed clogging and are considered to be troublesome from an operational standpoint [95].

The protozoa are the predominant small animals with all forms from the Phytomastigophora to Suctoria. The free-swimming ciliates predominate at the slime surface, while the stalked ciliates predominate in the lower regions. The primary function of protozoa is not to stabilize the waste, but to control the bacterial population in the system.

Higher animals including worms, snails, and insect larvae feed on the lower forms of microorganisms in the microbial slime system and live in the upper aerobic areas. As a result, these higher animals can help to keep the bacterial population in a state of high growth or rapid food utilization. Filter flies are nuisance organisms in the trickling filters. Much of the early study of trickling filter populations was aimed at the control of these nuisance organisms by flooding, chlorination, and the use of various pesticides.

A comprehensive description of the organisms found in trickling filters has been presented by Cooke [94], whose listing of various organisms can be found from Ref. (103).

Trickling filters and similar microbial slime systems are short-term retention devices that should not be expected to act as effective reduction devices for *S. typhosa*, *S. paratyphi*, and *M. tuberculosis*. This is true also of pathogenic protozoa, such as *Entamoeba histolytica* [96].

When loaded lightly with carbonaceous substances, the trickling filter does some nitrification because of the presence of *Nitrosomonas* and *Nitrobacter*. However, when heavily loaded with carbonaceous matter, nitrification in a trickling filter or similar system may be absent or nominal.

5 Theories and Mechanisms

5.1 Transfer of Oxygen in Slime Layer and Liquid Film

The relationships of the supporting medium, the biological slime film, the waste liquid, and the atmosphere, as illustrated in Fig. 3.12, have challenged the kinetic and mathematical skill of environmental engineers for many years. It was generally believed [36, 103] that the slime layer had an aerobic zone and an anaerobic zone, defined by the depth of penetration of oxygen in the slime layer. Further studies [72, 73] have reported that the biological slime film does not always consist of an aerobic layer and anaerobic layer. With dilute substrate, the respiration was found to be substrate-limited, and high concentrations of oxygen were found throughout the slime mass. More concentrated substrates increased oxygen utilization; however, oxygen concentrations simply approached zero in the depth of the slime but were still above zero. A new term "anoxic" has been used by many environmental engineers to define an environment in which the dissolved oxygen concentration ranges from 0 to about 1 mg/L. Accordingly it is believed that there would only be an aerobic zone in the slime layer when the substrate concentration is low and there could be an aerobic zone and an anoxic zone in the slime layer when the substrate concentration in the liquid film is so high that oxygen concentrations are below 1 mg/L in the depth of the slime.

For a specific wastewater flow rate, the oxygen concentration gradient in a slime layer is a function of the dissolved oxygen concentration at the air-liquid interface and the substrate concentration. The latter affects the oxygen requirements of the slime layer. At low substrate concentrations in the liquid film, there will be a decrease in the oxygen requirements of the slime layer that should increase the oxygen concentration at the slime-liquid interface and reduce the mass flux of oxygen across the interface [97]. The reduced oxygen requirement and increased oxygen concentration at the interface will then result in an increased depth of penetration of oxygen in the slime layer and probably an aerobic zone throughout the full depth of the slime layer [72, 73]. At high substrate concentration in the liquid film, the oxygen requirements of the slime layer approach a constant; then the mass flux of oxygen across the slime-liquid interface will be constant, and the thickness of the active portion (i.e., aerobic zone) of the slime layer for the entire bed depth will also be constant assuming this condition exists at any depth in the process bed [97].

The effect of hydraulic loadings on the transfer of oxygen in the slime layer in the liquid film has also been studied [90]. As specified by Maier [104] and verified by Jank [97], a range for hydraulic loadings is normally encountered in full-scale operation of a trickling filter. For laminar flow, an increase in flow rate will result in an increase in liquid velocity and an increase in the mass flux of oxygen across the air-liquid and slime-liquid interfaces. The increased supply of oxygen at the slime-liquid interface will result in a greater depth of oxygen penetration in the slime layer or a thicker active slime layer at a specific applied organic loading.

The depth of oxygen penetration is dependent on the molecular diffusion coefficient of oxygen in the slime layer, the rate of oxygen utilization, and the oxygen concentration at the slime-liquid interface. Jank and Dryman [90] have reported that for a specific wastewater flow rate and substrate concentration, an oxygen concentration gradient will be established within the slime layer that is directly related to the mass flux across the slime-liquid interface. Although the supply of oxygen at the slime-liquid interface can be either contributed by the dissolved oxygen content of the influent wastewater or transferred from the air to the liquid film as the wastewater flows across the slime layer, Jank [97] discovered that the quantity of oxygen contributed by the influent wastewater was negligible when compared to the oxygen requirement of the slime layer.

5.2 Transfer of Substrate in Liquid Film and Slime Layer

A conceptual illustration of the substrate concentration gradient within a slime layer (i.e., biofilm) as shown in Fig. 3.15 has been proposed by Williamson and McCarty [105, 106]. It is assumed that the rate of reaction is limited by a single substrate S. Let the substrate concentration outside the biofilm in the bulk liquid be S_0; at the biofilm surface, S_s; within the biofilm cellular matrix, S_c; and deep within the biofilm a

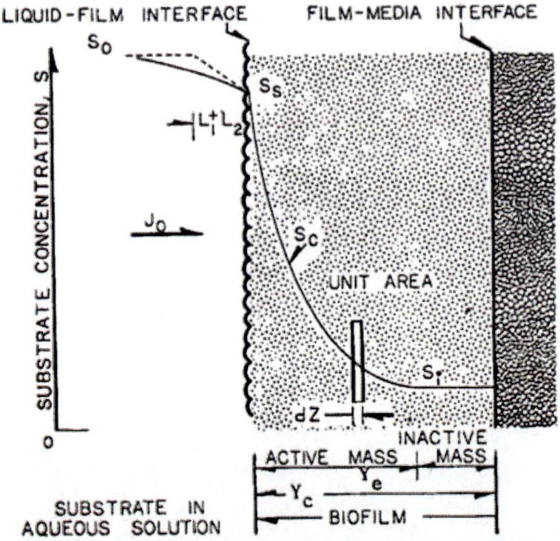

Fig. 3.15 Substrate profiles within a biofilm. (Source: WPCF)

constant limiting value, S_i. The substrate concentration gradient (dS_c/dz) at the slime-liquid interface where $z = 0$ has been shown by Whalen et al. [72] to be intermediate between low values (metabolism-limited case) and high values (diffusion-limited case). The biofilm depth to the point at which $S_c = S_i$ is termed the effective depth, Y_e, and will contain those microorganisms actively metabolizing the substrate. The surface flux of the chemical species can be calculated from Fick's Law [107] as:

$$J_0 = A_c D_w (S_0 - S_s)/Y \tag{3.29}$$

in which:

A_c = biofilm area, cm^2
D_w = diffusion coefficient of the chemical species through water, cm^2/d
Y = the depth of a stagnant liquid layer outside the slime-liquid interface, cm
J_0 = surface flux of the chemical species, mg/d

Under steady-state conditions, the substrate concentration gradient will be retained in the biofilm, and the mass flux of substrate across the slime-liquid interface will be equal to the total mass of substrate utilized by the active slime layer. If a biofilm is not metabolism-limited, then the substrate concentration within the depths of the biofilm will reach a minimum value of S_i, at which point bacterial metabolism stops. This situation occurs only in relatively thick biofilms. On the other hand, if the biofilm depth is restricted by either sloughing or hydraulic shear, then metabolism of the substrate may occur throughout the entire slime layer – this is a typical metabolism-limited case.

The mass transfer of substrate within the biofilm per unit area can also be described by Fick's Law as [105]:

$$\frac{\partial S_m}{\partial t} = -A_c D_c (\partial S_c / \partial z) \tag{3.30}$$

in which:

$\partial S_m / \partial t$ = rate of substrate mass transfer, mg/d
$\partial S_c / \partial z$ = substrate concentration gradient perpendicular to the surface plane, mg/cm^4
D_c = diffusion coefficient within the biofilm, cm^2/d

The rate of substrate utilization at any point within the biofilm is assumed to follow the Monod relationship [13, 108]:

$$-(dS_c/dt) = kS_c X_c/(S_c + K_s) \tag{3.31}$$

in which:

$-(dS_c/dt)$ = utilization rate of the rate-limiting substrate, mg/L
k = maximum utilization rate of the rate-limiting substrate, mg/d/mg
K_s = Monod half-velocity coefficient, mg/L
S_c = rate-limiting substrate concentration, mg/L
X_c = bacterial concentration within the biofilm, assumed to be constant with depth, mg/L

Through application of Eqs. (3.30) and (3.31) to the differential element of width dz shown in Fig. 3.15, and combination of the mass transfer terms, a steady-state equation is derived:

$$\frac{d^2 S_c}{dz^2} = \frac{kS_c X_c}{D_c (S_c + K_s)} \tag{3.32}$$

Equation (3.32) is a second-order nonlinear ordinary differential equation. Although it does not possess an explicit solution, it can be solved for the two limiting cases of the Monod equation [13]. The biofilm surface flux and the biofilm substrate concentration for the limiting cases at $S_s \gg K_s$ and $S_s \ll K_s$ are presented elsewhere [105, 106]. The biofilm model may be used to describe the utilization rate of any substrate by a biofilm if that substrate is both flux and substrate-limiting. Technical terms used in this chapter are all defined in the Nomenclature Section.

6 Types of Trickling Filters

6.1 General Description

A trickling filter is a packed bed of media covered with slime over which wastewater is passed so that it "trickles" downward as a thin laminar film. Oxygen and organic substrates diffuse through the liquid film into the slime film where biological oxidation and synthesis occur. End products and sludges appear in the filter effluent.

6.2 *Low-Rate, High-Rate, and Super-Rate Filters*

Trickling filters are classified by hydraulic and organic loading as low-rate, high-rate, and super-rate (Figs. 3.16, 3.17, and 3.18). Super-rate filters are also referred to as roughing filters. There is no rigid dividing line between the various rate trickling filters [109]. However, some rather general ranges are given in Table 3.5 to differentiate the various types of trickling filters by the hydraulic and organic loadings. Hydraulic loading is the rate of application of wastewater to the surface, usually expressed in millions of gallons per day per acre of surface area (mgad) gallons per day per square foot (gal/d/ft^2) or cubic meters per day per square meter (m^3/d/m^2). Organic loading is in pounds of BOD per acre-foot per day lb/ac.ft/d, pounds of BOD per cubic yard per day (lb/yd^3/d), pounds of BOD per 1000 cubic foot per day (lb/1000 ft^3/d), or kilograms of BOD per cubic meter per day (kg/m^3/d).

Accordingly, a low-rate filter is one designed for the applied loadings of not more than 0.4 kg BOD/m^3/d (25 lb BOD/1000 ft^3/d) and 4.08 m^3/m^2/d (100 gal/ft^2/d). The

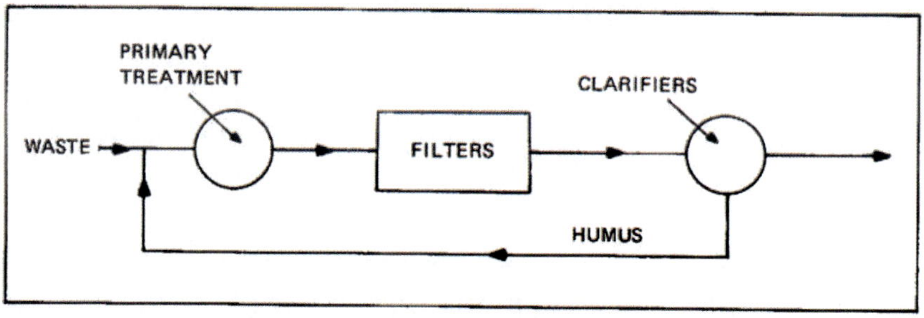

Standard-rate trickling filters.

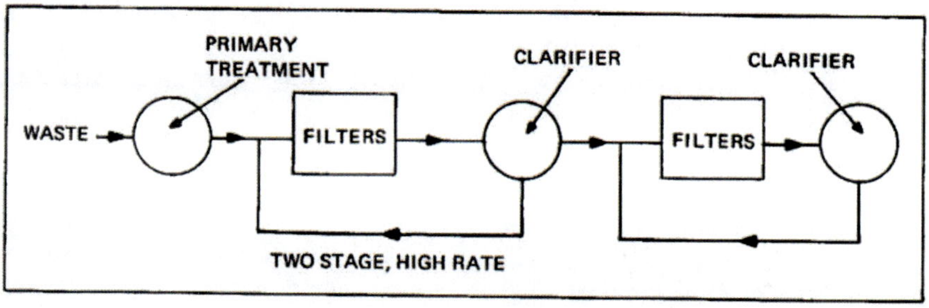

High-rate trickling filters.

Fig. 3.16 Standard (low-rate) and high-rate trickling filters. (Source: US EPA)

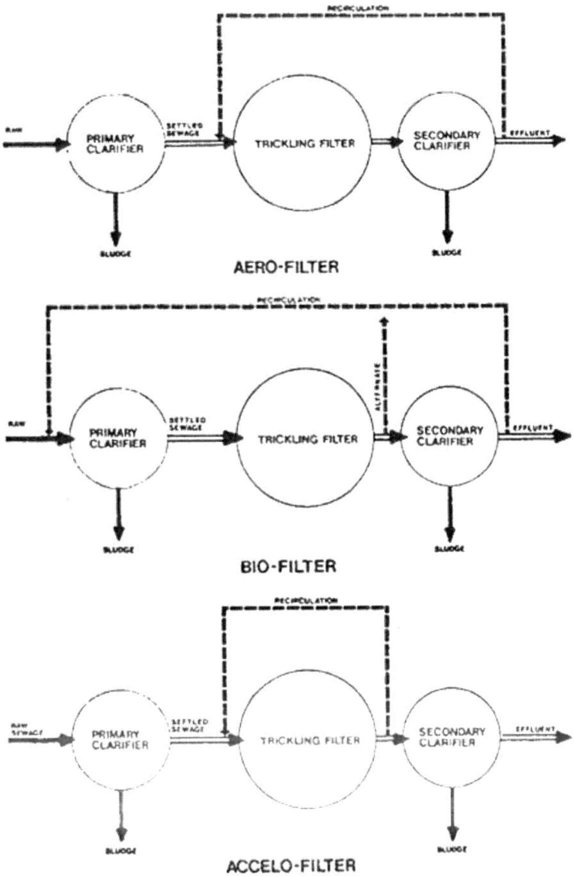

Fig. 3.17 Trickling filter variations. (Source: US EPA)

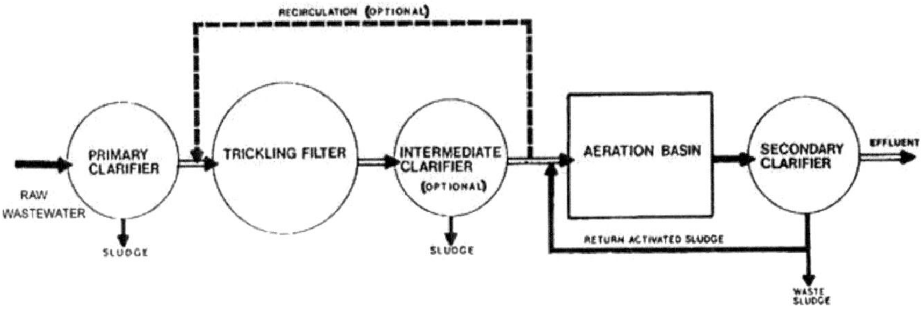

Fig. 3.18 Typical roughing filter installation. (Source: US EPA)

low-rate filter, which is also referred to as standard-rate filter, was the backbone of secondary biological treatment for over 50 years. The low-rate biological filter is about 1.5–2.5 m (5–8 ft) deep. Its rock media vary from 3.8 to 6.4 cm (1.5 to 2.5 in) nominal diameter and can be dosed with either fixed nozzle distributors or rotary

Table 3.5 Design comparison of different rate filters

Design	Trickling filter classification				
Characteristics	Standard (low) rate	Intermediate rate	High rate (rock media)	High-super rate (plastic media)	Super rate (for roughing)
Hydraulic loading					
$gal/ft^2/d$	25–100	100–230	230 to 1000	350–2100 [a]	1400–4200
$m^3/m^2/d$	1.0–4.1	4.1–9.4	9.4–40.8	14.3–85.7	57.1–171.4
mgad	1.1–4.4	4.4–10	10.0–43.5	15.2–91.4	61.0–182.9
Organic loading					
lb BOD_5/1000 ft^3/d	5–25	20–30	25–300	Up to 300	100 plus
kg BOD_5/d	0.08–0.40	0.32–0.48	0.40–4.80	Up to 4.80	1.60 plus
lb BOD_5/ac-ft/d	218–1089	871–1307	1089–13,068	Up to 13,068	4356 plus
Recirculation (ratio)	Minimum (0)	Usually (0.5–3)	Always (0.5–3)	Usually	Not required
Filter flies	Many	Varies	Few	Few	Few
Sloughing	Intermittent	Varies	Continuous	Continuous	Continuous
Depth of bed, ft	5–8	5–8	4–8	UP to 40	3–20
BOD removal, %	80–85	50–70	65–80	65–85	40–65
Effluent nitrification	Well	Some	Nitrites	Limited	No

Note: 1 $gal/ft^2/d = 0.0408\ m^3/m^2/d = 0.04354$ mgad $= 407.52\ m^3/ha/d$
1 lb $BOD_5/1000\ ft^3/d = 0.016$ kg $BOD_5/m^3/d = 43.56$ lb BOD_5/ac.-ft/d $= 0.027$ lb/$BOD_5/yd^3/d$
[a]Not including recirculation

distributors. Low-rate filters commonly have a dosing period of 3 min and a rest period of 6 min. The underdrains are usually sized to flow half-full at the design flow rate for adequate ventilation. Head loss through the filter may be 1.52–3.05 m (5–10 ft), which may be disagreeable if the plant site is too flat to permit gravity flow. Since the low-rate filter has a greater detention time and lower hydraulic and organic loadings, it produces a more highly nitrified effluent than the high-rate filter. The possible drawbacks of low-rate filters are:

(a) Odor problems caused by the septic sewage when weather is warm
(b) The presence of filter flies
(c) The occurrence of filter ponding

For solving the odor problem, the low-rate filters should be located where the odors would not create a nuisance or where the weather is not warm. When the filter fly persists as a nuisance, these nuisance organisms can be controlled by flooding, chlorination, or the use of various pesticides [110, 111]. Ponding of filters occurs when strong wastes are applied at low hydraulic loading rates. These difficulties are believed to be related to excessive amounts of bacterial growth that clog the trickling filters and can be controlled by higher flow rates to keep surfaces flushed. However, if both high organic and high hydraulic loadings are applied to a trickling filter, then the filter can no longer to be classified as a low-rate filter.

High-rate filters are normally designed for substantially higher organic and hydraulic loadings than low-rate filters. There are three principal types of high-rate filters depending on rate of feeding, recirculation, or loading. The three types are the biofilter, the accelo-filter, and the aero-filter, as indicated in Fig. 3.17.

The biofilter is a relatively shallow filter, generally 1.2–1.5 m (4–5 ft) in depth, which utilizes recirculation of a portion of the filter effluent to the primary sedimentation basin for a second passage through the filter. Organic loadings of biofilters are in the order of 0.88–1.11 kg BOD/m^3/d (55–69 lb BOD/1000 ft^3/d) based on the strength of the primary tank effluent, with hydraulic loadings ranging from 9.38 to 28.15 m^3/m^2/d (230–690 gal/ft^2/d). If additional treatment is necessary to lower the BOD content in the effluent, a second-stage filter may be provided. By appropriate selection of flow patterns and recirculation ratios, it is possible to satisfy the desired degree of treatment.

The aero-filter, which has a relatively deep media bed of 1.83–2.74 m (6–9 ft), utilizes a low momentary rate of wastewater application to the filter by means of a special type of distributor designed for frequent "raindrop" applications over a maximum area of the filter at one time. Recommended organic loadings are from 1.11 to 1.19 kg BOD/m^3/d (69–74 lb BOD/1000 ft^3/d), and the hydraulic loading rate is more than 9.38 m^3/m^2/d (230 gal/ft^2/d) with recirculation if necessary to maintain this rate. It should be noted that recirculation is used only during periods of low wastewater flow or only in amounts necessary to ensure proper operation of the distributor. BOD removal of single-stage treatment ranges from 63% to 78%. If additional treatment is desired, a second-stage filter may be provided, and with very strong organic wastewater, intermediate clarification may be used. When two-stage treatment is used, the organic loading of first-stage filter is 1.65–1.80 kg BOD/m^3/d (2.75–3.0 lb BOD/yd^3/d), and the loading of second-stage filter is about 60% of the first-stage loading. Some of the arrangements used in the aero-filter plants are shown in Fig. 3.17.

The accelo-filter, which is normally 1.83–2.44 m (6–8 ft) deep, utilizes direct recirculation of unsettled filter effluent back to the inlet of the distributor. The recommended organic loadings are in the same range as the biofilter, in the order of 0.88–1.11 kg BOD/m^3/d (55–69 lb BOD/1000 ft^3/d) based on the strength of the primary clarifier effluent. Recirculation is used as in the biofilter to affect the desired degree of treatment. As with the biofilter, a large variety of flow patterns is possible, including use of a primary high-rate filter and a secondary low-rate filter.

Early investigators considered an applied hydraulic loading of 9.38 m^3/m^2/d (230 gal/ft^2/d) or more is necessary to flush organic solids from the filter media and prevent clogging. It was found later that serious clogging did not occur when dosing rates ranged from 4.08 to 9.38 m^3/m^2/d (100–230 gal/ft^2/d). Many trickling filters were then designed to operate in this so-called intermediate-rate range.

The super-rate filter is defined by its high hydraulic and organic loadings, as indicated in Table 3.5. The plastic media filters can be operated at the high/super-rate range for secondary treatment.

Either plastic media filters or rock media filters can be used as the so-called roughing filters. A roughing filter (see Fig. 3.18) is used to reduce the organic load in which subsequent treatment may be applied to the effluent or where intermediate treatment is required [112]. Therefore, the roughing filler is generally installed advantageously ahead of activated sludge or any other secondary process. The filter design differs from other biological filters mainly because the determining factor is the high hydraulic loading as well as the high organic loading of certain wastewaters that are to be handled. Although roughing filters may give a high weight per unit volume of organic load removal, their settled effluent will still contain substantial organics in terms of BOD_5.

Most of the primary filters with plastic packing in a two-stage system may be operated at high hydraulic and organic loadings because of the plastic media's light weight, requiring less support structure, less cost, and freedom from corrosion. Plastic super-rate filters generally have deep beds with plastic media that have large void spaces and are dosed continuously. The highest hydraulic loading found in the literature approached 469.2 $m^3/m^2/d$ (11,500 $gal/ft^2/d$) which, however, was applied to a shallow bed filter [103]. The depth of the super-rate filter depends on the type of medium employed; however, the limitations in the medium available for use in super-rate filters were overcome by considerations outlined by Pearson [113]. Recirculation is normally practiced to maintain efficiency and keep the slime film in a wetted condition.

6.3 *Single- and Multi-stage Trickling Filter Plants*

A single-stage plant is one in which wastewater is passed through a single trickling filter; if there are two or more trickling filters, they would be operated in parallel. Wastewater may be, and usually is, recirculated through single-stage filters. Useful design charts were prepared by Baker and Graves [114] for single-stage filters using various design formulas. Although the single-stage filter is generally adequate and satisfactory for treating domestic wastewater, recent trends require an increase in detention time for handling occasional or constant overloaded conditions. The limitations of increasing depth are that the conventional rock filter media may not be able to support a very deep bed and a significant foundation structure must be provided. With the development of plastic media that are lighter in weight and possess proper ventilation characteristics, single-stage trickling filter plants using deep beds are being considered. Figures 3.16 and 3.17 show the flow diagrams of single-stage treatment.

In a two-stage filtration plant (i.e., double filtration plant), the effluent from a primary trickling filter, after the portion that is to be recirculated has been withdrawn for return to it directly or through the primary clarifier, passes through a secondary

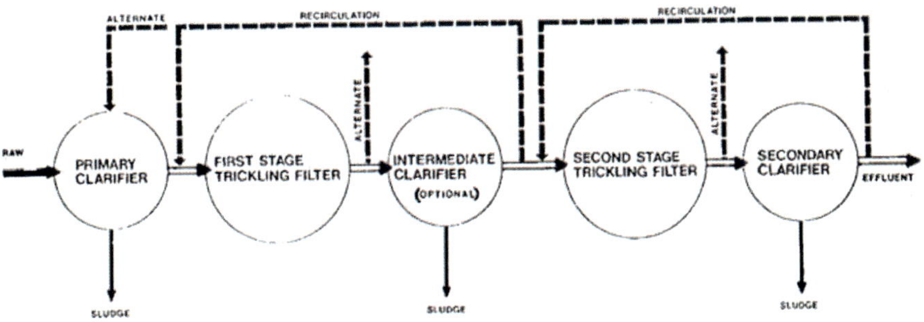

Fig. 3.19 Staging of filters. (Source: US EPA)

trickling filter. In other words, two-stage filtration means two biological trickling filters in series with or without intermediate clarifiers, followed by final clarification as shown in Fig. 3.19. The development of the two-stage filtration plant grew out of necessity from overloaded conditions at treatment plants in the pre-World War II period. A typical example of overloaded condition would be an existing single-stage trickling filter with a relatively small volume and receives a strong waste high in BOD. In this case, the installation of a two-stage trickling filter plant could have been the solution to this problem. Many publications [36, 110, 114–119] present the formulas for designing two-stage trickling filter plants. The authors will present and summarize them in the next section. As mentioned earlier, with the development of synthetic media, the use of a super-rate filter ahead of the existing trickling filter with rock media has been a popular practical solution.

Three-stage or tertiary treatment usually connotes activities concerned with nutrient control, primarily phosphorus and nitrogen. Although the tertiary filtration may be 80–100% superior to double filtration based on hydraulic advantages [120], it may be necessary only in exceptional cases. A factor in favor of the three-stage system is the development, under heterogeneous population of microorganisms, of selected strains of microorganisms in each filter stratum [49]. Wastewater treatment plants having three-stage trickling filter processes always requires high capital investment, which is undesirable.

7 Performance Models and Design Procedures

Although there are a variety of microbial slime systems, almost all performance models were developed for the design of conventional biological filters [80, 83–91, 101, 121–166, 173–177]. Different design results, almost infinite in number, can be obtained for removing a given amount of biochemical oxygen demand (BOD) from a waste stream in accordance with a performance model when different values of bed depth, surface area, recirculation rate, hydraulic loading, and wastewater temperature are assumed. The authors present the most common performance models in the following sections.

7.1 National Research Council Models

The National Research Council (NRC) compiled data for 34 operating rock trickling filters over an 8-month period during World War II [80]. The range of BOD removal in the field operations was between 75% and 95%. These field data are presented graphically in Fig. 3.20. The average performance of a single-stage rock trickling filter based on these field data can be expressed by the empirical models proposed by the NRC in 1946.

$$E_1 = \frac{100}{1 + K_{\mathrm{nrc}}(W/VF)^{0.5}} = L_0 - L_e/L_0 \tag{3.33}$$

$$F = \frac{1+R}{(1+0.1R)^2} \tag{3.34}$$

$$W = L_0 Q \tag{3.35}$$

$$R = Q_r/Q \tag{3.36}$$

in which:

$K_{\mathrm{nrc}} = 0.0085$
E_1 = % efficiency of BOD removal for process, including recirculation and sedimentation
L_0 = BOD of influent, mass/volume, lb/MG
L_e = BOD of effluent, mass/volume, lb/MG
W = BOD loading to filter, mass/time, lb/d

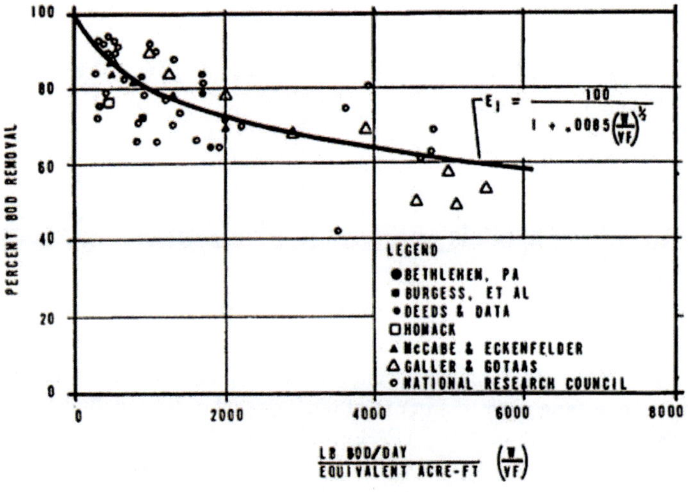

Fig. 3.20 Comparison of trickling filter operating data with NRC equation. (Source: US EPA)

V = volume of filter media, volume, ac.-ft
F = recirculation factor
R = recirculation ratio
Q = influent wastewater flow rate through the trickling filter, volume/time, MGD
Q_r = recirculation flow rate, volume/time, MGD

For the second-stage filter, the NRC model is:

$$E_2 = \frac{100}{1 + K_{nrc}(W'/VF)^{0.5}(1 - E_1)^{-1}} \tag{3.37}$$

$$W' = (1 - E_1)W \tag{3.38}$$

in which:

E_2 = % efficiency of BOD removal for second-stage filtration, including recirculation and settling
W' = BOD loading to second-stage filter, mass/time, lb/d

7.2 *Velz Model*

In 1948 Velz [30] postulated that the BOD removal per unit depth of trickling filter was proportional to the BOD remaining.

$$dL/dD = -K_e L \tag{3.39}$$

which integrates to:

$$L_D/L = e^{-K_e D} = 10^{-K_{10} D} \tag{3.40}$$

in which:

D = the depth of the trickling filter, length, ft
L_D = the BOD remaining in the effluent at depth D, mass/volume, mg/L
L_0 = BOD of untreated wastewater, mass/volume, mg/L
L = applied BOD (mass/volume, mg/L) which is removable, not over 0.90 L_0
K_e = rate of BOD removal, base e
K_{10} = rate of BOD removal, base 10

When recirculation is used, the applied BOD_L may be determined from Eq. (3.41):

$$L_a = \frac{QL_0 + Q_r L_e}{Q + Q_r} = \frac{L_0 + RL_e}{1 + R} \tag{3.41}$$

where:

L_a = applied BOD (mass/volume, mg/L) after dilution by recirculation
L_e = effluent BOD, mg/L
R = recirculation ratio = Q_r/Q
Q_r = recirculation flow rate, volume/time, MGD
Q = influent wastewater flow rate through the trickling filter, volume/time, MGD

7.3 Upper Mississippi River-Great Lakes Board Model

The standards established by the Upper Mississippi River-Great Lake Board [87] in 1952 specify a maximum daily trickling filter loading of 1.77 BOD_5/m^3 (110 lb BOD_5/1000 ft^3) of filter, a filter depth of not less than 1.52 m (5 ft) nor more than 2.13 m (7 ft), a filter influent BOD concentration not to exceed three times the effluent BOD concentration, and a hydraulic loading of not less than 9.88 $m^3/m^2/d$ (10 mgad) nor more than 29.65 $m^3/m^2/d$ (30 mgad). The standards were formulated into a performance model by Rankin in 1953 [121].

$$E = \frac{1 + \frac{Q_r}{Q}}{1.5 + \frac{Q_r}{Q}} \tag{3.42}$$

where E is the fractional efficiency of BOD removal.

7.4 Howland Models

In 1958 Howland [85] proposed that the rate of BOD removal was a function of contact time (t), giving the performance model:

$$L_e/L_0 = e^{-k/t} = \exp\left(-k't\right) \tag{3.43}$$

$$t = \frac{k''D}{Q^n} \tag{3.44}$$

in which n, k', and k'' are constants. Therefore, the remaining BOD in the effluent is obtained by substituting Eq. (3.44) into Eq. (3.43), yielding:

$$L_e/L_o = \exp\left[-k_T(D/Q^n)\right] \tag{3.45}$$

in which k_T is the reaction rate at the wastewater temperature T and n was determined to be 2/3. Besides, Howland [85] also introduced the effect of wastewater temperature on the reaction rate, k_T, in the BOD reduction equation:

$$k_T = k_{20}\,\theta^{(T-20)} \tag{3.46}$$

in which T is the wastewater temperature, in degrees Celsius; k_{20} is the reaction rate at 20 °C; and θ is the temperature coefficient equal to 1.035 according to Howland [85]. The value of θ has been reported to vary from 1.020 to 1.072 by Eckenfelder [122].

7.5 *Eckenfelder Models*

Eckenfelder [86, 116–118] modified the Howland [85] and Schultze [123] models in 1961 to evaluate the effect of a decreasing amount of BOD removal per unit of depth with increasing trickling filter depth, resulting in a series of his performance models.

In a manner analogous to activated sludge under plug flow conditions, BOD removal can be related to the available biological slime surface and to the time of contact of wastewater with that surface.

$$L_e/L_o = \exp\left(-kX_v t\right) \tag{3.47}$$

in which:

L_e = BOD remaining, mass/volume, mg/L
L_o = BOD in raw wastewater, mass/volume, mg/L
k = removal rate constant
X_v = volatile biological solid concentration, mass/volume
t = residence time, time

In a trickling filter, the mean residence time is defined as:

$$T = CD^m/q^n \tag{3.48}$$

where:

D = trickling filter depth, length, ft
q = hydraulic loading, volume/area/time, mgad
C, m, n = constants which are a function of the filter media and specific surface
m – 1 or 2 in most applications

The concentration of biological volatile solids, X_V, is a function of the specific surface of slime, a:

$$X_V = f(a) \tag{3.49}$$

where a = specific surface area (area of slime/volume of filter media). Therefore, the basic equation for BOD removal by a trickling filter with no recycle becomes:

$$L_e/L_o = \exp\left(-k'aD^m/q^n\right) \tag{3.50}$$

For a specific filter packing where, a, is known to be constant, Eq. (3.50) becomes:

$$L_e/L_o = \exp\left(-KD^m/q^n\right) \tag{3.51}$$

for a trickling filler with no recycle.

Equation (3.52) can be used for a trickling filter system with recycle.

$$L_e/L_a = \exp\left(-KD^m/q^n\right) \tag{3.52}$$

in which:

L_a = BOD in raw wastewater following dilution with recycle flow, mg/L

$$K = k'a \tag{3.53}$$

When circulation is used, the influent BOD is diluted by recirculation flow. By a material balance, the BOD applied to the trickling filter (L_a) can be calculated by Eq. (3.41). Since the trickling filter performance is a function of wastewater temperature, consideration must be given to temperature variation by adjustment of the reaction rate constant k or K according to Eq. (3.46) when Eckenfelder's models are used for filter design.

7.6 Galler and Gotaas Model

In 1964 Galler and Gotaas formulated an empirical performance model for trickling filter design from a multiple regression analysis of data from pilot plants and existing trickling filter plants with effluent BOD, L_e, as the dependent variable [88, 125, 129]:

$$L_e = \frac{0.46L_a^{1.19}(1+R)^{0.28}(Q/A)^{0.13}}{(1+D)^{0.67}T^{0.15}} \tag{3.54}$$

in which:

L_a = applied BOD (mass/volume, mg/L) after dilution by recirculation [see Eq. (3.41)]
Q = influent wastewater flow rate through the trickling filter, volume/time, MGD
Q_r = recirculation flow rate, volume/time, MGD
A = trickling filter area, ac.
D = trickling filter depth, length, ft
T = wastewater temperature, °C

7.7 Biofilm Model

Figure 3.15 illustrates the biofilm model. Suspended organic wastes may be adsorbed onto the biofilm surface and hydrolyzed into smaller soluble substances that, together with other dissolved organics, diffuse through a relatively stagnant liquid layer into the biofilm. The oxygen required for biochemical oxidation of the organics must also diffuse into the biofilm at a rate proportional to the microorganisms' need. As oxygen and organics diffuse past the microorganisms in the biofilm, the microorganisms consume the organic wastes at a rate that is either a function of oxygen concentration or organic substrate concentration, depending upon which is limiting.

The flux of substrate into the biofilm (J, mg/cm²/d) can be closely approximated by an equation of the form:

$$J = k_T S^p \tag{3.55}$$

where:

S = substrate concentration (organics, ammonia or oxygen), mg/L
p = coefficient generally equal to 0.91 for oxygen, 0.94 for organic substrate, and 0.97 for ammonia k_T = rate coefficient (mg/cm²/d) at wastewater temperature T °C
T = temperature, °C expressed by Eq. (3.46), in which θ is equal to 1.039
k_{20} = 0.054 for organic substrate, 0.05 for ammonia, and 0.21 for oxygen, mg/cm²/d

From a mass balance assuming plug flow through the reactor, Eq. (3.55) can be integrated. The surface area in the biofilm (A_c, cm²) is estimated by:

$$A_c = aV \tag{3.56}$$

where a is the media surface per unit volume, cm²/cm³, and V is the volume of the attached-growth media, cm³. The integration of Eq. (3.55) results in the following equation for substrate concentration in the reactor effluent:

$$S_e = \left[S_o^{(1-p)} - (1-p) k_T a V / Q \right]^{[1/(1-p)]} \tag{3.57}$$

where S_e is the effluent substrate concentration, mg/L; S_o is the bulk liquid substrate concentration, mg/L; p is a coefficient not equal to 1; and Q is the flow rate, cm^3/d.

7.8 US Army Design Formulas

In 1972, the US Army Corps of Engineers (US ACE) studied water problems in urban areas in addition to its traditional sanitary science role at recreation sites. Six years later, new design equations were developed by the US ACE for design of the plastic media trickling filters [147]. Equations [58] to [60] are presented below for calculating the filter depth, filter surface area, media volume, and sludge production, respectively [69]:

$$D = -\frac{q^n}{a K_{ace}} \ln \frac{L_e + L_e R}{L_o + L_e R} \tag{3.58}$$

where D = depth of filter, ft; q = hydraulic loading, gpm/ft^2; L_e = desired effluent BOD_5, mg/L; R = recirculation ratio = Q_r/Q; L_o = influent BOD_5, mg/L; a = specific surface area of the media, ft^2/ft^3; n = media factor, determined from laboratory; and K_{ace} = reaction rate constant ranging from 0.0015 to 0.003, determined in the laboratory.

$$A = \frac{10^6 Q}{1440 q} \tag{3.59}$$

where A = surface area of the filter, ft^2, and Q = average daily wastewater flow, MGD.

$$V = AD \tag{3.60}$$

where V = volume of filter media, ft^3.

$$P_s = 8.34 Q\, L_o\, F_s \tag{3.61}$$

where P_s = sludge produced, lb/d; L_o = influent BOD_5, mg/L; and F_s = sludge production factor, lb solids/lb BOD_5. F_s value ranges from 0.42 to 0.65 lb solids/lb BOD_5.

7.9 *US Environmental Protection Agency Model*

A general EPA model has been presented in a US EPA report [129] for designing all attached-growth systems primarily to assess the removal phenomenon as a function of hydraulic loading rate per unit volume.

$$\frac{L_e}{L_o} = \exp - K_p \left(\frac{V}{695Q}\right)^{0.5} \tag{3.62}$$

where:

V = attached-growth media volume, ft^3
Q = wastewater design flow excluding recycle flow, MGD
L_e = reactor effluent BOD_5, mg/L
L_o = reactor influent BOD_5, mg/L

$$K_p = \text{performance measurement parameter} = 0.265 + \ln \frac{q_w}{20} \tag{3.63}$$

q_w = wastewater surface application rate (wetting rate), gpm/ft^2

The values of the performance measurement parameter (K_p) and the applicable wetting rates are presented below:

Filter media	q_w, gpm/ft^2	K_p
Rock	0.1	0.15
Rock	0.2	0.18
Rock	0.3	0.20
Rock	0.4	0.22
Plastic	0.75	0.23

8 Design and Construction Considerations

A trickling filter has three principal components: (a) the distribution system that applies the wastewater to the filter media, (b) the filter medium that provides surface area for the microorganisms to grow, and (c) the underdrain system that supports the medium and provides drainage of the waste flow to a collection channel while permitting air circulation. The hydraulic considerations of three principal components as well as related minor components are described in this section.

Wastewater may be distributed over the trickling filter by rotary distributors or other suitable devices permitting uniform distribution to the surface area. At design average wastewater flow, the deviation from a calculated uniformly distributed daily volume per unit area (such as gal/ft^2/d) shall not exceed ±10% at any point.

Installations of motor-driven, rotary-type distributors have been used in filters ranging in size from 7.6 to 46 m (25–150 ft) in diameter.

There is a single conduit to convey the wastewater from primary settling tank to the distributor. Methods of conveying wastewater include gravity feed, dosing siphons, and pumping. When the filter is not designed for continuous dosing, the distributor is usually preceded by a pump or dosing tank and siphon.

All hydraulic factors involving proper distribution of wastewater on the filters should be carefully calculated. A minimum head of 0.6 m between low water level in siphon chamber and center of distributor arms is desirable. The head requirements of distributors are set by the manufacturers. The major head loss is the difference in elevation from the lowest water surface in the main underdrain channel. A minimum clearance of 0.15 m between distributor arms and filter media shall be provided. The head loss approximates 2.4 m (8 ft) for a filter 1.9 m (6 ft) deep and can be considerably greater for a deeper synthetic media filter.

It is important to know that bead losses chargeable to the trickling filter usually exceed the sum of all other head losses in the entire wastewater treatment plant. Compared with the activated sludge process, the trickling filter process requires much higher drop in static head, but requires less power.

Most trickling filter plants constructed in this country are circular and have rotary distributors. The majority of filters have reinforced concrete walls around the circumference, usually 0.2–0.3 m thick. The side walls are extended about 1 m to provide wind breaks.

The filter media may be crushed rock, slag, redwood, or specially manufactured synthetic plastic material. The media shall be durable, be resistant to spalling or flaking, and be chemically and biologically inert. The characteristics of various filter media are shown in Table 3.6 [130].

Underdrains with semicircular inverts or equivalent should be provided, and the underdrainage system shall cover the entire floor of the trickling filter. Inlet openings into the underdrains shall have an unsubmerged gross combined area equal to at least 15% of the surface area of the filter. The slope of underdrains shall be at least 1%. Effluent channels shall be designed to produce a minimum velocity of 0.6 m/s (2 ft/s) at average daily wastewater flow to the filter.

Ventilation of filter is important in maintaining the aerobic conditions necessary to secure effective biological treatment; therefore, adequate passageways at the bottom of the filters must be provided to permit free flow of air. Installation of vent stacks on the filter periphery, ventilating manholes, and discharge of filter effluent to the subsequent sedimentation basin in an open channel or partly filled pipes are methods employed to insure adequate ventilation. The Ten State Standards [87] require that inlet openings into the filter underdrains have an unsubmerged gross combined area equal to at least 15% of the surface area of the filter and that the size of drains, channels, and pipe be such that no more than 50% of their cross-sectional area will be submerged under the design hydraulic loading. Synthetic media manufacturers often recommend 0.1 m^2 (1 ft^2) of ventilating area for each 3–4.6 m of plastic media filter's tower periphery for domestic wastewater. Consideration may also be given to a forced ventilation system.

Table 3.6 Summary of BOD removal characteristics of various media treating settled wastewater

Description	Specific surface ft^2/ft^3	Temperature range °C	Influent BOD range mg/L	Depth ft	Hydraulic loading range mgad	n^a	K,[a] at 20 °C
1 1/2 in. flexi rings	40.0	2–26	65–90	8	12.5–26.9	0.39	0.46
1 in. clinker	61.5	7–17	220–320	8	0.96–1.2	2.56	0.865
2 1/2 in. clinker	37.4	7–17	220–320	6	0.96–1.2	0.84	0.685
1 in. slag	60.0	7–17	220–320	6	0.96–1.2	0.30	0.865
2 1/2 in. slag	33.0	7–17	220–320	6	0.96–1.2	0.75	0.640
1 in. rock	43.3	7–17	220–320	6	0.96–1.2	2.36	0.74
2 1/2 in. rock	27.6	7–17	220–320	6	0.96–1.2	3.80	0.645
1 in. rounded gravel	44.5	7–17	200–320	6	0.96–1.2	3.00	0.625
2 1/2 in. rounded gravel	19.7	7–17	220–320	6	0.96–1.2	5.40	0.57
Surfpac	28.0	24	200	21.6	31–350	0.50	0.395
Surfpac	28.0	24	200	12	62–250	0.45	0.33
2 1/2 and 4 in. rock filter	15.0	24	200	12	31–94	0.49	0.275
1 1/2 and 2 1/2 in. slag	42.0	7-1 7	112–196	6	5–12.5	1.0.	0.87
1 to 3 in. granite	29.0	16–18	186–226	6	2–16	0.4	0.312
3/4 in. Raschig rings	75.8	16–18	186–226	6	2–16	0.7	0.55
1 in. Raschig rings	52.2	16–18	186–226	6	2–16	0.63	0.42
1 1/2 in. Raschig rings	35.0	16–18	186–226	6	2–16	0.306	0.28
2 1/4 in. Raschig rings	22.7	16–18	186–226	6	2–16	0.276	0.25
Straight block	28.2	16–18	186–226	6	2–16	0.345	0.2

[a]n and K are constants for Eq. (3.31); m is equal to 1

9 Recent Advances in Biological Waste Treatment Processes

Dissolved gas flotation (DGF) is a process involving pressurization of gas at 25–95 psig for dissolving gas into water and subsequent release of pressure (to one atm) under laminar flow hydraulic conditions for generating extremely fine gas bubbles (20–80 microns) which become attached to the impurities to be removed and rise to the water surface together. The impurities or pollutants to be removed which are on the water surface are called float or scum which are scooped off by sludge collection means. The clarified water is discharged from the flotation clarifier's bottom. The gas flow rate is about one percent of influent liquid flow rate. The attachment of gas bubbles to the impurities can be a result of physical entrapment, electrochemical attraction, surface adsorption, and/or gas stripping. The specific gravity of the bubble-impurity agglomerate is less than one, resulting in buoyancy or non-selective flotation (i.e., Save-All). Dissolved air flotation (DAF) is one of the dissolved gas flotation (DGF) processes when air is used for the generation of gas bubbles. DGF or DAF is highly efficient for clarification of both water or wastewater [101, 130–160].

The first wave of evolutionary development in potable water treatment has been replacement of sedimentation clarifiers with dissolved air flotation (DAF) clarifiers using commercial DAF products, such as Supracell, Sandfloat, Clari-DAF, AquaDAF, KAMWT, etc. [101, 131, 141, 155–160]. This evolution has been extremely successfully. The largest city in the world, New York City, has adopted DAF for potable water treatment. The readers are referred to the Glossary section for all technical terms and commercial product names.

The second wave of evolutionary development in wastewater treatment has started [130, 132–140, 142–154]. Typical examples are shown in Fig. 3.21 (A suspended-growth activated sludge system uses primary flotation clarification and secondary flotation clarification) and Fig. 3.22 (An attached-growth trickling filter system uses primary flotation clarification and secondary flotation clarification) [101, 157], in which traditional primary sedimentation clarification is replaced by primary flotation clarification and traditional secondary sedimentation clarification is replaced by secondary flotation clarification.

A few newly developed biological processes are listed below:

1. *Innovative biological wastewater treatment system:* It normally includes (a) preliminary treatment units (i.e., screen, comminutor, grit chamber, etc. for the removal of sand, gravel, cinders, coffee grounds, small stones, cigarette filter tips, logs, cans, and other large-sized unwanted materials from raw wastewater), (b) primary flotation clarification for removing mainly total suspended solids from preliminary effluent, (c) secondary biological treatment units (such as activated sludge aeration or equivalent plus secondary flotation clarification) for removing dissolved organic/inorganic pollutants from primary effluent, and (d) tertiary treatment plant units (i.e., filtration, granular activated carbon

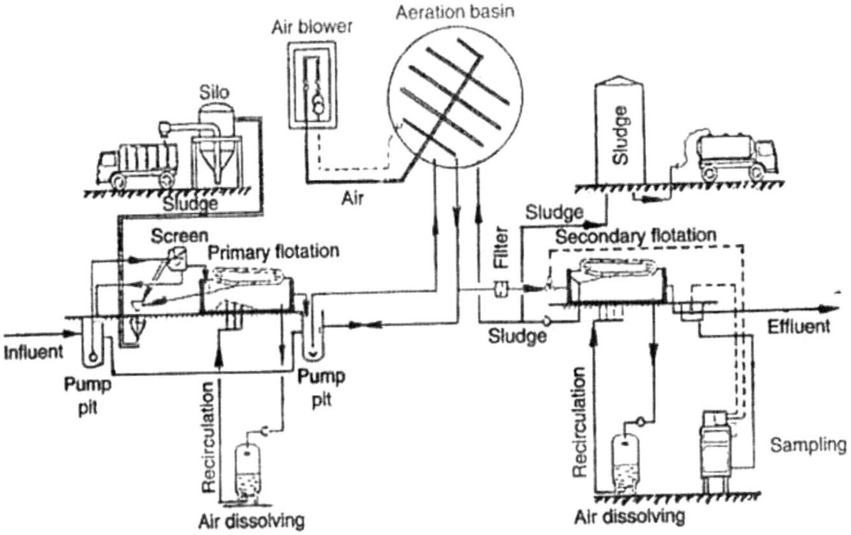

Fig. 3.21 A suspended-growth activated sludge system uses primary flotation clarification and secondary flotation clarification

adsorption, ion exchange, oxidation, nitrification, denitrification, and/or disinfection) for final polishing the secondary effluent in order to meet the effluent discharge standards.

2. *Innovative single-sludge activated sludge system:* It is a biological activated sludge process system involving the use of one suspended activated sludge solid going through aerobic and anoxic zones for carbonaceous oxidation, nitrification, denitrification, and phosphorus removal. The single-sludge activated sludge system is commonly a continuous flow system using sedimentation clarification. The Lenox Institute of Water Technology (LIWT) has developed three innovative single-sludge activated sludge systems all for carbonaceous oxidation, nitrification, denitrification, and phosphorus removal: (a) continuous flow single-sludge system using flotation clarification; (b) sequencing batch reactors (SBR) using sedimentation for carbonaceous removal and nutrient removal; and (c) SBR using flotation for carbonaceous removal and nutrient removal.
3. *Denitrification followed by flotation clarification:* The process involves the reduction of nitrates and nitrites to nitrogen gas through the action of facultative heterotrophic bacteria. In suspended-growth separate-stage denitrification processes, nitrified wastewater containing primarily nitrates is passed through a mixed anaerobic vessel containing denitrifying bacteria. Since the nitrified feedwater contains very little carbonaceous materials, a supplemental source of carbon (such as methanol, sugar, acetic acid, ethanol, or other compounds) is required to maintain the denitrifying biomass. This supplemental energy is provided by feeding methanol, for instance, to the biological reactor along with

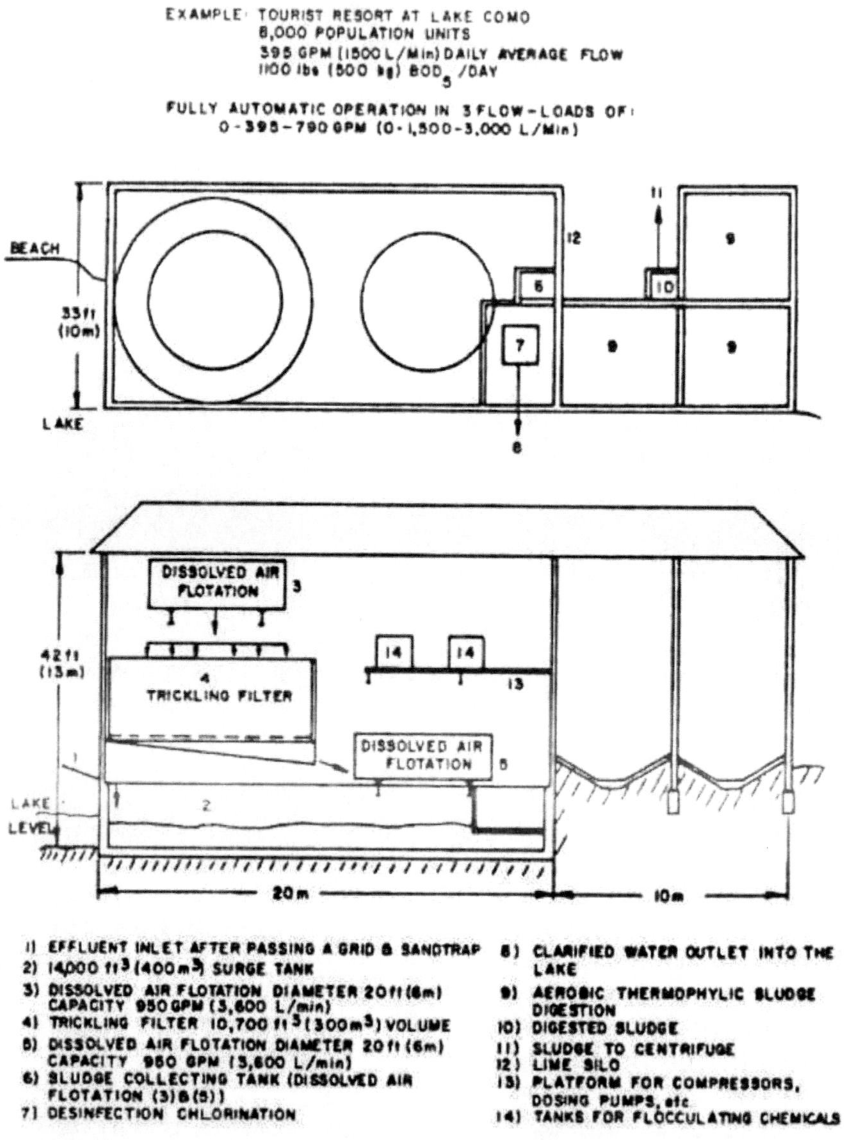

Fig. 3.22 An attached-growth trickling filter system uses primary flotation clarification and secondary flotation clarification

the nitrified wastewater. Mixing in the anaerobic denitrification reaction vessel may be accomplished using low-speed paddles analogous to standard flocculation equipment. Flotation clarification follows the denitrification step with the floated sludge being either returned to the head end of the denitrification system or wasted. An intermediate aeration step for stabilization (much less than 40 min) may be needed to guard against carryover of carbonaceous materials. The denitrification reactor may be covered but not air-tight to assure anaerobic conditions.

4. *Denitrification followed by membrane clarification:* Same as "denitrification followed by flotation clarification" except that the flotation clarification is replaced by a membrane filtration reactor.
5. *Sequencing batch flotation (SBF):* A sequencing batch process (either biological or physicochemical) involving the use of separate batch process steps in sequence of filling water/wastewater, reacting with chemicals or biomass, floating lightweight insoluble flocs and/or biomass, discharging treated effluent, and wasting floated insoluble scums for water or wastewater treatment is also called sequencing batch flotation (SBF) process (US Patent 5354458).
6. *Biological sequencing batch reactor (SBR):* A biological SBR system performs all the necessary functions of carbonaceous and nutrient removals in a single tank with variable water levels and timed aerobic aeration or anoxic mixing. This SBR system requires a minimum of three tanks and advanced automation equipment to control the cycle times and phases. The SBR control systems allow the operation to be configured to operate as almost any other suspended-growth biological reactor by adjusting the cycle phases between fill phase, react phase (either aerobic aeration or anoxic mixing), separation phase (either sedimentation or flotation), and decant phase. The biological SBR system involving the use of flotation in the separation phase is invented by Dr. Lawrence K. Wang, Dr. Lubomyr Kurylko, and Dr. Mu-Hao Sung Wang in 1994 (US Patent 5354458).

Glossary [158–166]

Activated sludge process: It is a continuous flow, biological treatment process characterized by a suspension of aerobic microorganisms, maintained in a relatively homogeneous state by the mixing and turbulence induced by aeration. The microorganisms are used to oxidize soluble and colloidal organics to carbon dioxide and water in the presence of molecular oxygen. The process is generally, but not always, preceded by primary clarification (either primary sedimentation clarification or primary flotation clarification). The mixture of microorganisms and wastewater formed in the aeration basins, called mixed liquor (ML), is transferred to secondary clarification (either secondary sedimentation clarification or secondary flotation clarification) for liquid-solid separation. The major portion of the microorganisms separated from the secondary clarifiers (secondary sedimentation clarifiers or secondary flotation clarifiers) is recycled to the aeration basins to be mixed with incoming wastewater, while the excess, which constitutes the waste activated sludge (WAS), is sent to the sludge handling facilities. The rate and concentration of the returned activated sludge (RAS) returned to the aeration basins determines the mixed liquor suspended solids (MLSS) level developed and maintained in the aeration basins. During the oxidation process, a certain amount of the organic material is synthesized into new cells, some of which then undergoes auto-oxidation (self-oxidation or endogenous respiration) in the aeration basins, the remainder forming net growth or excess sludge. Oxygen is required in the process to support the oxidation and synthesis reactions. In the conventional activated sludge plant, the wastewater is commonly aerated for a period of 4–8 hours (based on average daily flow) in a plug flow hydraulic mode. Either surface or submerged aeration systems can be employed to transfer oxygen from air to wastewater.

Aerobic digestion: The process equipment of aerobic digestion is similar to that of the activated sludge process. Aerobic digestion is not a suspended-growth biological process in principle and is for sludge treatment (not for wastewater treatment). It is a suspended-destruction biological process. In the presence of oxygen, the biodegradable matter in the activated sludge (biosolids) is oxidized, destroyed, and converted to carbon dioxide and water. Oxygen must be supplied in direct proportion to the mass of volatile solids (VS) destroyed. Sludge stabilization is not complete until there has been an extended period of primarily endogenous respiration (15–20 days).

Aerobic: An environmental condition in which free and dissolved oxygen is available in an aqueous environment (for instance, nitrification is an aerobic process).

Alkalinity: The capacity of water to neutralize acids by the water's content of carbonates, bicarbonates, hydroxide, and other compounds, also known as buffering capacity.

Ammonia nitrogen: Elemental nitrogen present in the form of ammonia (NH_3).

Ammonification: A biochemical or chemical process in which soluble organic nitrogen is converted to ammonia.

Anaerobic: An environmental condition in which free and dissolved oxygen is not available (dissolved oxygen = 0) in an aqueous environment (for instance, anaerobic digestion is an anaerobic process).

Anaerobic digestion: This is a biological destruction process for destroying biodegradable organic sludge (biosolids) in the absence of molecular oxygen or under anaerobic conditions. It is a three-stage sludge stabilization process for sludge volume reduction: (a) hydrolysis stage, breakdown of particulate matter and large macromolecules; (b) acid formation stage, fermentation of the soluble organic matter formed in the first reaction to volatile acids; and (c) methanogenesis stage, conversion of the volatile acids to the stable end produces, such as methane gas, carbon dioxide gas, and water. Four anaerobic digestion operational modes are available: (a) standard-rate digestion mode, or one stage, unheated, and unmixed; (b) high-rate digestion mode, one stage, heated, and mixed; (c) two-stage digestion mode; and (d) anaerobic contact process. Anaerobic digestion can occur in two temperature ranges: mesophilic (80–110 degree F) and thermophilic (113–149 degree F). Mesophilic range operation is more stable, while thermophilic range operation offers improved sludge dewatering. The process reduces volatile solids and thereby reduces the overall mass of the sludge, lowering subsequent land disposal costs.

Anaerobic: An environmental condition in which free, dissolved, and combined oxygen is unavailable in an aqueous environment.

Anoxic: A environmental condition in which oxygen is almost only available in a combined form, such as nitrate (NO_3), nitrite (NO_2), or sulfate (SO_4), in an aqueous environment. Dissolved oxygen (DO) concentration is in the range of 0 to 0.1 mg/L.

AquaDAF: It is a rectangular dissolved air flotation (DAF) clarifier commercially available from SUEZ Water Technologies and Solutions, Richmond, VA 23229, USA.

Assimilation: The biological or microbiological process in which nitrogen is taken in by cells to be used for growth and reproduction.

Attached-growth biological processes: In an attached-growth biological process system, the biomass responsible for biochemical reactions are attached on the surface of rocks, sands, woods, disks, screens, etc. where the wastewater pass through, and the organic pollutants and nutrients are removed. They include trickling filters, denitrification filter using coarse media, denitrification filter using fine media, slow sand filters, rotating biological contactors, etc. The effluent from an attached-growth bioreactor discharges to a water-solid separator (sedimentation clarifier, or flotation clarifier, or membrane filter) for further processing.

Autotrophic organisms: Organisms that use carbon dioxide, sunlight, or other inorganic substances for their cell growth (nitrifying bacteria are autotrophic).

Bar screens: There are two types of bar screens (or racks). The most commonly used, and oldest technology, consists of hand-cleaned bar racks. These are generally used in smaller wastewater treatment plants (WWP). The second type of bar screen is the type that is mechanically cleaned, which is commonly used in larger facilities.

Biochemical oxygen demand (BOD): A biochemical quantitative measure of the amount of oxygen used in the biochemical oxidation of organic matter in a specified time period (usually 5 days), at a specified temperature (usually 20 °C), and under specified conditions (in the dark).

Biological nutrient removal (BNR): The removal of nitrogen and/or phosphorus by the use of proliferation and selection of certain microbiological populations.

Biological phosphorus removal (BPR): The biological process of phosphorus removal in wastewater that relies on the proliferation and selection of a microbiological population capable of storing phosphorus in excess of their normal growth requirements. Also called biological nutrient removal (BNR) or enhanced biological phosphorus removal (EBPR).

Biosolids: It is a general term for any organic solids containing carbon, such as activated sludge, septic tank sludge, manure, etc.

Carbonaceous biochemical oxygen demand (CBOD): A chemical quantitative measure of the amount of dissolved oxygen required for the biological oxidation of carbon-containing compounds in a sample.

Chemical oxygen demand (COD): An environmental chemical test used to measure indirectly organic compounds in water or wastewater.

Chlorination: One of the disinfection processes involving the use of chlorine gas. Disinfection with chlorine is accomplished by adding chlorine gas to the water or wastewater. Sometimes application of sodium hypochlorite (liquid) for disinfection is also called chlorination. Academically disinfection using sodium hypochlorite should be called hypochlorination.

Clari-DAF: It is rectangular dissolved air flotation (DAF) clarifier commercially available from Xylem Water and Wastewater, Zelienople, PA 16063, USA.

Comminuting: It is a grinding or shredding operation for reducing the particle size of objects or debris in the influent wastewater. They are installed with a screen directly in the influent wastewater flow's channel, with the shredded particles returned to the flow downstream of the screen. The influent flow is channeled to and through these units. The debris is collected against the screen, or outside drum, and the teeth which penetrate this screen cut up the solids. When the solids are reduced to the size of the screen or drum openings, they pass through and on for additional downstream wastewater treatment. The barminutor is a comminuting device that incorporates revolving cutters that move up and down the upstream face of a bar screen, shredding and cutting whatever debris has accumulated against the screen. The screenings are transported to the cutting device, shredded, and then allowed to fall back into the influent channel downstream of the bar screen.

Complete mix activated sludge process: It is an activated process in which aeration tank (with about 3–5 hours DT) is usually circular and the influent wastewater and the mixed liquor are completely mixed during aeration operation. It is most useful for treating wastewater with fluctuating organic strength because it is least susceptible to shock loads among all activated sludge processes. A mechanically stirred reactor best simulates a complete mix condition. When diffused aeration in a rectangular aeration tank is used, the mixture of wastewater influent and the return sludge should enter the aeration tank at several points in a central location with the effluent going into channels on the sides of the aeration tank.

Contact stabilization process: It is a modification of the activated sludge process. In this modification, the adsorptive capacity of the floc is utilized in the contact tank to adsorb suspended, colloidal, and some dissolved organics. The hydraulic detention time in the contact tank is only about 30–60 minutes (based on average daily flow). After the biological sludge is separated from the wastewater in the secondary clarifier (either secondary sedimentation clarifier or secondary flotation clarifier), the concentrated sludge is separately aerated in the stabilization tank with a detention time of 2–6 hours (based on sludge recycle flow). The adsorbed organics undergo oxidation in the stabilization tank and are synthesized into microbial cells. If the detention time is long enough in the stabilization tank, endogenous respiration will occur, along with a concomitant decrease in excess biological sludge production. Following stabilization, the re-aerated sludge is mixed with incoming wastewater in the contact tank, and the cycle starts anew. Volatile compounds are driven off to a certain extent by aeration in the contact and stabilization tanks. Metals will also be partially removed, with accumulation in the sludge. This process requires smaller total aeration volume than the conventional activated sludge process. It also can handle greater organic shock and toxic loadings because of the biological buffering capacity of the stabilization tank and the fact that at any given time the majority of the activated sludge is isolated from the main stream of the plant flow. Generally, the total aeration basin volume (contact basin plus stabilization basin) is only 50–75% of that required in in the conventional activated sludge system. An equalization basin is generally recommended for equalizing the wastewater influent flow.

Conventional biological wastewater treatment system: It normally includes (a) preliminary treatment units (i.e., screen, comminutor, grit chamber, etc. for the removal of sand, gravel, cinders, coffee grounds, small stones, cigarette filter tips, logs, cans, and other large-sized unwanted materials from raw wastewater), (b) primary sedimentation clarification for removing mainly total suspended solids from preliminary effluent, (c) secondary biological treatment units (such as activated sludge aeration or equivalent plus secondary sedimentation clarification) for removing dissolved organic/inorganic pollutants from primary effluent, and (d) tertiary treatment plant units (i.e., filtration, granular activated carbon adsorption, ion exchange, oxidation, nitrification, denitrification, and/or disinfection) for final polishing the secondary effluent in order to meet the effluent discharge standards.

Dechlorination: It is a process for the removal of excess amount of residual chlorine in a chlorination disinfection process. Dechlorination is accomplished by using a reducing chemical such as sulfur dioxide (gas), sodium bisulfate (liquid), or sodium metabisulfite (liquid).

Denitrification filter: It is an attached-growth biological process filter using wood, plastic, rock, granular activated carbon, or sand as filtration media for denitrification (conversion of nitrate to nitrogen gas) of wastewater under anoxic/anaerobic condition.

Denitrification followed by flotation clarification: The process involves the reduction of nitrates and nitrites to nitrogen gas through the action of facultative heterotrophic bacteria. In suspended-growth separate-stage denitrification processes, nitrified wastewater containing primarily nitrates is passed through a mixed anaerobic vessel containing denitrifying bacteria. Since the nitrified feedwater contains very little carbonaceous materials, a supplemental source of carbon (such as methanol, sugar, acetic acid, ethanol, or other compounds) is required to maintain the denitrifying biomass. This supplemental energy is provided by feeding methanol, for instance, to the biological reactor along with the nitrified wastewater. Mixing in the anaerobic denitrification reaction vessel may be accomplished using low-speed paddles analogous to standard flocculation equipment. Flotation clarification follows the denitrification step with the floated sludge being either returned to the head end of the denitrification system or wasted. An intermediate aeration step for stabilization (much less than 40 minutes) may be needed to guard against carryover of carbonaceous materials. The denitrification reactor may be covered but not air-tight to assure anaerobic conditions.

Denitrification followed by membrane clarification: Same as "denitrification followed by flotation clarification" except that the flotation clarification is replaced by a membrane filtration reactor.

Denitrification followed by sedimentation clarification: The process involves the reduction of nitrates and nitrites to nitrogen gas through the action of facultative heterotrophic bacteria. In suspended-growth separate-stage denitrification processes, nitrified wastewater containing primarily nitrates is passed through a mixed anaerobic vessel containing denitrifying bacteria. Since the nitrified feedwater contains very little carbonaceous materials, a supplemental source of carbon (such as methanol, sugar, acetic acid, ethanol, or other compounds) is required to maintain the denitrifying biomass. This supplemental energy is provided by feeding methanol, for instance, to the biological reactor along with the nitrified wastewater. Mixing in the anaerobic denitrification reaction vessel may be accomplished using low-speed paddles analogous to standard flocculation equipment. Following the reactor, the denitrified effluent is aerated for a short period (5–10 minutes) to strip out gaseous nitrogen formed in the previous step which might otherwise inhibit sludge settling. Sedimentation clarification follows the gas stripping step with the collected sludge being either returned to the head end of the denitrification system or wasted. An intermediate aeration step for stabilization (about 50 minutes) between the denitrification reactor and the stripping step may be used to guard against carryover of carbonaceous materials. The denitrification reactor may be covered but not air-tight to assure anaerobic conditions.

Denitrification: The biochemical reduction reaction of reducing nitrate nitrogen to nitrogen gas in an anoxic environment.

Disinfection: Potable water disinfection is used in an attempt to destroy all pathogenic agents in drinking water for public health protection. Wastewater disinfection is used in an attempt to destroy pathogenic agents in a wastewater stream and protect the best uses of the receiving stream or groundwater. Environmental disinfection is used in an attempt to destroy pathogenic agents in rooms, houses, swimming pools, buildings, restaurants, streets, parks, etc., to prevent the spread of pathogenic agents, such as pathogenic bacteria and coronavirus in the environment. Typical disinfection techniques used include chlorine (gas), sodium hypochlorite (liquid), bromine (gas), bromine dioxide (gas), UV (light), ozone (gas), and cationic surfactant (water-soluble solid).

Dispersion: A uniform and maximum separation of extremely fine particles, often of colloidal size.

Dissolved air flotation (DAF): One of the dissolved gas flotation (DGF) processes when air is used for the generation of gas bubbles. See dissolved gas flotation (DGF).

Dissolved gas flotation (DGF): It is a process involving pressurization of gas at 25–95 psig for dissolving gas into water and subsequent release of pressure (to one atm) under laminar flow hydraulic conditions for generating extremely fine gas bubbles (20–80 microns) which become attached to the impurities to be removed and rise to the water surface together. The impurities or pollutants to be removed which are on the water surface are called float or scum which are scooped off by sludge collection means. The clarified water is discharged from the flotation clarifier's bottom. The gas flow rate is about one percent of influent liquid flow rate. The attachment of gas bubbles to the impurities can be a result of physical entrapment, electrochemical attraction, surface adsorption, and/or gas stripping. The specific gravity of the bubble-impurity agglomerate is less than one, resulting in buoyancy or non-selective flotation (i.e., Save-All).

Dissolved oxygen (DO): Molecular oxygen dissolved in water or wastewater.

Enhanced biological phosphorus removal (EBPR): The process of phosphorus removal in wastewater that relies on the proliferation and selection of a microbiological population capable of storing phosphorus in excess of their normal growth requirements. Also called biological nutrient removal (BNR).

Eutrification: Nutrient over-enrichment of a body of water, causing increased growth of algae and rooted aquatic plants.

Extended aeration activated sludge process: A long aeration time (over 18 hours DT) is employed in an extended aeration system. It is applicable only to small wastewater treatment plants of less than 3785 m^3/day or 1.0 MGD. Normally no primary clarifier is provided, and no exercise of optimum return sludge control is attempted although the return activated sludge (RAS) flow is about 75% to 100% of plant flow. The objective is to simplify the process both in construction and in operation. Sludge production is very small because of prolonged endogenous oxidation, which minimizes the problem of sludge treatment and disposal. The aeration effluent's sludge normally has an inferior settling characteristic. Therefore, a secondary flotation clarifier is better for an extended aeration plant.

Fermentation: The process in which bacteria degrades organic matter under anaerobic conditions, such as in a collection system, primary clarifier, anaerobic selector, or fermenter tank.

Filamentous organism: Microorganisms, such as bacterial, algal, or fungal species that grow in thread-like colonies, or a biological mass that may interfere with settling in clarifiers or may interfere with drainage through a filter.

Filtration: It is usually a granular medial filtration process which involves the passage of wastewater or water through a bed of filter media with resulting deposition of suspended solids. Eventually the pressure drop across the bed becomes excessive, or the ability of the bed to remove suspended solids is impaired. Cleaning is then necessary to restore operating head and effluent quality. The time in service between cleanings is termed the filter run time or run length. The head loss at which filtration is interrupted for cleaning is called the terminal head loss, and this head loss is maximized by the judicious choice of media sizes. Dual media filtration involves the use of both sand and anthracite as filter media, with anthracite being placed on top of the sand. Gravity filters operate either by using the available head from the previous treatment unit or by pumping to a flow split box after which the wastewater flows by gravity to the filter cells. Pressure filters utilize pumping to increase the available head. A filter unit generally consists of a containing vessel; the filter media; structures to support the media; distribution and collection devices for filter influent, effluent, and backwash water flows; supplemental cleaning devices; and necessary controls for flows, water levels, and backwash sequencing. Backwash sequences can include air scour or surface wash steps. Backwash water can be stored separately or in chambers that are integral parts of the filter unit. Backwash water can be pumped through the unit or can be supplied through gravity head tanks.

Flotation thickening: In a dissolved air flotation (DAF) sludge thickening process, air is introduced into liquid sludge that is being held at an elevated pressure. When the sludge is depressurized, the dissolved air is released as finely divided air bubbles carrying the solids to the top, where they are compacted and removed.

Gravity thickening: Its tank design is similar to a conventional sedimentation tank. Dilute sludge is fed to a center feed well. The feed sludge is allowed to settle and compact, and the thickened sludge is withdrawn.

Grinding: It is a unit operation for reducing the particle size of objects or debris in the influent wastewater, also termed shredding or comminuting. These devices may be installed with a screen directly in the wastewater flow or separately out of the wastewater flow, with the shredded particles returned to the flow downstream of the screen. Only those shredding and grinding devices that are installed directly in the influent channel are termed comminuting devices.

Grit chamber: It is a grit removal device that is designed to allow the settling out of this material. Grit removal is an important process for several reasons: (1) to prevent cementing effects at the bottom of sludge digesters and primary clarification tanks; (2) to reduce the potential for clogging of pipes and sludge hoppers; (3) to protect moving mechanical equipment and pumps from unnecessary wear and abrasion; (4) to reduce accumulations of materials in aeration tanks and sludge digesters which would result in a loss of usable volume; and (5) to reduce accumulations at the bases of mechanical screens. There are two types of grit chambers. The velocity controlled grit chambers limit the velocity in the rectangular channels to a maximum of 1 foot per second (fps). This velocity is low enough to allow the grit to settle but fast enough to maintain a majority of the organic material in suspension. The aerated grit chambers are normally sized on the basis of both detention time and volume of air. Typically, the detention time is in the range of 2–5 minutes, and the air flow is in the range of 0.04–0.06 cu ft/gallon of wastewater. The constant head type of system is normally designed using an overflow rate of 15,000 gallons per day per square foot and a 1 minute detention time at peak day flows.

Grit: It includes sand, gravel, cinders, coffee grounds, small stones, cigarette filter tips, and other large-sized unwanted materials in wastewater.

Head loss: It is an indirect measure of loss of energy or pressure of flowing water. It is measured as the difference in elevation between the upstream water surface and the downstream water surface.

Heterotrophic: Microorganisms that use organic matter (carbon) for energy and growth and can grow in both aerobic and anoxic environments using both dissolved and chemically bound oxygen (nitrates).

Hydraulic retention time (HRT): The given time it takes wastewater, including any return flows, to pass through a certain area.

Hypochlorination: Application of sodium hypochlorite (liquid) for disinfection is also called hypochlorination. It achieves the same result as that of chlorination using chlorine gas. Its use is generally limited to smaller water or wastewater treatment facilities or where there are significant health and safety concerns regarding the handling and use of chlorine gas.

Incineration: It is dry combustion of sludge or solid wastes to produce an inert ash. This ash can then be beneficially used or disposed of in a sanitary landfill. The process must dry the sludge cake or small solid wastes, destroy the volatile content by burning, and finally produce a sterile ash. A variety of incinerator configurations exist. The most common include (a) multiple hearth furnace incinerator and (b) fluidized bed incinerator.

Innovative biological wastewater treatment system: It normally includes (a) preliminary treatment units (i.e., screen, comminutor, grit chamber, etc. for the removal of sand, gravel, cinders, coffee grounds, small stones, cigarette filter tips, logs, cans, and other large-sized unwanted materials from raw wastewater), (b) primary flotation clarification for removing mainly total suspended solids from preliminary effluent, (c) secondary biological treatment units (such as activated sludge aeration or equivalent plus secondary flotation clarification) for removing dissolved organic/inorganic pollutants from primary effluent, and (d) tertiary treatment plant units (i.e., filtration, granular activated carbon adsorption, ion exchange, oxidation, nitrification, denitrification, and/or disinfection) for final polishing the secondary effluent in order to meet the effluent discharge standards.

Innovative physicochemical flotation wastewater treatment system: It includes (a) preliminary treatment units (i.e., screen, comminutor, grit chamber, etc. for the removal of sand, gravel, cinders, coffee grounds, small stones, cigarette filter tips, logs, cans, and other large-sized unwanted materials from raw wastewater), (b) primary flotation clarification for removing mainly total suspended solids from preliminary effluent, (c) secondary physicochemical treatment units (such as chemical precipitation/coagulation or equivalent plus secondary flotation clarification) for removing dissolved organic/inorganic pollutants from primary effluent, and (d) tertiary treatment plant units (i.e., filtration, granular activated carbon adsorption, ion exchange, oxidation, nitrification, denitrification, and/or disinfection) for final polishing the secondary effluent in order to meet the effluent discharge standards. In the nitrification and denitrification steps, only tertiary flotation clarification will be used for solid-water separation.

Innovative physicochemical flotation-membrane wastewater treatment system: It includes (a) preliminary treatment units (i.e., screen, comminutor, grit chamber, etc. for the removal of sand, gravel, cinders, coffee grounds, small stones, cigarette filter tips, logs, cans, and other large-sized unwanted materials from raw wastewater), (b) primary flotation clarification for removing mainly total suspended solids from preliminary effluent, (c) secondary physicochemical treatment units (such as chemical precipitation/coagulation or equivalent plus secondary membrane clarification) for removing dissolved organic/inorganic pollutants from primary effluent, and (d) tertiary treatment plant units (i.e., filtration, granular activated carbon adsorption, ion exchange, oxidation, nitrification, denitrification, and/or disinfection) for final polishing the secondary effluent in order to meet the effluent discharge standards. In the nitrification and denitrification steps, only tertiary membrane clarification will be used for solid-water separation.

Ion: A charged atom, molecule, or radical that affects the transport of electricity through an electrolyte or gas. An atom or molecule that has lost or gained one or more electrons.

KAMET: It is the abbreviation of Krofta Advanced Municipal Effluent Treatment (KAMET) and is a circular package plant consisting of both Supracell and Sandfloat with Supracell on the top. KAMET system can be used for either water treatment or wastewater treatment, but is advertised for wastewater treatment.

KAMWT: It is the abbreviation of Krofta Advanced Municipal Water Treatment (KAMWT) and is a circular package plant consisting of both Supracell and Sandfloat with Supracell on the top. KAMWT system can be used for either water treatment or wastewater, but is advertised for water treatment.

Kraus activated sludge process: When nitrogen deficiency occurs in biological wastewater treatment and exogenous supply of nitrogen is expensive, the Kraus process can be useful in that an internal or endogenous supply of nitrogen is used to maintain the growth of active biomass. The Kraus process has a regular aeration tank (conventional, step aeration, or complete mix) plus an extra reaeration tank. The reaeration tank receives (a) return activated sludge from a secondary clarifier, (b) digester supernatant, (c) digested sludge, and (d) possible other nitrogen source, if needed, and aerates the mixture. When the reaeration tank effluent is introduced into the aeration tank with the wastewater influent, the nitrate released from the reaeration tank serves as the supplemental nitrogen source for synthesis.

Lagoon, aerated: An aerated lagoon is a holding basin (3–10 days DT; 6–20 ft depth) in which air is mechanically introduced to speed up aerobic decomposition of organic pollutants in wastewater. It is essentially the same as the extended aeration activated sludge biological process, except that an earthen basin is used for the aerobic suspended-growth bioreactor. The downstream of an aerated lagoon can be a facultative stabilization pond (or facultative lagoon), a sedimentation clarifier, or a dissolved air flotation clarifier, each incorporating the recycle of biological solids (or biosolids). The oxygen required by the aerated lagoon process is supplied by surface or diffused aerators. As with other suspended-growth systems, the turbulence created is used to maintain the contents of the basin in suspension.

Lagoon, anaerobic: Anaerobic lagoons are capable of treating high-strength wastewater. They are relatively deep (8–20 ft depth; 20–50 days DT) ponds with steep sidewalls in which anaerobic conditions are maintained by keeping wastewater organic loading so high that complete deoxygenation is prevalent. Although some oxygenation is possible in a shallow surface zone, once grease form an impervious surface layer, complete anaerobic conditions develop. Waste treatment or stabilization results from thermophilic anaerobic digestion of organic wastes. The treatment process is analogous to that occurring in single-stage untreated anaerobic digestion of sludge in which acid-forming bacteria break down organics. The resultant acids are then converted to carbon dioxide methane, cells, and other products. In a typical anaerobic lagoon, raw wastewater enters near the bottom of the pond (often at the center) and mixes with the active microbial mass in the sludge blanket which is usually about 6 ft. deep. The effluent discharge is located near one of the sides of the pond, submerged below the liquid surface. Excess undigested grease floats to the top, forming a heat-retaining and relatively air-tight cover. Wastewater flow equalization and heating are generally not practiced. Excess sludge is washed out with the lagoon effluent. Recirculation of waste sludge is not required.

Lagoon, facultative: Facultative lagoons are intermediate-depth (3–8 ft depth; 20–180 days DT) ponds in which the wastewater is stratified into three zones. These zones consist of an anaerobic bottom layer, an aerobic surface layer, and an intermediate zone. Stratification is a result of solids settling and temperature-water density variations. Oxygen in the surface stabilization zone is provided by reaeration and photosynthesis. This is in contrast to aerated lagoons in which mechanical aeration is used to create aerobic surface conditions. In general the aerobic surface layer serves to reduce odors while providing treatment of soluble organic by-products of the anaerobic processes operating at the bottom. Sludge at the bottom of facultative lagoons will undergo anaerobic digestion producing carbon dioxide, methane, and cells. The photosynthetic activity at the lagoon surface produces oxygen diurnally, increasing the dissolved oxygen during daylight hours, while surface oxygen is depleted at night. It is used for treating raw, screened wastewater, or primary effluent, secondary effluent, or weak biodegradable industrial wastewater.

Membrane bioreactor (MBR): Currently it is one of the suspended-growth biological processes (such as activated sludge process) involving the use of a bioreactor (such as an aeration tank) for carbonaceous oxidation, nitrification, and/or denitrification and the use of membrane filtration for subsequent bioreactor effluent treatment. The membrane filter discharges the membrane filter effluent (as the product water), recycles certain amount of activated sludge to the bioreactor, and wastes the excess amount of activated sludge. In the future a MBR may also be a combination of an attached-growth bioreactor and a membrane filter.

Mixed liquor suspended solids (MLSS): The concentration of suspended solids present in activated sludge mixed liquor, usually expressed as milligrams per liter (mg/L).

Mixed liquor volatile suspended solids (MLVSS): The fraction of the suspended solids in biological reactor's mixed liquor that can be burned off by combustion at 550 ° C. or the organic portion of the solids (an estimation of the microorganisms and food). Usually expressed as milligrams per liter (mg/L).

Mixed liquor: The water containing biosolids in a biological reactor.

Nitrate (NO_3): An oxygenated form of nitrogen.

Nitrification: It is a biological process by which ammonia in wastewater is converted by *Nitrosomonas* and *Nitrobacter* to nitrite and then to nitrate in the presence of oxygen. The biological reactions involved in these conversions may take place during activated sludge treatment or as a separate stage following the removal of carbonaceous materials. Separate-stage nitrification may be accomplished via suspended-growth or attached-growth unit processes. In either case, the nitrification step is preceded by a pretreatment sequence (i.e., preliminary treatment, primary treatment, and secondary treatment). Possible secondary treatment may be (a) biological secondary treatment, such as activated sludge, trickling filter, roughing filter, plus secondary clarification, or (b) physicochemical secondary treatment, such as chemical precipitation/coagulation plus secondary clarification. Low BOD (i.e., BOD_5/TKN ratio of less than 3) in the secondary effluent will assure a high concentration of nitrifiers (*Nitrosomonas* and *Nitrobacter*) in the nitrification biomass. The most common separate-stage nitrification process is the plug flow suspended-growth configuration with clarification. In this process, pretreatment effluent (i.e., nitrification influent) is pH adjusted as required, and aerated, in a plug flow mode. Because the carbonaceous demand is low, nitrifiers predominate. A clarifier (either sedimentation clarifier or flotation clarifier) follows aeration, and nitrification sludge is returned to the aeration tank. A possible modification is the use of pure oxygen in place of conventional aeration during the plug flow operation.

Nitrification: The process of oxidizing ammonia nitrogen to nitrate in wastewater by chemical or biological reactions.

Nitrifier (nitrifying bacteria): Bacteria that are capable of oxidizing nitrogenous material, such as *Nitrobacter* and *Nitrosomonas*.

Nitrite (NO_2): It is an intermediate oxygenated form of nitrogen.

Nitrite lock: Incomplete nitrification resulting in excess nitrite levels that react with chlorine. Also known as "chlorine sponge."

***Nitrobacter*:** It is a group of nitrifying bacteria that oxidize nitrite to nitrate.

Nitrogen (N): N is an essential nutrient that is often present in wastewater as ammonia, nitrite, nitrate, and organic nitrogen. The sum of these is expressed as total nitrogen.

Nitrogen cycle: It is a chemical transformation cycle of nitrogen through various stages of decomposition and assimilation.

Nitrogen gas (N_2): Gaseous form of nitrogen, N, also called atmospheric nitrogen. It comprises approximately 79% of atmospheric gas.

Nitrogenous biochemical oxygen demand (NBOD): The quantitative measure of the amount of oxygen required for the biological oxidation of nitrogenous material, such as ammonia nitrogen and organic nitrogen (measured by subtracting carbonaceous BOD from total BOD values).

***Nitrosomonas*:** A genus of nitrifying bacteria that oxidize ammonia to nitrite.

***Nocardia*:** It is a group of irregularly bent, short-branching filamentous organisms that cause dense dark foam in aeration basins. Associated with high fats, oils, and greases. Easily identified under a microscope.

Nutrient: A substance that is taken in by organisms and promotes growth. Carbon, nitrogen, and phosphorus are essential nutrients to most aquatic organisms.

Organic nitrogen: It is an organic nutrient containing nitrogen which is chemically bound in organic molecules such as proteins, amines, and amino acids. It is calculated by subtracting the ammonia nitrogen analysis results from total Kjeldahl nitrogen analysis results.

Organic phosphate: (a) It is a molecule of phosphate combined with an organic compound. Total phosphorus includes orthophosphate, polyphosphate, and organic phosphate. (b) It is an organic phosphorus originating from organic sources, such as body and food waste, and sometimes industrial sources. Some can biologically decompose into inorganic orthophosphate; other types are non-biodegradable and will pass through the wastewater system without treatment.

Organic: A volatile, combustible, or biodegradable compound containing carbon bound with other elements.

Orthophosphate: It is an inorganic nutrient containing phosphorus which is required for plant and animal growth. An inorganic, soluble form of phosphorus that is readily available to plants and animals without further breakdown and accounts for 70–90% of total phosphorus in wastewater. The easiest form of phosphorus that can be treated chemically. Total phosphorus includes orthophosphate, polyphosphate, and organic phosphate.

Oxic: It is an environmental condition in which an aquatic environment contains dissolved oxygen, usually in the range of 1.5–2.0 mg/L.

Oxidation ditch: It is a ring-shaped channel suspended-growth biological process system similar to extended aeration process for mainly carbonaceous oxidation and nitrification. It was developed to minimize waste activated sludge production through endogenous decay of the sludge mass. It is equipped with mechanical aeration devices, such as mechanical brush aerators surface aerators and jet aerator devices, to aerate and pump the wastewater. Screened wastewater enters the ditch is aerated and circulates at about 0.8–1.2 ft/sec. The economics of oxidation ditches appear most favorable when the solids retention time (SRT) is long, particularly where nitrification is required. When both nitrification and denitrification are required, certain portion of the ditch must be operated under anoxic condition (instead of normal aerobic oxidation condition); the process should be renamed as "oxidation-denitrification ditch."

Oxidation-reduction potential (ORP): The potential required to transfer electrons from the oxidant to the reductant; the quantitative measure, in mV, of the state of oxidation in wastewater treatment.

Oxidation: The addition of oxygen, removal of hydrogen, or removal of electrons from a compound (for instance, organic matter may be oxidized to a more stable compound).

Particulate: Solids suspended in water, wastewater, or air that can vary widely in shape, size, density, and charge.

Phosphate: PO_4^{3-} ion

Phosphorus (P): P is an essential element and nutrient for all life forms. Occurs as orthophosphate, polyphosphate, and organic phosphates, the sum of which is total phosphorus.

Phosphorus-accumulating organisms (PAOs): Microorganisms (bacteria) that are capable of uptaking and storing orthophosphate in excess of their biological requirements.

Polyhydroxyalkanoates (PHAs): Energy-rich carbon polymers inside a bacterial cell, which are converted from readily available organic molecules, such as volatile fatty acids (VFAs) in the wastewater. PHAs are the intracellular energy storage of the phosphorus-accumulating organisms (PAOs). The PAOs utilize PHA as an energy source to uptake phosphorus from the wastewater in the aerobic zone of the biological phosphorus removal process.

Polyphosphate (Poly-P): Inorganic phosphorus derived from synthetic detergents. May be hydrolyzed into orthophosphates.

Polyphosphate: It is a large molecule containing many individual molecules of orthophosphate. Total phosphorus includes orthophosphate, polyphosphate, and organic phosphate.

Post-aeration: It is one of the tertiary treatment processes. Because of the more stringent water quality standards being adopted by various regulatory agencies, the practice of post-aeration has increased substantially in recent years. The introduction of water quality-based effluent standards and permits that include high dissolved oxygen levels has made it necessary for many wastewater treatment plants to post-aerate the plant effluent before discharge.

Pre-aeration: Pre-aeration is one of the preliminary treatment processes. It promotes a more uniform distribution of suspended and floating solids. Aerating wastewater prior to primary clarification (sedimentation or flotation) can also improve its treatability; provide grease separation, odor control, grit removal, and flocculation; and increase BOD removals. It is now common to combine grit removal with pre-aeration as one unit process. Since dissolved air flotation (DAF) will provide aeration for grease separation, odor control, and oxidation, pre-aeration is not needed if DAF is used for primary flotation clarification (instead of sedimentation clarification).

Preliminary effluent: The effluent from a preliminary treatment system (i.e., bar screen, comminutor, and grit chamber) by which most of large objects, such as rocks, logs and cans, grit, etc., in raw wastewater have been removed.

Preliminary treatment: It is the first treatment step or preliminary step of either a conventional wastewater treatment system or an independent physicochemical treatment system. Preliminary treatment consists of bar screen, comminutor, and grit chamber mainly for removing large objects, such as rocks, logs and cans, grit, etc., from raw wastewater. Comminutor is an option depending on the nature and characteristics of raw wastewater.

Primary effluent: The effluent from a primary treatment system (either primary sedimentation clarification or primary flotation clarification) by which most of total suspended solids in wastewater have been removed.

Primary flotation clarification: It is a unit process or unit operation for the removal of mainly total suspended solids (settleable solids and floatable solids) from screened wastewater using a primary flotation clarifier.

Primary flotation clarifier: A dissolved air flotation (DAF) reactor is used to float total suspended solids (TSS) from screened wastewater by decreasing their apparent density. DAF consists of saturating a portion or all of the wastewater feed or a portion of recycled effluent with air at a pressure of 25–90 lb/square inch (gage). The pressurized wastewater is held at this pressure for 0.5–3 minutes in a retention tank and then released to atmospheric pressure to the flotation chamber. The sudden reduction in pressure results in the release of microscopic air bubbles which attach themselves to TSS and oil particles in the wastewater in the flotation chamber. This results in agglomeration which, due to the entrained air, has greatly increased vertical rise rates of about 0.5–2 ft/min. The floated materials rise to the surface to form a froth layer (float). Specially designed scrapers or other skimming devices continuously remove the froth (or float). The retention time in the flotation chambers is usually about 20–60 minutes for rectangular flotation clarifier and about 3–15 minutes for circular flotation clarifier using zero-horizontal velocity design. The effectiveness of dissolved air flotation depends upon the attachment of bubbles to the suspended solids and/or oil which are to be removed from the waste stream. The attraction between the air bubble and particle is primarily a result of the particle surface charge and bubble size distribution. The more uniform the distribution of water and micro-bubbles, the shallower the flotation clarifier can be. Generally, the depth of effective flotation units is between 3 and 9 feet.

Primary sedimentation clarification: It is a unit process or unit operation for the removal of mainly total suspended solids (settleable solids and floatable solids) from screened wastewater using a primary sedimentation clarifier.

Primary sedimentation clarifier: It is a tank used to settle mainly total suspended solids (TSS) from screened raw wastewater by gravity. The main objectives of a primary sedimentation clarifier are removal of settleable solids by settling them to the clarifier bottom and removal of floatable solids by skimming them from the clarifier's wastewater surface. In a rectangular sedimentation clarifier, the wastewater flows from one end to the other, and the settled sludge is moved to a hopper at one end, either by scrapers called "flights" set on parallel chains or by a single bottom scraper set on a traveling bridge. Floating materials, such as grease and oil, are collected by a surface skimmer and then removed from the rectangular sedimentation clarifier. In a circular sedimentation clarifier, the wastewater usually enters in the middle and flows toward the outside edge. Settled sludge is pushed to the hopper that is in the middle of the circular clarifier's tank bottom. Floating material is removed by a surface skimmer connected to the sludge collector.

Primary treatment: It is an important wastewater treatment step (either primary sedimentation clarification or primary flotation clarification mainly for removing total suspended solids from preliminary treatment effluent) after the preliminary treatment (i.e., bar screen, comminutor, and grit chamber mainly for removing large objects from raw wastewater), but before secondary treatment (either biological treatment or physicochemical treatment mainly for removing dissolved organic/inorganic pollutants from primary effluent).

Protozoan: Small, one-celled animals such as amoebae, ciliates, and flagellates.

Rapid mixing: A water treatment unit process of quickly mixing a chemical solution uniformly through the process water.

Reduction: A chemical reaction involving the addition of electrons to a chemical entity.

Return activated sludge (RAS): (a) Settled activated sludge, returned from the bottom of final sedimentation clarifiers, to mix with incoming raw or primary effluent; (b) floated activated sludge, returned from the top of flotation clarifiers, to mix with incoming raw or primary effluent.

Rotating biological contactor (RBC): In a RBC system, a series of closely spaced circular polystyrene or polyvinyl chloride disks is partially submerged in wastewater. Biological growth (biomass) attached to the surface of the disks. The biomass is kept in aerobic conditions (if carbonaceous oxidation and nitrification are intended) or anoxic conditions (if denitrification is intended) by rotation of the disks. Aerobic RBC is opened, so biomass may utilize the oxygen in the air, while anoxic RBC for denitrification is normally closed to block off the air. The RBC effluent discharges to a clarifier (either a sedimentation clarifier or a flotation clarifier) for solid-water separation, returning sludge, wasting sludge, and clarifier effluent discharge.

Sandfloat: It is combined circular dissolved air flotation and filtration (DAFF) clarifier designed by the Lenox Institute of Water Technology (LIWT) and manufactured by Krofta Engineering Corporation (KEC).

Secondary effluent: The effluent from a secondary treatment step which may be either (a) biological treatment, such as activated sludge aeration or equivalent plus secondary clarification, or (b) physicochemical treatment, such as chemical precipitation/coagulation plus secondary clarification. Secondary treatment step removes most of dissolved organic/inorganic pollutants from primary effluent. Since secondary clarification (either secondary sedimentation clarification or secondary flotation clarification) is the final step of secondary treatment, the secondary effluent is also the secondary clarification effluent.

Secondary flotation clarification: It is a unit process or unit operation for the removal of the bio-oxidation process generated activated sludge and/or the chemical precipitation/coagulation process generated chemical sludge using a secondary flotation clarifier.

Secondary flotation clarifier: A dissolved air flotation (DAF) reactor is used to float biological sludge (activated sludge) from biologically oxidized wastewater (such as aeration tank effluent) and/or chemical sludge from chemically coagulated/flocculated wastewater (such as flocculator effluent) by decreasing the sludge's apparent density. DAF consists of saturating a portion or all of the wastewater feed or a portion of recycled effluent with air at a pressure of 25–90 lb/square inch (gage). The pressurized wastewater is held at this pressure for 0.5–3 minutes in a retention tank and then released to atmospheric pressure to the flotation chamber. The sudden reduction in pressure results in the release of microscopic air bubbles which attach themselves to suspended solids and other particles in the wastewater in the flotation chamber. This results in agglomeration which, due to the entrained air, has greatly increased vertical rise rates of about 0.5–2 ft/min. The floated materials rise to the surface to form a froth layer (float). Specially designed scrapers or other skimming devices continuously remove the froth (or float). The retention time in the flotation chambers is usually about 20–60 minutes for rectangular flotation clarifier and about 3–15 minutes for circular flotation clarifier using zero-horizontal velocity design. The effectiveness of dissolved air flotation depends upon the attachment of bubbles to the suspended solids which are to be removed from the waste stream. The attraction between the air bubble and particle is primarily a result of the particle surface charge and bubble size distribution. The more uniform the distribution of water and micro-bubbles, the shallower the flotation clarifier can be. Generally, the depth of effective flotation units is between 3 and 9 feet.

Secondary membrane clarification (or secondary membrane separation or secondary membrane filtration): A water-solid separation/clarification process uses membrane device instead of conventional sedimentation clarification in the secondary wastewater treatment step.

Secondary sedimentation clarification: It is a unit process or unit operation for the removal of the bio-oxidation process generated activated sludge and/or the chemical precipitation/coagulation process generated chemical sludge using a secondary sedimentation clarifier.

Secondary sedimentation clarifier: It is a tank used to settle the chemical precipitation/coagulation process generated chemical sludge and/or the bio-oxidation process generated activated sludge for removing dissolved organic/inorganic substances from wastewater. The main objectives of a secondary sedimentation clarifier are removal of settleable chemical and/or biological sludge solids by settling them to the clarifier bottom. In a rectangular sedimentation clarifier, the wastewater flows from one end to the other, and the settled sludge is moved to a hopper at one end, either by scrapers called "flights" set on parallel chains or by a single bottom scraper set on a traveling bridge. In a circular sedimentation clarifier, the wastewater usually enters in the middle and flows toward the outside edge. Settled sludge is pushed to the hopper that is in the middle of the circular clarifier's tank bottom.

Secondary treatment: It is a wastewater treatment step after primary treatment (either primary sedimentation clarification or primary flotation clarification). Secondary treatment may be either biological treatment (such as activated sludge aeration plus secondary clarification) or physicochemical treatment (such as chemical precipitation/coagulation plus secondary clarification) mainly for removing dissolved organic/inorganic pollutants from primary effluent.

Selector hydraulic retention time (HRT): The given time it takes wastewater, including any return flows, to pass through a tank (selector).

Selector: A zone in a biological treatment process with specific environmental conditions that allow for the growth or lack of growth of certain microorganisms (such as an anoxic or anaerobic zone).

Sequencing batch exchanger (SBE): A sequencing batch process involving the use of separate batch process steps in sequence of filling water/wastewater, reacting with ion exchanger, settling insoluble spent ion exchanger, discharging treated effluent, and recycling the settled spent ion exchanger for regeneration, resulting in removing ionic pollutants from water or wastewater (US Patent 5354458).

Sequencing batch flotation (SBF): A sequencing batch process (either biological or physicochemical) involving the use of separate batch process steps in sequence of filling water/wastewater, reacting with chemicals or biomass, floating light-weight insoluble flocs and/or biomass, discharging treated effluent, and wasting floated insoluble scums for water or wastewater treatment is also called sequencing batch flotation (SBF) process (US Patent 5354458).

Sequencing batch reactor (SBR), biological: A biological SBR system performs all the necessary functions of carbonaceous and nutrient removals in a single tank with variable water levels and timed aerobic aeration or anoxic mixing. This SBR system requires a minimum of three tanks and advanced automation equipment to control the cycle times and phases. The SBR control systems allows the operation to be configured to operate as almost any other suspended-growth biological reactor by adjusting the cycle phases between fill phase, react phase (either aerobic aeration or anoxic mixing), separation phase (either sedimentation or flotation), and decant phase. The biological SBR system involving the use of flotation in the separation phase is invented by Dr. Lawrence K. Wang, Dr. Lubomyr Kurylko, and Dr. Mu-Hao Sung Wang in 1994 (US Patent 5354458).

Sequencing batch reactor (SBR), physicochemical: A physicochemical SBR system performs all the necessary functions of organic and inorganic pollutants removals in a single tank with variable water levels and timed mixing, coagulation/flocculation or clarification (either sedimentation or flotation). This SBR system requires a minimum of three tanks and advanced automation equipment to control the cycle times and phases. The SBR control systems allows the operation to be configured to operate as almost any other suspended floc physicochemical reactor by adjusting the cycle phases between fill phase, react phase (chemical precipitation/coagulation, flocculation), separation phase (either sedimentation or flotation clarification), and decant phase. The physicochemical SBR system involving the use of either sedimentation or flotation in the separation phase is invented by Dr. Lawrence K. Wang, Dr. Lubomyr Kurylko, and Dr. Mu-Hao Sung Wang in 1994 (US Patent 5354458).

Short circuiting: Shortening of hydraulic detention time, such as when plant flow exceeds design flow.

Shredding: It is a unit operation for reducing the particle size of objects or debris in the influent wastewater, also termed grinding or comminuting. These devices may be installed with a screen directly in the wastewater flow or separately out of the wastewater flow, with the shredded particles returned to the flow downstream of the screen. Only those shredding and grinding devices that are installed directly in the influent channel are termed comminuting devices.

Sidestreams: Sources of inflow from within the wastewater treatment facility, such as supernatant return from digesters, centrate from centrifuge thickening, filtrate from filter presses, etc. These sidestreams may be high in solids, BOD, or nutrients and may add to the plant organic or hydraulic loading.

Single-sludge activated sludge system: It is a biological activated sludge process system involving the use of one suspended activated sludge solids going through aerobic and anoxic zones for carbonaceous oxidation, nitrification, denitrification, and phosphorus removal. The single-sludge activated sludge system is commonly a continuous flow system using sedimentation clarification. The Lenox Institute of Water Technology (LIWT) has developed three innovative single-sludge activated sludge systems all for carbonaceous oxidation, nitrification, denitrification, and phosphorus removal: (a) continuous flow single-sludge system using flotation clarification; (b) sequencing batch reactors (SBR) using sedimentation for carbonaceous removal and nutrient removal; and (c) SBR using flotation for carbonaceous removal and nutrient removal.

Sludge age (SA): SA is the length of time a particle of activated sludge stays in the biological treatment plant, measured in days. In a biological phosphorus removal plant, sludge age is the amount (pound or kilogram) of mixed liquor suspended solids in all the biological reactors divided by the suspended solids withdrawn from the system per day (pound per day of waste activated sludge or kilogram per day of waste activated sludge).

Sludge dewatering: Sludge dewatering is the second step of sludge handling, after sludge thickening. Sludge dewatering is required for all facilities that do not dispose of their sludge in liquid form. It is a physical (mechanical) process to reduce the moisture content of sludge. The purpose is to reduce sludge volume, in turn, to reduce subsequent sludge digestion costs and/or final sludge disposal (land disposal or incineration) costs. Dewatering is usually required prior to sludge incineration. There are a variety of dewatering processes available: (a) vacuum filtration; (b) centrifuge; (c) belt filter press; (d) plate and frame press; (e) sludge drying beds or infrared sludge dryer; and (f) sludge lagoons.

Sludge drying beds: It is a sludge dewatering process using solar energy to evaporate water from the sludge slurries.

Sludge handling: It includes every required processes or management regarding waste sludge, such as sludge thickening, sludge dewatering, sludge digestion, sludge storage, sludge transportation, sludge disposal, etc.

Sludge lagoons: It is an earthen basin used for holding and dewatering sludge slurries by solar energy.

Sludge thickening and dewatering combined, belt filter press: Conditioned sludge is first placed on a gravity drainage section where it is allowed to thicken. In this section, the bulk of the free water is removed from the sludge by gravity. On some units, this section is equipped with a vacuum assist, which enhances drainage and may help to reduce odors. Following gravity drainage, pressure is applied in a low-pressure section, where sludge is squeezed between opposing porous cloth belts. On some units, the low-pressure section is followed by a high-pressure section where the sludge is subjected to shearing forces that induce the release of additional quantities of water from the sludge. The final dewatered sludge cake is removed from the belts by scrapper blades.

Sludge thickening/dewatering centrifuge, basket centrifuge: They are suitable for small plants. It is a semi-batch type operation. Sludge cake is collected on the sides of the spinning bowl, while the centrate (dilute effluent stream from the centrifuge) overflows the bowl rim. Once solids have built up to a maximum thickness, the feed sludge is stopped, and scrapper blades peel the sludge from the walls. The process is then resumed.

Sludge thickening/dewatering centrifuge, solid bowl centrifuge: Sludge is fed at a constant flow rate into the rotating bowl where it is separated into a dense cake and a dilute stream called centrate. The units can be used with no prior chemical conditioning, but solids' centrate quality are improved considerably when the sludge is conditioned with polymer.

Sludge thickening: Sludge thickening is the first step of sludge handling. It is employed prior to subsequent sludge dewatering processes to increase the efficiency of the sludge dewatering equipment. There are at least five types of sludge thickening processes: gravity thickening, flotation thickening, centrifuge thickening, gravity belt thickening, and rotary drum thickening.

Sludge volume index (SVI): SVI is the ratio of the volume (in mL) of sludge settled from a 1000 mL sample in 30 minutes to the concentration of mixed liquor (in mg/L) multiplied by 1000. It indicates settling capabilities and compaction, and it is an indication of filamentous organism overgrowth and several other indicators of process problems.

Sodium hypochlorite: It is a liquid sodium hypochlorite solution to be used as a disinfectant for water or wastewater disinfection.

Solids retention time (SRT): The theoretical length of time, usually in days, that solids are retained in an aeration basin, clarifier, or other structure. SRT is used to calculate wasting rates.

Soluble BOD (SBOD): Soluble BOD is a sample that has been filtered through a 0.45 μm filter. Soluble BOD includes volatile fatty acids (VFAs) and organic material that will readily ferment to create more VFAs in an anaerobic selector.

Soluble: Dissolved in a solution and more readily available for biochemical or chemical reactions. For instance, soluble BOD or soluble COD are readily available as food to microorganisms in a biochemical reaction. Soluble BOD includes volatile fatty acids (VFAs) and organic material that will readily be available to microorganisms in a biological process.

Step aeration activated sludge process: It is an improved plug flow activated process system in which (a) return activated sludge (RAS) enters the head end of the aeration tank with a portion of the wastewater influent, (b) the wastewater influent piping is so arranged that an increment of sewage is discharged into the aeration tank (3–5 hours DT) at subsequent steps. In doing this, the waste load is more uniformly spread over the length of the aeration tank, resulting in better utilization of the oxygen supplied, and the biological process system is much less susceptible to shock load, and (c) the biological sludge is maintained highly active throughout the aeration tank so that more organics are removed in a shorter contact time.

Struvite: It is magnesium ammonium phosphate ($MgNH_4PO_4(H_2O)_6$). Struvite forms hard, very insoluble, white, yellowish-white, or brownish-white crystals.

Subnatant: The water or liquid remaining below a floated sludge (float) after flotation.

Substrate: The food or chemical substance on which organisms depend for growth. The organic matter in wastewater (as measured by the BOD_5 test) is a substrate for the microorganisms in a biological process. For instance, in biological phosphorus removal systems, volatile fatty acids are a readily available substrate used by phosphorus-accumulating organisms (PAOs).

Supernatant: The water or liquid remaining above a sediment or precipitate after sedimentation.

Supracell: It is a circular dissolved air flotation (DAF) clarifier designed by the Lenox Institute of Water Technology (LIWT) and manufactured by Krofta Engineering Corporation (KEC).

Suspended-growth biological processes: In a suspended-growth biological process system, biomass or microorganisms are suspended in the biological reactors for aerobic and/or anoxic biochemical reactions. They include all kinds of activated sludge processes as well as oxidation ditch process, such as (a) conventional activated sludge process using diffused aeration; (b) conventional activated sludge process using mechanical aeration; (c) high-rate activated sludge process using diffused aeration; (d) pure oxygen activated sludge process using covered reactor; (e) pure oxygen activated sludge process using uncovered reactor; (f) extended aeration activated sludge process with nitrification; (g) activated sludge process separate reactors for carbonaceous oxidation, nitrification, and denitrification; (h) single-sludge activated sludge process for carbonaceous oxidation, nitrification, and denitrification; (i) sequencing batch reactor (SBR); (j) membrane bioreactor (MBR); (k) oxidation ditch process; (l) aerated lagoon; etc. The effluent of the suspended-growth bioreactor (aeration tank, or oxygenation tank, or mixing tank) discharges to a solid-water separator (sedimentation clarifier, flotation clarifier, or membrane filter) for additional processing.

Tertiary membrane clarification (or tertiary membrane separation or tertiary membrane filtration): A water-solid separation/clarification process uses membrane device instead of conventional sedimentation clarification in the tertiary wastewater treatment step, such as nitrification and denitrification.

Tertiary treatment: It is a wastewater treatment step after secondary treatment, meaning tertiary treatment is for treating the secondary effluent in that most of organic substances have been removed. Normally tertiary treatment includes nitrogen removal processes, phosphorus removal processes, granular activated carbon, granular media filtration, membrane filtration, and post-aeration.

Thickening, sludge: Sludge thickening is employed prior to subsequent sludge dewatering processes to increase the efficiency of the sludge dewatering equipment. There are at least five types of sludge thickening processes: gravity thickening, flotation thickening, centrifuge thickening, gravity belt thickening, and rotary drum thickening.

Total Kjeldahl nitrogen (TKN): The combined amount of organic and ammonia nitrogen determined by a laboratory analysis.

Total nitrogen (TN): All forms of nitrogen including ammonia nitrogen, nitrate nitrogen, nitrite nitrogen, and organic nitrogen.

Total phosphorus (TP): All forms of phosphorus, including orthophosphate, polyphosphates, and organic phosphorus.

Volatile fatty acids (VFAs): Fatty acids containing fewer carbon atoms and are dissolvable in water. When organic material undergoes fermentation reactions, smaller and more readily available organic molecules are formed known as volatile fatty acids (VFAs). VFAs typically found in wastewater are acetic acid (acetate) and propionic acid. VFAs provide the food for PAOs.

Waste activated sludge (WAS): Solids removed/wasted from a biological process, such as activated sludge process, to prevent an excessive buildup in the system

References

1. U.S. EPA, EPA History: Federal Water Pollution Control Act Amendments of 1972, U.S. Environmental Protection Agency, http://www2.epa.gov/aboutepa/epa-history-federal-water-pollution-control-act-amendments-1972. Accessed 2015
2. Ruchhoft E (1939) Public Health Rep 54(12)
3. Mickinney (1960) Proc. 3rd Conf. on Biological Waste Treatment. Manhattan College

4. Stainer (1963) The Microbial World, 2nd edn. Prentice-Hall
5. Mickinney RE (1962) Microbiology for sanitary engineers. McGraw-Hill, New York
6. McCorty (1972) In: Mitchell R (ed) Water pollution microbiology. Wiley-Interscience, Hoboken
7. Stumm (1966) Proceedings of the 3rd international conference on water pollution research. Munich
8. Dirasian (1968) Electrode potentials—significance in biological systems, part 2: Experience in waste treatment, water and sewage works, October
9. Herbert D (1956) J Gen Micro 14.
10. Lawrence AW (1963) J San Engr Div, Proceedings ASCE. June.
11. Wang LK, Shammas NK, Hung YT (eds) (2005) Advanced biological treatment processes. The Humana Press, Inc., Totowa
12. Sarioğlu M, Orhon D, Görgün E, Artan N (2004) Design procedure for carbon removal in contact stabilization activated sludge process. Water Sci Technol v(11):285–292
13. Shammas NK, Wang LK (2005) Principles and kinetics of biological processes. In: Wang LK, Shammas NK, Hung YT (eds) Advanced biological treatment processes. The Humana Press, Inc., Totowa
14. Vesilind A (2003) Wastewater treatment plant design. Water Environment Federation and IWA Publishing, Alexandria
15. Wang LK, Poon CPC, Wang MHS (1977) Control tests and kinetics of activated sludge process. Water Air Soil Pollut 8:315–351
16. Beer C, Wang LK (1978) Activated sludge systems using nitrate respiration – design considerations. J Water Pollut Control Fed 50(9):2120–2131
17. Wuhrmann K (1963) In: Eckenfelder WW Jr, McCabe J (eds) Advances in biological waste treatment
18. Eckenfelder WW Jr, O'Connor DJ (1961) Biological waste treatment. Pergamon Press, New York
19. Federation of Sewage and Industrial Wastes Associations (1952) Air diffusion in sewage works. *Manual of Practice* 5
20. Water Pollution Control Federation (1967) Sewage treatment plant design. *Manual of Practice* 8, Washington, DC
21. Wang MHS, Wang LK (1979) Computer-aided analysis of biochemical oxygen demand and reaction constant. Int Soc Ecol Model J 1:61–67
22. Shammas NK (1986) Interactions of temperature, pH and biomass on the nitrification process. J Water Pollut Control Fed 58(1):52–59
23. Morrison SM, Newton CC (1972) The mechanism of waste treatment at low temperature. Completion report. Environmental Resources Center. Colorado State University, USA
24. Eckenfelder WW Jr (1967) Industrial water quality. The McGraw-Hill Co., NY. ISBN 9780071548663
25. Zanoni AE (1969) Secondary effluent deoxygenation at different temperatures. J Water Pollut Control Fed 41:640–659
26. Friedman AA (1972) Temperature effects on growth and yield of activated sludge. J Water Pollut Control Fed 44:1429
27. Novak JT (1974) Temperature-substract interactions in biological treatment. J Water Pollut Control Fed 46:984
28. Alleman JE (2015) The genesis and evolution of activated sludge technology. School of Civil Engineering. http://bridge.ecn.purdue.edu/~alleman/w3-class/456/article/article-aswisconsin.html, West Lafayette
29. Shammas NK, Wang LK (2005) Emerging suspended growth biological processes. In: Wang LK, Shammas NK, Hung YT (eds) Advanced biological treatment processes. Humana Press, Inc., Totowa
30. Jenkins D (1968) Control of activated sludge by mean cell residence time. J Water Pollut Control Fed 40, Part I.

31. Toerber ED (1974) Comparison of completely mixed and plug flow biological systems. J Water Pollut Control Fed 46:1995
32. Fazadkia M, Mahvi AH (2004) Comparison of extended aeration activated sludge process and activated sludge with lime addition method for biosolids stabilization. Pak J Biol Sci 7(12):2061–2065
33. KEE Group of Companies. Extended aeration. Waste Water Solutions International Inc., Ellicott City. http://www.keeprocess.com/html/extended_aeration.html. Accessed 2015
34. USFilter. Biological treatment-extended aeration. US Filter Corporation, Warrendale. http://www.usfilter.com/water/Corporate/Technologies/biological_extended_aeration_technologies. Accessed 2015
35. Rich LG (1963) Unit processes of sanitary engineering. Wiley, New York
36. Metcalf and Eddy (2003) Wastewater engineering treatment and reuse, 4th edn. McGraw Hill, New York
37. Balmer P (1967) J Water Pollut Control Fed 39:1021
38. Wastewater Information Website. Oxidation ditch - activated sludge treatment process. http://waterfacts.net/Unit_Processes/Oxidation_Ditch/oxidation_ditch.html. Accessed 2015
39. Bertanza G (1997) Simultaneous nitrification-denitrification process in extended aeration plants: pilot and real scale experience. Water Sci Technol 35(6):53–61
40. Collivignarelli C, Bertanza G (1999) Simultaneous nitrification-denitrification processes in activated sludge plants: performance and applicability. Water Sci Technol 40(4/5):187–194
41. Zipperm T, Fleischmann N, Haberl R (1998) Development of a new system for control and optimization of small wastewater treatment plants using oxidation-reduction potential. Water Sci Technol 38(3):307–314
42. Charpentier J, Godart H, Martin G, Mogno Y (1989) Oxidation-reduction potential (ORP) regulation as a way to optimize aeration and C, N and P removal: experimental basis and various full-scale examples. Water Sci Technol 21:1209–1223
43. Jones RM, Sen D, Lambert R (1998) Full scale evaluation of nitrification performance in an integrated fixed activated sludge process. Water Sci Technol 38(1):71–78
44. Barros PR, Carlsson B (1998) Iterative design of a nitrate controller using an external carbon source in an activated sludge process. Water Sci Technol 38(1):71–78
45. Wang LK, Shammas NK (2005) Single-sludge biological systems for nutrients removal. In: Wang LK, Shammas NK, Hung YT (eds) Advanced biological treatment processes. Humana Press, Inc., Totowa
46. Liu Y, Tay JH, Lin YM, Hung YT (2005) Biological nitrification and denitrification processes. In: Wang LK, Pereira NC, Hung YT (eds) Biological treatment processes. Humana Press, Inc., Totowa
47. Gerardi MH (2001) Nitrification and denitrification in the activated sludge process. Wiley-Interscience, New York
48. Wang LK, Menon R (2005) Membrane bioreactor. In: Wang LK, Shammas NK, Hung YT (eds) Advanced biological treatment processes. Humana Press, Inc., Totowa
49. Stephenson T, Judd S, Jeferson B, Brindle K (2000) Membrane bioreactors for wastewater treatment. IWA Publishing, London
50. Urbain V, Mobarry B, de Silva V, Stahl DA, Rittmann BE, Manem J (1998) Integration of performance, molecular biology and modeling to describe the activated sludge process. Water Sci Technol 37(3):223–229
51. Rosenberger S, Kruger U, Witzig R, Manz W, Szewzyk U, Kraumea M (2002) Performance of a bioreactor with submerged membranes for aerobic treatment of municipal waste water. Water Res 36:413–420
52. Yang J (2015) Membrane bioreactor for wastewater treatment, Bookboon.com, London, U.K. http://bookboon.com/en/membrane-bioreactor-for-wastewater-treatment-ebook#download
53. Achilli A, Cath TY, Marchand EA, Childress AE (2009) The forward osmosis membrane bioreactor: a low fouling alternative to MBR processes. Desalination 239(1–3):10–21

54. Zhi Huang Z, Ong SL, Ng HY (2012) Submerged anaerobic membrane bioreactor for low-strength wastewater treatment: effect of HRT and SRT on treatment performance and membrane fouling. Water Res 45(2):705–714
55. Kovalova L, Siegrest H, Singer H, Wittmer A, McArdel CS (2012) Hospital wastewater treatment by membrane bioreactor: performance and efficiency for organic micropollutant elimination. Environ Sci Technol 46(3):1536–1545
56. U.S. EPA (2013) Emerging technologies for wastewater treatment and in-plant wet weather management. U.S. Environmental Protection Agency, EPA 832-R-12-011 Washington, D.C., USA
57. Wang LK, Pereira N, Hung YT (eds) (2009) Biological treatment processes. The Humana Press, Inc., Totowa, NJ, USA
58. Wang LK, Shammas NK, Hung YT (2009) Advanced biological treatment processes. The Humana Press, Inc., Totowa, NJ, USA
59. Shammas NK, Wang LK (2005) Aerobic and anaerobic attached growth biotechnology. In: Wang LK, Hwa TJ, Ivanov V, Hung YT (eds) Environmental biotechnology. The Humana Press, Inc., Totowa, NJ, USA
60. U. S. EPA Trickling Filter-Wastewater Technology Fact Sheet, U. S. Environmental Protection Agency, EPA 832-F-00-014, Office of Water, Washington, DC, USA, September (2000)
61. Pennsylvania Department of Environmental Protection (2014) Wastewater treatment plant operator training module 20: trickling filters. Bureau of Water Supply and Wastewater Management, Harrisburg, PA, USA
62. Eckenfelder WW, O'Connor DJ (1964) Biological waste treatment. Pergamon Press, New York, USA, pp 221–247
63. UNEP (2014) Trickling filtration, International source book on environmentally sound technologies for wastewater and stormwater management. United Nations Environment Programme, Geneva, Switzerland
64. Dai Y, Constantinou A, Griffiths P (2013) Enhanced nitrogen removal in trickling filter plant. CH2M Hill, Milton, Queensland, Australia
65. Beer C, Wang LK, Bergenthal J (1976) Kinetics and control tests of nitrification-denitrification process, pall II: respiration and control tests. Proceedings of the Institute of Environmental Sciences 22:501–508
66. Wang LK, Wang MHS, Poon CPC, Bergenthal J (1978) Chemistry of nitrification-denitrification process. J Environ Sci 21:23–28
67. Christensen DR, McCarty PL (1975) Multi-process biological treatment model. J Water Pollut Control Fed 47:2652
68. Wang LK, Vielkind P, Wang MHS (1978) Mathematical models of dissolved oxygen concentration in fresh water. Ecol Model 5:115
69. Wang LK, Wang MHS, Poon CPC (1980) Theoretical and empirical developments of attached growth biological systems, part I. US Dept. of Commerce, National Technical Information Service, VA, USA, Tech. Report No. PB81-130577, 86 pp, March
70. Wang LK, Bergenthal J, Wang MHS (1981) Predicting the dispersion of airborne effluents for environmental management. J Environ Sci 24(1):39
71. Krofta M, Wang LK (1984) Development of innovative sandfloat system for water purification and pollution control. ASPE J Eng Plumbing 0(1):1–16
72. Bungay HR III, Harold DB Jr (1971) Simulation of oxygen transfer in microbial slimes. Biotechnol Bioengineering 8:569
73. Pierce DM (1978) Upgrading trickling filters. US Environmental Protection Agency, Washington, DC, USA, 62 pp
74. Imboff K, Fair GM (1956) Sewage treatment, 2nd edn. Wiley, New York, USA
75. Antonie RL (1970) Application of the bio-disc process to treatment of domestic wastewater. Technical paper presented at the 43rd annual conference of the water pollution control federation, Boston, MA, USA, October 4–9

76. Antonie RL (1972) Factors affecting BOD removal and nitrification in the bio-disc process. Technical paper presented at the Central States Water Pollution Control Association Annual Meeting, Milwaukee, Wisconsin, USA, p. 14–16 June
77. Antonie RL (1972) Three-step biological treatment with the bio-disc process. Technical paper presented at the New York Water Pollution Control Association, Montank, USA, 12–14 June
78. Torpey WN (1971) Biological treatment of wastewater. US Patent No. 3575849, Official Gazette, USA, 20 April, p 594
79. U. S. National Research Council (1946) Sewage Work J 18:791
80. Allen TS (1971) The design, construction, and operation of plastic media fillers. Technical paper presented at the Missouri Water and Sewage Conference, Jefferson City, MO, USA, 29 Sept
81. Hemming ML (1979) General biological aspects of wastewater treatment, including the deep-shaft process. Water Pollution Control 78:3
82. Public Works Journal Corp (1970) Handbook of trickling filter design. PWJC, Ridgewood
83. Velz CJ (1948) A basic law for the performance of biological filters. Sewage Work J 20:607
84. Howland WE (1958) Flow over porous media as in a trickling filter. Engineering Bulletin Extension Series No, 94. In: Proceedings of the 12th industrial waste conference, 1957, Purdue University, Lafayette, Ind., USA, 42, 435–65
85. Eckenfelder WW Jr (1961) J Sanitary Eng. 87, No. SA4, Proc. Paper 2860, 33
86. UMRGLB (1971) Recommended standards for sewage works. Great Lakes-Upper Mississippi River Board of State Sanitary Engineers, USA
87. Galler WS, Gotaas HB (1964), Sanitary Eng. 90, No. SA6, Proc. Paper 4194, 59
88. Gloyna EF, Eckenfelder WW (1968) Advances in water quality improvement. University of Texas Press, Austin, TX, USA, pp 204–214
89. Jank BE, Drynan WR (1973) Substrate removal mechanism of trickling filters. J Environ Eng 3:187
90. Gotaas HB, Caller WS (1973) Design optimization for biological filter models. J Environ Eng EE6:831
91. McKinney RE (1962) Microbiology for sanitary engineers. McGraw-Hill, New York, USA, pp 199–212
92. Rogovskaya TI, Lazareva MF Sb Ved Pr Vvs Sk Chemickotechnolo Pardubice 6(1):43
93. Cooke WB (1959) An ecological life history of Aureobasidium pulluans. Ecology 40:273. (1962)
94. Hawkes HA (1963) The ecology of waste water treatment. Pergamon Press, London, 203 pp
95. Buras N, Kott YJ (1969) A simple test for differentiation between E.coli and A.aerogenes water research, Elsevier, NY, USA
96. Jank BE (1971) Substrate removal mechanism of trickling filters operated at high organic and hydraulic loadings. PhD Thesis, University of Waterloo, Ontario, Canada
97. Wang MHS, Wang LK, Poon CPC (1982) Theoretical and empirical developments of attached growth biological systems, part II. US Dept. of Commerce, National Technical Information Service, USA. Tech. Report No. PB82-216409, 38 pp, June
98. NSW Plastics (2015) SESSIL trickling filter media-vertical plastic strip media, Roanoke. http://www.nswplastics.com/environmental/pdfs/sessil_flyer.pdf
99. Brentwood Industries (2015) Brentwood trickling filter media installation guidelines, Reading. http://www.brentwoodprocess.com/pdfs/tfig.pdf
100. Tsuno H, Hidaka T, Nakamoto M (2004) Development of pre-coagulation and bio-filtration process for advanced treatment of sewage. Water Sci Technol Water Supply 4(1):13–22
101. Wang LK (1984) Investigation and design of a denitrification filter. Civ Eng Pract Des Eng 3:347–362
102. The Dow Chemical Co. (1972) A literature search and critical analysis of biological trickling filter studies, vol 4 and II. Environmental Protection Agency, Washington, DC, USA, 702 pp
103. Maier WJ (1966) Mass transfer and growth kinetics on a slime layer, a simulation of the trickling filter. PhD Thesis, Cornell University, Ithaca, NY, USA

104. Williamson K, McCarty PL (1976) J Water Pollut Control Fed 48(1):9
105. Williamson K, McCarty PL (1976) A model of substrate utilization by bacterial films. J Water Pollut Control Fed 48(2):281
106. Frank-Kamenetskii DA (1955) Diffusion and heat exchange in chemical kinetics. Princeton University Press, NJ, USA
107. Monod J (1942) Recherches sur Ia Croissance des Cultures Bacteriennes. Herman, Paris
108. Lakeside Equipment Corporation (2015) Biological treatment options with trickling filter. http://www.lakeside-equipment.com/Product_Pages/trickle.htm
109. Scott HG (1961) Filter fly control at sewage plants. Sanitation (Los Angeles, California) 24(1):14
110. Hawkes HA (1998) Origin and development of the biological monitoring working party score system. Water Research, Elsevier, NY, USA
111. Poon CPC, Wang LK, Wang MHS (1986) Activated sludge processes. In: Wang LK, Pereira NC (eds). Biological treatment processes. Humana press, Totowa, NJ, USA, pp. 229–303
112. Poon CPC, Wang LK, Wang MHS (1986) Waste stabilization ponds and lagoons. In: Wang LK, Pereira NC (eds). Biological treatment processes. Humana press, Totowa, NJ, USA, pp. 305–359
113. Wang LK, Wang MHS, Poon CPC (1986) Frickling filters. In: Wang LK, Pereira NC (eds). Biological treatment processes. Humana press, Totowa, NJ, USA, pp. 361–425
114. Wang MHS, Wang LK, Poon CPC (1986) Rotating biological contactors. In: Wang LK, Pereira NC (eds). Biological treatment processes. Humana press, Totowa, NJ, USA, pp. 427–448
115. Thackston EL, Eckenfeider WW (1972) Water quality engineering, new concepts and developments. Jenkins Publishing Co., New York, USA, pp 115–120
116. Eckenfelder WW, Ford DL (1970) Water pollution control experimental procedures for process design. Jenkins Publishing Co., New York, USA, pp 173–183
117. Clark JW, Viessman W Jr, Hammer MJ (1977) Water supply and pollution control. International Textbook Co., New York, USA, pp 529–563
118. The City of Austin and the University of Texas, Austin, Texas (1971) Design guides for biological wastewater. U. S. Government Printing Office, Washington, DC, USA, pp 145–178
119. Meltzer D (1958) Experimental investigations into biological filtration of sewage at Klipspruit sewage purification works. Water Pollut Abstract 31:1230
120. Rankin RS (1953) Performance of biofiltration plants by three methods. In: Separate proceedings, ASCE, USA, No. 336
121. Eckenfelder WW Jr (1966) Industrial water pollution control. McGraw-Hill, New York
122. Shulze KL (1960) Load and efficiency of trickling filters. J Water Pollut Control Fed 32:245
123. U. S. EPA (1977) Aerobic biological wastewater treatment facilities. US Government Printing Office, Washington, DC, USA
124. WPCF (1977) Wastewater treatment plant design. Water Pollution Control Federation, Washington, DC, USA, 560 pp
125. Culp GL, Heim NF (1978) Field manual for performance evaluation and troubleshooting at municipal wastewater treatment facilities. US Government Printing Office, Washington, DC, USA, 397 pp
126. U. S. Army (1978) Design of wastewater treatment facilities major systems. US Environmental Protection Agency, Washington, DC, USA
127. U. S. EPA (1978) Innovative and alternative technology assessment manual. US Environmental Protection Agency, USA
128. Bonjes HH, Faisst JA, Lineck TS (1979) Capital and O & M cost estimates for biological wastewater treatment processes. US Environmental Protection Agency, Cincinnati, OH, USA
129. Shammas NK, Hahn HH, Wang MHS, Wang LK (2021) Fundamental of chemical coagulation and precipitation. In: Wang LK, Wang MHS and Shammas NK (eds) Environmental flotation engineering. Springer Nature, Switzerland, pp. 95–142

130. Shammas NK, Wang LK, Hahn HH (2010) Fundamental of wastewater flotation. In: Wang LK, Shammas NK, Selke WA, Aulenbach DB (eds) Flotation technology. Humana Press, Totowa, NJ, USA, pp 121–164
131. Wang LK, Shammas NK, Wu BC (2010) Electroflotation. In: Wang LK, Shammas NK, Selke WA, Aulenbach DB (eds) Flotation technology. Humana Press, Totowa, NJ, USA, pp 165–198
132. Shammas NK, Pouet MF, Grasmick A (2010) Wastewater treatment by electrocoagulation-flotation. In: Wang LK, Shammas NK, Selke WA, Aulenbach DB (eds) Flotation technology. Humana Press, Totowa, NJ, USA, pp 199–220
133. Shammas NK, Wang LK, Landin M (2010) Treatment of paper mill whitewater, recycling and recovery of raw wastewater. In: Wang LK, Shammas NK, Selke WA, Aulenbach DB (eds) Flotation technology. Humana Press, Totowa, NJ, USA, pp 221–268
134. Wang LK, Shammas NK (2010) Ozone-oxygen oxidation flotation. In: Wang LK, Shammas NK, Selke WA, Aulenbach DB (eds) Flotation technology. Humana Press, Totowa, NJ, USA, pp 269–326
135. Shammas NK (2010) Wastewater renovation by flotation. In: Wang LK, Shammas NK, Selke WA, Aulenbach DB (eds) Flotation technology. Humana Press, Totowa, NJ, USA, pp 327–346
136. Shammas NK (2010) Flotation-filtration system for wastewater reuse. In: Wang LK, Shammas NK, Selke WA, Aulenbach DB (eds) Flotation technology. Humana Press, Totowa, NJ, USA, pp 347–362
137. Shammas NK, Wang LK, Selke WA, Aulenbach DB (2010) Laboratory simulation and testing of air flotation and associated processes. In: Wang LK, Shammas NK, Selke WA, Aulenbach DB (eds) Flotation technology. Humana Press, Totowa, NJ, USA, pp 593–518
138. Shammas NK, Wang LK, Selke WA (2010) Completely closed water systems in paper mills. In: Wang LK, Shammas NK, Selke WA, Aulenbach DB (eds) Flotation technology. Humana Press, Totowa, NJ, USA, pp 401–428
139. Aulenbach DB, Shammas NK, Wang LK, Kitter RDJ (2010) Lake restoration using DAF. In: Wang LK, Shammas NK, Selke WA, Aulenbach DB (eds) Flotation technology. Humana Press, Totowa, NJ, USA, pp 429–456
140. Wang LK, Aulenbach DB, VanDyke JP (2010) Jiminy peak, Hancock, Massachusetts wastewater treatment plant: the first RBC-flotation-UV wastewater treatment plant in the USA. In: Wang LK, Shammas NK, Selke WA, Aulenbach DB (eds) Flotation technology. Humana Press, Totowa, NJ, USA, pp 457–484
141. Wang LK, Shammas NK, Aulenbach DB, Selke WA, Guss DB (2010) Pittsfield water treatment plant: once the world's largest flotation filtration plant. In: Wang LK, Shammas NK, Selke WA, Aulenbach DB (eds) Flotation technology. Humana Press, Totowa, NJ, USA, pp 485–501
142. Shammas NK, Wasowski JP, Wang LK, Selke WA (2010) Pretreatment of meal-processing waste. In: Wang LK, Shammas NK, Selke WA, Aulenbach DB (eds) Flotation technology. Humana Press, Totowa, NJ, USA, pp 502–566
143. Wang LK, Aulenbach DB, Shammas NK (2010) Treatment of seafood processing wastewater. In: Wang LK, Shammas NK, Selke WA, Aulenbach DB (eds) Flotation technology. Humana Press, Totowa, NJ, USA, pp 567–592
144. Aulenbach DB, Wang LK, Selke WA (2010) Algae removal by flotation. In: Wang LK, Shammas NK, Selke WA, Aulenbach DB (eds) Flotation technology. Humana Press, Totowa, NJ, USA, pp 363–400
145. Shammas NK, Bennett GF (2010) Principles of air flotation technology. In: Wang LK, Shammas NK, Selke WA, Aulenbach DB (eds) Flotation technology. Humana Press, Totowa, NJ, USA, pp 1–48
146. Wang LK, Shammas NK, Selke WA, Aulenbach DB (2010) Gas dissolution, release, and bubble formation in flotation systems. In: Wang LK, Shammas NK, Selke WA, Aulenbach DB (eds) Flotation technology. Humana Press, Totowa, NJ, USA, pp 49–84

147. Bennett GF, Shammas NK (2010) Separation of oil from water by air flotation. In: Wang LK, Shammas NK, Selke WA, Aulenbach DB (eds) Flotation technology. Humana Press, Totowa, NJ, USA, pp 85–120
148. Krofta M, Wang LK (2000) Flotation engineering. Lenox Institute of Water Technology, Lenox, MA, USA. Lenox/01-05-2000/368, January 2000, 255 pages
149. Shammas NK, Wang LK, Wang MHS (2018) Removal of heavy metals from industrial effluents by chemical precipitation and chemical coagulation. In: Wang LK, Wang MHS, Hung YT, Shammas NK, Chen JP (eds) Handbook of advanced industrial and hazardous wastes management. CRC Press, Boca Raton, FL, USA, pp 1129–1140
150. Wang LK, Wang MHS, Shammas NK (2018) Treatment of wastewater, storm runoff and combined sewer overflow by dissolved air flotation and filtration. In: Wang LK, Wang MHS, Hung YT, Shammas NK, Chen JP (eds) Handbook of advanced industrial and hazardous wastes management. CRC Press, Boca Raton, FL, USA, pp 577–610
151. Palanlandy P, Adlan HMN, Aziz HA, Murshed MF, Hung YT (2018) Dissolved air flotation (DAF) for wastewater treatment. In: Wang LK, Wang MHS, Hung YT, Shammas NK, Chen JP (eds) Handbook of advanced industrial and hazardous wastes management. CRC Press, Boca Raton, FL, USA, pp 657–694
152. Pankivskyi YI (2000) Combined physical-chemical and biological treatment of municipal wastewater under cold weather conditions., Master of Engineering Thesis, Lenox Institute of Water Technology, USA; May 2000, 67 pages. In: Wang LK, Wang M-HS, Shammas NK (eds) Environmental Science, Technology, Engineering, and Mathematics (STEM), vol 2000, No. 5. Lenox Institute Press, Newtonville, NY, USA; 67 p. https://doi.org/10.17613/k22f-bz03
153. Aulenback DB, Wang LK, Wang MHS, Shammas NK (2021) Basic hydrology, water resources, and DAF boat plant for lake restoration. In: Wang LK, Wang M-HS, Hung Y-T, Shammas NK (eds) Environmental and natural resources engineering. Springer, Nature Switzerland, pp 229–298
154. Wang LK, Wang MHS, Wong JM (2021) Innovative circular gravity flotation and fiber detection for fiber separation. In: Wang LK, Wang MHS, Shammas NK (eds) Environmental flotation engineering. Springer Nature Switzerland, pp 167–188
155. Wong JM, Farmerie JE, Wang LK (2021) Operation and performance of Clari-DAF® system for water purification. In: Wang LK, Wang MHS, Shammas NK (eds) Environmental flotation engineering. Springer Nature Switzerland, pp 339–366
156. Wong, JM, Hess, RJ Wang, LK (2021) Operation and performance of the AquaDAF® process system for water purification. In: Wang MHS, Shammas NK (eds) Environmental flotation engineering. Springer Nature Switzerland, pp 301–338
157. Wang LK (2007) Merging flotation technologies. In: Wang LK, Hung YT, Shammas NK (eds) Advanced physicochemical treatment technologies. Humana Press, Totowa, NJ, USA, pp 449–484
158. Wang LK (2021) Humanitarian engineering education of the Lenox Institute of Water Technology and its new potable water flotation processes. In: Wang LK, Wang MHS, Shammas NK (eds) Environmental flotation engineering. Springer Nature Switzerland, pp 1–72
159. Wang LK, Wang MHS, Fahey EM (2021) Innovative dissolved air flotation potable water filtration plant in Lee, Massachusetts, USA. In: Wang LK, Wang MHS, Shammas NK (eds) Environmental flotation engineering. Springer Nature Switzerland, pp 93–94
160. Shammas NK, Wang LK (2016) Water Engineering: hydraulics, distribution and treatment. Wiley, New York, USA, 806 pages
161. Holtorff MS (1999) Municipal waste treatment by innovative physical-chemical and biological processes. Master Engineering thesis, Lenox Institute of Water Technology, Lenox, MA, USA
162. Wang LK (2010) Conversion factors for environmental engineers. In: Wang LK, Shammas NK, Selke WA, Aulenbach DB (eds) Flotation technology. Humana Press, Totowa, NJ, USA, pp 618–666

163. Wang MHS, Wang LK (2014) Glossary and conversion factors for water resources engineers. In: Wang LK, Yang CT (eds) Modern water resources engineering. Humana Press, Totowa, NJ, USA, pp 759–851
164. Wang MHS, Wang LK (2016) Glossary of land and energy resources engineering. In: Wang LK, Wang MHS, Hung YT, Shammas NK (eds) Natural resources and control resources. Springer, NYC, USA, pp 493–623
165. Wang MHS, Wang LK (2015) Environmental water engineering glossary. In: Yang CT, Wang LK (eds) Advances in water resources engineering. Springer, New York, USA, pp 471–556
166. Wang LK, Wang MHS (2015) Environmental engineering glossary, 2nd edn. Lenox Institute Press, Newtonville, NY, USA

Chapter 4
Removal of Endocrine Disruptors for Environmental Protection

Nazih K. Shammas, Lawrence K. Wang, and Mu-Hao Sung Wang

Contents

1 Introduction 171
2 Endorine System and Endocrine Disruptors 174
  2.1 Endocrine System 174
  2.2 Endocrine Disruptors 174
3 Descriptions of Specific EDCs 175
  3.1 Pesticide Residues 176
  3.2 Highly Chlorinated Compounds 180
  3.3 Alkylphenols and Alkylphenol Ethoxylates 182
  3.4 Plastic Additives 182
4 Water Treatments for EDC Removal 184
  4.1 Granular Activated Carbon (GAC) 184
  4.2 Powdered Activated Carbon (PAC) 186
  4.3 Coagulation/Filtration 186
  4.4 Lime Softening 187
5 Point-of-Use/Point-of-Entry Treatments 187
6 Water Treatment Techniques for Specific EDC Removal 188
  6.1 Methoxychlor 188
  6.2 Endosulfan 188
  6.3 DDT 189
  6.4 Diethyl Phthalate 189
  6.5 Di-(2ethylhexyl) Phthalate (DEHP) 189
  6.6 PCBs 190
  6.7 Dioxin 190
  6.8 Alkylphenols and Alkylphenol Ethoxylates 190
Glossary 191
References 192

N. K. Shammas · L. K. Wang (✉) · M.-H. S. Wang
Lenox Institute of Water Technology, Newtonville, NY, USA

Krofta Engineering Corporation, Lenox, MA, USA

L. K. Wang, M. -H. S. Wang, Y. -T. Hung, N. K. Shammas (eds.),
Environmental and Natural Resources Engineering, Handbook of Environmental Engineering 19, https://doi.org/10.1007/978-3-030-54626-7_4

Abstract The endocrine system regulates all biological processes in the body from conception through adulthood and into old age, including the development of the brain and nervous system, the growth and function of the reproductive system, as well as the metabolism and blood sugar levels. Endocrine disruptors are chemicals that either mimic endogenous hormones, interfere with pharmacokinetics, or act by other mechanisms resulting in adverse effects such as compromised reproductive fitness, functional or morphological birth defects, cancer, and altered immune functions.

Recently, public concern has focused on the possible hormonal effects of some environmental pollutants on wildlife and humans. These chemicals, referred to collectively as endocrine disruptors, comprise a wide range of substances including pesticides (methoxychlor), surfactants (nonylphenol), plasticizers (diethylphthalate), and organohalogens (polychlorinated biphenyls or PCBs and dioxin).

The feasibility of using the various techniques for the removal of the endocrine disruptors will depend on the size of the system and the cost-effectiveness. The two major concerns regarding technologies for small systems are affordability and technical complexity (which determine the needed skills for the system operators). The removal processes of these chemicals from water supply include granular activated carbon; powdered activated carbon; coagulation/filtration; and lime softening.

Keywords Endocrine system · Endocrine disruptors · Environmental pollution · Hormonal effects · Removal processes · Granular activated carbon (GAC) · Powdered activated carbon (PAC) · Coagulation/filtration · Lime softening

Acronym

AP	Alkylphenols
APE	Alkylphenol ethoxylate
BAT	Best available technology
CWA	Clean Water Act
DEHP	Di-(2ethylhexyl) phthalate
DEP	Diethyl phthalate
EDC	Endocrine-disrupting compounds
EDSP	Endocrine Disruptor Screening Program
GAC	Granular activated carbon
MCL	Maximum contaminant level
NP	Nonylphenol
NPDWR	National Primary Drinking Water Regulations
NPE	Nonylphenol ethoxylates
PAC	Powdered activated carbon
PCB	Polychlorinated biphenyls
POE	Point-of-entry

POU	Points-of-use
TCDD	Dioxin (2,3,7,8-tetrachlorodibenzo-p-dioxin)
TOC	Total organic carbon
USDA	US Department of Agriculture
USEPA	US Environmental Protection Agency

Nomenclature

Q_e	equilibrium capacity of the carbon for the target compound, μg/g
C_e	equilibrium liquid-phase concentration of the target compound, μg/L
K	Freundlich coefficient in $(\mu g/g)(L/\mu g)^{1/n}$
$1/n$	Freundlich coefficient, dimension-less units

1 Introduction

The emerging pollutants endocrine-disrupting compounds (EDCs) are defined as "chemicals that either mimic endogenous hormones, interfere with pharmacokinetics or act by other mechanisms" [1]. Such adverse effects as compromised reproductive fitness, functional or morphological birth defects, cancer, and altered immune functions, among others, have been reported in the scientific literature [1–3]. The term "endocrine disruptors" is used to describe substances that are not produced in the body but act by mimicking or antagonizing natural hormones. Endocrine systems, also referred to as hormone systems, are found in all mammals, birds, fish, and many other types of living organisms. They are made up of:

- Glands located throughout the body
- Hormones that are made by the glands and released into the bloodstream or the fluid surrounding cells
- Receptors in various organs and tissues that recognize and respond to the hormones

Hormones are released by glands and travel throughout the body, acting as chemical messengers. Hormones interface with cells that contain matching receptors in or on their surfaces. The hormone binds with the receptor, much like a key would fit into a lock. The hormones, or keys, need to find compatible receptors, or locks, to work properly. Although hormones reach all parts of the body, only target cells with compatible receptors are equipped to respond. Once a receptor and a hormone bind, the receptor carries out the hormone's instructions by either altering the cell's existing proteins or turning on genes that will build a new protein. Both of these actions create reactions throughout the body. Researchers have identified more than 50 hormones in humans and other vertebrates.

The endocrine system regulates all biological processes in the body from conception through adulthood and into old age, including the development of the brain and nervous system, the growth and function of the reproductive system, as well as the metabolism and blood sugar levels. The female ovaries, male testes, and pituitary, thyroid, and adrenal glands are major constituents of the endocrine system.

The Endocrine Disruptor Screening Program (EDSP) focuses on the estrogen, androgen, and thyroid hormones. Estrogens are the group of hormones responsible for female sexual development. They are produced primarily by the ovaries and in small amounts by the adrenal glands. Androgens are responsible for male sex characteristics. Testosterone, the sex hormone produced by the testicles, is an androgen. The thyroid gland secretes two main hormones, thyroxine and triiodothyronine, into the bloodstream. These thyroid hormones stimulate all the cells in the body and control biological processes such as growth, reproduction, development, and metabolism.

It is thought that EDCs may be responsible for some reproductive problems in both women and men as well as for the increases in the frequency of certain types of cancer. EDCs have also been linked to developmental deficiencies and learning disabilities in children. Because hormone receptor systems are similar in humans and animals, effects observed in wildlife species raise concerns of potential human health effects. During fetal development and early childhood, low-dose exposure to EDCs may have profound effects not observed in adults such as reduced mental capacity and genital malformations. Evaluating potential low-dose effects of environmental estrogenic compounds has been identified as a major research priority.

There is growing evidence that artificial chemicals in the environment can disrupt hormones by sending erroneous signals or blocking legitimate signals. Since the hormones are part of the endocrine system, the hormone disrupters are also called endocrine disrupters. And since the concern originated with estrogen, the "female" hormone, they are sometimes called estrogen mimics.

The strongest evidence that endocrine disrupters are damaging human or animal health comes from the animal kingdom. In the 1980s and 1990s, fish and beluga whales with horrible malformations showed up in the Great Lakes region, with cancers, ulcers, and other deformations. In the 1990s, an epidemic of misshapen reproductive organs in Florida alligators was blamed on a pesticide spill into the lake. Other reproductive abnormalities in gulls, mink, eagles, and other animals have been blamed on chemicals that mimic hormones. Furthermore, in the laboratory, tiny concentrations of hormone mimics lock onto cell receptors, causing the cells to reproduce in a phenomenon suspiciously like cancer.

The controversy about endocrine disrupters began with concern about chemicals that disrupt estrogen, but now it has expanded to cover chemicals that interfere with androgens male hormones. Keeping in mind that males have female hormones, and vice versa, it is immediately realized that the endocrine story is a lot more complicated than once thought.

Several studies have found a worldwide lowering of sperm counts and blamed it on the rising concentrations of estrogen mimics in the environment. Some scientists say estrogen mimics could also explain the growing incidence of breast cancer and perhaps prostate cancer as well [4–6].

The putative endocrine disrupters have structures akin to real hormones and seem to include breakdown products of several pesticides that are now banned, such as (a) DDT; (b) PCBs, a persistent group of chemicals still found in electrical equipment that pollutes lake and stream sediments in many industrial regions; (c) dioxins, a group of toxic chemical by-products from paper production and incineration; (d) chemicals found in the epoxy lining of "tin" cans, plastics used for storing food, dental sealants, and (e) vinclozolin, a fungicide used on fruit.

A growing body of scientific research indicates that man-made industrial chemicals and pesticides may interfere with the normal functioning of human and wildlife endocrine systems. A hormone is defined as any substance in the body that is produced by one organ then carried by the bloodstream to have an effect in another organ. The primary function of hormones, or the endocrine system, is to maintain a stable environment within the body; this is often referred to as homeostasis. The endocrine system also controls reproduction and growth.

Recently, public concern has focused on the possible hormonal effects of some environmental pollutants on wildlife and humans. These chemicals, referred to collectively as endocrine disruptors, comprise a wide range of substances including pesticides (methoxychlor), surfactants (nonylphenol), plasticizers (diethylphthalate), and organohalogens (polychlorinated biphenyls or PCBs and dioxin). Endocrine disruptors may be found in many everyday products – including plastic bottles, metal food cans, detergents, flame retardants, food, toys, cosmetics, and pesticides. Many industrial chemicals and pesticides have undergone extensive toxicological testing; however, since the purpose of this testing was not to find some subtle endocrine effects, these potential effects may not have been revealed. The persistence of some pesticides in the aquatic environment may pose a threat to the human population, especially if such substances occur in the nation's drinking water sources. As a result of this growing concern, the 1996 Safe Drinking Water Act (SDWA) Amendments and the Food Quality Protection Act require EPA to develop a screening and testing program to determine which chemical substances have possible endocrine disrupting effects in humans [7].

One example of the devastating consequences of the exposure of developing animals, including humans, to endocrine disruptors is the case of the potent drug diethylstilbestrol (DES), a synthetic estrogen. Prior to its ban in the early 1970s, doctors mistakenly prescribed DES to as many as five million pregnant women to block spontaneous abortion and promote fetal growth. It was discovered after the children went through puberty that DES affected the development of the reproductive system and caused vaginal cancer. Since then, Congress has improved the evaluation and regulation process of drugs and other chemicals.

Many of the potential EDCs may be present in surface water or groundwater. A number of drinking water treatment processes are available and may be used to remove many of the potential EDCs. This chapter presents treatment processes for large municipalities as well as small communities to remove specific EDCs from drinking water.

2 Endorine System and Endocrine Disruptors

2.1 Endocrine System

The endocrine system consists of glands, hormones, and receptors. Endocrine glands include the hypothalamus, pineal, pituitary, thyroid, parathyroid, thymus, adrenal, ovaries, prostate, and testes [8].

The pituitary gland acts as the control center, telling the other glands when to send their signals and how much hormone to send. The pituitary gets its cues from the hypothalamus, which acts as a regulator, telling the pituitary to increase hormone production or to slow it down and shut it off. These messages travel back and forth continuously throughout all parts of an organism, keeping everything balanced and coordinated.

The glands produce hormones, such as adrenocorticotropic hormone (ACTH), corticosteroid, adrenaline, estrogen, testosterone, androgen, insulin, triiodothyronine, and thyroxin.

Hormones are involved in just about every biological function. They are better known as the body's chemical messengers because they travel through the bloodstream and cause responses in other parts of the body. The amount of hormone that an animal's body produces depends upon the stimuli that its body receives. They also can work at astonishingly low concentrations – in parts per billion or even trillion. Hormones regulate [8]:

1. Reproduction and embryo development
2. Growth and maturation
3. Energy production, use, and storage
4. Electrolytes – the balance and maintenance of water and salt
5. Reaction to stimuli, such as fright and excitement
6. Behavior of human beings and animals

   Receptors, which are in the cells of various target organs and tissues, recognize and respond to the hormones. Receptors are a part of a complex biological feedback system that regulates the response. Any disruption to the balance can cause changes to take place in these reactions.

2.2 Endocrine Disruptors

Endocrine disruptors are synthetic or naturally occurring chemicals that interfere with the balance of normal hormone functions in animals, including humans. This imbalance can cause various abnormalities of the reproductive system, such as the feminization of males and the masculinization of females. Among other abnormalities, they also can cause enlargement of the thyroid gland, birth defects, behavioral changes, depressed immune systems, and an increased vulnerability to disease.

An endocrine disrupting chemical can affect the endocrine system of an organism in a number of ways, but they typically affect animals in three specific ways. They can mimic, block, or trigger a hormone response [8].

Mimic Chemicals Mimic chemicals respond like normal hormones inside the body. A good example of a mimicking endocrine disruptor is the potent drug diethylstilbestrol (DES), a synthetic estrogen. Doctors prescribed DES to as many as five million pregnant women to block spontaneous abortion prior to DES being banned in the early 1970s. When doctors first began prescribing DES, they believed that it would prevent miscarriage and promote fetal growth. However, researchers discovered that after the children went through puberty, DES affected the development of the reproductive system of the daughters of the mothers given DES, and it caused vaginal cancer. In addition, these women have an increased risk of developing endometriosis. Sons born to mothers given DES have an increased frequency of undescended testes (cryptorchidism), congenital birth defects, hypospadias (urethra opening on the underside of the penis), and decreased adult sperm count.

Blocker Chemicals The second group of disrupting chemicals is hormone blockers. These interfere with how naturally occurring hormones function. Blockers bind to the same protein receptors as the real hormone but do not stimulate any action. They sit in the way of the natural hormone and prevent it from sending its message. An example of a blocker is how DDE (a metabolic breakdown product of the pesticide DDT) blocked action of testosterone in male alligators in Lake Apopka, Florida, which led to undersized penises. Testosterone, a male hormone, is needed for proper reproductive development.

Trigger Chemicals Triggers are the third category of disruptors. They attach to protein receptors and then trigger an abnormal response in the cell. These triggers cause growth at the wrong time, an alteration of metabolism or synthesis of a different product. The best-known triggers are dioxin and dioxin-like chemicals. Dioxin acts through a hormone-like process to initiate entirely new responses.

3 Descriptions of Specific EDCs

In this section, the potential EDCs are grouped by chemical class (see Table 4.1). Descriptions of the EDCs provide a brief description of the chemical, its major uses, the major human exposure routes, health effects, water solubility, environmental persistence, occurrence/detection in water sources, drinking water standards, and statutes that regulate the substance in water. The best available technology (BAT) as determined by laboratory testing for removal of specific EDCs from water is indicated when this has been determined.

Table 4.1 Environmental Endocrine Disruptors [8]

Chemical	Group	Source
Polychlorinated dioxins, polychlorinated biphenyls	Polychlorinated compounds	Incineration, landfill and industrial production
DDT, dieldrin, lindane	Organochlorine pesticides	Agricultural runoff and atmospheric transport
Atrazine, trifluralin, permethrin	Pesticides	Agricultural runoff
Tributyltin	Organotins	Harbors (from antifoulants used to paint the hulls of ships)
Nonylphenol	Alkylphenols	Surfactants found in industrial and municipal effluents
Dibutyl phthalate, butylbenzyl phthalate	Phthalates	Plasticizers found in industrial effluents
Estradiol, estrone and testosterone; ethynyl estradiol	Natural hormones and synthetic steroids	Municipal effluents and agricultural runoff
Isoflavones, lignans, coumestans	Phytoestrogens	Plant material found in pulp mill effluents

3.1 Pesticide Residues

A number of pesticides have been implicated as endocrine disruptors, primarily in aquatic and wildlife species. Agricultural runoff is responsible for the presence of most pesticides found in surface waters. The pesticide concentrations in surface waters tend to be highest after the first storm following application. Pesticides may also enter source water from accidental spills, in wastewater discharges, or as runoff from urban and suburban areas. Because pesticides are known to be potentially highly toxic compounds, the maximum contaminant level (MCL) has been established for each of these substances. These limits were originally established on the basis of known toxicologic effects; however, in the future the MCLs may be set at even lower concentrations if adverse endocrine effects are detected due to their presence [9]. Again, this chapter does not infer that the reader is obligated to attain an MCL; rather this information is presented to demonstrate how future research on EDCs may eventually impact some MCLs.

3.1.1 DDT

DDT is an organochlorine insecticide used mainly to control mosquito-borne malaria. It is the common name of the technical product that is a mixture of three isomers of DDT and contains 65–80% p,p'-DDT. It is very soluble in fats and most organic solvents and practically insoluble in water. In the USA, DDT is currently used only for public health emergencies as an insecticide under Public Health Service supervision and by the US Department of Agriculture (USDA) or military for health quarantine. US Environmental Protection Agency (USEPA) banned the use of DDT in food in 1972 and the use in nonfoods in 1988 [9]. At present no US companies are producing DDT. The primary supporting evidence for adverse health effects in humans comes from an epidemiological study performed by Rogan in North Carolina [10] in which blood levels of DDE (a metabolite of DDT) were determined in pregnant women. Once the blood levels were determined for each woman, neurologic testing was then performed on the infants that were born from these pregnancies. A very strong correlation was found linking increased blood levels of DDE with poor performance of the neurologic tests by these infants [10, 11]. Strong correlation of maternal serum levels of DDE, a metabolite of DDT, with defects in muscular tone and hyporeflexia was observed in their children. More convincing evidence of endocrine effects has been observed in an ecological setting [12–14]. The initial reports were of egg shell thinning in bald eagles as well as vitellogenin (a protein that is normally only produced in the livers of female amphibians and fish) production in male African clawed frogs [12]. Primary exposure routes for humans are inhalation, ingestion, and dermal contact.

In spite of the 1972 ban of DDT in the USA, human exposure to DDT is potentially high due to its prior extensive use and the persistence of DDT and its metabolites in the environment. DDT has been detected in air, rain, soil, water, animal and plant tissues, food, and work environment [9]. Breakdown products in the soil environment are DDE and DDD, which are also highly persistent. Due to its extremely low solubility in water, DDT is mainly retained by soils and soil fractions with higher proportions of soil organic matter. While it is generally immobile or only very slightly mobile, DDT may leach into groundwater over long periods of time. DDT may reach surface waters primarily by runoff, atmospheric transport, drift, or direct application. DDT has been widely detected in ambient surface water sampling in the USA at a median level of one ng/L (nanogram per liter equal part per trillion). DDT is regulated by EPA under the Clean Water Act (CWA). Effluent discharge guidelines and water quality criteria have been set under the CWA [9].

3.1.2 Endosulfan

Endosulfan is a chlorinated hydrocarbon insecticide which acts as a poison for a wide variety of insects and mites on contact. Although it may be used as a wood preservative, it is used primarily on a wide variety of food crops, including tea, coffee, fruits, and vegetables, as well as on rice, cereals, maize, sorghum, or other grains. Human exposure to endosulfan is primarily through breathing air, drinking water, eating food, or working where endosulfan is used. Exposure to endosulfan mainly affects the central nervous system [15]. The effects of long-term/low-dose exposure are unknown. The most convincing evidence of endocrine effects in mammals is taken from laboratory animal studies in which doses of 5 mg/kg/d resulted in reduced sperm counts and altered testicular enzyme levels in male rats [9].

Endosulfan has been found in at least 143 of the 1416 National Priorities List sites identified by the EPA [9]. Although not easily dissolved in water, when released to water, endosulfan isomers hydrolyze readily in alkaline conditions and more slowly in acidic conditions. Endosulfan has been detected at levels of 0.2–0.8μg/L in groundwater, surface water, rain, snow, and sediment samples. Large amounts of endosulfan can be found in surface water near areas of application. The USEPA recommends that the amount of endosulfan in lakes, rivers, and streams should not be more than 74 ppb [9]. Humans can become exposed to endosulfan by drinking water contaminated with it.

3.1.3 Methoxychlor

Methoxychlor is an organochlorine insecticide that is effective against a wide range of pests encountered in agriculture, households, and ornamental plants. It is registered for use on fruits, vegetables, and forage crops. The use of methoxychlor has increased significantly since DDT was banned in 1972 [9]. It is similar in structure to DDT, but it has a relatively low toxicity and relatively low persistence in biological systems. Methoxychlor is not highly soluble in water. Methoxychlor is highly toxic to fish and aquatic invertebrates. Levels of methoxychlor can accumulate in algae, bacteria, snails, clams, and some fish, but it is usually transformed into other substances and rapidly released from their bodies. The most probable routes of exposure for humans are inhalation or dermal contact during home use and ingestion of food or drinking water contaminated with methoxychlor. Short-term exposure above the MCL causes central nervous system depression, diarrhea, and damage to liver, kidney, and heart tissue. Evidence suggests that high doses of technical methoxychlor or its metabolites may have estrogenic effects.

The risk of human exposure via groundwater should be slight, but it may be greater if application rates are very high or if the water table is very shallow. At present the strongest evidence of endocrine effects due to methoxychlor is taken from laboratory studies in which the relatively low dose of 0.5μg/kg/d caused reduced fertility in mice [16, 17].

In an USEPA pilot groundwater survey, methoxychlor was found in a number of wells in New Jersey and at extremely low concentrations in water from the Niagara River, the James River, and an unnamed Lake Michigan tributary [9]. Methoxychlor will most likely reach surface waters via runoff. Methoxychlor was detected in drinking water supplies in rural South Carolina. USEPA set a limit of methoxychlor in drinking water at 0.04 mg/L (see Table 4.2). USEPA advises that children should not drink water containing more than 0.05 mg/L for more than one day and that adults should not drink water containing more than 0.2 mg/L for longer periods of time [9].

Table 4.2 Regulated Chemicals That Cause Endocrine Disruption (U. S. EPA, 8)

Chemical	MCLG mg/L	MCL mg/L	Source	Health Effect
Atrazine	0.003	0.003	Runoff from herbicide used on row crops	Cardiovascular system; reproductive problems
Benzo(a)pyrene (PAHs)	0.0	0.0002	Leaching from linings of storage tanks and distribution lines	Reproductive difficulties; increased risk of cancer
Carbofuran	0.04	0.04	Leaching of soil fumigant used on rice and alfalfa	Problems with blood, nervous system and reproductive system
2,4-D	0.07	0.07	Runoff from herbicide used on row crops	Kidney, liver and adrenal gland problems
1,2-Dibromo-3-chloropropane (DBCP)	0.0	0.0002	Runoff/leaching from soil fumigant used on soybeans, cotton, pineapple and orchards	Reproductive difficulties; increased risk of cancer
Di(2-ethylhexyl) adipate	0.4	0.4	Discharge from chemical factories	General toxic effects; reproductive difficulties
Di(2-ethylhexyl) phthalate	0.0	0.006	Discharge from rubber and chemical factories	Reproductive difficulties; liver problems; increased risk of cancer
Dinoseb	0.007	0.007	Runoff from herbicide used on soybeans and vegetables	Reproductive difficulties
Dioxin (2,3,7,8-TCDD)	0.0	0.00000003	Emissions from incineration; discharge of chemical factories	Reproductive difficulties; increased risk of cancer
Ethylene dibromide	0.0	0.00005	Discharge from petroleum refineries	Problems with liver1 stomach, reproductive system and kidneys; increased risk of cancer
Glyphosate	0.7	0.7	Runoff from herbicide use	Kidney problems; reproductive difficulties
Hexachlorobenzene	0.0	0.001	Discharge from metal refineries and agricultural chemical factories	Liver or kidney problems; reproductive difficulties; increased risk of cancer
Methoxychlor	0.04	0.04	Runoff/leaching from insecticide used on fruits/vegetables/livestock	Reproductive difficulties
Polychlorinated biphenyls (PCBs)	0.0	0.0005	Runoff from landfills; discharge of waste chemicals	Skin changes; thymus gland problems; immune deficiencies; reproductive or nervous system difficulties; increased risk of cancer

(continued)

Table 4.2 (continued)

Chemical	MCLG mg/L	MCL mg/L	Source	Health Effect
Toxaphene	0.0	0.003	Runoff/leaching from insecticide used on cotton and cattle	Kidney, liver, or thyroid problems; increased risk of cancer
1,2,4-Trichloro-benzene	0.07	0.07	Discharge from textile finishing factories	Changes in adrenal glands

3.2 Highly Chlorinated Compounds

3.2.1 Polychlorinated Biphenyls (PCBs)

Polychlorinated biphenyls are a group of manufactured organic compounds that include 209 different chemical forms known as congeners. This high number of many different chemical forms is possible because from one to ten chlorine atoms can attach to the carbon atoms that make up the basic chemical structure of this family of compounds. PCBs are thermally stable and resistant to oxidation, acids, bases, and other chemical agents. PCBs tend to be more soluble in lipid-based solvents than in water; however, among the 209 congeners, there is a wide range of water solubility and lipid solubility with the lesser chlorinated congeners being more water soluble [9].

In the environment, PCBs can be contaminated with dibenzofurans, dioxins, and polychlorinated naphthalenes. Since 1974, all PCB manufacturing has been banned and previous use in electrical capacitors and transformers has been greatly reduced. Because of their chemical-resistant properties, PCBs have persisted in the environment in large quantities despite the manufacturing ban. The primary routes of potential human exposure to PCBs are ingestion of food and water as well as through dermal contact. There is extensive human data which shows a strong association of low birth weights and shortened gestation with PCB exposure in humans [18, 19]. In addition, extensive neurologic testing of children who experienced exposure to PCBs prior to birth revealed impaired motor function and learning disorders [20]. Studies have indicated that PCBs concentrate in human breast milk.

PCB releases from prior industrial uses and the persistence of the compounds in the environment have resulted in widespread water and soil contamination. They have been found in at least 383 of the 1430 National Priorities List sites identified by the USEPA [9]. The PCBs with a high degree of chlorination are resistant to biodegradation and appear to be degraded very slowly in the environment. PCB concentrations in water are higher for the lower chlorinated PCBs because of their greater water solubility. PCBs have been found in runoff, sediments, soil, creek water, leachate, underground oil-water layer, and pond effluents. Concentrations in these locations have ranged from 4 to 440,000μg/L [9]. In water, small amounts of PCBs may remain dissolved, but most adhere to organic particles and sediments. PCBs in water bioaccumulate in fish and marine mammals and can reach levels several orders of magnitude higher than levels found in the water [21–23] . USEPA regulates PCBs under the CWA and has established water quality criteria and toxic pollutant effluent standards. Based on the carcinogenicity of PCBs, USEPA published a MCL Goal (MCLG) for PCBs at zero and the MCL of 0.5μg/L (0.5 ppb) under the SDWA (see Table 4.2).

3.2.2 Dioxin (2,3,7,8-tetrachlorodibenzo-p-dioxin TCDD)

Dioxin is considered an EDC on the basis of its effects that occur during pregnancy which result in many malformations observed in the offspring of many species including humans [24, 25]. Dioxin is a contaminant formed during the manufacture of 2,4,5-trichlorophenoxyacetic acid (2,4,5-T), an herbicidal compound that comprised about 50% of the defoliant Agent Orange, and 2,4,5-T derivatives, as well as other chemicals synthesized using 2,4,5-trichlorophenol. Dioxins may also be formed during incineration of chlorinated industrial compounds such as plastic and medical waste.

Dioxin is one of the most acutely toxic compounds synthesized by modern chemistry. TCDD is the most toxic member of the 75 dioxins that exist and is the one most studied. It is almost insoluble in water. TCDD is stable in water, dimethylsulfoxide, 95% ethanol, or acetone. It can undergo a slow photochemical and bacterial degradation, though normally it is extremely stable. Dioxin is degraded when heated in excess of 500 °C or when exposed to ultraviolet radiation under specific conditions. TCDD has no known commercial applications but is used as a research chemical. TCDD has been found in at least 91 of 1467 National Priorities List sites identified by the USEPA [9]. Dioxins are widespread environmental contaminants. They bioaccumulate throughout the food web because of their lipophilic properties and slow metabolic destruction. The primary source of dioxin exposure to humans is from food.

3.2.3 Furan

Furan is classified as cyclic, dienic ether; it is a colorless, flammable liquid. It is insoluble in water but is soluble in alcohol, ether, and most common organic solvents. Furan is used primarily as an intermediate in the synthesis and production of other organic compounds, including agricultural chemicals (insecticides), stabilizers, and pharmaceuticals [9]. Furan is a carcinogen [26] and its primary route of potential human exposure is inhalation.

Furan was detected in 1 of 63 industrial effluents at a concentration of less than 10μg/L [9]. Furan was detected in a creek in the Niagara River watershed and in the Niagara River.

3.3 Alkylphenols and Alkylphenol Ethoxylates

Nonylphenol (NP) and octylphenol are the largest volume alkylphenol products manufactured in the USA. Alkylphenols (APs) such as nonylphenol and octylphenol are mainly used to make alkylphenol ethoxylate (APE) surfactants. These surfactants are the primary active ingredients in industrial chemicals that are used as cleaning and sanitizing agents. Nonylphenol ethoxylates (NPEs) account for approximately 80% of total APE use with total US production exceeding 500 million pounds per year [9]. Alkylphenols are also used as plasticizers, in the preparation of phenolic resins, polymers, heat stabilizers, antioxidants, and curing agents. APEs do not break down completely in wastewater treatment plants or in the environment. The most widely used NPEs have nine- or ten-member carbon chains attached to the ethoxylate group. Thus, the great majority of NPEs in use are easily dissolved in water. Human exposure to APs and APEs may occur through contaminated drinking water that has been extracted from polluted waters. At present there is no conclusive evidence that APs or APEs cause adverse health effects in humans; however, there are many reports of alkylphenols causing production of a female-associated liver protein, vitellogenin, in male fish [27].

Investigations of NP levels in rivers have found values varying between 2μg/L in the Delaware River in Philadelphia to 1000μg/L in the Rhine and 1000μg/L in a tributary of the Savannah River [9, 28]. Drinking water is frequently taken from rivers and can easily become contaminated with alkylphenols. Analysis of many drinking water samples in the USA has found an overall average concentration of alkylphenolic compounds of 1μg/L. Studies in the USA show NPE removal from wastewater ranging from 92 to 99% with minor seasonal variations. NPE concentrations in discharges after treatment are reportedly low, varying between 50 and 200μg/L. Draft USEPA water quality guidelines for nonylphenol in freshwater are 6.6μg/L water (four-day average) and 25μg/L (one-hour average), and in saltwater, they are 1.6μg/L (four-day average) and 6.2μg/L (one-day average).

3.4 Plastic Additives

3.4.1 Bisphenol A

Bisphenol A is an industrial chemical used to synthesize epoxy resins or polycarbonate plastic. Human exposure to the potential endocrine disrupting effects of bisphenol A may occur when this chemical leaches out of the plastic due to incomplete polymerization or breakdown of the polymer upon heating. Polycarbonates are commonly used for food and drink packaging materials, and infants are the subgroup of the population that is most highly exposed to this compound. Bisphenol A is also used in plastic dental fillings.

Bisphenol A is a solid which has low volatility at ambient temperatures. It has a water solubility of 120–300 mg/L. Its water solubility increases with alkaline pH values. Releases of bisphenol A into the environment are mainly in wastewater from plastics-producing industrial plants and from landfill sites that contain large quantities of plastics. Bisphenol A does not bioaccumulate in aquatic organisms to any appreciable extent. If released into acclimated water, bisphenol A would biodegrade. In untreated water, bisphenol A may biodegrade after a sufficient adaptation period, it may adsorb extensively to suspended solids and sediments, or it may break down upon exposure to light [9].

3.4.2 Diethyl Phthalate (DEP)

Diethyl phthalate is a synthetic substance that is commonly used to increase the flexibility of plastics used to make toothbrushes, automobile parts, tools, toys, and food packaging. It is also used in cosmetics, insecticides, and aspirin. DEP can be released fairly easily from these products since it is not part of the polymer. Plastic materials containing DEP in waste disposal sites constitute the major reservoir of DEP in the environment. If released to water, DEP is expected to undergo aerobic biodegradation. Humans are exposed to DEP through consumer products and plastics, contaminated air, or contaminated drinking water and foods.

There is evidence which shows a strong correlation with impaired reproductive performance in multigeneration studies in rodents [29]; however, endocrine effects associated with DEP exposure in humans have not been reported.

DEP has accumulated and persisted in the sediments of the Chesapeake Bay for over a century. DEP has been detected in surface water samples from Lake Ponchartrain and the lower Tennessee River, as well as other industrial river basins. Surface water samples collected along the length of the Mississippi River contained DEP in significant concentrations. DEP has been detected in groundwater in New York State public water system wells, near a solid waste landfill site in Norman, OK, and at sites in Fort Devens, MA; Boulder, CO; Lubbock, TX; and Phoenix, AZ. DEP has been identified in drinking water in the following cities: Miami, Philadelphia, Seattle, Lawrence, New York City, and New Orleans [9].

3.4.3 Di (2-ethylhexyl) Phthalate (DEHP)

Di-(2-ethylhexyl) phthalate is a manufactured chemical that is used primarily as one of several plasticizers in polyvinyl chloride (PVC) resins that make plastics more flexible. It is the most commonly used of a group of related chemicals called phthalates or phthalic acid esters. DEHP is also used in inks, pesticides, cosmetics, and vacuum pump oil. DEHP is everywhere in the environment because of its use in plastics in large quantities, but it evaporates into air and dissolves in water at very low rates. Colon et al. [30] identified the presence of phthalate esters in the serum of young Puerto Rican girls with premature breast development. The primary routes of potential human exposure to DEHP are inhalation, ingestion, and dermal contact in occupational settings and from air pollution, water pollution, food consumption, and food wrapped in PVC. It is easily dissolved in body fluids such as saliva and plasma. DEHP is biodegradable, but it tends to partition into sediment where it is relatively persistent. It also tends to bioconcentrate in aquatic organisms. Because of its low vapor pressure, human exposure to DEHP in either water or air appears to be minimal.

DEHP has been detected frequently in surface water, groundwater, and finished drinking water in the USA at concentrations in the low μg/L range. Groundwater in the vicinity of hazardous waste sites may be contaminated with DEHP. EPA regulates DEHP under the CWA and the SDWAA. DEHP is included on lists of chemicals for which water quality criteria have been established under the CWA. EPA [9] classifies DEHP as a water priority pollutant and has set the MCL Goal at zero. EPA has set the MCL at six parts DEHP per billion parts of drinking water (6μg/L).

4 Water Treatments for EDC Removal

Water suppliers use a variety of treatment processes to remove contaminants from drinking water. Individual processes may be arranged as series of processes applied in a sequence. Water utilities select a treatment train that is most appropriate for the contaminants found in the source water [31–35]. The most commonly used physicochemical processes include coagulation and flocculation, sedimentation, filtration, and disinfection for surface water [36]. Some treatment trains also include ion exchange and adsorption [37]. These conventional processes are inefficient for substantially reducing certain pesticide concentrations and other EDCs.

The processes described below and in the following sections can be used for removal of EDCs as specified either individually or as a class of compounds. The feasibility of using the various techniques will depend on the size of the system and the cost effectiveness. The two major concerns regarding technologies for small systems are affordability and technical complexity (which determine the needed skills for the system operators).

4.1 Granular Activated Carbon (GAC)

Activated carbon is similar to charcoal in composition, but its surface has been altered to enhance its sorption properties. Activated carbon is made from a variety of materials including wood, coal, peat, sawdust, bone, and petroleum distillates. Activated carbon produced from wood and coal is the most commonly used variety in drinking water treatment plants. The base carbon material is dehydrated then carbonized through slow heating in the absence of air. It is then activated by oxidation at high temperatures (200–1000 °C), resulting in a highly porous, high surface area per unit mass material [38]. The activation process is considered a two-step procedure in which amorphous material is burned off and pore size is increased. Typically, GACs have surface areas ranging from 500 to 1400 m^2/g.

GAC treatment removes contaminants via the physical and chemical process of sorption. The contaminants accumulate within the pores, and the greatest efficiency is attained when the pore size is only slightly larger than the material being adsorbed. Removal efficiencies for many organic contaminants are good to excellent. Water quality parameters such as dissolved organic matter, pH, and temperature can significantly affect the removal efficiency of GAC. However, for GAC treatment of drinking water, it is necessary to reduce the total organic carbon (TOC) of the treated water through the preliminary steps of coagulation/filtration [39] before treatment with GAC. Its removal efficiencies change drastically once the bed nears exhaustion, as contaminant breakthrough occurs. GAC beds can be reactivated by removing the granular carbon from the water treatment chambers, drying the material then placing it in large furnaces that heat the material to 1200–1400 °F. This heating process removes any residual of contaminants from the pores and again enlarges the pore size. This feature and the high temperatures needed to attain reactivation should be kept in mind when considering claims of some manufacturers that flushing point-of-use (POU) GAC filters with hot water will reactivate units or increase operating efficiency. The increased temperatures that are reached with hot water do not in any manner achieve reactivation.

The performance of GAC for specific contaminants is determined in the laboratory by trial runs and is performed one chemical at a time. The following description is presented to provide the reader with a basic understanding of how the relative capacity of activated carbon to remove a chemical from water (a liquid phase) is determined. Data are gathered within a laboratory setting and determined on the basis of one chemical at a time. The Freundlich equation can be used to indicate the efficiency of GAC/PAC treatment. The Freundlich equation is expressed as

$$Q_e = K \times C_e^n \tag{4.1}$$

where:

Q_e = equilibrium capacity of the carbon for the target compound, µg/g
C_e = equilibrium liquid-phase concentration of the target com/pound, µg/L
K = Freundlich coefficient in $(\mu g/g)(L/\mu g)^{1/n}$
$1/n$ = Freundlich coefficient, dimension-less units

Equation (4.1) can be converted into a linear form as shown below:

$$\log Q_e = \log K + \frac{1}{n} \log C_e \tag{4.2}$$

Plotting of Eq. (4.2) on a logarithmic paper will yield the value of K from the intercept and $1/n$ from the slope of the straight line. The K values that are determined for each chemical are a means of expressing the "ability" of a particular GAC to remove a chemical.

Typically when *K* values that are greater than 200 are attained the process is considered to be economically feasible [9]. In addition, the process of GAC can be fine-tuned, that is, certain basic parameters such as pH, temperature, or choice of carbon source can be altered to increase efficiency of the process when certain critical contaminants such as pesticides must be removed.

Careful monitoring and testing are required to ensure that all contaminants are removed. The carbon media must be replaced regularly. The replacement intervals depend on:

1. Type of contaminant
2. Concentration
3. Rate of water usage
4. Type of carbon used in the system

There is potential for bacterial growth on the adsorbed organic chemicals; routine maintenance must be performed. When POU devices are used for compliance for small systems, programs for long-term operation, maintenance, and monitoring must be provided by the water utility.

4.2 Powdered Activated Carbon (PAC)

Powdered activated carbon (PAC) also functions by adsorption of contaminants from water onto a solid phase material, in this case powdered carbon. PAC differs from GAC in that the powdered carbon is added to the water in a large tank, a period of time is provided for adsorption of the contaminants to occur, and then the powdered carbon is later removed in a filtration process [40]. This process also differs from GAC in that PAC needs to be added continually to the process; however, the process is less expensive and less technically demanding, but it is more labor intensive. PAC is more adaptable to short-term applications rather than as a continual use process. For contaminants such as pesticides which are mostly used during a six-week period in late spring and summer, PAC may be a particularly useful choice [9]. The water being treated comes into contact with much less carbon material per unit volume treated, so the process is not as efficient as GAC.

GAC is the BAT (best available technology) for removal of all of the selected EDCs that were discussed in the previous section. However, since other technologies are used in the multistep process of drinking water treatment, a brief discussion is included for those processes that enhance the performance of GAC.

4.3 Coagulation/Filtration

Coagulation/filtration processes involve the addition of chemicals like iron salts and aluminum salts, with and without anionic, cationic, or anionic-cationic polymers that coagulate and destabilize particles suspended in the water [39]. The suspended particles are ultimately removed via clarification and/or filtration. Conventional filtration includes pretreatment steps of chemical coagulation, rapid mixing, and flocculation, followed by floc removal via sedimentation or flotation [36]. After clarification, the water is filtered using common filter media such as sand, dual-media, and tri-media. Direct filtration has several effective variations, but all include a pretreatment of chemical coagulation, followed by rapid mixing. The water is filtered through dual- or mixed-media using pressure or gravity filtration units [31, 35, 36].

4.4 Lime Softening

In the lime-softening (LS) process, the pH of the water being treated is raised sufficiently to precipitate calcium carbonate and, if necessary, magnesium hydroxide to reduce water hardness. Some materials in solution are removed by incorporating them into the particles of calcium carbonate/magnesium hydroxide and coprecipitate, which are then separated by settling, filtration, or flotation [31, 35, 36].

5 Point-of-Use/Point-of-Entry Treatments

The SDWA identifies both point-of-entry (POE) and points-of- use (POU) treatment units as options for compliance technologies for small systems. A POU treatment device treats only the water at a particular tap or faucet, resulting in other taps in the facility serving untreated water. POU devices are typically installed at the kitchen tap. POU devices are listed as compliance technologies for inorganic contaminants, synthetic organic contaminants, and radionuclides [34]. POU devices are not listed for volatile organic contaminants because they do not address all routes of exposure. POE treatment units treat all of the water entering a facility (household or other building), resulting in treated water from all taps. POE devices are still considered emerging technologies because of waste disposal and cost considerations.

POE and POU treatment units often use the same technological concepts as those used in central treatment processes, but on a much smaller scale. Technologies that are amenable to the POU and POE scale treatment include activated alumina, GAC, reverse osmosis, ion exchange, and air stripping [37].

When POU and POE units are used by a public water system to comply with the National Primary Drinking Water Regulations (NPDWRs), the SDWA requires that the units be owned, controlled, and maintained by the public water system or by a person under contract with the public water system. This is to ensure that the units are properly operated and maintained to comply with the MCL or treatment techniques. This will also ensure that the units are equipped with the required mechanical warnings to automatically alert the customers to the occurrence of operational problems.

6 Water Treatment Techniques for Specific EDC Removal

The EDCs addressed in this chapter that are included in the NPDWRs as drinking water contaminants are methoxychlor, DDT and DDE, endosulfan, PCBs, DEP, and DEHP [41–49]. The EDCs in this section are grouped by chemical class. Removal techniques for the EDCs not listed in the NPDWRs will be based on removal of similar contaminants that are listed.

The treatment processes are described with considerations of advantages, limitations, and special considerations. The actual choice of a process to include in a treatment train will ultimately depend on:

1. The source water quality
2. The nature of the contaminant to be removed
3. The required quality of the finished water
4. The size of the drinking water system

6.1 *Methoxychlor*

The BAT for removal of methoxychlor from drinking water is GAC. Steiner and Singley [50] have tested a wide range of water treatment processes and found GAC to be the most efficient for removal of methoxychlor. They found that over a broad range of concentrations (ranging from 1 mg/mL to 25 mg/mL), the GAC process could remove sufficient quantities of methoxychlor so that the finished water met MCL requirements which is 0.1 mg/mL.

6.2 *Endosulfan*

The BAT for removal of endosulfan from drinking water is GAC. In the USEPA report, Carbon Adsorption for Toxic Organics [51], the following *K* values, as determined by the Freundlich equation and actual test, were determined:

Type of endosulfan	K (μg/g)(L/μg)$^{1/n}$
Alpha-endosulfan	6135
Beta-endosulfan	1990
Endosulfan sulfate	2548

Table 4.3 Isotherm Constants for Selected EDCs (U. S. EPA, 9)

	Isotherm Constants		Calculated Value
Chemical	K mg/g $(L/mg)^{1/n}$	$1/n$	μg/g $(L/\mu g)^{1/n}$
Alpha-endosulfan	194	0.50	6135
Beta-endosulfan	615	0.83	1990
Endosulfan sulfate	686	0.81	2548
DDT	332	0.50	10,499
DDE	232	0.37	18,000
Diethyl phthalate (DEP)	110	0.27	17,037
Diethylhexyl phthalate (DEHP)	11,300	1.50	8308
PCB-1221	242	0.70	1922
PCB-1232	630	0.73	4067
Nonylphenol	250	0.37	19,406

For small system compliance, GAC, POU-GAC, and PAC can be used to remove endosulfan from drinking water supplies (see Table 4.3) [9, 51, 52].

6.3 DDT

The BAT for removal of DDT from drinking water is GAC. In the USEPA report [42], the following *K* values, as determined by the Freundlich equation and actual test, were determined: DDT has a *K* value of 10,449μg/g $(L/\mu g)^{1/n}$ which is sufficiently above the cutoff point of 200μg/g $(L/\mu g)^{1/n}$ to be judged an effective treatment method and DDE (a DDT metabolite with endocrine activity) of 18,000μg/g $(L/\mu g)^{1/n}$.

6.4 Diethyl Phthalate

The BAT for removal of diethyl phthalate from drinking water is GAC. In the USEPA report [51], the following *K* value, as determined by the Freundlich equation and actual test for diethyl phthalate, yielded a *K* value of 17,037μg/g $(L/\mu g)^{1/n}$.

6.5 Di-(2ethylhexyl) Phthalate (DEHP)

The BAT for removal of DEHP from drinking water is GAC. In the USEPA report [42], the following *K* value, as determined by the Freundlich equation and the test, was determined. DEHP has a *K* value of 8308μg/g $(L/\mu g)^{1/n}$ which is one of the highest values established among the 130 compounds that they tested; GAC is very effective for the removal of DEHP from drinking water.

6.6 PCBs

In the USEPA report [51], two studies were reported for PCB-1221 and PCB-1232. The K value determined for PCB-1221 was 1922μg/g $(L/\mu g)^{1/n}$, and the K value for PCB-1232 was 4067μg/g $(L/\mu g)^{1/n}$. Both mixtures are among the lesser chlorinated groups containing 21 and 32% chlorine, respectively. Relative to other PCB mixtures, they are more hydrophilic and hence would have lower K values than the commercial PCB mixtures, Aroclor 1242, 1248, 1254, and 1260. The most troublesome PCB environmental mixtures tend to be derivatives of this later group of compounds; therefore, GAC should be a very effective method for removal of environmental PCB compounds from drinking water.

6.7 Dioxin

Dioxin is not water soluble; hence it is not likely to be present in untreated drinking water unless it would be attached to sediment in raw water. Because most conventional water treatment methodologies such as coagulation-sedimentation and filtration are effective in removing sediment, it is likely that these processes would be very effective in the removal of the contaminant, dioxin.

6.8 Alkylphenols and Alkylphenol Ethoxylates

GAC is best used for removal of these contaminants from drinking water. Previous laboratory-scale testing for removal of nonylphenol with GAC has yielded K values of 19,406 at a water pH of 7.0 [9]. For consistency of removal of synthetic organic chemicals, GAC, POU-GAC, and PAC are recommended for small system compliance. GAC devices include pour-through for treating small volumes, faucet-mounted for POU, in-line for treating large volumes at several faucets, and high volume commercial units for treating community water supply systems. Careful selection of the type of carbon is based on the specific contaminants in the water and the manufacturer's recommendations. Site-specific conditions may affect the percentage removal using these techniques, including the presence of "competing" contaminants. Source water-specific testing will be needed to ensure adequate removal. For GAC, surface waters may require pre-filtration. PAC is most applicable to those systems that already have a process train including mixing basins, flocculation, flotation clarification (or sedimentation clarification), and filtration.

Glossary [53–55]

Endocrine disrupting compounds (EDC's) They are chemicals that either mimic endogenous hormones, interfere with pharmacokinetics or act by other mechanisms. Such adverse effects as compromised reproductive fitness, functional or morphological birth defects, cancer, and altered immune functions, among others, have been reported.

Endocrine disruptors (a) They are substances that are not produced in the body but act by mimicking or antagonizing natural hormones. (b) They are chemicals that either mimic endogenous hormones, interfere with pharmacokinetics, or act by other mechanisms resulting in adverse effects such as compromised reproductive fitness, functional or morphological birth defects, cancer, and altered immune functions. (c) They are environmental pollutants with adverse hormonal effects on wildlife and humans, such as pesticides (methoxychlor), surfactants (nonylphenol), plasticizers (diethylphthalate), and organohalogens (polychlorinated biphenyls or PCBs and dioxin).

Endocrine systems They are also referred to as hormone systems and are found in all mammals, birds, fish, and many other types. The endocrine system regulates all biological processes in the body from conception through adulthood and into old age, including the development of the brain and nervous system, the growth and function of the reproductive system, as well as the metabolism and blood sugar levels.

Granular activated carbon adsorption Granular activated carbon (GAC) is used in wastewater substances GAC systems that generally consist of vessels in which the carbon is placed, forming a "filter" bed. These systems can also include carbon storage vessels and thermal regeneration facilities. Vessels are usually circular for pressure systems or rectangular for gravity flow systems. Once the carbon adsorptive capacity has been fully utilized, it must be disposed of or regenerated. Usually multiple carbon vessels are used to allow continuous operation. Columns can be operated in series or parallel modes. All vessels must be equipped with carbon removal and loading mechanisms to allow for the removal of spent carbon and the addition of new material. Flow can be either upward or downward through the carbon bed. Vessels are backwashed periodically. Surface wash and air scour systems can also be used as part of the backwash cycle.

Hormone A hormone is any member of a class of signaling molecules, produced by glands in multicellular organisms, that are transported by the circulatory system to target distant organs to regulate physiology and behavior.

Powdered activated carbon adsorption Powdered activated carbon (PAC) is used in both water and wastewater treatment. For water treatment, PAC systems generally involve of dosing powdered activated carbons in slurry form to rapid mixers and flocculators (for mixing and flocculation, and adsorption) before reaching clarifiers for PAC-floc separation and clarifier effluent discharge. For wastewater treatment, powdered activated carbon in slurry form is dosed to either mixing/flocculation chambers (if they exist) or to an aeration basin directly to be mixed with mixed liquor for adsorption reactions. The spent PAC again is recovered or removed by a subsequent clarifier (either sedimentation clarifier or flotation clarifier). When PAC is used in conjunction with a flotation clarifier, it is also called “adsorption flotation”.

References

1. U. S. FDA (2018) Endocrine Disruptor Knowledge Base (EDKB). U. S. Food and Drug Administration, Web Site http://edkb.fda.gov/
2. European Commission (2018) Endocrine disruptor web site, http://europa.eu.int/comm/environment/endocrine
3. IPCS (2018) Summary and details: scientific facts on endocrine disruptors, International Program on Chemical Safety, Green Facts, http://www.greenfacts.org/endocrine-disruptors/
4. Naz RK (ed) (1999) Endocrine disruptors: effects on male and female reproductive system. CRC Press, Boca Raton, FL, USA
5. Chhanda G (2000) Reproductive malformation of the male offspring following maternal exposure to estrogenic chemicals. Proc Soc Exp Biol Med 224:61–68
6. Triendl R (2001) Genes may solve hormone-disrupter debate, Robert. Nature 209
7. U. S. EPA (2017) Endocrine Disruptor Screening and Testing Advisory Committee (EDSTAC) final report. Office of Prevention, Pesticides and Toxic Substances, U. S. Environmental Protection Agency, January (1999). http://www.epa.gov/scipoly/oscpendo/history/finalrpt.htm
8. Jesperson K (2003) Endocrine disruptors: what are they doing to you, National Drinking Water Clearing House. *On Tap* 2(4):16–22, and 49
9. U. S. EPA (2001) Removal of endocrine disruptor chemicals using drinking water treatment processes, technology transfer report # EPA/625/R-00/015. The National Risk Management Research Laboratory, Office of Research and Development, U. S. Environmental Protection Agency, Cincinnati, OH, USA
10. Rogan W, Gladen B et al (1996) Neonatal effects of transplacental exposure to PCBs and DDE. J Pediatr 109:335–341
11. Stehr-Green PA (1989) Demographic and seasonal influences on human serum pesticide residue levels. J Toxicol Environ Health 27:405–421
12. Palmer BO, Palmer SK (1995) Vitellogenin production by xenobiotic estrogens in the red-eared turtle and African clawed frog. Environ Health Perspect 103(Suppl. 4):19–25
13. Colborn T, vom Saal FS, Soto AM (1993) Developmental effects of endocrine-disrupting chemicals in wildlife and humans. Environ Health Perspect 101:378–384
14. Guillette LJ, Crain DA (1996) Endocrine-disrupting contaminants and reproductive abnormalities in reptiles. Comments Toxicol 5:381–399
15. ATSDR (2018) Agency for toxic substances and disease registry ToxFAQs. Endosulfan (1995). http://www.atsdr.cdc.gov/tfacts41.html
16. Cummings AM, Gray LE (1989) Antifertility effect of Methoxychlor in female rats: dose- and time-dependent blockade of pregnancy. Toxicol Appl Pharm 97:454–462
17. vom Saal FS, Nagel SC (1995) Estrogenic pesticides: binding relative to estradiol in MCF-7 cells and exposure during fetal life on subsequent territorial behavior in male Mice. Toxicol Lett 77:343–350
18. Taylor PR, Lawrence CE, Hwang HL, Paulson AS (1984) Polychlorinated biphenyls: influence in birth weight and gestation. Am J Public Health 74:1153–1154
19. Patandin SC, Koopman-Esseboom MA, de Ridder N, Weisglas-Kuperus G, Sauer PJ (1998) Effects of environmental exposure to polychlorinated biphenyls and dioxins on birth size and growth in Dutch children. Pediatr Res 44:538–545

20. Jacobson JL, Jacobson SW, Humphrey HE (1996) Intellectual impairment in children exposed to polychlorinated biphenyls *In Utero*. N Engl J Med 335:783–789
21. Gladen BC, Rogan WJ (1991) Effects of perinatal polychlorinated biphenyls and dichlorodiphenyl dichloroethene on later development. J Pediatr 113:991–995
22. Huisman MC, Koopman-Esseboom V, Fidler M, Hadders-Algra CG, can der Paauw LG, Tinstra N, Weisglas-Kuperus PJ, Sauer BC, Boersma ER (1995) Perinatal exposure to polychlorinated biphenyls and dioxins and its effect on neonatal neurological development. Early Hum Develop 41:111–127
23. Brouwer A, Longnecker MP, Birnbaum LS, Cogliano J, Kostyniak P, Moore J, Schantz S, Winneke G (1999) Characterization of potential endocrine-related health effects at low-dose levels of exposure to PCBs. Environ Health Perspect 107(suppl 4):639–649
24. Tohyama C (2004) Mechanisms of dioxin toxicity: implications for assessing risk to humans, conference on Environmental Endocrine Disruptors. Colby-Sawyer College, New London, CT, USA, pp 6–11
25. U.S. Department of Health and Human Services (2001) 2,3,7,8-Tetracblorodibenzo p-Dioxin (TCDD); DIOXIN, CAS No.1746-01-6, Public Health Service National Toxicology Program, USA, 9th Report on Carcinogens, January
26. U. S. Department of Health and Human Services (2001) Furan, CAS No. 110–00-9. 9th Report on Carcinogens, Public Health Service National Toxicology Program, USA, January
27. Jobling S, Reynolds T, White R, Parker MG, Sumpter JP (1995) A variety of environmentally persistent chemicals, including some phthalate plasticizers, are weakly estrogenic. Environ Health Perspect 103(6):582–587
28. Maguire RJ (1999) Review of the persistence of nonylphenol and nonylphenol ethoxylates in aquatic environments. Water Qual Res J Can 34(1):37–78
29. Hauser R (2004) Phthalates: understanding human exposure and health risks, Conference on Environmental Endocrine Disruptors. Colby-Sawyer College, New London, CT, USA, pp 6–11
30. Colón I, Caro D, Bourdony CJ, Rosario O (2000) Identification of phthalate esters in the serum of young Puerto Rican girls with premature breast development. Environ Health Perspect 108:9
31. AWWA, ASCE (1998) Water treatment plant design, 3rd edn. McGraw-Hill Company, Inc, New York, USA
32. U. S. EPA (1990) Technologies for upgrading existing or designing new drinking water treatment facilities, EPA/625/4–89/023. U. S. Environmental Protection Agency, USA
33. U. S. EPA (1997) Water on tap: a consumer's guide to the nation's drinking water, EPA/815/K-97/002. U. S. Environmental Protection Agency, USA
34. U. S. EPA (1998) Small system compliance technology list for the non-microbial contaminants regulated before 1996, EPA/815/R-98/002. U. S. Environmental Protection Agency
35. U. S. EPA (1999) Drinking water treatment, EPA/810/F-99/013. U. S. Environmental Protection Agency, USA
36. Wang LK (2005) In: Hung YT, Shammas NK (eds) Physicochemical treatment processes. The Humana Press, Inc., Totowa, NJ, USA
37. Wang, L. K, Hung, Y. T. and Shammas, N. K. (eds.) Advanced physicochemical treatment processes, The Humana Press, Inc., Totowa, NJ, USA (2005)
38. Hung YT, Lo HH, Wang LK, Taricska JR, Li KH (2005) Granular activated carbon adsorption. In: Wang LK, Hung YT, Shammas NK (eds) Physicochemical treatment processes. The Humana Press, Inc., Totowa, NJ, USA
39. Shammas NK (2005) Coagulation and flocculation. In: Wang LK, Hung YT, Shammas NK (eds) Physicochemical treatment processes. The Humana Press, Inc., Totowa, NJ, USA
40. Hung YT, Taricska JR, Lo HH, Wang LK, Li KH (2005) Powdered activated carbon adsorption. In: Wang LK, Hung YT, Shammas NK (eds) Advanced physicochemical treatment processes. The Humana Press, Inc., Totowa, NJ, USA
41. Endocrine Disruptor knowledge Base (EDKB), Resources to Predict Estrogen and Androgen Activity, National Center for Toxicological Research, U.S. Food and Drug Administration, USA (2014)

42. NIEHS (2018) Endocrine disruptors. National Institute of Environmental Health Sciences, http://www.niehs.nih.gov/health/topics/agents/endocrine
43. WHO/UNEP (2012) State of the science of endocrine disrupting chemicals – an assessment of the state of the science of endocrine disruptors. The United Nations Environment Programme and World Health Organization
44. NRDC (2018) Endocrine disruptors. Natural Resources Defense Council, NRDC Home WebSite http://www.nrdc.org/health/effects/qendoc.asp
45. EWG (2018) Dirty dozen endocrine disrupters. Environmental Working Group, EWG Home WebSite, http://www.ewg.org/research/dirty-dozen-list-endocrine-disruptors
46. Diamanti-Kandarakis E, Bourguignon J-P, Giudice LC, Hauser R, Prins GS, Soto AM, Zoeller RT, Gore AC (2009) Endocrine-disrupting chemicals: an Endocrine Society Scientific Statement. Endocr Rev 30(4):293–342
47. Fagin D (2012) Toxicology: the learning curve. Nature Int Weekly J Sci 490:7421
48. U.S. FWS (2014) Endocrine (hormone) disrupters. U.S. Fish and Wildlife Service, Environmental Quality Home Site, Last updated 17 Apr
49. U. S. EPA (2018) Endocrine Disruptor Screening Program (EDSP). U. S. Environmental Protection Agency, EDSP Home WebSite, http://www.epa.gov/endo/pubs/edspoverview/whatare.htm
50. Steiner J, Singley J (1979) Methoxychlor removal from potable water. J Am Water Works Ass 17:284–286
51. Dobbs and Cohen (1980) Carbon adsorption for toxic organics, report # EPA/600/8-80/023. U. S. Environmental Protection Agency, Cincinnati, OH, USA
52. Kamrin M, Hayden N, Christian B, Bennack D, D'Itri F (1991) water quality: home water treatment using activated carbon, WQ-13, Cooperative Extension Service, Purdue University
53. Wang LK, Wang MHS, Hung YT, Shammas NK (eds) (2020) Environmental and resources engineering. Springer, New York
54. Wang LK, Wang MHS, Shammas NK, Aulenbach DB (eds) (2020) Environmental flotation engineering. Springer, New York
55. Wang LK, Wang MHS, Hung YT, Shammas NK (eds) (2020) Integrated natural resources management. Springer, New York

Chapter 5
Cooling and Reuse of Thermal Discharges

Yung-Tse Hung, James Eldridge, Jerry R. Taricska, Kathleen Hung Li, and William W. Shuster

Contents

1 Introduction 198
2 Cool Ponds 199
2.1 Mechanism of Heat Dissipation (Cooling) 199
2.2 Design of Cooling Ponds 200
3 Cooling Tower 211
3.1 Mechanism of Heat Dissipation in Cooling Towers 212
3.2 Type of Towers 213
3.3 Natural Draft Atmospheric Cooling Towers 215
3.4 Natural Draft, Wet Hyperbolic Cooling Towers 218
3.5 Example 1 220
3.6 Hybrid Draft Cooling Towers 220
3.7 Induced-Mechanical or Forced Draft Wet Cooling Towers 220
3.8 Cooling Tower Performance Problems 221
Glossary 226
References 227

Y.-T. Hung (✉)
Department of Civil and Environmental Engineering, Cleveland State University, Cleveland, OH, USA
e-mail: yungtsehung@gmail.com

J. Eldridge
Sales Department, Lantec Product, Agoura Hills, CA, USA
e-mail: eldridge@lantecp.com

J. R. Taricska
Environmental Consultant, Naples, FL, USA
e-mail: drjerrytaricska@gmail.com

K. H. Li
Technical Writing Consultant, Austin, TX, USA
e-mail: kathyli31@yahoo.com

W. W. Shuster
Department of Chemical and Environmental Engineering, Rensselaer Polytechnic Institute, Troy, NY, USA

L. K. Wang, M. -H. S. Wang, Y. -T. Hung, N. K. Shammas (eds.),
Environmental and Natural Resources Engineering, Handbook of Environmental Engineering 19, https://doi.org/10.1007/978-3-030-54626-7_5

Abstract Thermal pollution is produced by industries such as electric power plants, pulp and paper mills, chemical facilities, and other process industries that use water and subsequently discharge water with elevated temperature. The receiving streams, rivers, lakes, and other waters can dramatically alter the native environment. Often, elevated water temperature will be detrimental to native species of plants and animals. As water warms, the solubility of oxygen decreases. In addition to promoting competitive species of plants and animals, warmed waters may lead to aesthetic and odor problems if anaerobic conditions are created. For these and many other reasons, the discharge of warm process water is widely regulated by governmental agencies in developed nations. The authors introduce both cooling ponds and cooling towers in detailed. The cooling pond topics covered are mechanism of heat dissipation, cooling pond types and design, completely mixed cooling, recirculation cooling, surface temperature prediction, longitudinal mixing, heat dissipation evaluation, once-through cooling ponds, seasonal weather variation, pond location, cooling, etc. The cooling tower topics covered are mechanism of cooling tower's heat dissipation, cooling tower types and designs, natural draft atmospheric cooling towers, natural-draft wet hyperbolic cooling towers, design examples, hybrid draft cooling towers, induced, mechanical or forced wet cooling towers, cooling tower performance, operational problems and solutions, and reuse of thermal discharge.

Keywords Cooling ponds · Cooling towers · Reuse of thermal discharges · Heat dissipation mechanism · Types of cooling ponds · Types of cooling towers · Design · Surface heat transfer · performance · Natural draft atmospheric cooling towers · Natural-draft wet hyperbolic cooling towers

Nomenclature

a	contact area/tower volume, 1/ft or 1/m
A	surface area of pond, ft or m^2
β	slope of the saturated vapor pressure curve between the dew point temperature and the pond surface temperature, mm Hg °C
C1	Bowen's conduction-evaporation coefficient, 0.468 mm Hg °C
C_p	specific heat of water, 1.0 cal/g °C
d	average depth of flow path, ft or m
dq_t/dt	net rate of surface heat exchange, kcal/m^2 d
E	equilibrium temperature, °C
E	equilibrium temperature of the pond at zero added heat load, °C
e_a	air vapor pressure, mm Hg
e_s	saturated vapor pressure of water at pond surface, mm Hg
$f(U)$	evaporative wind speed function, Kcal/d-m^2-mm Hg
h_a	enthalpy of air-water vapor mixture at wet bulb temperature, J /kg dry air or Btu/lb dry air

h_w	enthalpy of air-water mixture at bulk water temperature, J /kg dry air or Btu/lb dry air
H_s	gross solar radiation, kcal/d m^2
K	surface heat coefficient or heat transfer coefficient at the pond surface, Kcal/d - m^2-°C
K	mass transfer coefficient, lb water/hr ft^2
KaV/L	tower characteristic
q	heat dissipated, Kcal/d
q_a	long-wave solar radiation, Kcal/day-m^2 or 400 Btu/day-ft^2
q_c	conductive-convective heat loss, Kcal/day-m^2 or 400 Btu/day-ft^2
q_d	Discharge of a heated effluent, Kcal/day-m^2 or 400 Btu/day-ft^2
q_e	evaporative heat loss, Kcal/day-m^2 or 400 Btu/day-ft^2
$q_{r\,j}$	heat flux, Kcal/d-m^2
q_s	short-wave solar radiation, Kcal/day-m^2 or 400 Btu/day-ft^2
q_t	net heat flux across the air-water interface, Kcal/day-m^2 or 400 Btu/day-ft^2
$q_{a\,r}$	reflected portion of q_a, Kcal/day-m^2 or 400 Btu/day-ft^2
$q_{b\,r}$	long-wave black radiation from water surface, Kcal/day-m^2 or 400 Btu/day-ft^2
$q_{s\,r}$	reflected portion of q_s, Kcal/day-m^2 or 400 Btu/day-ft^2
Q	volumetric flow rate, m^3/d
T	bulk water temperature, F or °C
T_{avg}	average temperature = $(T_d + T_s)/2$, °C
T_d	dew point temperature, °C
T_s	surface temperature of the pond, °C
T_x	the boundary condition of T at $X = X$
T_o	the boundary condition of T at $X = 0$
$T_{s\,m}$	completely mixed cooling pond's actual surface temperature, °C
$T_{s\,u}$	completely unmixed cooling pond's actual surface temperature, °C
T1	hot water temperature, °F or °C
T2	cold water temperature, °F or °C
u	average flow velocity, m/d
U	wind speed, km/hr
V	active cooling volume/plan area/, m or ft
X	distance from pond entry, m
α	mixing coefficient
α_m	α evaluated for the completely mixed cooling pond
α_u	α evaluated for the completely unmixed cooling pond
ρ	density of water, 1.0 g/cm^3
$\delta T/\delta X$	longitudinal temperature gradient, °C/m

1 Introduction

The discharge of water at elevated temperature is often described as *thermal pollution*. It is produced by industries such as electric power plants, pulp and paper mills, chemical facilities, and other process industries that use and subsequently discharge water. Even if the discharged water is merely elevated in temperature, allowing it to return to streams, rivers, lakes, and other waters can dramatically alter the native environment.

Often, elevated water temperature will be detrimental to native species of plants and animals. In such instances, alternate, invasive, and possibly undesirable species will be supported by the elevated water temperature. As water warms, the solubility of oxygen decreases. In addition to promoting competitive species of plants and animals, warmed waters may lead to aesthetic and odor problems if anaerobic conditions are created. For these and many other reasons, the discharge of elevated process water is widely regulated by governmental agencies in developed nations.

The fundamental principle of evaporation of water to achieve cooling is widely utilized in nature. In the human body, sweat glands produce perspiration (water), which cools the body by evaporation. The fact that this cooling is more effective during a 95 °F day in Arizona than an afternoon of the same temperature in Florida is widely understood. This concept is illustrated by the following example.

Problem
Given a pail full of water, change the water in the pail into vapor.

Solution
Several possible methods are possible to achieve this goal:

1. Let the pail stand until all the water in the pail has evaporated
2. Obtain a second, empty pail, and pour the water into the empty pail. Repeat this process again and again between the pails. In this way, one will intuitively increase the water/air interaction, so evaporation of the water will be enhanced.
3. If possible, blow air over the surface of the water in the pail. A simple house fan will suffice. Again, intuitively, evaporation of the water is enhanced, and the moving air will carry away the water vapor as fresh air is pushed across the water.
4. And, per the previous mention of Arizona vs. Florida, low humidity air will support the desired evaporation of water much more efficiently than high humidity air.

The purpose of this chapter is to discuss cooling ponds and cooling towers. Essentially, a cooling pond is choice "a" in the previous example. A cooling tower combines choices 2–4.

Cooling ponds are the simplest choice to lower the temperature of a warm water discharge stream and may be the least costly alternative. Such a solution to a water cooling problem relies entirely upon contact of the pond water surface with the surrounding environment. As such the warm water will slowly approach the desired outlet temperature. Ponds are designed to be either *recirculating* or *once-through type*.

A cooling pond may be a good choice if ample room for such a pond is available and land costs are low at a given location. The slow process most often requires a very large pond surface, perhaps several acres, for proper heat transfer to occur. Also, if they are located in a rural area and if the quality of the water being cooled is fairly good, such man-made ponds can offer recreational uses in addition to serving to lower the water temperature. An important note regarding cooling ponds is that this method of lowering the water temperature is most likely to limit evaporative losses of water vs. a cooling tower.

However, if the needed land area for a pond is not available and/or the slow cooling of the water is unacceptable, a cooling tower will be the design choice. Cooling towers are classified as *natural* or *mechanical draft*, as well as *wet* or *dry* types.

2 Cool Ponds

2.1 Mechanism of Heat Dissipation (Cooling)

A cooling pond permits a sufficient buildup of the process water so that the needed heat exchange may occur between the water and the surrounding environment. The warmed water entering the pond will lose and gain heat as it passes through the pond by the combined mechanisms of conduction, convection, radiation, and evaporation. Numerous factors affect the rates of these various mechanisms as well as the overall rate of heat transfer. If the heat inputs are equal to heat outputs, the pond will be at an equilibrium condition, and the surface temperature of the pond will be at some constant value.

The heat exchange between the pond and the surrounding environment will be governed by the temperature difference between the atmosphere and the pond surface, the surface area of the pond, meteorological conditions such as rain or sun, wind speed, humidity, geographical location, and so forth. Obviously these conditions will vary depending upon both time of day as well as time of year.

The various heat inputs and outputs have been identified and presented in typical ranges [1]. Expressed in both Kcal/day-m^2 and Btu/day-ft^2, these values are as follows.

Inputs

1. q_s = short-wave solar radiation
   Range of q_s from 1085–7588 Kcal/day-m^2 or 400–2800 Btu/day-ft^2
2. q_a = long-wave solar radiation
   Range of q_a from 6504–8672 Kcal/day-m^2 or 2400–3200 Btu/day-ft^2

Outputs

3. $q_{s\ r}$ = reflected portion of q_s
   $q_{s\ r}$ ranges from 109 to 542 Kcal/day-m^2 or from 40 to 200 Btu/day-ft^2
4. $q_{a\ r}$ = reflected portion of q_a
   $q_{a\ r}$ ranges from 190 to 325 Kcal/day-m^2 or from 70 to 120 Btu/day-ft^2
5. $q_{b\ r}$ = long-wave black radiation from water surface
   $q_{b\ r}$ ranges from 6504 to 9756 Kcal/day-m^2 or from 2400 to 3600 Btu/day-ft^2
6. q_e = evaporative heat loss
   q_e ranges from 5420 to 21,680 Kcal/day-m^2 or from 2000 to 8000 Btu/day-ft^2

Input or Output: Dependent upon Circumstances

7. q_c = conductive-convective heat loss
   q_c ranges from −867 to +1084 Kcal/day-m^2 or from −320 to +400 Btu/day-ft^2

The net heat flux across the air-water interface (pond surface) can now be expressed as the difference between inputs and output, or:

$$q_t = (q_s + q_a) - (q_{s\ r} + q_{a\ r} + q_{b\ r} + q_e) + q_c \tag{5.1}$$

Note that as indicated above, value for conductive-convective heat loss can be either positive of negative.

As stated previously, at equilibrium q_t is zero. The surface temperature of the pond at this condition is called the *equilibrium temperature*. Also, note that the values for $q_{b\ r}$, q_e and q_c are dependent upon the pond surface temperature. The variables q_s, q_a, $q_{s\ r}$, and $q_{a\ r}$ are independent of pond surface temperature. The discharge of a heated effluent into the pond will add another heat quantity to the input side of Eq. (5.1), q_d.

As the result of changing weather conditions, the surface temperature of most bodies of standing water (be these natural ponds, lakes, or man-made cooling ponds) approaches the equilibrium temperature asymptotically. Equilibrium temperature of such bodies of water is normally calculated on the basis of average daily temperatures.

2.2 *Design of Cooling Ponds*

Often the design of a cooling pond is dictated by a common rule of thumb. One such design rule widely accepted for cooling pond design is 1–2 ac/Mw of pond capacity or 200–400 Kcal of heat will be dissipated per hour per square meter of pond surface area. Such rules of thumb should always be double-checked with sound engineering principles. Local weather patterns and land formations are obvious limitations that must be considered as such factors will affect the rate of heat transfer possible from the pond's surface. Thus the size of the pond needed for the required cooling should also be considered [2].

Basic design equations are presented here for both recirculating cooling ponds as well as for once-through type ponds. The effects of longitudinal mixing and short circuiting of both types of cooling ponds are discussed. Seasonal and geographical considerations are also considered, as well as other miscellaneous considerations regarding cooling pond design.

2.2.1 Evaluation of Heat Dissipation of Completely Mixed and Recirculating Cooling Ponds

The primary parameter when designing a recirculating cooling pond is the amount of water surface area required to dissipate the excess heat of the water that enters the pond. An example would be a cooling pond at a thermal electrical power generation site. In the design, a difference will exist above the equilibrium surface temperature of the pond. The surface area of the pond needed to dissipate the surplus heat of the water in the pond is directly related to the mean surface heat transfer capacity of the pond [1].

Capacity of a Completely Recirculating Pond to Dissipate Added Heat Loads

The ability of a pond to dissipate a heat load from an influent has been demonstrated to be described as [1]:

$$q = KA\ (T_s - E) \tag{5.2}$$

where q is heat dissipated (Kcal/d); K is heat transfer coefficient at the pond surface (Kcal/d - m^2-°C); A is surface area of pond (m^2), T_s is surface temperature of the pond, (°C); and E is equilibrium temperature of the pond at zero added heat load, (°C). Equation (5.2) may also be expressed as

$$T_s = q/KA + E \tag{5.3}$$

Now writing the equation in terms of heat flux, $q_{\mathrm{r\,j}}$ Kcal/d-m^2, Eq. (5.3) becomes

$$T_s = q_{\mathrm{r\,j}}/K + E \tag{5.4}$$

Equation (5.4) expresses the relationship between the pond surface and the rise of the pond temperature above the equilibrium temperature. Equation (5.2) or some modification of this equation is used in cooling pond design to predict pond surface temperature.

Surface Heat Transfer Coefficient

An expression has been derived for the surface heat transfer coefficient, K, by means of a balance of the rates of heat exchange at the pond surface. This expression can be manipulated and several approximations made to arrive at the following definition of K [3]:

$$K = 92.5 + (C_1 + \beta) f(U) \tag{5.5}$$

where K is surface heat transfer coefficient (Kcal/d-m^2-°C), 92.5 is back radiation, C1 is Bowen's conduction-evaporation coefficient (0.468 mm Hg °C), β is slope of the saturated vapor pressure curve between the dew point temperature and the pond surface temperature (mm Hg °C), and $f(U)$ is evaporative wind speed function, (Kcal/d-m^2-mm Hg).

Equation (5.5) defines the surface heat transfer coefficient in terms of three components. These three components approximately represent the sum of heat dissipation from back radiation, conduction, and evaporation.

Relationship for the Slope of the Saturated Vapor Pressure Curve, β

Various values of β are determined by one of two methods, reference to vapor pressure tables or to a vapor pressure chart such as that in Fig. 5.1.

Fig. 5.1 Plot of the slope of the vapor pressure curve for water vs. the average dew point and the pond water surface temperature

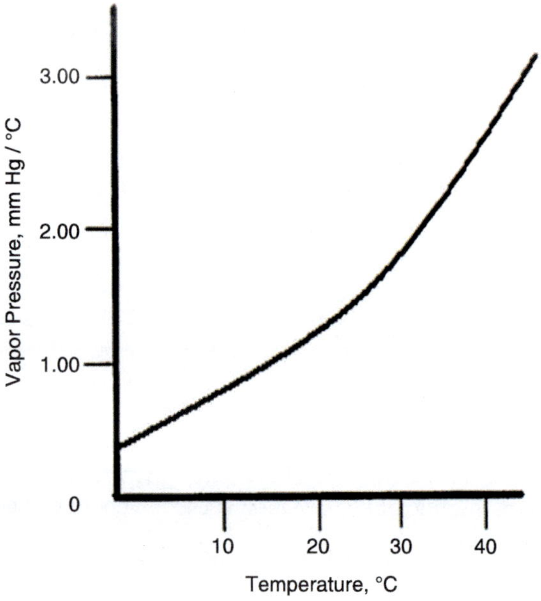

$$\beta = (e_s - e_a)/(T_s - T_d) \tag{5.6}$$

where β is slope of vapor pressure (mm Hg/°C), e_s is saturated vapor pressure of water at pond surface (mm Hg), e_a is air vapor pressure (mm Hg), T_s is surface temperature of pond water (°C), T_d is dew point temperature (°C), and T_{avg} is average temperature $[T_d + T_s)/2]$ (°C).

Relationship for the Wind Speed Function, f(U)

Brady, Graves, and Geyer [3] used curve fitting, plotting techniques, and multiple regression analysis of data collected at three cooling pond sites to develop an empirical relationship to predict the wind speed function. In metric units, this expression states:

$$f(U) = 189.8 + 0.784\, U^2 \tag{5.7}$$

where $f(U)$ is wind speed function (kcal/d m^2) and U is wind speed (km/hr). It is important to note that Eq. (5.7) predicts that the wind speed function, $f(U)$, is not sensitive to relatively low wind velocities. This has the effect of reducing the accuracy of wind speed data at low wind velocities. This also reduces the sensitivity of the heat transfer coefficient in calm wind conditions. A possible explanation of the apparent insensitivity of K to calm wind conditions is related to the cooling effect of vertical convection currents in the atmosphere [3].

Relationship for Equilibrium Temperature, *E*

An expression of pond equilibrium temperature has been derived by Brady, Graves, and Geyer by examining the balance of heat exchange rates at the pond surface. This expression can be manipulated and by several approximations, *E*, is expressed as [3]

$$E = T_d + (H_s/K) \tag{5.8}$$

where E is equilibrium temperature (°C), H_s is gross solar radiation (kcal/d m^2), K is heat transfer coefficient (kcal/ d m^2 °C), and T_d is dew point temperature (°C). A test of Eq. (5.8) was made by evaluation of values of "*E*" directly with data for dew point temperature and gross solar radiation at three recirculating cooling pond sites. Plots of this evaluation revealed that Eq. (5.8) was normally accurate within a few degrees Celsius if the heat transfer coefficient was greater than 135 kcal/ d m^2 °C [3].

2.2.2 Prediction of Surface Temperature, T_s, of a Completely Mixed Recirculating Cooling Pond

The surface heat transfer coefficient (K) and the equilibrium temperature (E) can be determined using known weather conditions and Eqs. (5.5, 5.6, 5.7, and 5.8). An iterative approach is used to account for the fact that the average surface temperature (T_s) is not known initially. The first estimate of T_s is corrected in successive approximations. Required weather conditions are the dew point temperature, gross solar radiation, and the wind speed. With these values known T_s can be calculated for a specific heating load [3].

Calculation of Cooling Pond Exit Temperature

Problem

Given the following information calculate the cooling pond exit temperature:

Application	Fossil fuel-fired power plant
Hot water source	Plant condensers
Type of cooling pond	Completely recirculating
Surface area of pond, A	4×10^6 m^2
Heat load	8×10^9 kcal/day
Hot water flow rate	9255 L/s
Location	Southern USA
Prevailing weather condition	July
Mean dew point, T_d	23 °C
Mean solar radiation	5750 kcal/ m^2·d
Mean wind speed	13.5 km/hr

Solution

Estimate the surface water temperature, T_s, to be 36 °C, and with the given mean dew point temperature, T_d, of 23 °C (given in conditions above), calculate the T_{avg}:

$$T_{avg} = (T_d + T_s)/2 = (36 + 23)/2 = 29.5^{\circ}\text{C}$$

Using Fig. 5.1 with the average temperature to determine the slope of the vapor pressure curve, $\beta = 1.77$ mm Hg °C.

At the given wind speed, $U = 13.5$ km/hr and using Eq. (5.7) determine

$$f(U) = 189.8 + 0.784\ U^2 \tag{5.7}$$

$$f(U) = 189.8 + 0.734\ (13.5)^2 = 323.6\ \text{kcal/m}^2\ \text{d}$$

The surface heat coefficient, K, is now determined from Eq. (5.5):

$$K = 92.5 + (C_1 + \beta) f(U) \tag{5.5}$$

$$K = 92.5 + (0.468 + 1.77)\ 323.6 = 816.5\ \text{kcal/m}^2 \cdot \text{d} \cdot {}^\circ\text{C}$$

Using the given solar radiation, $H_s = 5750$ kcal/m^2·d, the equilibrium temperature, E, can now be determined from Eq. (5.8):

$$E = T_d + (H_s/K) \tag{5.8}$$

$$E = 23 + (5{,}750/816.5) = 23 + 7.0 = 30^\circ\text{C}$$

The surface temperature is now determined using Eq. (5.3):

$$T_s = q/KA + E \tag{5.3}$$

$$T_s = \left\{8 \times 10^9 / \left[(816.5)\left(4 \times 10^6\right)\right]\right\} + 30.0$$

$$T_s = 2.5 + 30.0 = 32.5^\circ\text{C versus the assumed value of } 36^\circ\text{C}$$

Repeating this procedure now using 33 °C as T_s, from Fig. 5.1, $\beta = 1.65$ and previously calculated value for $f(U)$ of 323.6 kcal/m^2 d, recalculated K and E

$$K = 777.9\ \text{kcal/m}^2 \cdot \text{d} \cdot {}^\circ\text{C and } E = 30.4^\circ\text{C}$$

and

$$T_s = q/KA + E \tag{5.3}$$

$$T_s = \left\{8 \times 10^9 / \left[(777.9)\left(4 \times 10^6\right)\right]\right\} + 30.4$$

$$T_s = 2.6 + 30.4 = 33.0^\circ\text{C}$$

Because this calculated value for T_s equals the assumed value, no additional iterations are needed.

Design Chart

Figure 5.2 will assist in this type of cooling pond design calculation [3]. This figure gives the values for heat transfer coefficient, K (in kcal/m^2·d·°C), at the intersection of the wind speed, U, in km/h and average temperature, T_{avg} (in °C). The controlling local weather conditions may be found in standard references [4].

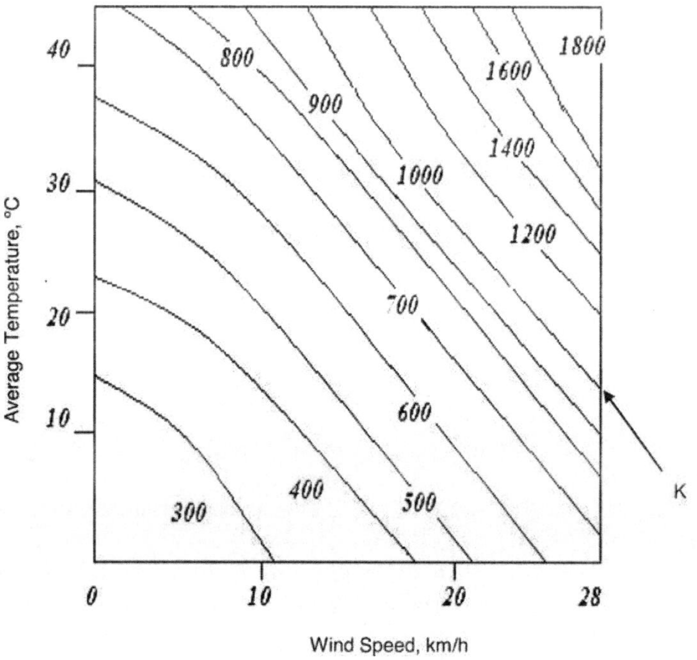

Fig. 5.2 Values of heat transfer coefficients in terms of wind speed and average temperature

2.2.3 Longitudinal Mixing Effects on Recirculated Cooling Pond Performance

In addition to dissipating heat from water flowing from an industrial process, the lowest possible intake temperature of the water re-entering the process is a prime objective of proper operation of the pond if that water is to be reused in the process. Dissipating heat and lowering the temperature of the water recirculated back to the process are not necessarily complementary objectives.

Pond exit temperature (process intake temperature) at a given set of weather conditions is directly related to the amount of longitudinal mixing that takes place within the cooling pond. Also affecting this temperature is the uniformity of flow through the pond. Short circuit flows of water within the pond will result in direct transport of warm water entering the pond to the pond discharge point. Such short circuit flow will return water to the process at elevated temperature.

If insufficient longitudinal mixing occurs, a number of techniques are available to alleviate the condition. Transverse baffles placed in the pond will increase the length of the flow path of water in the pond. Vertical skimmer walls may be placed in the pond to minimize wind-driven short circuits. This effect of wind-driven water currents is also minimized by a high length to width ratio of pond size. Relatively shallow pond design will minimize vertical water current formation in the pond.

2.2.4 Evaluation of Heat Dissipation in Once-Through Cooling Ponds

Various mechanisms of heat transfer of warm water have been previously described. Various models exist that allow for prediction of water discharge temperature from such cooling ponds. The needed pond surface area may be predicted using such models for a given temperature change of the water in the pond.

Edinger and Geyer [1] developed a method that predicts the design parameters of a cooling pond with a good balance of accuracy vs. facility. This method is also summarized in a government publication [5]. The model assumes complete (ideal) mixing in the depth and width of the water flow. No longitudinal mixing of the water, in the flow direction, is assumed. Further assumptions are constant cross-sectional flow and constant flow velocity. Absence of flow channel curvature is also assumed in this model. In this way, the water entering the pond is modeled as completely mixed at point of entry into the pond. The warm water then transfers heat to the surroundings exponentially as it proceeds as a "slug" of warm water passing through the pond. These assumptions and approximations allow for a mathematical expression that predicts the surface temperature as a function of longitudinal distance of the pond. In differential form, the net rate of heat exchange is

$$dq_t/dt = -\mathrm{K}(\mathrm{T_s} - E) \tag{5.9}$$

where dq_t/dt is net rate of surface heat exchange (kcal/m^2 d), K is heat transfer coefficient (kcal/m^2, °C), T_s is pond surface temperature (°C), and $E =$ equilibrium temperature, °C.

This heat exchange causes the water temperature to fall in relation to distance of water travel in the cooling pond. This may be expressed mathematically:

$$-K(T_s - E) = \rho\, C_p\, d\, u(\delta T/\delta X) \tag{5.10}$$

where ρ is density of water (1.0 g/cm^3), C_p is specific heat of water (1.0 cal/g °C) d is average depth of flow path, (m), u is average flow velocity (m/d), δT/δX is longitudinal temperature gradient (°C/m), and X is distance from pond entry (m). Integration of Eq. (5.10) with the boundary condition of $T = T_x$ at $X = X$ and that $T = T_o$ at $X = 0$ results in the expression

$$e^{-\alpha} = (T_x - E)/(T_o - E) = -KX/\left(e^{\rho\, C_p\, d\, u}\right) \tag{5.11}$$

where $\alpha\, i = KX/\rho\, C_p\, d\, u = KA/\rho\, C_p\, Q$ *and* $Q =$ volumetric flow rate, m^3/d. With the previously discussed weather conditions, the cooling pond area required for a given heat load may be calculated. Or alternatively, the exit temperature of water leaving a cooling pond of a known surface area may be determined.

Calculation of Pond Exit Temperature

In the previous example the following values were found to be appropriate:

$$K = 778.5\ \text{kcal/m}^2\ \text{d}^\circ\text{C}$$

$$C_p = 1.0\ \text{kcal/kg}^\circ\text{C}$$

and

$$\rho = 1.0\ \text{g/cm}^3 = 1{,}000\ \text{kg/m}^3$$

The volumetric flow, Q, in terms of m^3/d is determined as follows:

$$Q = 9{,}255\ \text{L/s} = (9{,}225)\ (60)\ (60)\ (24)/1{,}000 = 8.0 \times 10^3\ \text{m}^3/\text{d}$$

Assumed inlet temperature of water entering the power plant condensers is 28 °C. If the water is warmed 10 °C as it passes the condensers, the temperature of the water entering the cooling pond can be assumed to be 38 °C (T_o). Therefore, the following data applies:

As shown in Eq. (5.11), the value for α can be determined as follows:

$$\alpha = \text{KA}/\rho\ C_\text{p}\ Q$$

$$\alpha = (778.5)\ (4 \times 106)/[(1{,}000)\ (1.0)\ (8 \times 105)] = 3.89$$

Substituting values for E, T_o, and α into Eq. (5.11) and solving for the temperature water, T_x, exiting the cooling pond

$$e^{-\alpha} = (T_\text{x} - E)/(T_\text{o} - E) \tag{5.11}$$

$$e^{-3.89} = (T_\text{x} - 30.4)/(38 - 30.4)$$

$$T_\text{x} = 30.56^\circ\text{C}.$$

The effect of pond surface area on the exit temperature of the water may be evaluated by repeating this calculation for various values for the area of the pond (A). Alternatively, exit water temperature may be used to determine the required cooling pond surface area.

2.2.5 Relationship of a Completely Mixed to a Totally Unmixed Pond Once-Through Cooling Ponds

The purpose of once-through cooling ponds, both completely mixed and completely unmixed, is twofold: to dissipate heat from water rejected from some industrial process and to achieve the lowest possible effluent temperature of water exiting the pond. Both results are directly related to both longitudinal mixing and short-circuiting currents present within the pond.

Comparison Between Net Plant Temperature Rise of a Completely Mixed Pond vs. a Completely Unmixed Pond

The ratio of the net plant temperature rise in the water exiting the pond from a completely mixed vs. a completely unmixed pond has been determined [1] and is:

$$(T_{s\,m} - E)/(T_{s\,u} - E) = e^{\alpha}/(1 + \alpha) \tag{5.12}$$

where $T_{s\,m}$ is completely mixed cooling pond's actual surface temperature (°C) and $T_{s\,u}$ is completely unmixed cooling pond's actual surface temperature, °C.

At fixed flow rate, Q; surface area, A; and similar weather conditions, the completely unmixed pond will yield faster cooling of the water as compared to the completely mixed pond. This is caused as a consequence of the warm water in the unmixed pond being unmixed initially or otherwise diluted with cooler water upon entry into the pond. The driving force of cooling, T_s–E, is maintained at the highest possible value.

Area of Completely Unmixed Cooling Pond Needed to Provide Identical Surface Temperature as with a Completely Mixed Cooling Pond

The area of a completely unmixed cooling pond needed to provide the identical surface temperature as with a completely mixed cooling pond can be determined with Eq. (5.11). In this instance $T_{sm} = T_{su}$, so T_{sm}–$E = T_{su}$–E as well; thus

$$(T_{sm} - E)/T_{su}, = e^{\alpha}/(1 + \alpha_m) = 1 \tag{5.13}$$

where α_m is α evaluated for the completely mixed cooling pond and α_u is α evaluated for the completely unmixed cooling pond. The actual difference between α_m and α_u is the area of each respective cooling pond. The flow rate, Q, is identical for each pond. Additionally, the specific heat of the water, C_p; the water density, ρ; and the surface heat transfer coefficient, K, are the same for both or either ponds. Therefore the ratio of the two required areas for the two types of cooling ponds can be determined.

2.2.6 Effect of Seasonal Weather Variations on Cooling Pond Requirements and Performance

The equilibrium temperature, *E*, for most locations inside the continental USA reaches a low in early January and then reaches a high six months later in early July. The pattern of variation of *E* is actually quite smooth for most locations. The difference between seasonal high and low is more distinct at more northern (higher latitude) locations [4].

The surface heat exchange coefficient, K, varies in a relatively inversely proportional relationship to "*E*." Highest heat exchange coefficients are realized in summer, lowest in winter. The variation of "*K*" is not as regular as the variation "*E*" as the result of seasonal wind variations. As a result, the maximum value of "*K*" occurs in some locations (coastal, mountains, etc.) at times of the year other than mid-summer.

2.2.7 Effect of Geographic Location on Cooling Pond Performance

As mentioned above, geographic locations will affect both "*E*" and "*K*." At higher latitudes, solar radiation is reduced and significantly affects the equilibrium temperature of a cooling pond. Topography (hills, flat plain, etc.) will influence predominant wind speeds as well as wet bulb temperature and hence influence *K*. As a result, within the continental USA, the best conditions for cooling pond performance exist in the southern Great Plains between the Mississippi River and the Rocky Mountains (Kansas, Oklahoma, north Texas, eastern Colorado, etc.). Favorable conditions for cooling pond performance also exist along the Atlantic seaboard between Cape Cod and the Gulf of Mexico. Poor conditions for cooling pond performance will be found in the Great Basin region between the Rocky and Sierra Nevada Mountains (Nevada and Utah). It should be stressed that local "micro-climate" and local topographic conditions must be considered when determining cooling pond performance [4].

2.2.8 Other Factors Related to Cooling Pond Design

Topics pertinent to the design of cooling ponds are the following.

Dikes and Baffles

A cooling pond has traditionally been constructed by building dikes and/or other embankments to form a man-made pond. Therefore, cooling ponds are normally constructed where the local topography is amenable to such construction. If a pond must be excavated, construction costs will escalate [4].

Evaporation of Water

In some locations, particularly southern and western, evaporative loss of water from a cooling pond can be a serious problem. Such water must be replenished and if construction of a cooling pond is proposed for arid or semi-arid regions, sources of substitute water must be available.

Infiltration

This term refers to entry and movement of water through soils. If present, infiltration may affect the design and location of a cooling pond for two reasons: (1) if recharging of the local water table takes place, that water table may rise in elevation and affect surrounding land uses and (2) groundwater contamination may occur.

Fogging

A disadvantage of cooling ponds is the potential for fog formation. If a fog forms and obscures local visibility, this can be a problem, particularly if a highway or an airport runway is nearby.

Recreation

As previously mentioned, if the water being cooled in the pond is of sufficiently good quality, boating, fishing, and perhaps even swimming recreational opportunities may be an auxiliary benefit of pond construction.

Economics

If sufficient land is available at a reasonable cost, and the local topography is amenable to construction needs, a cooling pond may be a good solution to a cooling problem as the pond will cost less than a tower to construct.

3 Cooling Tower

To treat large volumes of warm water from an industrial process such as a coal or nuclear power plant, a cooling pond is not practical because of the enormous size of the pond needed. Indeed, the term lake would be more appropriate to describe such a body of water. As a result a *cooling tower* is the practical solution to the cooling needs of most large industrial plants. Such towers are often massive in size and

Fig. 5.3 The cooling system at the CHPP-5 (TEC-5) heating and power plant in Kharkiv, Ukraine. (Photo provided courtesy of Dr. Gennadiy K. Voronovskiy, General Director)

visible from a long distance. Note the size of the cooling system in Fig. 5.3. The design of cooling towers is now a specialized niche of environmental engineering. Therefore, it is important for engineers to understand the basic design of such towers, their advantages and disadvantages, operational considerations, and their environmental impact.

3.1 Mechanism of Heat Dissipation in Cooling Towers

The normal operation of a cooling tower involves both heat and mass transfer. Excellent explanations of the material and heat balances involved have been given in detail [6]. For the purposes of this discussion, the relative temperature and humidity gradients present within a tower are given in Fig. 5.4.

In this example, the top of the tower in Fig. 5.4 (a) contains water at a higher temperature vs. the dry bulb temperature of the air. The water is therefore being cooled by both evaporation and transfer of sensible heat from the water to the air. As evaporation of water is also taking place at the air-water interface, the humidity will change as well.

Possible relative conditions at the bottom of the cooling tower are sketched in Fig. 5.4 (b). Now the water temperature is above the wet bulb temperature but below the dry bulb temperature. As the water is being cooled, the temperature at the water-air interface is lower than the bulk temperature of the water. A temperature gradient will exist in the direction of the water-air interface. This is a consequence of the fact that the air temperature is higher than the temperature at the water-air interface. Sensible heat is being transfer from both water and air to the interface. The sum of these two heat transfers will equal the latent heat flow from the interface into the air as the consequence of evaporation.

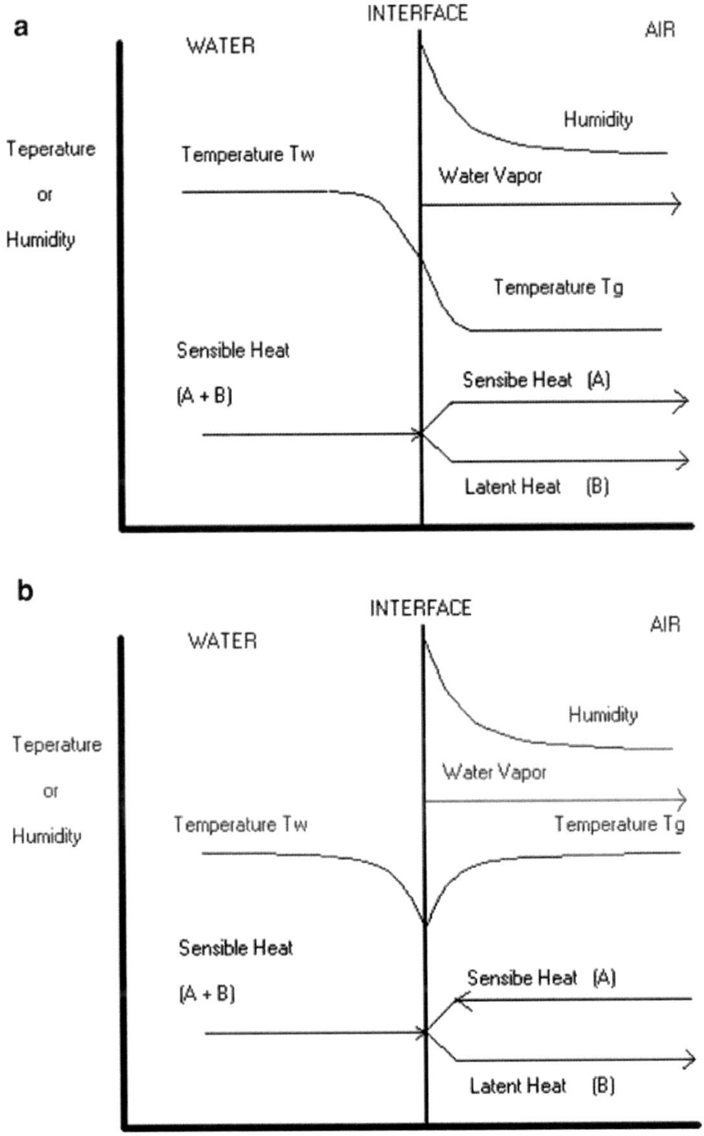

Fig. 5.4 (**a**) Relative temperature and humidity conditions at the top of a typical cooling tower. (**b**) Relative temperature and humidity conditions at the bottom of a typical cooling tower

3.2 Type of Towers

Cooling towers are normally classified as natural or induced draft types. Additionally, towers are either wet or dry types [7].

Induced (also called mechanical) draft towers employ a fan to move air through the tower. Natural draft towers rely upon the difference in density between the warm air exiting a tower and the cool air inside the tower to act as the driving force for air movement. Most towers at large industrial locations have direct contact between air and water and are hence considered to be "wet" towers. Some towers, however, are designed to keep the air and water separate and hence are considered to be "dry" towers.

Dry cooling towers are used in specialized applications where temperatures are very high. Convection is the primary mechanism of heat transfer. Since dry towers are markedly less efficient than wet towers, they are not discussed at length here.

Common types of cooling towers are presented in Figs. 5.5, 5.6, 5.7, 5.8, 5.9, 5.10, 5.11, 5.12, and 5.13 and are described in the following sections.

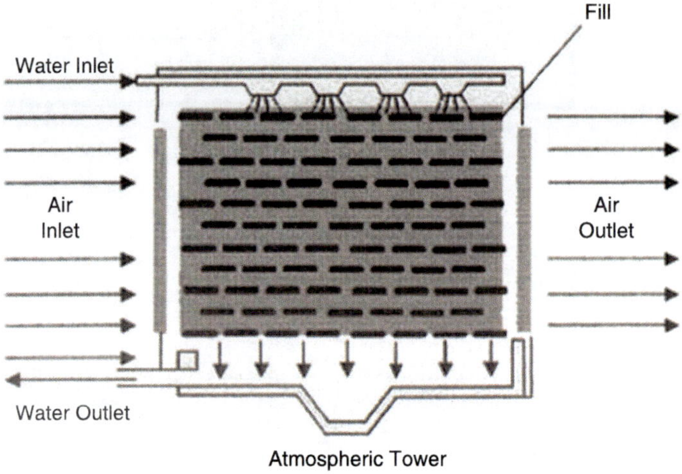

Fig. 5.5 Natural draft atmospheric cooling tower

Fig. 5.6 Counter-flow natural draft cooling tower

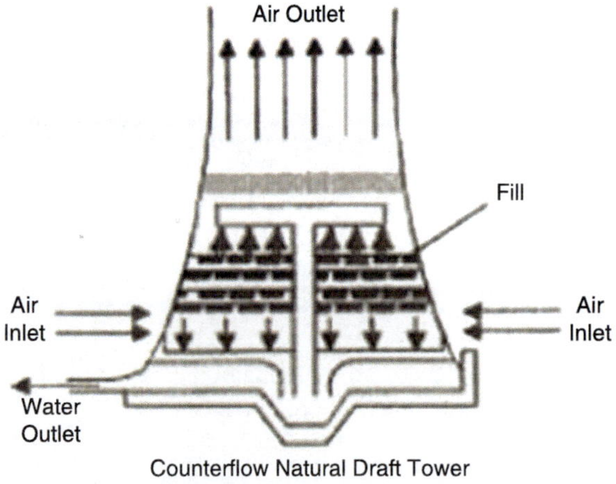

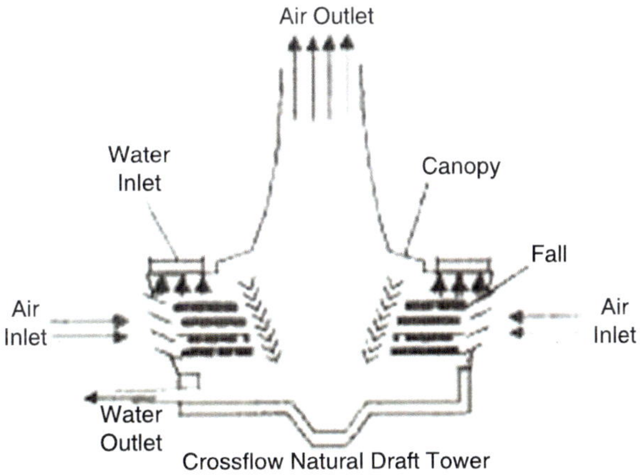

Fig. 5.7 Cross-flow natural draft cooling tower

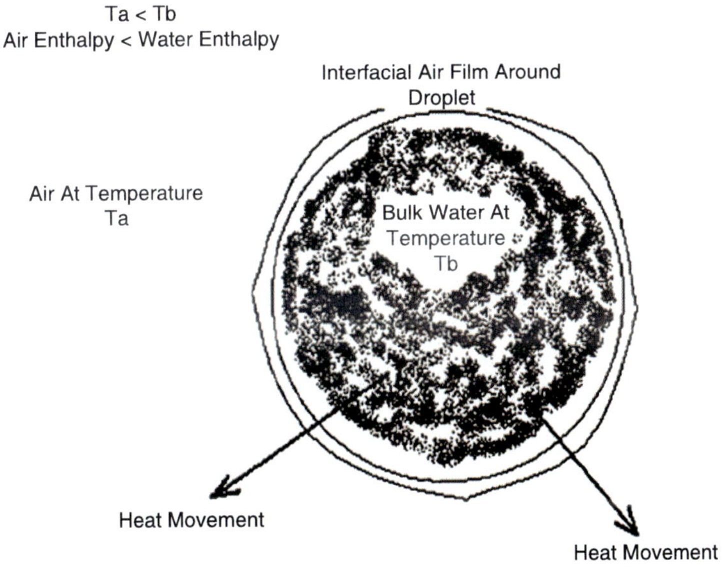

Fig. 5.8 Film of cool air surround a warm droplet of water, with movement of heat from the water to the air

3.3 Natural Draft Atmospheric Cooling Towers

The flow scheme for this design of cooling tower is presented in Fig. 5.5. Air enters the tower through open louvers on the sides of the tower. As the air passes through the tower, it comes in contact with water that is sprayed downward in the tower. Wind currents primarily drive the horizontal airflow through the tower. Such towers are sometimes open, containing only the falling water spray.

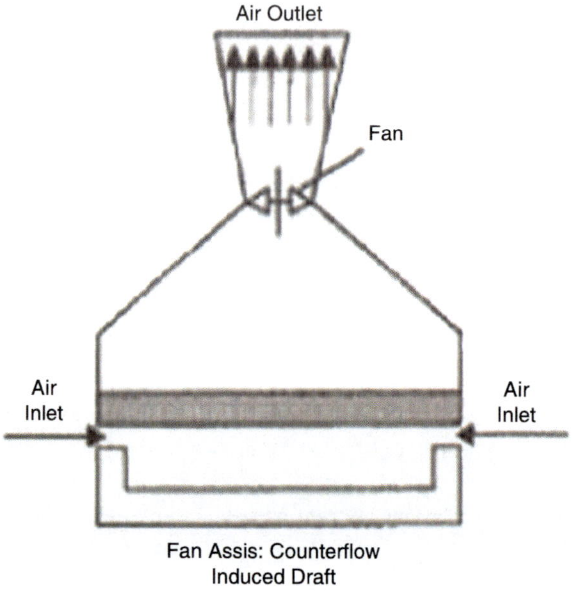

Fig. 5.9 Fan assist (draft or mechanical type) counter-flow induced draft cooling tower

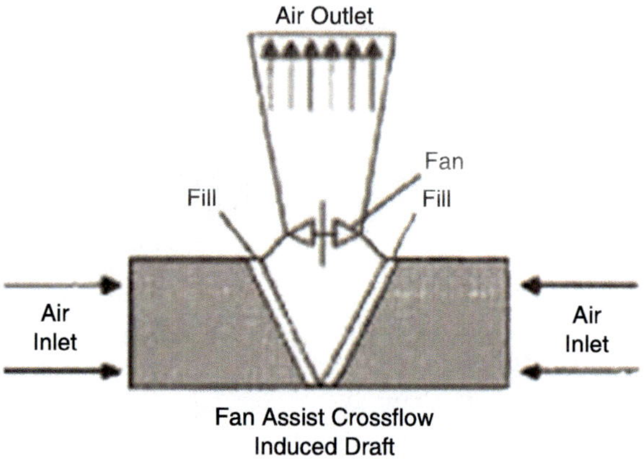

Fig. 5.10 Fan assist (induced or mechanical type) cross-flow induced draft cooling tower

An alternate design of this type of tower uses packing (or fill) with an open design. Splashboards, triangular PVC rods, or an open polyethylene design are among the several possibilities for the fill design. The basic purpose of these various fills is the same: to increase water/air contact surface with minimum added resistance to both air and water flows.

With the exception of large, hyperbolic cooling towers that are a special case (such as those in Fig. 5.3), this type of cooling tower is normally used for smaller installations such as hospitals, office buildings, or other relatively small facilities. This type of cooling tower is the typical of refrigeration units.

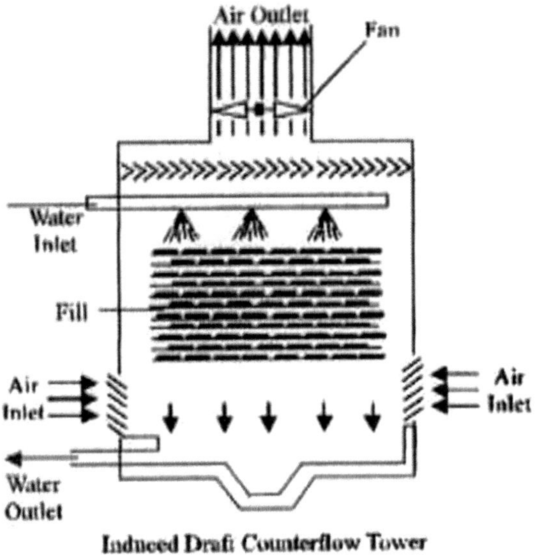

Fig. 5.11 Induced draft (mechanical) counter-flow cooling tower

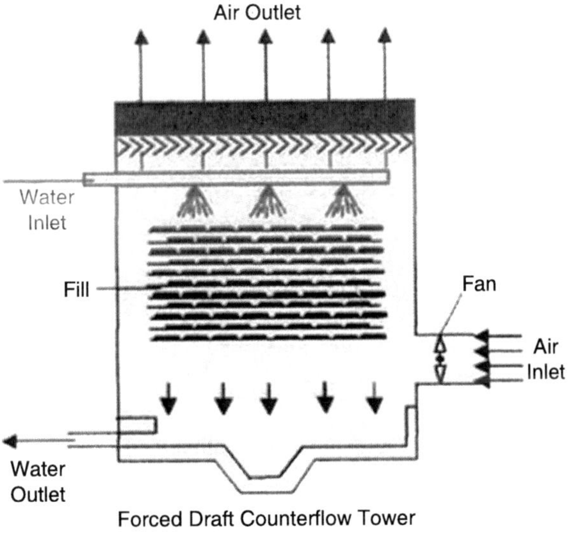

Fig. 5.12 Forced (mechanical) draft counter-flow cooling tower

Advantages and disadvantages of this type of cooling tower have been previously described [6] as

Advantages
No moving or other mechanical parts
Relatively little maintenance
No recirculation or air required

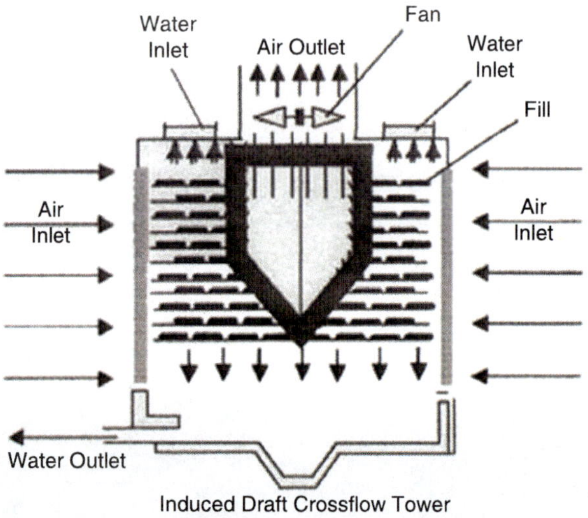

Fig. 5.13 Induced draft (mechanical) cross-flow cooling tower

Disadvantages
Comparatively large sizes needed for a given heat load
High initial capital cost
High pumping costs
Must have unobstructed location to take advantage of available winds
Cooling is dependent upon wind direction and velocity
Often had excessive drift loss

The last disadvantage is of concern. Drift droplets will likely contain chemical impurities and as such will be considered to be airborne emissions by governmental regulatory agencies [8, 9].

Ozone treatment of cooling tower water has recently been suggested by the US Department of energy as a method of treating cooling tower water that saves costs and reduces the need for chemical additives [10].

3.4 *Natural Draft, Wet Hyperbolic Cooling Towers*

As seen in Fig. 5.3, this type of tower is named for the distinctive, hyperbolic shape of the stack through which air passes after contacting warm water sprayed over packing inside the lower part of the tower. Often massive in size, hyperbolic cooling towers are often 150–180 m (490–515 ft) high and 100–135 m (330–450 ft) in diameter. Such towers are often made from reinforced concrete. Airflow through the tower is driven by the density difference between the heated (so lighter) air leaving the tower and the colder (heavier) air outside the tower. Wind velocity blowing across the tower opening may additionally aid this airflow; however, if the tower is correctly designed, the operation of the cooling tower will not be dependent upon wind flows. This type of cooling tower design is illustrated in Figs. 5.6 and 5.7.

The bottom 3–6 m (10–20 ft) contains fill (packing) that supports extended surface-to-surface contacts of air and water in both the counter and cross-flow designs. In a counter-flow design, warm water is distributed by spray nozzles over the fill material and flows downward as the air passes upward. If cross-flow design is utilized, warm water flows downward over the fill as air passes in a 90° plane to the water flow. The center of the tower is open and the airflow here turns upward to exit the tower.

Counter-flow design supports more efficient heat transfer as the coolest water will always be in contact with the coolest air and the minimum amount of fill is required. However, the geometry of the cross-flow design promotes improved air/water contact, and at the same capacity as the counter-flow tower, the cross-flow tower will operate at a lower head loss.

The purpose of the fill (packing) material, as mentioned above, is to support and enhance surface-to-surface contact of air and water to facilitate heat transfer. Transfer of heat may be modeled based upon a thin film of air surrounding a water droplet per Fig. 5.8. This movement of heat can be modeled using the Merkel equation [11, 12]:

$$K\,aV/L = \int_{T1}^{T2} dT/(h_w - h_a) \tag{5.14}$$

where KaV/L is tower characteristic, K is mass transfer coefficient (lb water/ft^2 or kg water/ m^2), a is contact area/tower volume (ft or m), V is active cooling volume/ plan area (ft or m), T1 is hot water temperature (°F or °C), T2 is cold water temperature (°F or °C), T is bulk water temperature (°F or °C), h_w is enthalpy of air-water mixture at bulk water temperature (J/kg dry air or Btu/lb dry air), and h_a is enthalpy of air-water vapor mixture at wet bulb temperature (J/kg dry air or Btu/lb dry air).

Further derivation of this equation is possible to determine the tower characteristic and can be found in standard reference texts. Here, it is used to demonstrate that the function of fill or packing is to maximize the surface of water available to the air to promote the most efficient heat transfer possible.

Natural draft hyperbolic cooling towers are attractive due to their relative lack of mechanical and electrical components. They can accommodate large quantities of water and are relatively efficient cooling units. However, the height of such units, which is needed for proper draft, is objectionable from a public relations viewpoint. The plume of condensed water vapor leaving a large hyperbolic tower is often viewed as pollution by the general public.

3.5 Example 1

Select the proper wet bulb temperature for a proposed cooling tower in southern California. When designing a cooling tower, the highest possible wet bulb temperature must be determined. This will assure that the cooling tower will operate at all times. The Los Angeles airport has the following maximum wet bulb temperatures [4]:

- 67 °F is exceeded during 2% of summer hours.
- 69 °F is exceeded during 0.5% of summer hours.
- 71 °F is exceeded during 0.1% of summer hours.

Depending upon how critical the application is, the designer may wish to design to a slightly higher temperature that 71 °F. Note that at 0.1% of the summer hours, if 71 °F is used for design purposes, the tower may be expected to fail over 20 h, or approximately 1 d:

$$(90\ \text{days/summer})\ (24\ \text{hours/day})\ (0.01) = 21.6\ \text{hours}$$

Of course these hours of poor performance will most likely be spread over the entire summer period, so the amount of safety factors needed is at the discretion of the design engineer.

3.6 Hybrid Draft Cooling Towers

Two hybrid types of draft cooling towers are seen in Figs. 5.9 and 5.10. Although still considered to be draft-type cooling towers, both designs employ a fan to assist airflow through the towers. The large size of the cooling tower limits the actual size of the fan that can be used, so the design of the tower narrows to a point where the fan is installed, and then the tower widens after the fan as the air exits the tower.

The advantage of these hybrid designs is that no packing or fill products is used. The lack of the packing to force additional water-air contact requires the fan to make up for this lost efficiency in heat transfer. The fan also adds to maintenance and operational costs of the tower. These designs do reduce initial capital outlay, as the towers will be considerably smaller than true natural draft cooling towers.

3.7 Induced-Mechanical or Forced Draft Wet Cooling Towers

As opposed to the hybrid-induced draft towers just discussed, cooling towers that employ a fan to drive air through a packed cooling tower are known as mechanical or

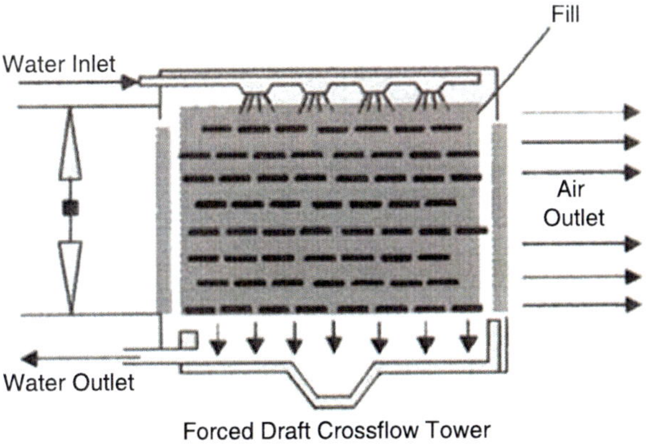

Fig. 5.14 Forced draft (mechanical) cross-flow cooling tower

induced draft type. Often a bank of such cooling towers are built together to satisfy the total cooling requirement, as the practical fan size limits the size of each individual cooling tower. Both counter and cross-flow air paths may be used.

In Figs. 5.11 and 5.13, it is seen that the fan is located behind the packed section of the tower. From the exit section of the tower, the fan is able to pull the air through the packing. The designs in Figs. 5.12 and 5.14 place the fan at the entrance to the tower to push air through the packed section and then out of the tower. The cooling towers in Figs. 5.11 and 5.13 are called *induced draft*, and the designs in Figs. 5.12 and 5.14 are called *forced draft* cooling towers.

The induced draft design normally utilizes larger fans than forced draft. This allows for higher air exit velocities and lessens problems with recirculation as a result. Air distribution is often superior using induced draft vs. forced draft design. As the fan is in the exit in an induced draft cooling tower, this design permits the warm, moist air to pass through the fan, increasing corrosion problems.

Both induced and forced draft cooling towers can be noisy. Both types will also obviously consume large amounts of electric power during normal operations and have more maintenance considerations than natural draft cooling towers. These designs are practical for rooftops. Also, if sufficient land is available, a bank of either type of design will not be the dominant feature of the landscape as with large natural draft hyperbolic cooling towers.

3.8 Cooling Tower Performance Problems

This discussion is intended to be brief, as many good sources of operational advice are now available particularly on the Internet.

Some of the causes of poor performance in a cooling tower [13–20] and suggested corrective actions are given below.

3.8.1 Scale Formation

In a cooling tower, water evaporates and potentially enables mineral deposits to form a "scale" on surfaces. If such a scale does form, it acts as a barrier to heat transfer, and the performance of the cooling tower will degrade. Possible sources of such scale are hard water and/or inadequate blow-down [21].

If hard water is present, water softeners combined with adjustment of the blow-down rate will likely prevent further formation of scale. The advice of a reputable industrial water treatment service company may prove valuable in such a situation.

Removing existing scale deposits can be difficult. Hard water deposits generally will dissolve in a mild acid backwash. But if scale is thick and/or has had the opportunity to dry, it may have to be manually chipped off surfaces.

3.8.2 Clogged Spray Nozzles

Again, hard water can cause deposits to form in the spray nozzles that introduce the water into the cooling tower. Alga or other solids may also cause deposits. Spray nozzles should be inspected as a result to verify that the proper flow pattern of water into the tower is present on a regular basis. If partial plugging occurs, the flow of water over the fill will be uneven and water/air contact surface will be lost. Consequently, heat transfer and tower performance will suffer. A proper water treatment program recommended by an experienced water treatment company should be followed.

3.8.3 Poor Airflow

Poor airflow through the tower will degrade tower performance. Various causes can partially interrupt airflow – debris in the tower exit, in the entrance, or in the packing will cause unwanted and undesirable air currents to form that will short circuit parts of the tower. Loose fan and motor mountings, poor fan alignment, damaged fan blades, gear box problems, and other various problems can all contribute to poor airflow as well. Thus, regular maintenance visual inspection of the cooling tower is important.

3.8.4 Poor Pump Performance

If pumps (recirculation, blow-down, etc.) are not performing properly, then water flow through the tower will be degraded as will tower performance. Again, preventative maintenance is preferable to an emergency shutdown. Perhaps most instructive are case studies of problem cooling tower installations [22–24].

3.8.5 Choosing the Correct Cooling Tower

As previously mentioned, the Internet offers a window to a wide source of technical information. As information is readily found on the World Wide Web, this chapter will not direct readers to specific sites for design advice. Using several search engines with proper key phrases will return several excellent results to consult regarding a cooling tower project. The basic data to gather at the start of such a project [25] are:

1. The heat load on the tower
2. Range (temperature difference of water in vs. water out)
3. Approach (temperature difference water leaving vs. air entering wet bulb temperature)
4. Approximate tower size

As can be seen, if Items 1–3 are known, one will be ready to consult with the engineering staffs of potential suppliers of a cooling tower to determine Item 4.

If the wet bulb temperature is not known, a quick estimate can be made by dipping a thermometer into the water and then reading the lowest temperature the thermometer registers as the water evaporates. This temperature is the wet bulb temperature. Several readings in the environmental conditions for which the cooling tower is proposed should be taken. Pertinent articles regarding cooling tower design [26] frequently appear in various trade magazines as well.

3.8.6 St. Johns River Power Park [27]

This 632,000 KW coal-fired electric power plant is jointly own by JEA and Florida Power Light. It is located in Jacksonville, Florida. Phase constructed was started in the 1980s and completed in 1988. Initially, the plant is equipment emission control units which include NOx burners, electrostatic precipitators, and wet limestone flue gas de-sulfurization units. A di-basic acid injection system to enhance SO_2 removal and selective catalytic reactors to further reduce NOx emissions was added in 204 and 2009, respectively. The facility has two hyperbolic-shaped cooling towers with fill (packing) each 460 feet high with base diameter of 370 feet. The cooling towers are equipment with bottom and top drift eliminators. The design criteria for these natural-draft counter-flow units are as follows (Tables 5.1, 5.2, and 5.3).

Table 5.1 Monthly dry bulb temperatures

Specific frequency level (%)	1	2.5	5	10
Temperature (°F)	95	93	92	90

Table 5.2 Monthly wet bulb temperatures

Specific frequency level (%)	1	2.5	5	10
Temperature (°F)	81	80	79	78

Table 5.3 Design requirements for cooling tower

Parameter	Unit	Value
Total water quantity cooled per tower	gpm	247,700
Min partial load operation	% of flow	60
Total cooling duty per tower	10^6 Btu/hr	2960
Design cooling range	°F	24.6
Temperature of water entering tower	°F	114.6
Temperature of water leaving tower	°F	90
Design approach to ambient atmospheric design wet bulb temperature	°F	10
Ambient atmospheric design wet bulb temperature	°F	80
Relative humidity for the above condition	%	58
Maximum drift loss at design conditions (including blow through with a 45 mph wind)	%	0.002
Design minimum ambient temperature	°F	10
Basin normal operating level range	ft	0–4
Basin depth	ft	5
Basin total water storage	ft^3	424,000
Cooling tower system blow-down, max	ggpm	17,200
Cooling tower system blow-down, max	°F	90

3.8.7 Legionnaires' Disease

Legionnaires' is a form of pneumonia. The bacterium that causes this disease will flourish in cooling tower water if conditions are favorable [18, 28]. As no amount of drift elimination will prevent sub-micron particles of water from leaving a cooling tower, it is important to stress that a proper water control program be followed for normal cooling tower operation. As previously noted, a qualified water treatment service company should be engaged to assure that such bacterial growth is prevented in the operations of a cooling tower. To prevent airborne Legionnaires' and other diseases, the cooling water can be treated by specially formulated chemicals [18], filtration, copper ionization, UV [15], ozone [10, 14–17], or dissolved air-ozone flotation [19]. Controlling hydrogen sulfide emissions from cooling tower water has been discussed by Nagl [20]. Center for Disease and Control and Prevention (CDC) has developed program to implementing control measures for corrective actions to control the development of legionella species in cooling towers and other water using units. These measures are implemented when temperature and disinfectant control levels are not met in the cooling towers [29].

3.8.8 Reuse of Thermal Discharges

Reclaimed water (reuse) from Miami-Dade County, Florida, will be used for cooling after additional treatment for the two new nuclear units at Turkey Point Power Plant.

Additionally, the existing combined-cycle natural gas unit at Turkey Point Power Plant (Unit 5) will use the reclaimed water for cooling. Miami-Dade County will provide up to 90 million gallons day (MGD) or average daily flow of 75 MGD. This reclaimed water would otherwise be injected in deep well for disposal [30–32].

The gas-fired Mattawomen Power Plant in Maryland uses reclaimed water for water supply for the cooling towers. Cycle of concentration (make-up volume/blow down volume) typically range 2–3 and can be improved with increase treatment of the makeup water. The water reclamation facility that provides reclaimed water to Mattawomen Power Plant improved its treatment (lowering total nitrogen and phosphorus concentrations). As a result, the cycle of concentration increased from 4–5 to 8–10 [32–34].

3.8.9 Recent Research

The current cooling tower research examined microbial effects on human health and corrosion. Other research examined the methods to improve operation, reduce emissions, and increase water supply.

The legionnaire's disease (LD) outbreak in New York, New York, in 2015 was examined. The outbreak causes a total of 138 adults with LD requiring 128 hospitalization and 16 deaths. The researcher test water samples for legionella DNA from 183 cooling towers, in which 162 of them were positive to for legionella. The result of their investigation indicated that aerosol containing outbreak legionella strain from a cooling tower infected person residing several kilometer away from the tower [28]. A new 45 minutes DNA test can detect and quantify legionella bacteria compare to the 10–14 day Petri dish method [35]. Researchers determined that air condition cooling towers have aquatic environments and have presents of pathogenic free living amoebae [36]. Boamah D.K. et.al describe the interactions between Legionella with protozoa which is host for the bacteria. The biology, evolution, and persistence of Legionella in the environment are connected to the protozoa [37]. The use of biocide, bronopol, considerably reduces metal loss from corrosive bacterial strains in cooling tower water system [38]. Researcher compared biofilm removal percentage and detachment rate for six-green bio-dispersants and two non-green bio-dispersants. Higher removal percentages and detachment rates were achieved by green bio-dispersant over the non-green bio-dispersants [39]. Experimental investigation showed that changing operating conditions and distribution system in cooling tower that emissions of PM_{10} (Particulate matter with particles less than or equal to 10 micrometers in diameter) can be reduced. Additionally, water drift with PM10 can also be reduced adding drift eliminator [40]. Coal-fired power plant that utilizes an integrated liquid absorbent-based CO_2 capture and desalination processes were examined to increase water supply. The investigators showed by modifying these processes, the amount and quality of water produced can be used for makeup water for flue gas desulfurization and/ or cooling tower at negligible energy consumption [41].

Acknowledgment This chapter is dedicated in memory of Dr. William Shuster, a former Chairman and Professor of Chemical and Environmental Engineering at the Rensselaer Polytechnic Institute in Troy, New York, USA. Dr. Shuster mentored over 50 doctoral students from around the world and was well-known in the field. The authors of this chapter salute Dr. Shuster for his lifelong contributions to chemical engineering education.

Glossary

Approach The difference in temperatures of water temperature exiting the tower vs. the dry bulb temperature of the air entering the tower. (ΔT, water out vs. dry bulb air)

Basin The bottom of the tower, water is collected here after passing the tower.

Blow-down some of the water collected in the basin is exited from the tower rather than recirculated within the tower. This prevents buildup of solids in the water. *(Note: the same term is used in wet scrubbing, as the objective is also the same.)*

Cooling Range The temperature difference between hot water entering the tower and cooled water exiting the tower.

Drift Also known as *windage*, small droplets of water, which are carried out of the tower by the airflow, normally expressed as a % of total water flow rate.

Drift Eliminators Baffles, or other specialized collection products, placed in the air exit path to capture water droplets. *(Note, per the previous mention of scrubbing, again, mist elimination in scrubbing is a similar application.)*

Dry Bulb Temperature Air temperature, measured by standard means.

Heat Load Amount of heat exchanged within the tower between water and air, expressed per unit of time. It can also be expressed as the product of the quantity of water circulated within the tower per unit time and the cooling range.

Make-up Water Water that is added to the tower to replenish water losses due to evaporation, drift, blow-down and other miscellaneous water losses.

Packing Also called *fill*, this is a product placed within the tower to promote uniform water flow distribution and to enhance water-air contacts. *(Note: Again, note the similarity to wet scrubbing applications.)*

Performance Normally expressed as the amount of cooling for a given quantity of water at a given wet bulb temperature.

Relative Humidity The ratio of the partial pressure of water vapor in the air to the vapor pressure of water at the temperature of the air, which contains the water vapor.

Wet Bulb Temperature The temperature measured by an ordinary thermometer as a thin film of water on the thermometer is evaporated into a surrounding air stream. *(Note: Recall the example earlier in this discussion of cooling in Arizona* vs. *Florida – if the air surrounding the thermometer is already saturated with water, no evaporation and hence no cooling of the water can take place. In such an instance the wet and dry bulb temperatures are equal.)*

References

1. Edinger JE, Geyer JC (1965) Publication No. 65–902. Edison Electric Institute, USA
2. Berman LD (1961) Evaporative cooling of circulating water. Pergamon Press, New York, USA
3. Brady DK, Graves WL Jr, Geyer JC (1969) Publication No. 69–901. Edison Electric Power Institute, USA
4. U.S. Department of Commerce (1968) Climactic atlas of the United States. US Government Printing Office, Washington, D.C., USA
5. U.S. Department of the Interior FWPCA (1968) Industrial waste guide on thermal pollution. US Government Printing Office, Washington, D.C., USA
6. McKelvey KK, Brooke ME (1959) The industrial cooling tower. Elsevier, Amsterdam
7. Editor, Power (1973)
8. US EPA (2002) Compilation of air pollutant emission factors, AP-42, Vol. 1, Sec. 13.4, 5th edn. Federal Register, USA
9. US EPA (2002) National pollutant discharge elimination system – proposed rules. Federal Register, USA
10. U.S. Department of Energy (1995) Federal technology alerts: ozone treatment for cooling Towers, USA
11. Rosaler RC (1995) The standard handbook of plant engineering, 2nd edn. McGraw-Hill, New York, USA
12. Green DW (ed) (1984) Perry's chemical engineers' handbook, 6th edn. McGraw-Hill, New York, USA
13. Washington State University Cooperative Extension Energy Program, Technical Brief WSUEEP98013, revised (1998) February, USA
14. WQA (2002) Ozone: a reference manual. Water Quality Association, Washington, D.C., USA
15. Roseman J (2003) Cooling tower disinfection. Water Qual Products 8(4):14–16
16. Wang LK (1984) Pretreatment and ozonation of cooling tower water, part I. U.S. Dept. of Commerce, National Technical Information Service, Springfield, VA, USA
17. Wang LK (1984) Pretreatment and ozonation of cooling tower water, part II. U.S. Dept. of Commerce, National Technical Information Service, Springfield, VA, USA
18. Wang LK (1984) Prevention of airborne legionnaires' disease by formulation of a new cooling water for use in central air conditioning systems. U.S. Dept. of Commerce, National Technical Information Service, VA, USA
19. Wang LK (1985) Proceedings of the Seventh Mid-Altantic Industrial Waste Conference, pp 207–216
20. Nagl G (1999) Environ Technol 9(7):19–24
21. Shuster WW (1986) Cooling of thermal discharges. In: Wang LK, Pereira NC (eds) Handbook of environmental engineering, volume 4, water resources and natural control processes. Humana Press, Inc., Totowa, NJ, USA, pp 107–138
22. Wayne Barber, Nucleonics Week, June 4, 1998
23. Thurman G (2001) Power Mag
24. Swanekamp R (2001) Power Mag
25. Terranova J (2001) Selecting the right cooling tower on a budget: balancing project needs and the owner's expectations. HPAC Engineering, V 73 N 11, Nov 2001, IIF – IIR, record ID 2002-1531
26. Jeanine Katzel (2000) Trends in cooling towers, Plant Engineering, 54(1):38–43
27. Taricska JR (2018) Personal communication with Hai Vu of JEA. Jacksonville Florida, USA
28. Lapierre P, Nazarian E, Zhu Y, Wroblewski D, Saylors A, Passaretti T, Tran A, Lin Y, Kornblum J, Morrison S, Mercante JW, Fitzhenry R, Weiss D, Raphael BH, Varma J, Zucker HA, Rakeman JL, Musser KA (2017) Legionnaires' disease outbreak caused by endemic strain of Legionella pneumophila New York, New York, USA, 2015, Center of Disease, Control and Prevention. Emerg Infect Dis, www.cdc.gov/eid 23(11)

29. CDC (2017) Developing a water management program to reduce Legionella growth & spread in buildings. A practical guide to implementing industry standards. US Department of Health and Human Services, Atlanta, GA, USA, 1–32 June
30. Editor (2018) Spartan Bioscience Announce World's First Onsite Onsite Legionella DNA Test, Plumb. Eng., TMB Publishing Inc., Niles
31. Soares SS, Souza TK, Berte FK, Cantarelli VV, Rott MB (2017) Occurrence of infected free-living amoebae in cooling tower of southern Brazil. Curr Microbio 74(12):1461–1468
32. Boamah DK, Zhou G, Ensminger AW, O'Conner TJ (2017) From many hosts, one accidental pathogen: the diverse Protozan Hosts of Legionella. Front Cell Infect Microbio 7:477
33. Narenkumar J, Ramesh N, Rajasekar A (2018) Control of corrosive bacterial community by bronopol in industrial water system. Biotech 8:55
34. DiPippo F, de Tora F, Di Gregorio L, Buccolini M, Capocecera R, Rossetti S, Tandol V (2017) Green bio-dispersant removal efficacy estimation for controlling biofilms in cooling towers. Ann Microbiol 67(11):779–784
35. J. Ruiz, A.S. Kaiser, M. Lucas, Experimental determination of drift and PM_{10} cooling tower emissions: influence of components and operation conditions, Environ Pollut, 230, 422–431, (2017)
36. Feron P, Thiruvenkatachari R, Cousins A (2017) Water production through CO_2 capture in coal-fired power plants. Energy Sci Eng 5(5):244–256
37. FPL, www.fpl.com/clean-energy/nuclear/turkey-point-conservation.html. Accessed 29 Jan 2018
38. Editor (2013) Wastewater won't be wasted at Turkey Point, NEI, News & Media, 13 Sept
39. Casey EB (2016) US power profile shift water opportunity. Water Online 12
40. Veil JA (2007) Use of reclaimed water for power plant cooling. US Department of Energy, USA
41. JEA (2005) Best management practice and guidance manual for cooling towers

Chapter 6
Basic Hydrology, Water Resources, and DAF Boat Plant for Lake Restoration

Donald B. Aulenbach, Lawrence K. Wang, Mu-Hao Sung Wang, and Nazih K. Shammas

Contents

1 Introduction ... 232
2 Special Features of Water and Their Relationship to Life on Earth ... 234
3 Hydrology ... 236
4 Evaporation ... 245
5 Transpiration ... 249
6 Evapotranspiration ... 249
7 Infiltration and Percolation ... 251
8 Runoff ... 251
9 Ground Water ... 266
10 Impact of Pollution on Lakes ... 275
11 Thermal Impacts on the Aquatic Environment ... 283
12 Toxics in Water Resources ... 287
13 Goals of Water Pollution Control ... 289
14 Dissolved Air Flotation Boat Plant for Lake Restoration ... 293
Glossary ... 294
References ... 296

The authors Donald B. Aulenbach and Nazih K. Shammas are deceased at the time of publication of this chapter

D. B. Aulenbach
Rensselaer Polytechnic Institute, Troy, NY, USA

Lenox Institute of Water Technology, Newtonville, NY, USA

L. K. Wang (✉)
Rensselaer Polytechnic Institute, Troy, NY, USA

Lenox Institute of Water Technology, Newtonville, NY, USA

Krofta Engineering Corporation, Lenox, MA, USA

M.-H. S. Wang · N. K. Shammas
Lenox Institute of Water Technology, Newtonville, NY, USA

L. K. Wang, M. -H. S. Wang, Y. -T. Hung, N. K. Shammas (eds.), *Environmental and Natural Resources Engineering*, Handbook of Environmental Engineering 19, https://doi.org/10.1007/978-3-030-54626-7_6

Abstract This chapter introduces the basic hydrology, water resources, and a dissolved air flotation (DAF) boat plant for lake water treatment. Specific coverage includes special features of water and their relationship to life on earth, hydrology, evaporation, transpiration, evapotranspiration, infiltration, percolation, runoff, ground water, impact of pollution on lakes, thermal impacts on the aquatic environment, toxics in water resources, goals of water pollution control, dissolved air flotation boat plant for lake restoration, glossary, and a list of authors' recent publications.

Keywords Donald B. Aulenbach · Lenox Institute of Water Technology · Nazih K. Shammas · Hydrology · Water resources · Dissolved air flotation · DAF · Boat water treatment plant · DAF boat plant · Lake restoration · Water pollution control

Nomenclature

A	The tributary drainage area in hectares or acres or mi^2
B	The ratio of heat loss by conduction to heat loss by evaporation
C	The coefficient of runoff, which is equal to the ratio of runoff to precipitation for the individual area at a specific time and usually averages about 0.3 (however, if there was a recent antecedent storm that has saturated the ground, the value may approach 0.9, and it may be greater than 1 with melting ice and snow)
c	The weir coefficient, which varies between 2.5 and 3.5
C_e	A coefficient which varies, being 15 for small shallow lakes and ponds and for leaves and grass and 11 for large and deep bodies of water
d	Drawdown in m (ft) in the vicinity of the pumping well in which the observation well is r m (ft) from the pumped well
e	The voids ratio, e, which is the ratio of the pore volume to the solid volume and equals:
E	The evaporation, in/mo
e_a	The vapor pressure of the air in millibars
e_s	The vapor pressure of the water surface in millibars
F	The recurrence interval or frequency in years that a stated intensity is equaled or exceeded
f	A coefficient equal to $T_m \times P_m/100$ and equal to the monthly consumptive use
H	The head above the spillway in m (ft)
H_v	The latent heat of vaporization which is assumed to be 536 cal/g
i	The average intensity of rainfall for the specified period of time and may be expressed as cm/h (in/h)
k	The monthly consumptive use coefficient, which equals 0.55 for garden vegetables, melons, and winter grain; 0.65 for grasses, brambles, leaves, peas, potatoes, tomatoes, peppers, and small grain; and 0.75 for corn, alfalfa, and orchard with cover

K	10
L	The length of flow in feet
L	The length (width) of the spillway in m (ft)
L_a	Leakage and refers to ground water flow into or out of mercury
n	The number of events (*m* is the order of magnitude of a particular event)
O	The excess rainfall in in/h
p	The porosity, which is equal to the ratio of the pore void or interstitial space to the total volume of the rock or soil
P_a	The atmospheric pressure
P_m	The monthly percent of daytime hours of the year
P_t	Precipitation
Q	The rate of runoff at a given location and time
Q	The discharge in m^3/d (gal/min)
Q	The rate of discharge in m^3/s (cfs)
Q_b	The net energy loss by long-wave radiation to the atmosphere
Q_r	The reflected solar and sky short-wave radiation incident at the water surface
Q_s	The solar and sky short-wave radiation incident at the water surface
Q_v	The net energy advected to the water which is the difference
Q_θ	The increase in energy stored in the water
R_0	Runoff
s	Slope in ft/ft
S	Storage
Sc	The coefficient of storage as a decimal
STEM	Science, technology, engineering, and mathematics
t	The duration of rainfall in minutes
T	The coefficient of transmissibility in m^3/d unit gradient (gal/day (ft/ft))
t	The time of pumping in days
T_a	The temperature of the air in °C
T_m	The mean monthly temperature in °F
T_s	The temperature of the water surface in °C
U	The consumptive use in inches per period (usually a month)
u	A tabulated value in the Theis method
V_a	The vapor pressure of the air or dewpoint temperature, in inches of mercury
V_s	The vapor pressure of the water surface in inches of mercury
w	The wind velocity 30 ft above the surface in mi/h
$W(u)$	Theis' well function value
ρ	The density of the water

1 Introduction

It has been said that pollution is mostly a resource out of place. That is, most natural things are subject to change and recycle, but when they are exposed to a new environment or are present in excessively high concentrations, they interfere with the natural system.

When *Homo sapiens* evolved, knowledge also evolved, with its application, thinking. At first there were only a few humans around, and they did not upset the balance of nature. However, by thinking they developed tools, created housing, and controlled fire. They developed agriculture and controlled animals. They seemingly overcame nature and were able to survive longer. They learned to control diseases and to ameliorate human imperfections. But their desire was not only health, longevity, and reproduction but also comfort. In order to provide these comforts, agriculture, mining, manufacturing, transportation, and communications were developed using chemicals that were either harmful to the environment in certain concentrations, or were persistent so that they would not deteriorate. We are now starting to realize the consequences of all human activities. Also, nearly all these activities involve water in some fashion.

Before disease transmission was understood, pollution control was most likely influenced by the presence of foul odors. This could be considered an aesthetic problem or an air pollution concern. Body wastes and other decomposing materials generally have an undesirable odor, thus creating an aversion to them. Thus decomposing matter was usually removed as far as possible from the cave or other housing. Unknowingly, this controlled the spread of many diseases.

However, humans' gregarious nature caused them to congregate into communities, leading to the large cities of today. This has resulted in human wastes where they are not desirable. Inadequate management of human wastes has been traced to many of the great plagues of the Middle Ages. The commonly accepted first incidence of relating human wastes to disease is attributed to studies of a cholera epidemic in London in 1849 by Dr. John Snow. Through an epidemiological study, he traced the incidence of cholera to a well that was contaminated by human wastes in a certain block of housing. Whether or not the story is true that he resolved the problem by removing the handle from the pump, Dr. Snow did show the relationship between human wastes and disease. These findings were independent of the development of the microscope by Pasteur in France only a few years later. This led to the removal of human wastes from homes and the modern concept of water carriage of wastes. This also resulted in gross water pollution problems downstream from the sewage discharge. Thus early pollution control efforts were related to public health.

Most frequently cities were located along bodies of water, which could be used as a source of food supply and as a source of power, and the water itself could be used for drinking, for manufacturing processes, for transportation of raw and manufactured goods, and as a recipient of the wastes from all these sources. It was found that the increased organic load exceeded the reaeration capacity of the stream or lake, creating conditions inhibitory to desirable fish and aquatic organisms. Depending upon the orientation, it could also contaminate the drinking water supply. Significant pollution control activities arose in the early 1900s with disinfection of drinking water by chlorination, coagulation and filtering of drinking water, and treatment of wastes to remove the organic load before discharge. Thus began the era of the sanitary engineer.

Although waste treatment greatly reduced disease transmission, the removal of the organic materials did not reduce the oxygen demand as expected. It was observed that inorganic nutrients, particularly nitrogen and phosphorus, were little reduced in the waste treatment process. These nutrients were available to stimulate aquatic growths, especially photosynthetic algae, which is new organic material. When these algae die, they decompose and exert an oxygen demand on the receiving body of water, not unlike the organic load that was so diligently removed by the treatment system. Although this was not a public health hazard, it did interfere with biological life in the receiving waters, including eliminating certain fish. As more and more interrelationships between waste treatment and its impact on the receiving water were observed, we had the development of the so-called environmental engineer.

Despite our concerted efforts to treat wastes to a degree that the effluent will have no deleterious impact on the environment, we are continually developing and finding new substances that are or can be deleterious. These include toxic and hazardous materials frequently used in manufacturing processes, persistent or refractory substances that may have been developed for this very purpose (viz., PCBs, various insecticides and herbicides), hormones from bioengineering processes, endocrine disruptors, and antibiotics from constant use and overuse in controlling diseases and infections. In addition to direct discharges of treated wastes, there are indirect discharges such as leaching from landfills, leakage from underground storage tanks, and condensation of volatile substances in the atmosphere during rainfall. Oxides of nitrogen and sulfur from combustion produce acid rain that may be carried great distances from their source, resulting in lakes of acidity high enough to interfere directly with aquatic life, particularly fish, or to dissolve inhibitory minerals from the surrounding rocks, with similar effects. Each problem requires its own solution. Thus it can be seen that as mankind progresses, pollution problems seem to increase.

Whereas we place greatest emphasis on the effects of pollution on humans, it must be realized that other parts of the environment, including both plants and animals, may also be impacted. Frequently, these may also indirectly impact humans. A commonly used example of this is the benefit of DDT in controlling insect-borne diseases. After considerable use and study, it was found that the DDT created thin shells in the eggs of birds, thereby interfering with their reproduction. Particularly, birds of prey were reduced, resulting in an increased population of rodents and other small animals with the subsequent destruction of many crops. The chain of impact is often diverse.

Although there are sources of pollution from the air and the soil, this chapter will be restricted to the study of water bodies. This is not to lessen the importance of these other sources; however, limits must be placed on the coverage in an individual chapter. This is just one part of humans learning to control their environment without destroying it.

2 Special Features of Water and Their Relationship to Life on Earth

Water is an unusual substance. It has many physical and chemical properties that do not follow the normally expected trends. Scientists have been able to explain them. However, this does not make them any less special.

The observation of the peculiar properties of water dates back at least as far as L. J. Henderson [1], in 1912, in which he describes the properties of water that are most important. First of all was that 3/4 of the earth's surface is covered by water and that, if it were not for the mountains of land, the entire earth would be covered by 4 kilometers (2.5 miles) deep of water.

Chemically, water is very stable; very few substances react directly with it, although many reactions take place within it. Perhaps its thermal properties are the most anomalous properties of water. The first of these is its specific heat or heat capacity. Water is the standard, having been assigned a value of 1.000 in the range of 0–1 °C. This is a relatively high value and enables water to maintain a relatively constant temperature. It is important in the stabilization of biological and chemical reactions in the aquatic environment and serves to moderate the earth's climate by providing a reservoir of heat in the tropical oceans and a reservoir of cold in the polar ice caps. These effects in turn help aquatic organisms maintain a constant temperature and reduce the amount of heat regulation necessary for mammals.

A second important thermal property is water's latent heat of fusion and vaporization. The latent heat of fusion, 88 cal/g, is quite high. The latent heat of vaporization, 536 cal/g, is the highest known for any substance. The latent heats represent the amount of heat that must be provided to a given weight of water to convert it from either the solid to the liquid or the liquid to the vapor state at the same temperature. This is very important in the evaporation of water and represents a large transfer of energy from the tropical oceans, where energy is available to evaporate the water, and this energy is released elsewhere on the earth in the form of condensation. The latter effect is also very important since it regulates or moderates temperatures throughout the world, preventing the tropics from becoming excessively hot and the poles extremely cold. It is similarly important in physiological heat regulation, in terms of both perspiration in animals and transpiration in plants.

The thermal conductivity of a substance represents the time rate of transfer of heat by conduction through a unit thickness across a unit area for a unit of temperature in a unit of time. The value for water of 0.0125 cal/s-cm-cm^2-°C is quite high compared to other liquids. Ice, however, has a higher thermal conductivity, which results in a lake's more rapid cooling once the first layer of ice has formed.

Probably the most significant thermal anomaly is thermal expansion. Because of orientation of the water molecules, the most compact configuration occurs at 4 °C. Thus, the most dense water will be 4 °C. This results in deep temperate climate lakes having a fairly constant temperature of 4 °C on the bottom. Furthermore, ice, the solid state of water, is less dense than water and therefore floats on the surface. Water is the only substance whose solid state floats on its liquid surface. This is very important in our environment since if this were not the case ice would form at the bottom of all deep lakes and this ice would never thaw. Such a circumstance would severely restrict biological activity in the bottom of the lake and result in a buildup of layers of sediments and ice in the bottom of all lakes.

There are several non-thermal properties of water that are also peculiar to it. The first one to be considered is its solvent power. More elements and compounds dissolve in water than in any other liquid. This is very important in making nutrients available to biological life in the water, whose food must be waterborne. In addition, the materials dissolved in the water help to maintain the osmotic pressure of the aquatic organisms.

Water also has a very high ionizing power as a result of its high dielectric constant, which helps to reduce the force of attraction between atoms or molecules in the liquid. This is important in providing soluble nutrients to aquatic organisms and also affects cell permeability.

Only mercury has a higher surface tension than water. The value for distilled water is 7.5×10^{-6} N/m (75 dynes/cm), whereas for mercury it is 43.6×10^{-6}N/m. This creates a tough, thin elastic film on the surface of the liquid and results in the formation of drops. It also affects capillarity, which controls the rise of liquid in the soil. The surface film of a lake actually forms a habitat for certain organisms known as pleuston. Microscopic pleuston are called neuston. Those living on the upper surface of the interface film are called epineuston, and those on the underside are called hyponeuston. The pleuston include the water striders, the mosquito larvae, certain snails, and other organisms. Some organisms that do not live in the surface film nonetheless deposit their eggs there. Thus the surface film of water is a very special habitat to certain organisms.

The viscosity or resistance to flow of water varies with its temperature. This affects lake mixing, particularly during the summer, when the surface of the lake is warm and more readily mixed by the wind. It also affects the flotation of organisms in the euphotic zone (the area through which sunlight penetrates). This is important in maintaining balanced algal growths and in differentiating between the species that may be present at certain times of the year as a function of temperature.

Water is considered a noncompressible fluid, although precise measurements show that it is very slightly compressible. For all intents and purposes, it may be considered a noncompressible fluid for either measurement purposes or for the aquatic organisms living therein.

From all these physical and chemical parameters, it may be seen that water is very important in sustaining life. It is the author's belief that water is the most important factor in maintaining life on earth and that if scientists want to determine whether or not life could exist on any other planet, they must measure its presence there. In short, all life is dependent upon water and its unique properties.

3 Hydrology

Hydrology is the science of the phenomena of water in all its states; its distribution and occurrence in the earth's atmosphere, on the earth's surface, and in the soil and rock strata; and the relationship of these phenomena to the life and activities of humankind. It is concerned with both the quality and the quantity of water available. Since water is essential to the preservation of life on earth, we need a good understanding of hydrology.

The amount of water present on the earth today is essentially the same as that which was present when the earth was formed. It may be in a different physical state and some minor additions may have occurred through chemical reactions and cosmic particle bombardment, but for the most part, the total quantity of water on the earth has not changed. It does, however, undergo many physical transformations from the vapor phase to the liquid and the solid phases. This is a constant and ongoing cycle, commonly called the hydrologic cycle. Figure 6.1, from Osborn and Harrison [2], depicts the various phases and locations of water on the earth. It may be seen that water vapor occurs in the atmosphere, condenses, and forms precipitation that falls over the face of the earth, both the land and the water. A portion of the water evaporates in falling; another amount evaporates after striking objects on earth such as trees or buildings. Another portion flows along the ground, forming streams and/or lakes, whereas still another part infiltrates the soil to become ground water. Both surface water and ground water flow toward the lowest level, ultimately reaching the ocean. Here a large amount of the evaporation occurs. Evaporation occurs from all moist ground surfaces and all open waters. In addition, transpiration occurs from all plant life. Usually in evaluating the amount of water vapor formed, evaporation and transpiration are combined as evapotranspiration since it is difficult to differentiate one from the other on the earth's surface.

Tremendous amounts of water pass through the hydrologic cycle. Some idea of this can be gained from Fig. 6.2 [3]. Large amounts of energy are utilized in the evapotranspiration process in converting water to water vapor. Also, the amount of precipitation onto the land and the ocean is similar when one considers that approximately 75% of the earth is covered by the ocean. This does not imply, however, that the distribution of rainfall over the earth's surface is uniform. There are many localized factors that control the distribution of precipitation across the earth's surface.

The main source of energy for evapotranspiration is the sun. Small increases in evaporation may occur as a result of human activities, particularly in the use of heat for power generation. However, this total amount is small compared to the energy of the sun and is significant only in localized areas. When water evaporates, the vapor is lighter than air; therefore, it rises and expands as the atmosphere rarifies. Expansion

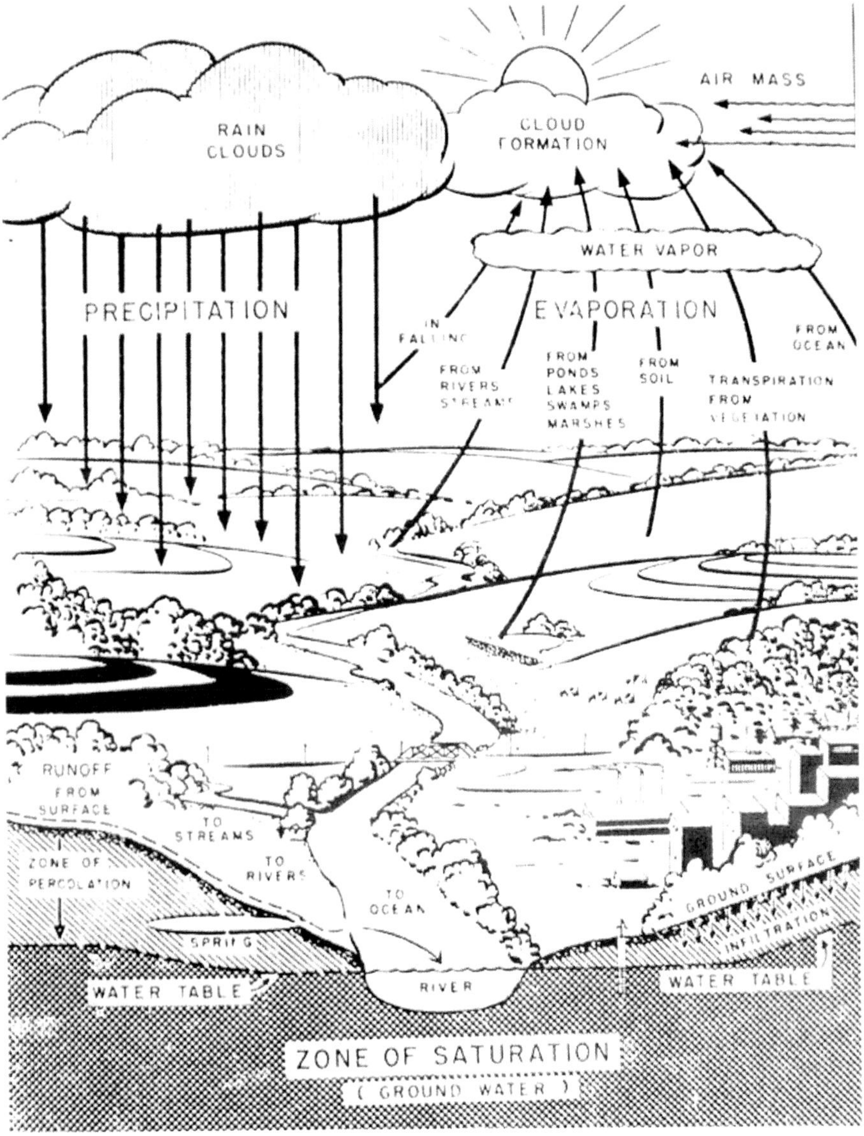

Fig. 6.1 The hydrologic cycle [2]

causes cooling, which ultimately results in condensation, releasing again the energy that was required to evaporate the water initially. Fortunately, the water vapors do not rise in the atmosphere indefinitely. About 50% of the water condenses before it reaches 1.6 km (1 mi) altitude. About 4% remains at 6 km (4 mi) and essentially none has been observed above 20 km (12 mi). Therefore, the water vapor does not escape the earth's atmosphere.

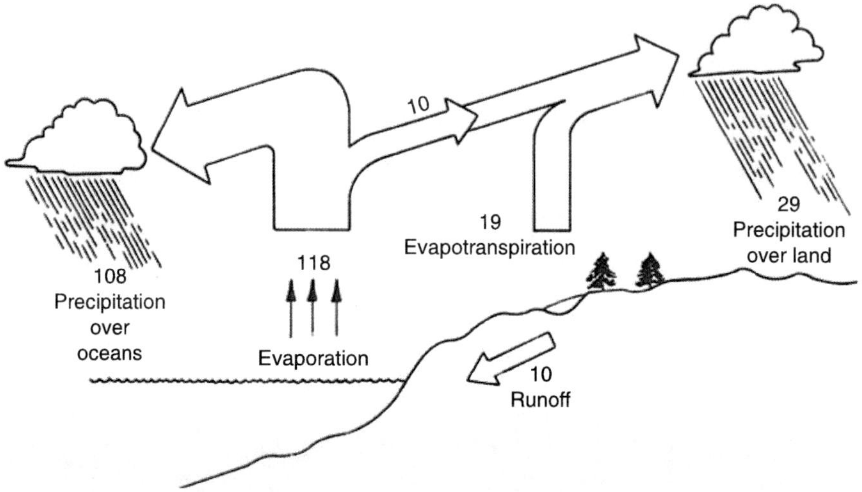

Fig. 6.2 World hydrologic cycle. Units are 10^5 gal/yr [3]

Precipitation occurs in many forms. Rain, in which droplets of liquid water fall to the earth, is the most common. Drizzle is the descent of very fine droplets of liquid water, usually with a diameter of less than 0.5 mm, which are small enough so that wind currents easily carry them both horizontally and occasionally upward. Snow is the precipitation of solid water, primarily in the form of hexagonal crystals. In North America, sleet is rain that falls through subfreezing air masses that cause the rain to freeze into ice pellets. In Britain, sleet is considered melting snow or a mixture of snow and rain. Freezing rain is caused when liquid rain falls on a subfrozen object coating it with ice. Whereas sleet occurs primarily in winter, hail is a form of frozen rain droplets that are found primarily in summer thunderstorms. In the formation of hail, the water droplets fall through a subfreezing air mass, forming small ice pellets. Depending upon the rising force of the air masses, these ice pellets may be carried upward into the warmer air mass where more condensation occurs, therefore, increasing their size. They fall again through the cold air mass, where the newly condensed water on the surface freezes forming a larger hailstone. Repeated rising, growing, falling, and freezing has been known to create hail stones as large as 5 cm (2 in). Dew forms directly from condensation on the ground, mainly during the night when the ground surface has been cooled by outgoing radiation. In some circumstances, dew can be a significant portion of the total precipitation.

All of precipitation is the result of cooling that causes the condensation of the water vapor. There are three main forms of cooling: dynamic, upward movement of air, and external cooling. Dynamic cooling is caused by the expansion of the air as it rises to a higher altitude. In general the temperature of the air above the earth's surface decreases with altitude. This change in temperature with altitude is called the lapse rate. In the first 3–5 km above the earth's surface, it is extremely variable. In some cases it is even negative (an increase in temperature with altitude), which is called an inversion. Generally in the range of 5–11 km of altitude, the decrease in temperature averages 0.7 °C/100 m altitude (3 °F/1000 ft). In the free atmosphere, where there are no other sources of heat or means of removal, the rising air cools at the dry adiabatic lapse rate. In the middle latitudes this decrease in temperature is 1°C/100 m (5.5 °F/1000 ft). This decrease in temperature occurs until the dewpoint is reached. It must be pointed out that the dewpoint also decreases at a rate of 0.2 °C/100 m (1.1 °F/1000 ft). After the dewpoint has been reached, the wet or, as it is sometimes called, retarded adiabatic lapse rate occurs. This results in a decrease in 0.6 °C/100 m (3.2 °F/1000 ft). The difference between the wet and dry adiabatic lapse rate is caused by the latent heat of condensation, which releases energy as condensation occurs.

Under adiabatic conditions, three scenarios can occur, as shown in Fig. 6.3. The first scenario represents stable air in which the existing lapse rate is less than both the wet and dry adiabatic lapse rate. Therefore any rising air would be cooled more rapidly than the surrounding air and, since it would be cooler, it would not rise. Under this condition there would be no convective rising. This is depicted in the first diagram in Fig. 6.3. The second scenario is one of unstable air. If the existing lapse rate is greater than both the wet and dry adiabatic lapse rates, any rising air will be cooled less rapidly than the surrounding air and thus would be relatively warmer than the surrounding air and would continue to rise. This creates the unstable conditions and is shown in the middle diagram in Fig. 6.3. If the existing lapse rate is less than the dry adiabatic lapse rate, but greater than the wet adiabatic lapse rate, the scenario is one of conditional stability. The condition would be stable only when no moisture is condensing. However, when moisture is condensing, unstable conditions would develop and the water vapor would continue to rise. This condition is useful in artificial rainmaking, which relies upon initiating the precipitation using dry ice, silver iodide, or some other nucleating agent.

Upward movements of air that cause precipitation generally fall under three categories. The first is convective movement, which is typified by tropical rains and thunderstorms. The surface air is heated rapidly by a warm sun during the day. The water vapor rises rapidly, expanding to the point at which it then cools, condenses, and precipitates. In tropical areas this is responsible for the routine daily precipitation that falls sometime in early afternoon. Thunderstorms may occur any time of the day or night, but are most common late in the afternoon as a result of the rapid daytime warming by the sun.

The second upward movement of air is called orographic precipitation. This occurs when horizontal currents of warm air pass over large bodies of water, such as the ocean, where they pick up moisture. When they reach land, the air is forced upward by the coastal mountains where it is cooled and forms precipitation. Since the moisture is removed in the upward motion that produces this cooling, the air mass becomes warmer and hence drier as it proceeds down the other side of the mountain. Thus on the windward side of the mountain, areas of heavy rainfall occur, whereas on the leeward side deserts occur. As a typical example, in the Pacific Northwest, 190–300 cm/yr of precipitation occurs (75–125 in/yr) producing rain forests on the windward sides of the mountain ranges. On the leeward sides, only

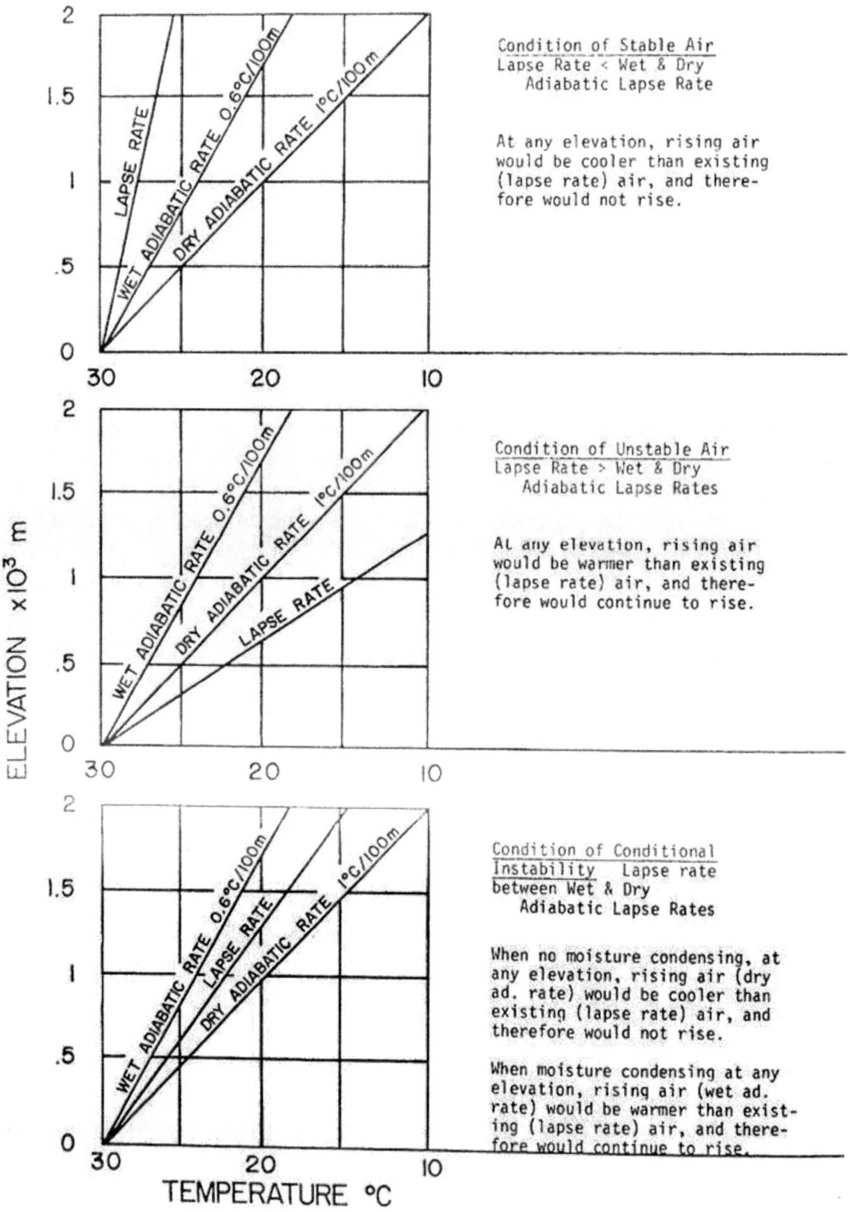

Fig. 6.3 Possible conditions of atmospheric stability

about 5% as much precipitation occurs. In some areas this is called the Chinook effect.

The third upward movement of air is caused by cyclonic motion. The air around low-pressure areas flows in a counterclockwise direction. On the surface of the earth,

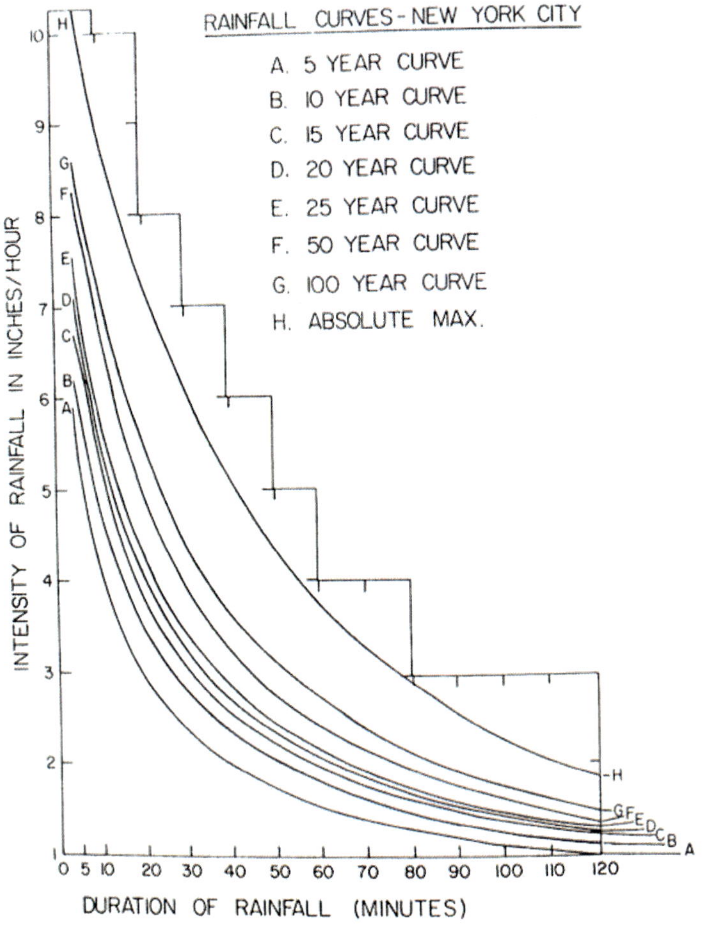

Fig. 6.4 Rainfall intensity-duration curves for New York City

the air moves from areas of high pressure to areas of low pressure. This creates areas of differential temperature, and where such cooling occurs precipitation also occurs.

The third scenario for precipitation is caused by external cooling, usually the presence of cold air masses. When cold air comes down out of the Arctic (from the Antarctic below the Equator), it comes in along the surface of the earth, forcing the warm moist air upward. This creates cooling that results in precipitation.

In addition to the variation of precipitation with location and means of formation, precipitation within a storm also varies with time. As a general rule, the longer the duration of precipitation, the less is the rate of precipitation. Figure 6.4 depicts the intensity duration curves for New York City. The figure also shows the recurrence interval or the number of years that a given precipitation would not be exceeded. For design parameters, the cumulative worst possible storm is also plotted. It is clearly

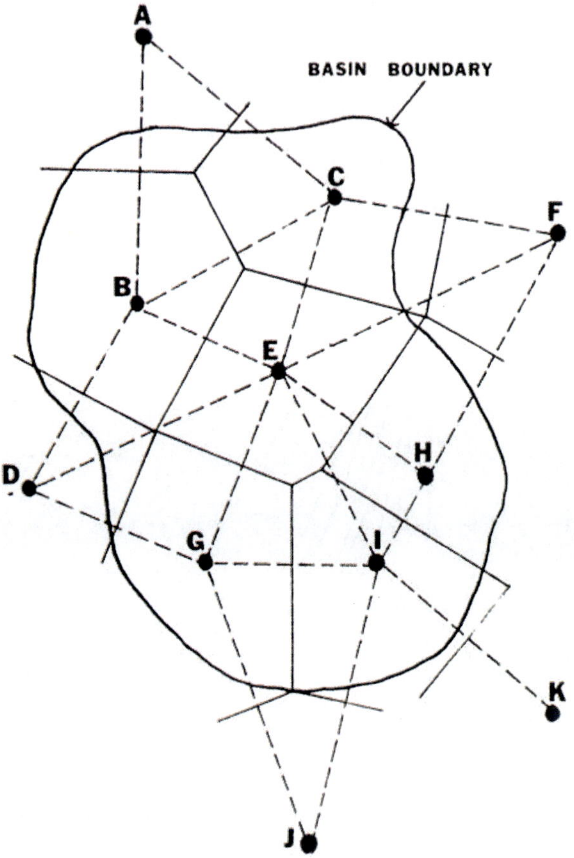

Fig. 6.5 Thiessen method for average precipitation over an area

evident from this figure that the longer the duration of precipitation, the less intense will be the storm.

In addition to the variation of precipitation with time, there is a very definite geographic variation in precipitation, particularly in considering a large area such as a drainage basin. Numerous methods have been presented to determine a representative value for precipitation over a large drainage basin area. The first method for determining a representative value would be to take the arithmetic mean of several precipitation stations within the basin. Values can be based on monthly or annual precipitation, but usually are not based on shorter periods of time. In the arithmetic method, only values from stations that are located within the stated drainage basin area are used. No results from adjacent stations are employed, even where a station may be very close to, but outside of, the drainage basin under study.

This has been corrected significantly by utilization of the Thiessen method [4] for determining precipitation over an area (Fig. 6.5). In this method stations located close to the basin boundary may be considered, but only a weighted average as a

Table 6.1 Calculations for Thiessen method

(1)	(2)	(3)	(4)	(5)	or (5a) omit (4)
Location	Ppt'n cm	Area km^{-2}	Fraction (or %) of total area[a]	Weighted ppt'n, cm (2) × (4)	Ppt'n × Area (2) × (3)
A	1.65	7	0.01	0.0165	11.55
B	3.71	120	0.90	0.7049	445.20
C	4.88	109	0.18	0.8784	531.92
D	3.91	20	0.03	0.1173	78.20
E	6.83	120	0.19	1.2977	819.60
F	7.16	0	–	–	–
G	7.57	92	0.15	1.1355	696.44
H	11.43	76	0.12	1.3716	868.88
I	12.70	82	0.13	1.6510	1041.40
J	4.44	0	–	–	–
K	4.95	0	–	–	–
Σ		626	1.00	7.17	4492.90/ 626 = 7.18

[a](4) = $A/\Sigma A$ = (3)/Σ(3) (×100 if using %)

function of the area within the given basin that is represented by that particular station. To put it another way, if a station adjacent to the basin is so close to the basin that the perpendicular bisector of a line between a station immediately outside of the basin and its closest station inside the basin falls within the basin, then the area closer to the outside station, but within the basin, is considered representative of the results of the station located outside the basin. In order to depict this, a stylized drainage basin is chosen as shown in Fig. 6.5. The points indicate the location of precipitation measuring stations within and adjacent to the basin. In order to determine the portion of the total area represented by each station, construction lines (shown as dotted lines in the figure) are drawn between each adjacent station. Then perpendicular bisectors of each of these dotted lines are drawn. The perpendicular bisectors are extended until they meet a similar perpendicular bisector of an adjacent line. With careful construction all lines should meet in points and it can be determined whether or not an area inside the basin is closer to a measuring station outside the basin. The perpendicular bisectors are drawn as solid lines on the figure. Then by means of a planimeter the area represented by each station may be measured. The total precipitation in the basin is then derived from the sum of all of the individual values of precipitation at the representative measuring stations multiplied by the fractions of the area of influence within the basin from that measuring station. A typical example of the calculations involved is shown in Table 6.1.

A third method for measuring precipitation in a large basin is the isohyetal method. Here the values of precipitation for each station are indicated on a map showing the location of each station. Smooth lines are drawn through areas of equal precipitation, as shown in Fig. 6.6. Now the area between each contour line is measured with a planimeter and the average precipitation between the two lines is multiplied by the ratio of that area to the total basin area, the sum of the individual

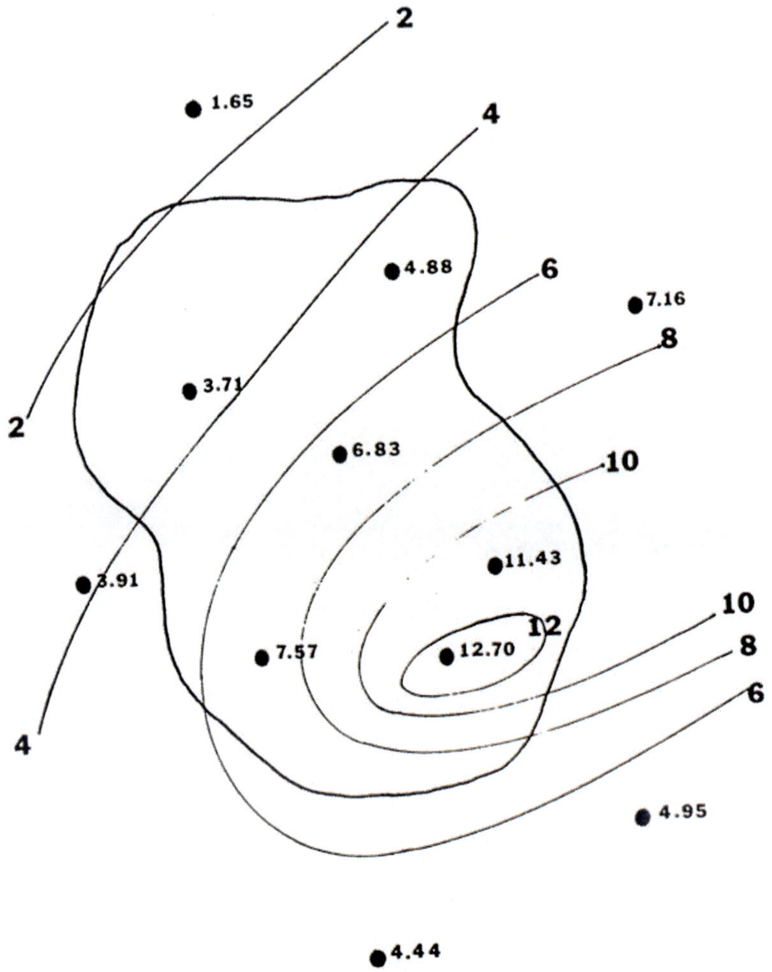

Fig. 6.6 Isohyetal method for average precipitation over an area

values giving the total precipitation over the drainage basin. An example of the calculation for Fig. 6.6 is shown in Table 6.2.

All of these methods have some sources of errors. However, the use of the isohyetal method can also take into account such known conditions as mountains, lakes, and so on. Thus, the method may be improved by some subjective interpretation of the data.

Of importance to the engineer is the frequency of recurrence of a storm of a given intensity or the maximum or minimum rainfall over a given period of time. This normally requires extended periods of record that frequently are not available. In lieu of extended records, information may be obtained from a similar location or a nearby location where long-term data are available. Where sufficient data are available, the

Table 6.2 Calculations for isohyetal method

(1) Isohyet, cm	(2) Area (*A*), km^2	(3) Fraction (or %) of total area $A/\Sigma A = (2)/\Sigma(2)$	(4) Avg. ppt'n, cm	(5) Weighted ppt'n, cm (3) × (4)
12	13	0.021	12.2	0.2562
10–12	77	0.123	11.0	1.3530
8–10	116	0.185	9.0	1.6650
6–8	145	0.232	7.0	1.6240
4–6	150	0.239	5.0	1.1950
2–4	105	0.168	3.0	0.5040
2	20	0.032	1.8	0.0576
Σ	626	1.000		6.6500

Alternative for column (5): (Σ(2) × (4))/Σ(2) = (5)

probability of any event occurring can be determined by plotting the individual values on probability paper. Either an arithmetic normality or a geometric normality may be plotted. This is accomplished by arranging the individual data points in ascending (or descending) order of magnitude. The probability of recurrence of any value then is the order of magnitude of that event divided by the total number of events plus 1. These values can then be plotted on a graph similar to Fig. 6.7. To determine the probability of any event exceeding (or being less than) a given value, one needs only to refer to the probability curve. The mean value is that which is exceeded every other year as a statistical average and this produces a mean value at the 50% occurrence. For a value that would be exceeded once in 10 yr, one would refer to the 90% level and for once every 100 yr refer to the 99-yr probability and so on. It must be remembered that these are statistical values. Figure 6.7 was plotted from 26 actually measured values. It may be seen that the upper end of the curve does not fit the data points. Furthermore, there is nothing to prevent a 100-yr occurrence from happening 2 yr in a row. The information merely provides data upon which the engineer can design a facility or a structure to handle the worst anticipated problem in a given period of time. The value of the structure and the value of the potential property damage must all be taken into account. For example, a dam should be constructed to handle the worst possible conditions, whereas a small storm drain could be designed to overflow on an average of once every 20 yr if there would be no significant damage caused by such flooding. Obviously, social concerns must also be considered, not just the financial aspect.

4 Evaporation

Evaporation is the formation of water vapor. Strictly speaking, it is the formation of water vapor from a liquid surface; however, for the sake of this discussion, evaporation here will also include evaporation from the solid state or ice, which technically is called sublimation.

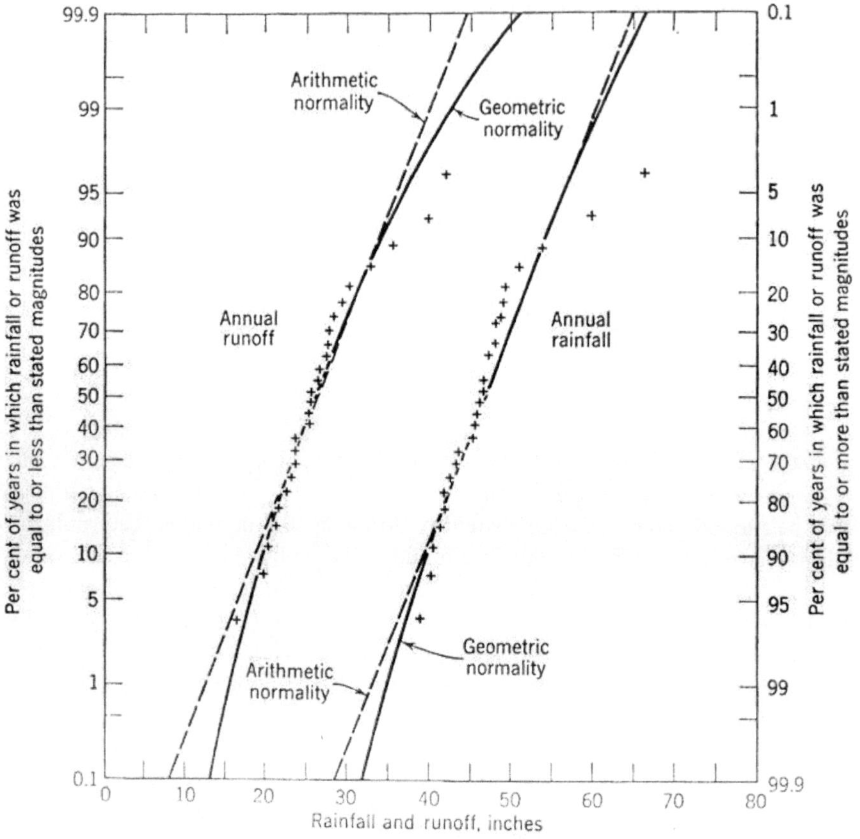

Fig. 6.7 Frequency distribution of annual rainfall and runoff

The most important factor causing evaporation is the amount of solar radiation. Indirectly, other factors affecting evaporation are also related to solar energy. The vapor pressure of the air controls the relative humidity; the lower the vapor pressure, the greater the evaporation. The temperature of both the water and the overlying air affects the vapor pressure. The higher the temperature, the greater will be the evaporation. Wind is very important in creating evaporation in that it carries away the saturated monolayer of water vapor immediately above a body of water. This brings in fresh air with a lower vapor pressure, thereby allowing increased evaporation to occur. An increase in the barometric pressure decreases evaporation. Thus, the lower pressure at greater elevations would be expected to result in increased evaporation. However, this effect is generally small and is exceeded by those from the temperature differential and the wind velocity, each of which has a greater effect upon evaporation at higher altitudes. The greater the quantity of dissolved salts in the water, the less evaporation occurs. Thus, slightly less evaporation occurs over oceans than over fresh water bodies. The difference becomes significant only in the evaporation of salt brine. The motion of water also increases the evaporation as a result of the increase in surface area created by waves and of course by the formation of spray droplets from waterfalls. Also important is the shape of the bottom of the body of water, particularly in lakes. The greater the relative area of shallow water to deep water, the higher will be the rate of evaporation.

Several means are available for determining the amount of evaporation from open water surfaces. The simplest is the use of the water budget, following the equation:

$$E = P_t - R_0 \pm S \pm L_a \tag{6.1}$$

in which

E = evaporation
P_t = precipitation
R_0 = runoff
S = storage
L_a = leakage and refers to ground water flow into or out of the given basin

The precipitation may be fairly readily measured by means of rain gauges. The runoff can be determined by a flow-measuring device on the outlet stream of the body of water. Storage merely refers to the change in elevation during the period of reference. It relates to the volume of water involved in any change in elevation. Normally on an annual cycle using the water year beginning October 1, the storage is approximately 0 from year to year. The ground water inflow or outflow is rather difficult to determine. In actuality, this equation is often used to determine the amount of leakage based upon some measured amounts of evaporation.

Evaporation (E) may also be determined by the use of an energy budget. In this case, the equation may be written as:

$$E = \frac{Q_s - Q_r - Q_b + Q_v - Q_\theta}{\rho H_v (1 + B)} \tag{6.2}$$

This equation gives E in cm when the energy values of Q_s, Q_r, Q_b, Q_v, and Q_θ are measured in cal/cm^2.

Q_s = the solar and sky short-wave radiation incident at the water surface
Q_r = the reflected solar and sky short-wave radiation incident at the water surface
Q_b = the net energy loss by long-wave radiation to the atmosphere
Q_v = the net energy advected to the water which is the difference between the energy of the inflow and the outflow of the given basin
Q_θ = the increase in energy stored in the water
ρ = the density of the water
H_v = the latent heat of vaporization which is assumed to be 536 cal/g
B = the ratio of heat loss by conduction to heat loss by evaporation and is commonly known as the Bowen ratio. The Bowen ratio (B) may be written as:

$$B = 0.61 \frac{T_s - T_a}{e_s - e_a} \left(\frac{P_a}{1000} \right) \tag{6.3}$$

in which

T_a = the temperature of the air in °C
T_s = the temperature of the water surface in °C
e_s = the vapor pressure of the water surface in millibars
e_a = the vapor pressure of the air in millibars
P_a = the atmospheric pressure

There are numerous mass transfer equations that can determine evaporation mathematically. Most of these are empirical and have been designed to fit a specific location. Typical among these is Meyer's equation:

$$E = C_e \left(V_s - V_a \right) \left(\frac{1 + w}{K} \right) \tag{6.4}$$

in which

E = the evaporation, in/mo
V_s = the vapor pressure of the water surface in inches of mercury
V_a = the vapor pressure of the air or dewpoint temperature, in inches of mercury
w = the wind velocity 30 ft above the surface in mi/h
K = 10
C_e = a coefficient which varies, being 15 for small shallow lakes and ponds and for leaves and grass and 11 for large and deep bodies of water

Evaporation may also be measured by means of a pan. Here the basic principle is to place water in an appropriate pan and to measure the amount of water evaporated over a given period of time. There are numerous sizes, shapes, and depths of pans. However, they come under three basic types. First is the sunken pan, which is usually placed in an indentation on the shore of the body of water whose evaporation is to be determined. By placing these in the ground, there is less heat loss through the sides, and the pans may be somewhat larger; however, they have a higher cost and are not portable. The second type is the floating pan, which is placed directly in the body of water. These most nearly simulate evaporation from that water, since there are no significant heat losses in this system. However, this type presents problems particularly from splashing and wave action, which can change the volume of water inside or outside of the pan. Also, it is difficult to read the water level when the water in the pan may be sloshing back and forth, and they are difficult to maintain in place. The third type of pan is a surface pan and is the most common type used. It is portable and relatively inexpensive. The standard weather bureau class A pan is 4 ft (1.2 m) in diameter and 10 in. (25 cm) deep. It is filled to 8 in. (20 cm) and evaporated to 7 in. (17.5 cm). It is placed above the ground adjacent to the body of water whose evaporation is to be measured. Nomographs and, of course, computer programs can be used to determine the pan coefficient, which relates the evaporation in the pan to the actual evaporation in the body of water.

In addition to evaporation from a water surface, there is evaporation of moisture from soil. This is usually a function of the evaporation of recent precipitation, but where the ground water level is close to the ground surface, capillary action may maintain moisture at the surface of the ground, where the moisture may evaporate. As a general rule, ground cover reduces the amount of evaporation from the soil surface. Compared to the evaporation from bare ground, areas covered by vegetation may result in only 0.7–0.8 as much evaporation, and a dense forest may result in only 0.2–0.4 as much. In general, the same factors that affect evaporation from an open water surface also have similar effects upon evaporation of the soil moisture.

5 Transpiration

Transpiration is the vaporization of water from the breathing pores of leaves and other plant surfaces. This amount can be significant in forested areas and can be as much as 2 million pounds of water per acre per year (300,000 kg/ha-yr). The amount of transpiration increases significantly with an increase in temperature, and sunlight is necessary for the process to occur. It is relatively difficult to measure transpiration alone, but it can be determined experimentally in special plots, greenhouses, and so on. However, it is difficult to compare these experimental studies with actual field conditions. In general under field conditions, it is almost impossible to separate transpiration from evaporation from the soil surface.

6 Evapotranspiration

Since it is difficult to determine transpiration alone, and since the total amount of vaporization of water is really the important parameter, evaporation and transpiration are usually combined in a term called evapotranspiration.

Evapotranspiration may be estimated by the inflow-outflow method, which is essentially the same as the method for measuring evaporation, where the two are combined and called evapotranspiration.

Considerable attention has been paid to the total of evapotranspiration by soil scientists. Meyer [5] has designed a series of charts that show the estimated amount of evaporation as a function of the rainfall during a month and the potential transpiration over the same period of time. These charts are available for different latitudes, but are basically designed for use in the middle latitudes. From the charts for evaporation and transpiration, the total evapotranspiration for a given area may be estimated.

Another method for determining evapotranspiration is the Blaney-Criddle method [6]. This method provides the potential consumptive use: that is, the amount of evapotranspiration that would occur if sufficient precipitation were available. It can be modified to provide the actual use based on the precipitation and a balance sheet. The basic equation for this method is:

$$U = \Sigma(f \times k) \tag{6.5}$$

in which

U = the consumptive use in inches per period (usually a month)

f = a coefficient equal to $T_m \times P_m/100$ and equal to the monthly consumptive use factor

T_m = the mean monthly temperature in °F

P_m = the monthly percent of daytime hours of the year

k = the monthly consumptive use coefficient, which equals 0.55for garden vegetables, melons, and winter grain; 0.65 for grasses, brambles, leaves, peas, potatoes, tomatoes, peppers, and small grain; and 0.75 for corn, alfalfa, and orchard with cover

Obviously with this method the operator must know the areas of each crop under production in the area being studied. The time of growth must be known and information must be available about the possibility of a second crop. This is a function of the length of the growing season for the particular area. The method gives more precise data, but obviously considerable calculation is involved in determining the areas of the various crops produced. Since U is merely the potential consumptive use, a balance sheet must be made up over a complete growing season or year, as appropriate. When precipitation exceeds potential evapotranspiration, the excess moisture may be stored. When evapotranspiration exceeds precipitation, the potential consumptive use may be fully utilized if there is sufficient storage from the previous month when there was excess precipitation. This may continue until all of the stored water has been utilized. Beyond this point, evapotranspiration cannot exceed the precipitation. The typical situation for the Northern latitudes is to have excess precipitation during the spring and insufficient precipitation in summer to allow for all of the potential consumptive evapotranspiration. One of the uses of this method is to determine when and how much additional irrigation would be needed to provide the optimum crop growth.

A more detailed method, based upon the potential evapotranspiration, for determining evapotranspiration has been developed by Thornthwaite [7]. The formula involves the relationship between temperature, latitude, and potential evaporation, and the actual use also depends upon the climate, soil moisture supply, plant cover, and land management. Since this method also determines potential evapotranspiration, a balance sheet similar to the Blaney-Criddle method must be utilized in order to determine the actual evapotranspiration over a growing season.

7 Infiltration and Percolation

The term infiltration represents the entrance of surface water into the ground at the soil-water or soil-air interface. It is affected by topography, the surface conditions of the ground, the characteristics of the surface material, and the rate of precipitation. It is greatest with plowed fields, with vegetative cover, and on flat ground.

Percolation represents the flow of water within the soil in both the horizontal and vertical directions. There is a fine line of distinction where infiltration ends and percolation begins. Percolation is controlled by gravity and the type of soil through which the water is flowing. Percolation will continue in a downward movement until an impermeable barrier is reached. This may be a very fine material such as clay or an impermeable substance such as rock. From this point the water will flow in more of a horizontal direction, but still down slope along the top of the impermeable barrier. The velocity of flow is a function of the soil material and the amount of water being applied to force the system in a downward direction as a function of gravity. This zone of water above the first confining layer is called the free or unconfined ground water. Since at various locations on the earth there are intermittent layers of porous and nonporous materials, water may enter a lower level and in flowing downward may pass under an impermeable layer and become a confined aquifer. Confined aquifers usually flow under pressure as a function of the connection to the free ground water surface. Thus there may be numerous aquifers as one proceeds deeper into the earth. However, as a general rule, the deeper the aquifer, the more dissolved solids are present and the poorer the quality of the water. Since the free ground water has the potential to be contaminated from the surface, the most desirable water from the standpoint of domestic use is usually that found in the first confining aquifer. A more detailed discussion of ground water is given in a separate section on ground water.

8 Runoff

Precipitation which does not evapotranspire back into the atmosphere or infiltrate into the soil as ground water runs across the surface of the earth as runoff, forming streams and lakes. These may be used as a source of water for various uses including navigation, recreation, waterpower both directly and in the generation of electricity, and many industrial uses including production of materials and cooling water. Surface waters serve as a water supply for many people particularly for large metropolitan areas. The animals in the water provide food for many people. When used for irrigation the water may be used for growing crops on land. Runoff represents the earth's freshwater resources. Once the water flows to the ocean, it becomes saline and takes on significantly different characteristics.

There are numerous factors that affect stream flow. These may be listed as follows:

1. Precipitation, particularly the amount of water equivalent, its distribution across a given watershed, and the extent and duration of storms.
2. Temperature is important particularly with regard to the accumulation of ice and snow during the winter months in temperate and polar climates.
3. Topography, specifically the slope and character of the ground.
4. Geology with regard to the relative imperviousness of the soil and the bottom of a stream. This represents the difference between surface water runoff and ground water.
5. Ground surface covering, specifically the extent, character, and type of vegetation.
6. The storage of water, whether it be naturally in lakes, ponds, swamps, or marshes, or under artificial storage in dams, in reservoirs, or within dykes.
7. The watershed itself, particularly the size, the prevailing winds, and the orientation of the watershed in relation to storms and mountains.
8. Wind intensity which will carry precipitation into or out of a basin and which, under unusual conditions, may even control the outflow of a river.
9. Erosion and silt, which tend to fill the stream bottom causing the channels to change and to increase flooding potential with subsequent storms.

When studying a given watershed, items 3–9 are normally considered to be constant. There can be some allowable changes in the character and type of vegetation, and artificial storage can be provided. Within limits, erosion and silt pollution can be controlled. Also in studying runoff, conditions during the nonfreezing time of the year are usually considered. However, it must be recalled that the worst potential for flooding in a stream occurs when a heavy rainfall coincides with spring snowmelt. This leaves the amount and distribution of precipitation as the most important factor to consider when evaluating storm runoff in a given drainage basin.

In evaluating storm runoff, the greatest concern is normally the peak or maximum rate of discharge. Normally, the peak discharge occurs at the time of concentration, which is the longest time for flow from the farthest point in time in any given drainage basin to reach a specific location downstream. The development of the time of concentration, t_c, is shown in Fig. 6.8. An idealized drainage basin is depicted with a stream flowing through it. The point of determination of flow is designated as point A. The units of time are designated as the numbers 1–5. This indicates that in the first unit of time all of the water falling within line B will reach point A. These lines are not necessarily arcs of a circle, but may be contoured as a function of the slope and the character of the ground. However, for convenience they are shown as relatively smooth lines on the figure. By the end of the second time period, all areas within line C will contribute flow to point A. Continuing this analogy, finally, after somewhat more than five time periods, any rainfall that occurs within the given drainage basin area will flow to point A during that given amount of time. The time needed for the rainfall that comes down the farthest distance from point A in the basin to reach that point A is considered t_c. As the area contributing to point A

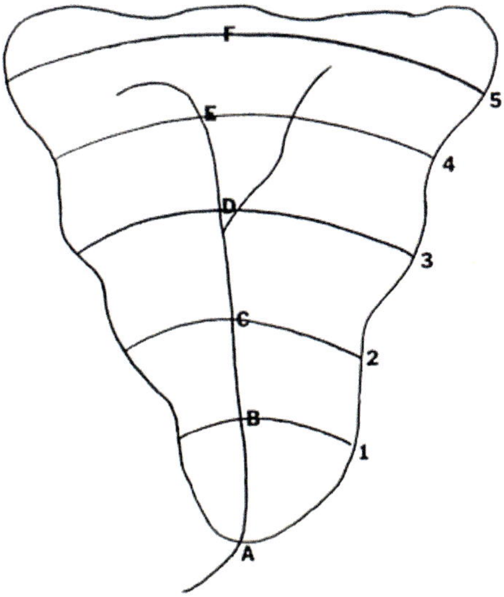

Fig. 6.8 Illustration of time of concentration, t_c, in a drainage basin

increases, the amount of runoff reaching point A increases significantly. However, it must be recalled from Fig. 6.5 that as the time of precipitation increases, the intensity of precipitation decreases. Thus, up until the time of concentration, there are two factors that affect the rate of runoff: they are the increasing area with time and the decreasing rate of precipitation with time. However, after t_c, the area contributing is constant and the only factor affecting runoff is the duration of precipitation, which results in a decreasing rate of precipitation within time. Thus, a curve of runoff vs. time during a storm event is depicted in Fig. 6.9, which shows that the initial increase in runoff is affected by both the increasing area contributing and the decreasing intensity of precipitation up to t_c and, thereafter, the decrease in runoff is caused by the decrease in intensity of the storm. As a general rule, for a given basin, the shorter the t_c, the higher will be the peak runoff rate; or conversely, anything done to increase t_c, will reduce the peak flow.

The value of t_c can be determined for a small drainage area, such as that going into a storm sewer or culvert, or for a large area to determine the flow in a stream. The main difference is the time factor, whether this is in minutes or days. An empirical equation to calculate t_c has been given by Homer and Flint [8]:

$$t_c = 4.68L^{0.332}O^{-0.675}s^{-0.281} \tag{6.6}$$

in which

L = the length of flow in feet
O = the excess rainfall in in/h
s = slope in ft/ft

For L in m and O in cm/h, the constant becomes 3.62. Excess rainfall is defined as that which produces runoff. It may be seen that this is also a function of prior wetting or the wetness of the ground surface at the beginning of the storm. The highest runoff

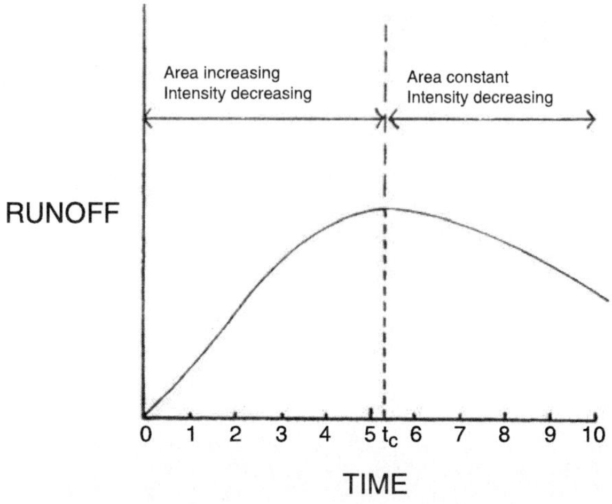

Fig. 6.9 Factors influencing a storm hydrograph

occurs when a storm occurs shortly after a previous storm that has saturated the ground with water. Then a greater portion of the new precipitation occurs as runoff instead of infiltration.

There are numerous methods available for estimating storm runoff. The statistical method implies a Type III Pearsonian family curve. This is valid when large amounts of data are available from previous runoff data. Frequently insufficient data are available and other methods must be used to estimate the storm runoff.

The most common method for measuring runoff is the rational method using the equation:

$$Q = CiA \tag{6.7}$$

Q = the rate of runoff at a given location and time.

C = the coefficient of runoff, which is equal to the ratio of runoff to precipitation for the individual area at a specific time and usually averages about 0.3. However, if there was a recent antecedent storm that has saturated the ground, the value may approach 0.9, and it may be greater than 1 with melting ice and snow.

i = the average intensity of rainfall for the specified period of time and may be expressed as cm/h (in/h). It may also be calculated by the equation:

$$i = \frac{KF^x}{a + t^n}$$

in which $K = 5–50$, $a = 0–30$, $n = 0.5–1.0$, and $x = 0.1–0.5$. All of the values K, a, n, and x vary with geographical location. F is the recurrence interval or frequency in years that a stated intensity is equaled or exceeded. Thus, F equals $(n + 1)/m$ where n is the number of events and m is the order of magnitude of a particular event; t is the duration of rainfall in minutes. The values expressed all present i in units of in/h.

A = the tributary drainage area in hectares or acres or mi^2. For ready conversion factors, there are 640 ac/mi^2. Also, if English units of area in acres are combined with precipitation in in/h, the value of Q becomes ac-in/h and this is approximately equal to 1 ft^3/s (cfs) (actual value, 1.002). If metric units of hectares are used for area, the value Q is given in ha-cm/h, which can be converted since ha-cm = 100 m^3. The intensity of the rainfall can also be determined by a precipitation gauge and knowing the duration of precipitation. This will provide the average intensity of rainfall during the time of precipitation.

Numerous empirical relationships based upon the rational method have been proposed as shown in Fair et al. [9] and Gray [10]. As may be expected, the empirical equations apply most validly to the specific drainage basin being studied by the researchers who published their results.

The rational method is useful for mean storm runoff values over the period of time of a storm. For the maximum flow, the time of precipitation used for the equation should be the time of concentration, t_c. To determine runoff at any particular time during and after a storm occurrence, various values of time would have to be chosen and the total flow integrated over a longer period of time.

In order to avoid the multiple calculations in determining storm flow by the rational method, Sherman [11] has devised a system called the unit hydrograph. The hydrograph applies only to a specific drainage basin and is derived from data pertaining to runoff over a period of one complete storm event. Based on this storm hydrograph, the anticipated flow from other storms in that same drainage basin can be estimated using four basic assumptions. These are:

1. The effects of all physical effects of a basin are reflected in the shape of the storm runoff hydrograph. That is, all runoff in that specific basin will have the same characteristic shape of curve.
2. For a given duration of rainfall, the duration of runoff is essentially constant and independent of the magnitude of the rainfall. This indicates that the curve is the same length for the same time of precipitation.
3. The ratio of added runoff for a certain duration of rainfall is proportional to the total volume of storm runoff. This means that an increased magnitude of precipitation over the same duration of rainfall will increase proportionately all portions of the hydrograph curve.
4. Precipitation extending over several durations of time, with or without interruption, creates a hydrograph that is a composite of all the hydrographs from unit hydrographs starting with the first duration of time of rainfall. The total runoff is a sum of the individual runoffs based upon the unit hydrograph values.

The unit hydrograph is the hydrograph of direct runoff from any storm that produces exactly 1.0 in of net rain [12]. The duration of rainfall varies as a function of the size of the basin; for stormwater inlets it may be in the order to 10 min,

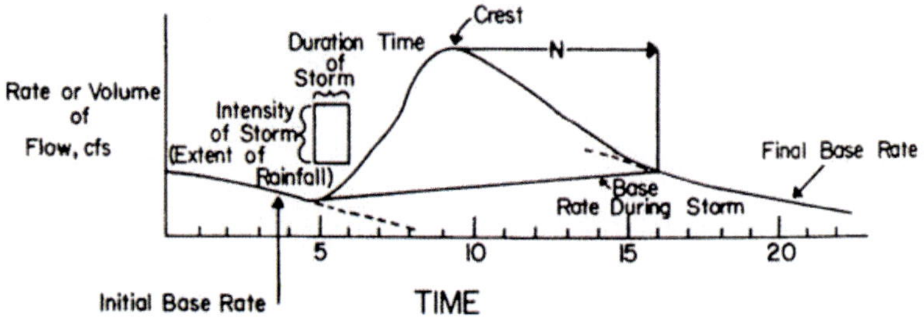

Fig. 6.10 Stream hydrograph showing a complete storm, the hydrograph, and the means for separating base flow

Table 6.3 Values of *N* as a function of drainage basin size

Area, mi^2	N	$N = A^{0.2}$
100	2	2.5
500	3	3.5
2000	4	4.5
5000	5	5.5
10,000	6	6.3

whereas in large basins it may be in the order of 1 d. Frequently a 6-h duration of precipitation is a convenient value for calculating the unit hydrograph for a river.

In the use of the concept of the unit hydrograph, the first process is to determine stream flow as the result of an individual storm event. When possible, the beginning time for this sequence should be long enough after a significant pre-occurring precipitation so that the stream will reach its base flow or relatively constant flow. The onset of precipitation and its total amount are recorded and correlated with the stream flow resulting from this storm. When possible, the flow data are continued until the flow returns to the initial base flow. This is depicted in Fig. 6.10. After the storm flow is plotted, the base flow is subtracted from the storm flow. Base flow is considered to be the value at the lowest point of the runoff curve immediately prior to the start of the increase in flow caused by the storm. The point at which the flow returns to base flow is determined by N as in Fig. 6.10. For large drainage basin area N is usually measured in terms of days. The value of N may be determined by one of several means, most of which are empirical. Two of these are shown in Table 6.3. Using a determined value of N, which is time from the peak of the runoff until it is considered that the stream has returned to base flow, a line can be drawn representing base flow. Individual values of base flow are then subtracted from the storm hydrograph to provide the unit hydrograph as a function of the specific storm as shown in Fig. 6.11. There are numerous alternatives for determining base flow, but in general the use of a straight line as indicated in Fig. 6.10 provides results that are just as valid as any others. Figure 6.11 represents a hydrograph for the duration of the storm studied. Also the figure represents the amount of runoff from the measured amount of precipitation. To convert this hydrograph to a unit hydrograph of 1.0 in of

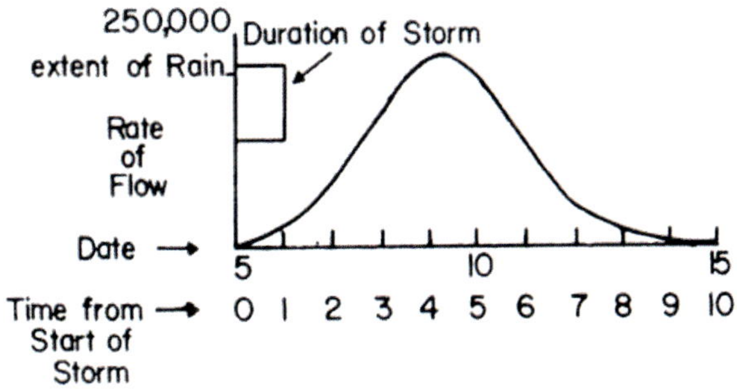

Fig. 6.11 Unit hydrograph

precipitation (or 1 cm), this may be calculated by multiplying or dividing the individual values in Fig. 6.11 by the ratio of 1 in to the total inches of storm precipitation. This makes the unit hydrograph convenient in calculating the impact of future storms. For example, using a unit hydrograph for 1 cm of precipitation, if the actual precipitation were 2 cm, each individual value on the unit hydrograph would be multiplied by a factor of 2 to determine the actual storm runoff. If the precipitation continues for longer than the unit of time for which the hydrograph has been drawn, a subsequent curve starting 1 unit of time later can be calculated and a series of curves drawn until the end of the storm. The total runoff then caused by the storm would be the sum of the individual values of each of the unit time hydrographs. It must be kept in mind that, to determine the total stream flow, the value for the base flow must be added to the total flow in the stream.

The information obtained from this method is useful in predicting storm runoff from any intensity and duration of storm. One precaution, however, must be taken when using the unit hydrograph for determining peak flow. This is that the values obtained in the initial storm hydrograph may represent different conditions of ground water moisture content at the start of the storm. This directly affects the ratio of the precipitation to the runoff. If the ratio of the total precipitation to the runoff is determined for the initial storm hydrograph, proportionate calculations can be made for any other ratio of precipitation to runoff.

The amount of flow and the rate of flow are important in the design of flood control facilities and for the design of dam spillways. The maximum height of a flood or the adequacy of a spillway design can be determined by a method called flood routing, per Creager et al. [13] and Pickels [14]. Basically this compares the amount of inflow to a reservoir during any short period of time with the rate of outflow through any control systems such as culverts, penstocks, or the spillway. When the inflow exceeds the outflow, the reservoir elevation will rise. As the reservoir elevation rises, the rate of discharge through the control systems also increases. Thus, by plotting the incremental increases in water level vs. the rate of discharge through the outlet devices, the maximum flood level and the duration and amount of runoff downstream can be calculated. This sequence can be determined utilizing a computer program or by a graphical method, as shown here.

The first step in flood routing is to determine the initial reservoir stage at the start of the design flood. As a general rule, if the volume of the flood runoff is high compared to the storage capacity of the reservoir, it is assumed that the reservoir would be full at the beginning of the design flood. However, if the capacity of the reservoir is large compared to the volume of flood runoff, then the initial reservoir elevation would be that expected at the beginning of the design flood or the elevation obtained by a flood having an average frequency of once every 25 yr. The rate and volume of inflow into the system may be determined utilizing the unit hydrograph and projecting to the worst conceivable storm. The worst conceivable storm can be estimated if sufficient long-term data are available for the watershed in question. The other information needed concerns the nature of the discharge system. This may consist of any combination of regulating outlets, penstocks, consumer demand, and so on. In a small dam, the discharge may be considered through a culvert. To determine the rate of flow through a culvert, the equation used is:

$$Q = KA2gh \tag{6.8}$$

in which

Q = the discharge in m^3/s (cfs)
K = the coefficient of discharge
g = gravity, 9.81 m/s^2 (32.2 ft/s^2)
h = the effective head on the center of the orifice in m (ft)

K may be calculated as:

$$\frac{1}{\left(1+2gL/C_c{}^2R_h\right)}$$

in which

C_c = the Chezy coefficient.
L = the length of the conduit in m (ft).
R_h = the hydraulic radius which = A/w_p = the area (A) divided by the wetted perimeter (w_p). Once the pipe is full, the value of R_h = pipe diameter/4.

For discharge over a spillway, a sharp crested weir with no approach velocity may be assumed. Here

$$Q = cLH^{3/2} \tag{6.9}$$

in which

Q = the rate of discharge in m^3/s (cfs)
c = the weir coefficient, which varies between 2.5 and 3.5
H = the head above the spillway in m (ft)
L = the length (width) of the spillway in m (ft)

The next information needed to evaluate the flood routing is the volume of storage in the reservoir at any elevation. This may be determined by measuring the area under water at a series of different elevations. The volume then becomes the sum of the calculations of the mean area times the difference in elevation between any two area measurements. The calculations for the storage are shown in Fig. 6.12. The upper portion of the figure indicates the method for calculating the volume at a given elevation, and this may be applied to the total storage available at any given elevation, as shown in the lower diagram. In this figure, English units of area and volume are indicated because of the simplified relationship between ac-ft and cfs.

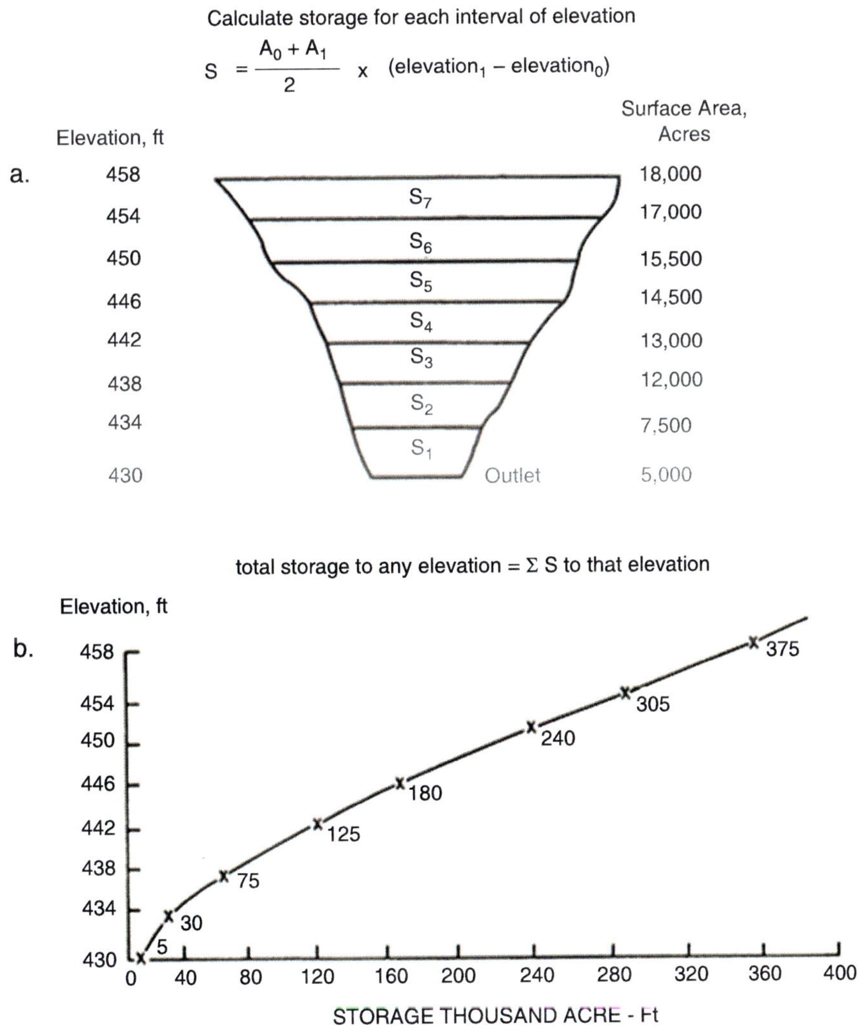

Fig. 6.12 Illustration of calculation of reservoir storage at any depth

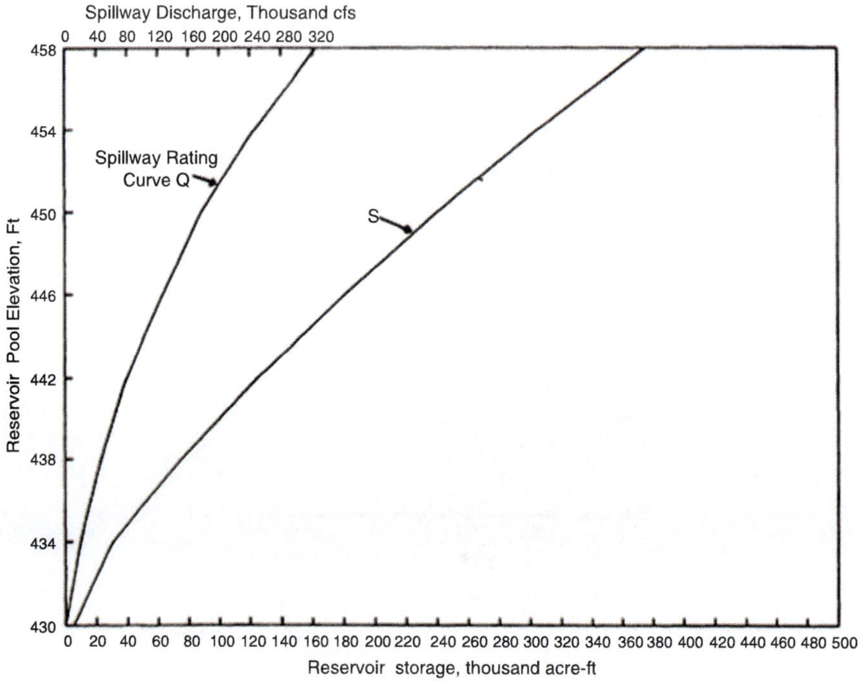

Fig. 6.13 Storage and discharge rate at any elevation

Obviously, metric units may be used with the appropriate conversion factors. For convenience, the discharge through the control devices may be plotted on the same curve as the storage vs. elevation, with the scale of the discharge in cfs double the scale for storage in ac-ft, as shown in Fig. 6.13. The reason for this correlation of scales is that 1 cfs is approximately equal to 2 ac-ft/d. However, in working with mean rate of discharge, it is desirable to work with half of the mean rate of discharge at the beginning of the time period and half of the mean rate of discharge at the end of the time period. That this is equivalent to the mean rate of discharge is shown by the comparison:

$$\frac{Q_1}{2} + \frac{Q_2}{2} = \frac{Q_1 + Q_2}{2} = Q_M \quad \text{or} \quad \overline{Q} \tag{6.10}$$

which represents the mean rate of discharge over a given time period. Also, one normally works with only a 12-h period, which is one-half of a day. Thus, to determine half of the outflow rate at a particular time with an inflow of 12 h (1/2 d), the calculation would be:

$$\frac{cfs}{2} \times 2\frac{ac-ft}{d-cfs} \times \frac{12\text{h}}{24\text{h/d}} = \frac{ac-ft}{2}$$

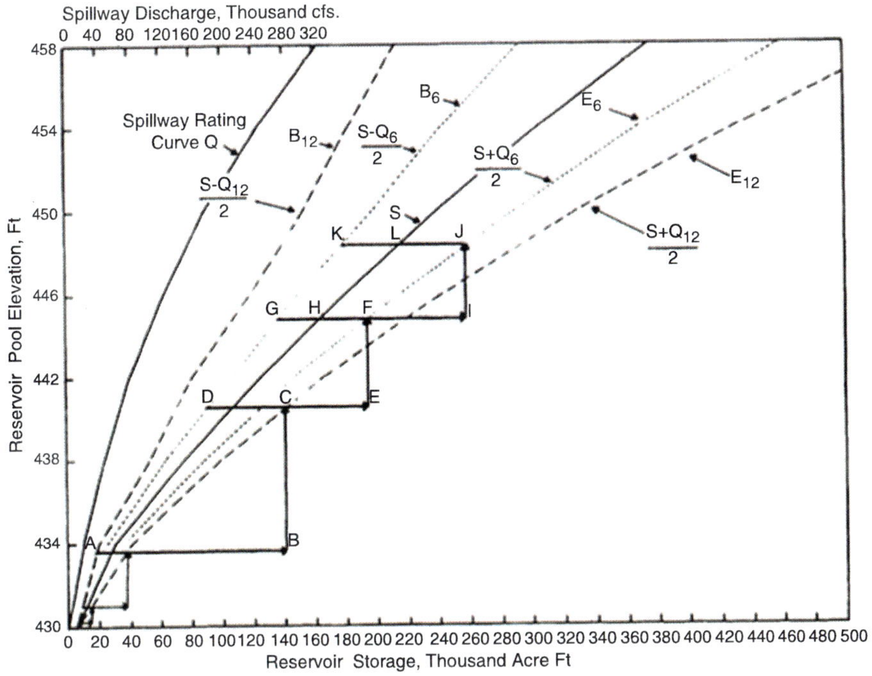

Fig. 6.14 Worksheet (partial) for demonstrating reservoir elevation and discharge with a given inflow over a specified time period

Thus by setting the scale of discharge (in cfs) in Fig. 6.12 to 1/2 that of the ac-ft, we have effectively correlated the ac-ft half day and half of the mean discharge over that time period. For a 6-h interval of inflow, which is used over the peak period, we would have to divide the cfs by another factor of 2.

Next, on a curve such as Fig. 6.13 are plotted the values of the storage plus and minus half of the outflow over a 12-h period and plus and minus half of the storage over a 6-h period (Fig. 6.14). The storage minus half the outflow in a 12-h period is designated as:

$$S - \frac{Q_{12}}{2}$$

or B_{12} indicating the beginning curve for a 12-h period, and the plot

$$S + \frac{Q_{12}}{2}$$

is indicated by the E_{12} curve or the ending line for a 12-h period. The similar lines for a 6-h period are indicated as the B_6 and the E_6 curves. Since the discharge curve in its upper scale is equal to half of the discharge in a 12-h period at any elevation, a compass can be used to draw the B_{12} and E_{12} curves. The value of the spillway discharge Q is, respectively, added to and subtracted from the storage curve in ac-ft. The B_6 and E_6 curves then are the midpoints between the respective B_{12} curve and the storage curve and the E_{12} curve and the storage curve. These lines are indicated in Fig. 6.14.

The next step is to choose the inflow hydrograph and determine the values of inflow during a 12- or 6-h period. An example of computation for this is shown in Table 6.4. The first column indicates the corresponding lines that have been drawn in Fig. 6.14. The first few influent calculations are not indicated as lines because the lines would be too short and would not show in the figure. The values for the first five columns are the values to be used to enter the Fig. 6.14. It may be seen that 12-h intervals may be chosen for the noncritical portion of the curve; however, a 6-h interval should be chosen over the peak of the curve. This time interval may vary for different drainage basin sizes. The values for column 3 are the instantaneous rates of inflow in cfs determined from the storm hydrograph; and the values for column 4 are merely the mean rate over the time period. Column 5 simply converts these values of cfs to ac-ft during the time interval. As may be seen during the 12-h time interval, these are exactly the same values as for cfs and for the 6-h period are half these values. Then, choosing the predetermined elevation of the reservoir at the start of the flood (for this example, the reservoir was considered to be empty), the amount of runoff from column 5 is added to the appropriate B (B_{12}) line in Fig. 6.14. For example, line AB occurring 48 h after the beginning of the flood represents an inflow of 112,700 ac-ft and it is added to the B_{12} line at an elevation of 433.6 ft, which is the elevation at the end of the previous time period. The length of line AB represents the 112,700 ac-ft inflow. Since this line extends beyond the E_{12} or ending curve for a 12-h period, line BC must be drawn in an upward direction in order to meet the E_{12} curve. This establishes a new reservoir elevation at the end of the time period and a new discharge rate that is equivalent to the Q value from the Q curve at the new elevation. From this point, a 6-h time interval was used and line DE therefore is drawn as the volume of flow of 98,600 ac-ft beginning with the B_6 curve and extending to the right. Again, this line extends beyond the E_6 curve and therefore must be extended as line EF upward to the E_6 curve. In all instances, when working with a 12-h influent, one must begin and end with a 12-h curve, and correspondingly, when working with a 6-h influent, one must begin and end with a 6-h curve. This system is continued until the point where a line designating the influent does not extend beyond the ending curve, and then a downward line would have to be drawn to the corresponding ending curve, indicating a drop in elevation of the reservoir and a decrease in the discharge rate. It may be seen from the values of Table 6.4, column 6, that the maximum reservoir elevation reached was 451.8 ft. at which time the spillway discharge equaled the volume of inflow.

The final results of the calculations in Table 6.4 are shown in Fig. 6.15, which plots the inflow vs. the outflow and also shows the stage or elevation at the corresponding time periods. Note that the units for stage are in feet of elevation, whereas the inflow and outflow units are given in terms of flow. It may be seen that the peak of the stage occurs at the point where the inflow and outflow curves cross.

Table 6.4 Sample flood-routing computations

Line on Fig. 6.14	Time from start of flood, h (1)	Length of time interval *t, h* (2)	Instantaneous inflow rate I, cfs (3)	Mean inflow rate during interval $(I_1 + I_2)/2$ cfs (4)	Vol. of inflow during interval (4) × (2) × *t*/24 ac-ft (5)	Reservoir elev. at end of interval, ft (6)	Spillway discharge rate at end of interval, cfs (7)
	0	12	2000	–	–	430.0	0
	12	12	3800	2900	2900	430.3	500
	24	12	10,400	7100	7100	431.0	2000
	36	12	52,400	31,400	31,400	433.6	12,000
AB	48	12	173,100	112,700	112,700	440.5	64,000
DE	54	6	221,400	197,200	98,600	444.6	106,000
GI	60	6	245,800	233,600	116,800	448.2	147,000
KJ	66	6	237,500	241,600	120,800	450.6	178,000
	72	6	211,800	224,600	112,300	451.8	193,000
	78	6	180,000	195,800	97,900	451.8	193,000
	84	6	148,600	164,200	82,100	451.1	184,000
	96	12	92,800	120,700	120,700	448.1	146,000
	108	12	54,100	73,400	73,400	444.5	105,000
	120	12	29,700	41,900	41,900	441.0	69,000
	132	12	16,500	23,100	23,100	438.2	44,000
	144	12	8800	12,600	12,600	436.1	28,000
	156	12	4000	6400	6400	434.5	17,000
	168	12	2500	3250	3250	433.4	11,000
	180	12	2500	2250	2250	432.6	8000
	192	12	2000	2000	2000	432.0	5000

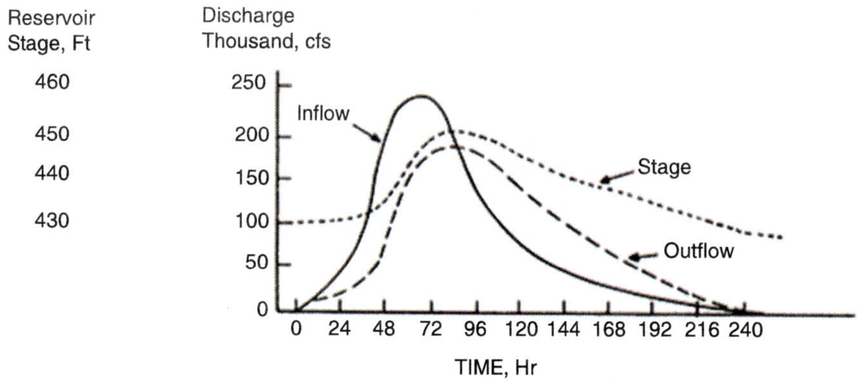

Fig. 6.15 Summary of the results of routing a flood through a reservoir

Also the peak flow has been reduced by this reservoir. Increasing the time of concentration and reducing the peak flow both combine to reduce the possible flood damage downstream. The calculation will also indicate the maximum stage, and if this exceeds a predetermined elevation, it will enable the design engineer to adjust the size of the spillway discharge to decrease the maximum stage. Conversely, the maximum stage can determine the limit of flooding and provide information on restrictions on building in the flood plain.

It should be noted that this type of calculation lends itself to a computer solution, since the rate of outflow at any elevation can be entered and the rate of inflow can be applied. The storage capacity would also have to be entered into the program. Once the storage vs. elevation and the rate of discharge are determined for any particular system, the impact of any potential storm can be readily determined.

In some instances, not the maximum flood but the maximum dependable supply is desired. This is essential where a dependable supply is required for any purpose such as drinking water supply, navigation, or hydropower. For this purpose a dam may frequently be constructed to create a reservoir for storage of water for periods when the flow is less than normal. The parameters in question include the stream flow, the demand flow, and the storage that can be provided. At any point in time, the demand flow cannot exceed stream flow plus the storage capacity. Storage capacity must be provided prior to the demand's exceeding the stream flow. The limit of demand approaches the mean stream flow over an extended period of time.

The relationship between runoff, demand, and storage may be shown in a mass diagram using the Rippl method [15]. In this system, a long-term cumulative plot of runoff vs. time is first made. It is desirable to keep the time integral for runoff to a maximum of a month, because if annual plots were made, a severe drought occurring during 11 months of a year could be missed if all of the annual precipitation occurred during 1 month. From the plot of the long-term data, the period of the most severe drought can be extracted and plotted on an expanded scale. A typical mass diagram of flow vs. time is shown in Fig. 6.16. Using this figure, several factors can be determined. If a desired draft is known, the rate of this draft can be plotted on the

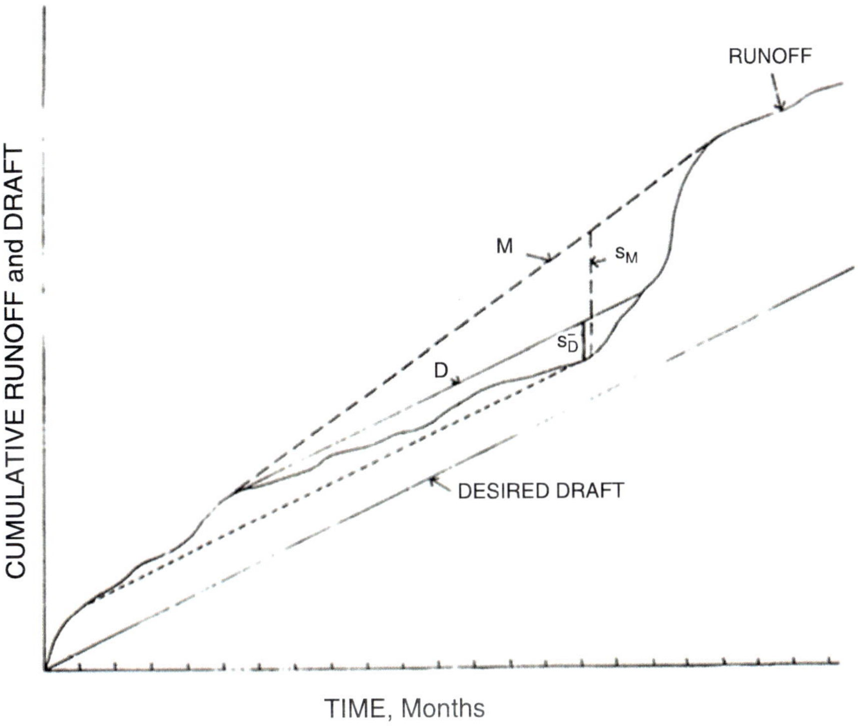

Fig. 6.16 Rippl method for determining reservoir storage

same scale as the mass diagram. In order to determine the practicality of this rate, a line parallel to the draft is drawn on the mass diagram tangent to the point at which the maximum drought begins. This line is shown as line D on Fig. 6.16. If this line intersects the mass diagram of the stream flow later in time, then this demand can be satisfied providing that a reservoir is constructed that represents the maximum difference between the stream flow and the desired draft. This is represented as the line S_D in Fig. 6.16. In order to determine the maximum draft that can be exerted on this stream during this drought, a line tangent to the start of the drought and tangent at the end of the drought could be drawn. This is designated as line M on Fig. 6.16, representing the maximum draft when the assumption is made that there are no physical limits on the size of the dam or the reservoir. In this case, the maximum storage to be provided is the maximum difference between this draft and the runoff and is depicted as S_M. Any demand that exceeds this rate could not be supplied because it would be impossible to build a storage reservoir large enough to provide this water. Obviously, the maximum draft may also be limited by the maximum size of the dam and/or the storage reservoir. If this is the controlling factor, then the potential available storage should be drawn on the figure and a line tangent at the start of the drought and touching the top of the maximum storage capacity should be drawn. Thus, this simple diagram can be used to determine the maximum allowable draft and the maximum size of the reservoir needed to provide this draft. One other useful bit of information can be derived from this figure. If a line is drawn parallel to the chosen draft, but tangent at the lowest point of the drought and extended backward in time until it again meets the runoff curve, this will indicate the time at which the reservoir would have had to have been built in order for it to have filled prior to the onset of the design drought. Thus it may be seen that this figure represents a simplified balance between stream runoff as it varies, the amount of withdrawal from the stream, and the storage that must be provided prior to the withdrawal exceeding the stream flow. It must be pointed out that the demand curve does not have to be a constant value; however, at no time can the total demand exceed the sum of the inflow and the storage available.

9 Ground Water

Ground water represents a significant source of water supply. In the order of 30 billion gal/d (110 million m^3/d) of ground water is used in the USA. It is the prime source of water for drinking in rural areas. Many medium-sized cities and numerous industries use ground water as their source of supply for drinking, processing, and cooling. One principal advantage of ground water is its relatively cool and constant temperature. Furthermore, the soil acts as a filter, removing particulates including bacteria. Thus, under most circumstances, ground water is not subject to contamination as is surface water. Potential sources of contamination include industrial discharges, landfills, and chemical dump sites. These are discussed in more detail in a later section.

Ground water results from the infiltration of precipitation into the soil. The water will proceed downward vertically until it reaches the first confining layer. Above this layer there is a saturated layer of soil containing the ground water. The upper edge of this saturated area is called the ground water table. This will vary with the amount of precipitation and the time since the last precipitation. There is flow in this saturated area in a downhill gradient. At some point the water in the saturated zone may flow under a confining zone. This now produces a confined aquifer. An aquifer is defined as a geologic formation that contains water and transmits it from one point to another in quantities sufficient to permit economic development. This usually defines a sand or gravel soil. An aquiclude is a formation that contains water, but cannot transmit it rapidly enough to furnish a significant supply. This is usually represented by clay, which holds much moisture but does not transmit it. An aquifuge is a formation such as rock that neither holds nor transmits water. There may be many intermittent levels of aquifers, aquicludes, and aquifuges through the soil. Water may occur at any depth and there are few locations where there is no water whatsoever beneath the soil surface. However, in general, as the depth increases, the total dissolved salts increase, with saline water being found at great depths. Thus, the useable quality of the water decreases with depth.

The volume of water in a saturated formation is a function of the pore space of the soil. The porosity, p, is equal to the ratio of the pore void or interstitial space to the total volume of the rock or soil as:

$$\rho = \frac{\text{Void volume}}{\text{Total volume}} \tag{6.11}$$

Concurrently, the voids ratio, e, is the ratio of the pore volume to the solid volume and equals:

$$e = \frac{\rho}{1 - \rho} \tag{6.12}$$

The safe yield of an aquifer is the rate at which water can be withdrawn from it without depleting the supply to such an extent that further withdrawal at this rate is no longer economically feasible. The safe yield thus is a function of the velocity of flow through the aquifer and the source of recharge of the aquifer. The velocity of flow through an aquifer is a function of the permeability of the soil. According to Darcy's law $V = ks$, in which s is the slope of the hydraulic grade line in vertical units divided by horizontal units (m/m, or ft/ft) and k is a constant as a function of the type of soil. Some relationships between soils and various parameters are shown in Fig. 6.17 and Table 6.5. The quantity of flow through the soil, Q, is a function of

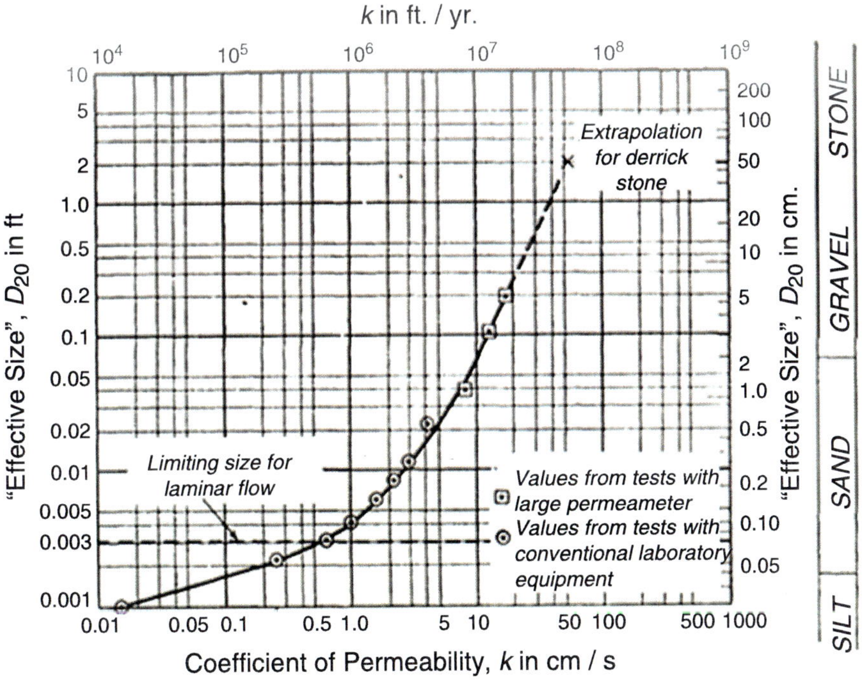

Fig. 6.17 Effect of transition from laminar to turbulent flow on the permeability coefficient

Table 6.5 Characteristics of soils

	Size		Porosity	Permeability	
Type	in	mm	%	gal/ft²-d	Velocity
Quartzite, granite				0.1	
Limestone and shale			5	1	
Sandstone			15	700	
Gravel	>0.08	1.5–8	25	100,000	1000 mi/yr
Very coarse sand	0.04–0.08	0.8–0.3	30	65,000	600 mi/yr
Coarse sand	0.02–0.04	0.5–2	35	8500	80 mi/yr
Medium sand	0.01–0.02	0.25–0.5	40	2000	20 mi/yr
Fine sand	0.005–0.01	0.05–0.25	45	80	1 mi/yr
Very fine sand	0.003–0.005	0.005–0.05	50	2	100 ft/yr
Silt	<0.003	0.005–0.05	58	1	50 ft/yr
Clay		0.005–0.05	70	0.002	1 ft/10 yr
Colloidal clay		10 Å–0.01	90	2×10^{-5}	1 ft/1,000 yr

$Q = Ps$, in which Q is the flow per unit area and P is the permeability. Standard English units for permeability have been gal/ft^2-d at a slope of 1 ft/ft at 60 °F; the equivalent metric units would be m^3/m^2-d at a slope of 1/1 at 15 °C. To handle large areas, a standard field coefficient of permeability is sometimes used in which the slope is extended to 1 ft/mi. In metric units this translates to 1 m/km. It must be noted that the permeability P is through a unit area of cross-section. To determine the coefficient of transmissibility, $T = Pt_a$, in which t_a is the thickness of the aquifer. T then has units of volume/day/unit of area of the aquifer for its entire depth. This is useful information in terms of determining the amount of water available.

It is useful to know the direction of flow of ground water and the slope. This can be determined by the use of three or more observation wells. The water level of each well is measured relative to some reference elevation. Frequently, mean sea level is used, but this is not necessary. The three (or more) observation wells are plotted in plan to scale (see Fig. 6.18). Choosing three adjacent wells, a line is drawn between the well with the highest water level elevation and the well with the lowest water level elevation. This line is then subdivided into equal increments assuming that for the relatively short distance between these two points the slope is constant. This line will then represent equal increments of slope such as meters or feet. A line is then drawn connecting the intermediate level well to its equivalent elevation on the line drawn between the highest and the lowest water levels. This is then considered a line of equal elevation. Any line perpendicular to this line will indicate the direction of flow from the highest to lowest elevation. Further if this perpendicular line is extended on the diagram to the location of the well with the highest water level, the distance from the well to this line can be determined. Knowing the change in elevation and the distance, the slope can be determined as the change in elevation with distance. In large areas where more than three wells are located, similar sets of wells are studied and the direction and slope may be determined for each area. In

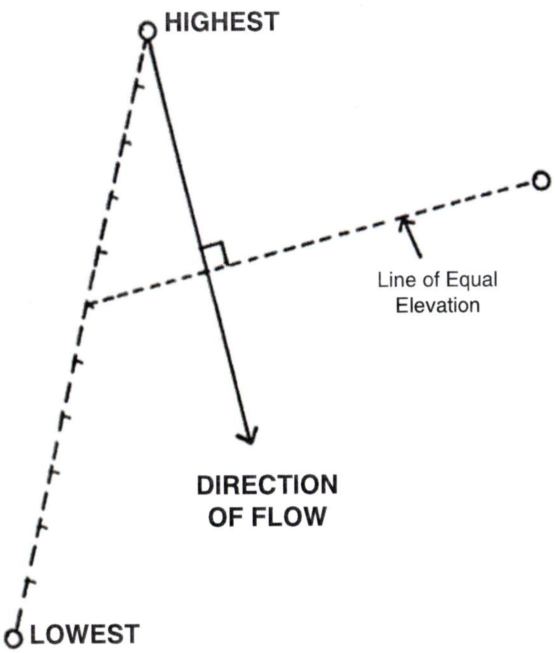

Fig. 6.18 Method for determining direction of flow and slope of ground water

some instances both the slope and direction of the flow may change in a relatively small distance.

It is also useful to know the time of flow so that the velocity between any two points may be established. This can be most readily accomplished by means of various tracers, but the dye Rhodamine WT has been found to be very useful for monitoring ground water flow (Aulenbach and Clesceri [16]). The dye or other tracer is placed in the well with the highest water level elevation. Samples are then obtained continuously or periodically (usually daily) until the dye appears in the next downstream well. Depending upon the means and location of the injection of the dye, either the first appearance or the peak concentration of the dye in the observation well is used. Knowing the time until the tracer is observed in the observation well and the actual distance within the soil, the velocity may be determined as the distance divided by time. It may be seen that the slope of the ground water table is especially essential in this case since the flow in a nonhorizontal direction will increase the distance from one point to another.

Since wells are utilized to obtain water from the ground, it is necessary to determine how much water can be obtained from the soil without exceeding the safe yield. Separate calculations must be made for wells in an unconfined aquifer and those in a confined aquifer. Those in an unconfined aquifer are frequently referred to as gravity or nonpressure wells, whereas those in a confined aquifer are referred to as pressure wells. If water is removed from a well at a constant rate until the conditions within the soil reach equilibrium, measurements can be made to determine the permeability of the soil under a set of operating conditions. Knowing the permeability, the maximum safe yield can be determined. For both nonpressure and pressure wells, at least one observation well is needed within the zone of influence of the pumped well. If measurements of water level can be made in the pumped well, this can serve as the second observation well. If it is not possible to determine the water level in the pumped well, then a second observation well is needed for the determination.

For the nonpressure well in the surface aquifer, the relationship between the flow and the permeability may be determined by the equation:

$$Q = \frac{\pi K_p \left(h_1^2 - h_w^2\right)}{\ln \frac{r_1}{r_w}} \tag{6.13}$$

The terms of h_1, h_w, r_1, and r_w for the equation are shown in Fig. 6.19, in which one observation well and the pumped well are used. K_p is a coefficient. If two observation wells are used (see Fig. 6.20), then $h_1$and $r_1$become the observation well farther from the pumped well, and h_2 and r_2 for the second observation well closer to the pumped well may be substituted for h_w and r_w, respectively.

For pressure wells in a confined aquifer, the equivalent equation is:

$$Q = \frac{2\pi K_p t (h_1 - h_w)}{\ln \frac{r_1}{r_w}} \tag{6.14}$$

The representative parameters for this equation are shown in Fig. 6.20. It may be seen that, in working with a pressure (artesian) well, the water level in the observation well will rise above the confining layer. This water level represents the piezometric surface or the elevation or pressure head to which the water will rise with no confinement. In this case, however, there is no actual cone of depression surrounding the pumped well in the confined aquifer. There is only a reduction in the piezometric head in the neighborhood of the pumped well.

The previous two equations assume that the zone of influence is circular. Under actual conditions this may not be true, since there is some horizontal flow in the aquifer and, therefore, the cone of depression will be lower on the downstream side of the pumped well. Also, under certain conditions, it may take a long time (up to 10 yr) to establish equilibrium. Thus, other methods for determining the yield of an aquifer are necessary.

One common method is the Theis method [17]. This considers the fact that an aquifer is somewhat elastic, so that removing water from it causes a lower pressure in the area of the pumped well that causes more water to enter into the area. The Theis method is based upon four assumptions: (a) the soil of the aquifer is homogeneous; (b) the coefficient of transmissibility, T, is constant; (c) the coefficient of storage, S_c, is constant, representing the yield of water per unit area of aquifer per unit drop in hydraulic gradient; and (d) the water is released from the aquifer instantaneously.

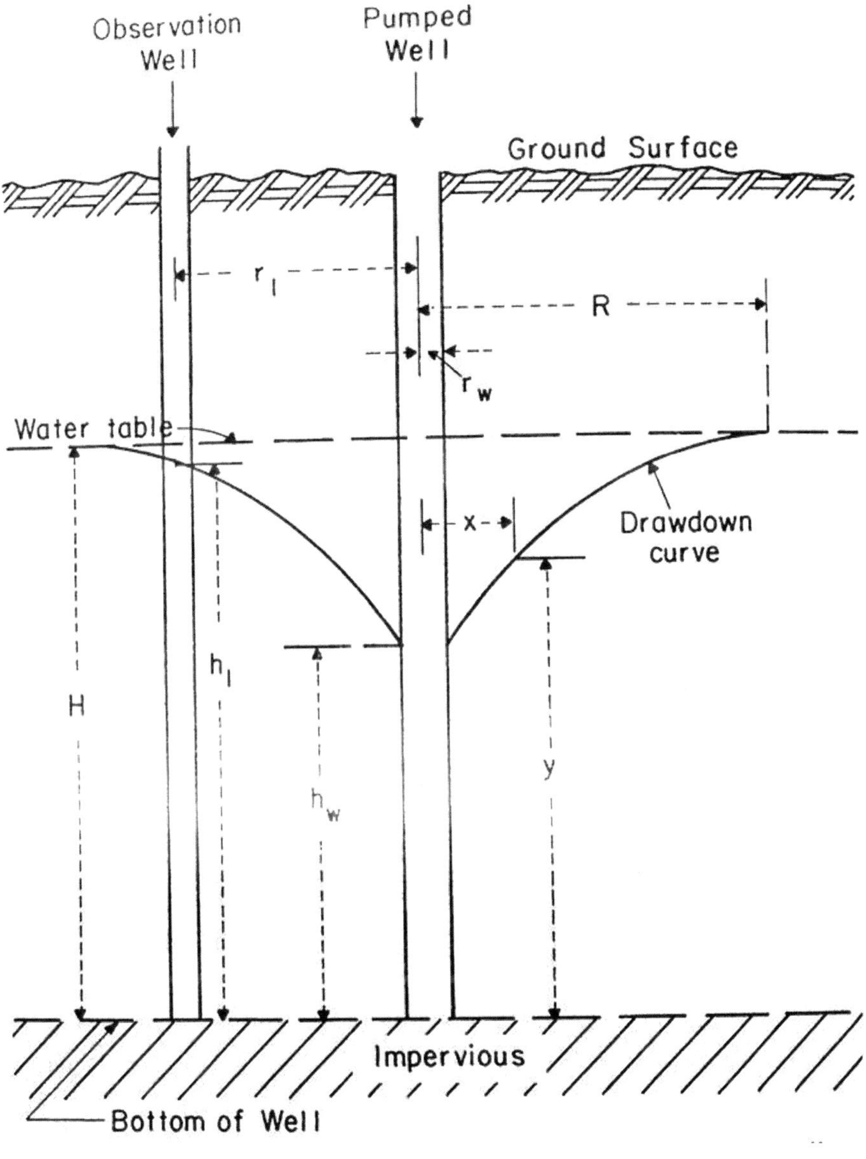

Fig. 6.19 Diagram illustrating the hydraulics of a nonpressure well

Actually under normal conditions there is a lag in the release of the water from the aquifer. The Theis method makes use of the well function *W(u)*.

$$W(u) = \int_{u}^{\infty} \frac{e^{-u}}{u} du \quad (6.15)$$

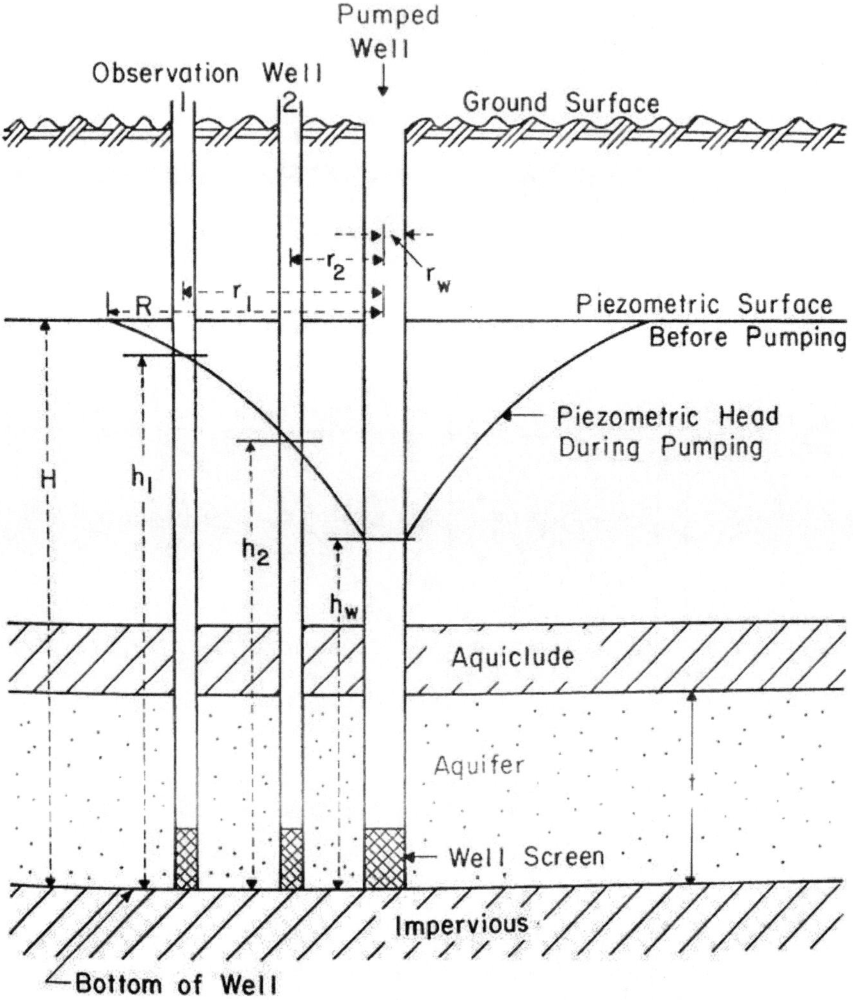

Fig. 6.20 Diagram illustrating the hydraulics of a pressure (artesian) well

For metric units

$$d = \frac{QW(u)}{4\pi T} \tag{6.15a}$$

$$u = \frac{r^2 S_c}{4Tt} \tag{6.15b}$$

in which Q is in m^3/d.

For English units

$$d = \frac{114.6QW(u)}{T} \tag{6.15c}$$

$$u = \frac{1.87r^2S_c}{Tt} \tag{6.15d}$$

In these equations

W(u) = Theis' well function value
d = drawdown in m (ft) in the vicinity of the pumping well in which the observation well is *r* m (ft) from the pumped well
Q = the discharge in m^3/d (gal/min)
T = the coefficient of transmissibility in m^3/d unit gradient (gal/day (ft/ft))
Sc = the coefficient of storage as a decimal
t = the time of pumping in days
u = a tabulated value in the Theis method

In English units of measurement, the constants convert directly from time of pumping in days to discharge in units of minutes. Theis has prepared a table of values for *u* and *W(u)*; however, it is much easier to plot these values on a log-log scale, as shown in Fig. 6.21. This diagram is for English units of measurement and the time *t* must be in terms of days. To solve this system, a well is pumped at a constant rate and the level of water in an observation well is determined on a daily basis. On log-log paper with the same scale as the type curve (Fig. 6.21), values of *d* in ft and r^2/t in ft^2/d are plotted. Then, overlaying this plot on the type curve and keeping the ordinates parallel, the one curve is moved over the other until they coincide or produce a curve of best match. Note that the two lines in the type curve (Fig. 6.21) are really one continuous line merely divided into two different ranges. With the two curves coinciding, any point on the two curves may be chosen and the values for *u*, *W(u)*, *d*, and r^2/t can be determined. Using these values, the transmissibility *T* can be determined as:

$$T = \frac{QW(u)}{4d} \tag{6.16}$$

and

$$S_c = \frac{4T + u}{r^2} \tag{6.17}$$

For English units

$$T = 114.6\frac{QW(u)}{d} \tag{6.16a}$$

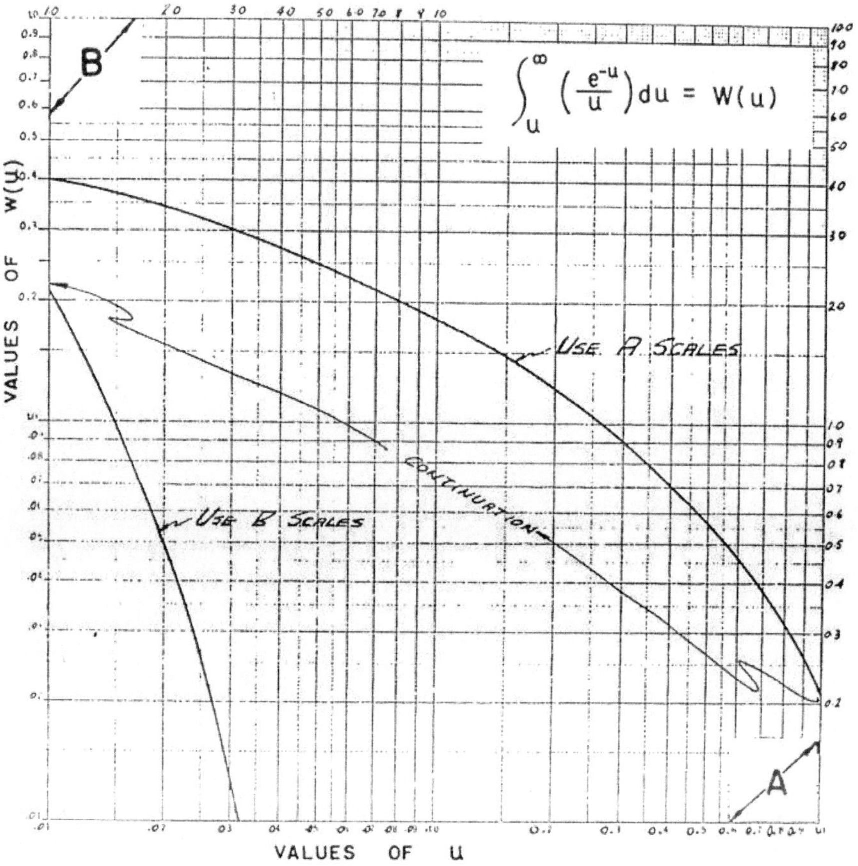

Fig. 6.21 Theis type curve for well data

and

$$S_c = \frac{uT}{1.87(r^2/t)} \tag{6.17a}$$

Jacob modified the Theis method to simplify it. His calculations are

$$T = \frac{264Q \log (t_2/t_1)}{d_2 - d_1} \tag{6.18}$$

and

$$S_c = \frac{0.37t_0}{r^2} \tag{6.19}$$

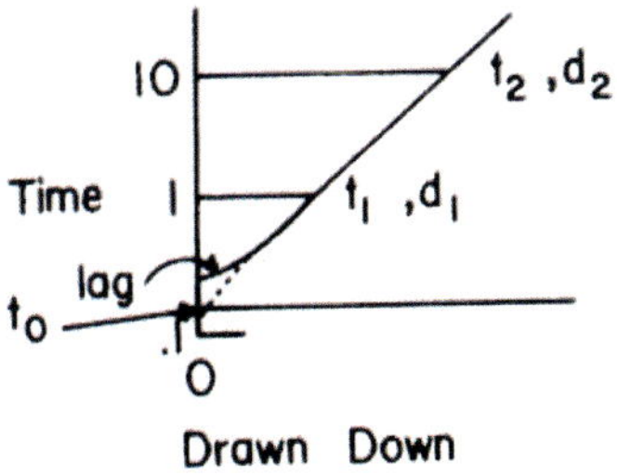

Fig. 6.22 Jacob's modification for determining storage and transmissibility of ground water

In Jacob's equations (Fig. 6.22),

d = the drawdown, ft, in the observation well at time t, in days, from the start of pumping
Q = in gal/min

Values of the log t are plotted vs. the drawdown, d. This may be done by plotting t on the log scale of semi-log paper. To simplify the calculation, corresponding values of d over one log cycle of t may be taken, as shown in Fig. 6.22. This simplifies the calculations because t_2/t_1 will always be 10 and the log of 10 = 1. To determine the transmissibility, T, the values of the drawdown at the two times then may be substituted into the equation and t determined. For the calculation of the storage, t_0 must be determined. This is determined by extending a curve of best fit of a straight line intersecting the origin of the drawdown line. This gives a projected value of t at $d = 0$. The equation then may be used for determining the storage of water in the aquifer.

The International Ground Water Modeling Center at Colorado School of Mines, 1500 Illinois Street, Golden, CO 80401, USA, www.mines.edu/Research/Centers/igwmc, has an annotated database of ground water models from all over the world.

10 Impact of Pollution on Lakes

It has frequently been stated that all lakes are born to die. This implies that over an extended period of time lakes will slowly fill in from erosion, and increase in nutrient content to support ever greater biological life, which will also die and settle to the bottom, increasing the rate of sedimentation and filling in of the bottom. As the lake becomes shallower, rooted aquatic plants can grow in the nutrient-rich sediment, thereby increasing the production of organic matter and increasing the rate of sedimentation by both the dying plant life and increasing the sedimentation of erosion by slowing the velocity of the flow of water in the area where the rooted aquatic plants occur. Ultimately the lake will fill in and may become a bog in which very little water is found, but high productivity occurs. Ultimately the large rooted terrestrial plants take over in the moist soil and a high-moor or climax bog is produced as the climax of the lake. Whereas all lakes may eventually reach this termination naturally, the rate at which this process proceeds is often influenced by humans, whose activities frequently result in increasing the erosion process that speeds the filling of the lake with sediment. Also, the discharge of sewage effluent and runoff from fertilized land may add to the nutrients in the lake, thereby increasing the rate of biological growth. This also adds to the increasing rate of sedimentation. Whereas little can be done to control the natural cycle of a lake, much can be done to prevent the increased rate of eutrophication that occurs from anthropogenic sources.

In terms of the natural rate of eutrophication in lakes, the morphology of the lake frequently has the greatest controlling influence. The morphology may even limit or restrict the impact of human activities until the morphology has been changed. As a general rule, a deep lake with steep sloping sides will have a much slower rate of eutrophication from all causes than a lake of similar volume that is shallow and has large shallow shoreline areas. Also the larger the total volume of the lake, the slower will be the rate of eutrophication.

Lakes in a temperate climate normally go through a stratification pattern that results in dimictic conditions (equal mixing throughout the lake twice a year). This is primarily caused by the anomalous temperature-density relation of water. It may be recalled that the most dense phase of water is 4 °C; thus, water at this temperature will tend to settle on the bottom of the lake, assuming that the external air temperature reaches 4 °C or lower sometime during the year. During the winter, ice will form on the surface of the lake, with the temperature of the water immediately below the ice at or near 0 °C. However, with a deep lake the temperature at the bottom approaches 4 °C. As spring comes and the sun warms the water, the ice melts and then begins to warm the surface of the lake. When all of the lake is warmed to 4 °C, even a gentle wind will circulate the water from top to bottom of the lake creating the spring overturn. As spring continues and the sun warms the surface of the lake, the warmer surface waters will tend to float on the colder lower-temperature water. If this heating occurs during a period of strong wind, there may still be complete mixing of the lake and the entire lake is warmed to the temperature of the surface. However, if warming occurs on a calm day, the surface of the lake will become significantly warmer than the lower level. With several days of warming and little wind, a point is reached at which the wind does not have sufficient energy to mix the upper warmer layers of lower-density and lower-viscosity water with the colder lower levels of higher-density and greater-viscosity water. Thus a period of stratification begins. Frequently, in large temperate lakes, the level of stratification is established near 10-m depth. In addition to the combination of warming and wind during the establishment of stratification, the shape and orientation of the lake with the wind have a great influence on the depth of the upper mixed zone. During the summer, there is an upper layer that is equally mixed by the wind, then a zone of rapidly decreasing temperature with depth, and then a third layer at the bottom with relatively constant and cold temperatures. The upper layer is called the epilimnion, the zone of great temperature vs. depth change is called the thermocline or metalimnion, and the lower layer is called the hypolimnion.

Once this summer stratification is established, there is little to no mixing in the hypolimnion. This also restricts the reaeration of the hypolimnion from the surface. Thus if large amounts of organic material reach the hypolimnion, the oxygen available will be utilized and the area will become anaerobic. If on the other hand there is little productivity in the lake, or if the lake is so clear that photosynthesis can occur in the hypolimnion, then there will not be oxygen depletion in the hypolimnion and this area will support a fine crop of cold-water fish. Also keeping the hypolimnion aerobic helps to stabilize the nutrients that settle to the bottom and are normally precipitated under aerobic conditions. It can be seen that this is a critical condition, for once the hypolimnion becomes anaerobic the nutrients that are soluble under anaerobic conditions are released to the water column, allowing for the growth of more biological materials that will in turn die, settle to the bottom, and utilize more dissolved oxygen. Thus, in order to maintain the lake in an oligotrophic condition, management efforts should be made to maintain the hypolimnion in an aerobic state.

As fall approaches, the surface of the water is cooled and the cooler water circulates to a depth of equal temperature and/or density. This tends to lower the thermocline until the lake again becomes uniform in temperature and even a gentle wind will mix the lake from top to bottom. This is called the period of fall turnover. As the air temperature reaches 4 °C and becomes colder, the surface of the lake will approach 0 °C, but the warmer 4 °C water will remain on the bottom. After ice is formed, there is no wind effect on the lake to cause mixing. Thus the bottom of a deep lake never freezes. However, again during ice cover there is no opportunity for reaeration of the lake from the surface, and thus the biological activity in the water must be such that it does not utilize the available oxygen in the take. Because of the cold temperature, this is usually fairly easy to maintain; however, in certain eutrophic lakes, oxygen depletion may occur under the ice. This period of stratification under the ice is called the winter stratification period. In evaluating a lake it may be seen that it is essential to evaluate it under all conditions of stratification and nonstratification.

A lake contains many biological communities. Within the water column are numerous organisms of microscopic size. These are generally termed plankton, which are microscopic floating organisms. The plankton may be subdivided into two general groups: the phytoplankton, which are the plant life and include the algae, the fungi, and the pollen that fall into a lake, and the zooplankton, which represent the animal forms. In another category, the plankton may be broken down into the nekton or free swimming organisms and the benthon, which are the microscopic organisms that exist on the bottom.

A prime concern is the algae, the microscopic green plants floating in the water column. These organisms represent the base of the food chain in that they can convert simple inorganic materials into organic materials with the aid of sunlight. In addition, during the daytime or sunlight hours, these plants undergo photosynthesis, which is a process in which oxygen is liberated in the water. It has been estimated that ¾ of the world's supply of oxygen is generated by algae in the ocean. In terms of the food chain, the algae are normally consumed by the zooplankton, which are in turn consumed by larger animal forms, which may be consumed by small fish, which may be consumed by large fish, which may be consumed by larger vertebrates, including humans. But it is the small algae that are the base of this food chain.

All biological systems require the presence of the proper balance of nutrients in order to grow and reproduce. For the larger organisms, the smaller organisms provide these nutrients. However, the algae, as the base of the food chain, must gain their nutrients in the inorganic form from the surrounding water. Organisms that rely on inorganic matter for growth are termed autotrophic, whereas those that require organic matter as a source of food are called heterotrophic. Growth is a function of the nutrients available (plus other limiting factors, such as temperature, light, and so on), but these nutrients may vary both with location and time within a lake. Specific organisms may vary in their specific nutrient requirements; however, in general these organisms have a certain demand for the essential elements. Carbon may be obtained from the solution of carbon dioxide. Hydrogen may be formed by hydrolysis of water, or from bicarbonates dissolved in the water. Oxygen may be produced in photosynthesis or may be secured from dissolved atmospheric oxygen. Nitrogen is secured from dissolved nitrogenous materials including both ammonia and nitrates. Phosphorus is usually derived from soil and animal excreta. Sulfur is usually derived from the soil, but is also present in animal excreta. The needed ratio of each individual material varies, but there is a general demand for approximately 60 parts of carbon to 15 parts of nitrogen to 1 part of phosphorus in most cellular material. In addition to these stated growth factors, there may be many other trace elements that are required in order to support adequate growth. Most frequently, the requirements for these elements are so low that there is an adequate amount available. Certain specific organisms may have special requirements. A typical example of this is the diatoms, which require the presence of silicon in order to manufacture their cell case, called a frustule.

Normally organisms such as algae will grow until one of the nutrients becomes limiting. Then growth may be retarded or eliminated entirely. The limits are in the ratio of the requirements; therefore, the limit is different for each element. This is referred to as Liebig's law of the minimum, which states that biological systems will grow until they are limited by the nutrient that is present in the limiting concentration. In lakes the most common limiting nutrients are nitrogen and phosphorus; however, in a few instances carbon has also been shown to be limiting. Most frequently, phosphorus is the limiting constituent; however, this is not exclusively so and there are many lakes in which nitrogen is the limiting factor.

Productivity is a measure of the utilization of inorganic material to produce plant growth. Very frequently, productivity in lakes is measured by the amount of fish available in the lake. Since the number of fish is indirectly related to the growth of the other organisms lower in the food chain and, ultimately, to the amount of algal growth, it may be seen that the productivity of fish is merely an indirect measure of the amount of algae growth in the first place. Whether or not productivity is desirable is a function of individual taste. A lake that is low in productivity will be clear and conversely will have a small fish population. On the other hand, a productive lake will be turbid because of the high concentration of plankton, but this will support a large fish population. However, this fish population usually represents different species. A lake low in productivity tends toward the more preferred game fish such as trout and salmon, whereas a lake that is high in productivity is more commonly inhabited by bass, pickerel, and catfish. The term oligotrophic has been used to describe lakes of low productivity and eutrophic for lakes of high productivity. In between is a region called mesotrophic, indicating that there is a gradual transition between oligotrophic and eutrophic. These terms are not intended to imply that all eutrophic lakes are undesirable or that all oligotrophic lakes are desirable. The desirability of a specific level of productivity is a function of the specific use of the lake and a balance of various types is needed. The long-range problem is that as lakes age the nutrients accumulate within the lake. New nutrients are brought into the lake by allochthonous inputs, which are substances entering the lake from outside of the water body. This includes siltation, organic matter from decaying leaves, and, of course, anthropogenic sources such as sewage or treated sewage effluent. Autochthonous inputs are those that are generated within the lake. This mostly represents a recycling of the nutrients within the lake since these are relatively constant. Thus, as there is an increase in allochthonous sources, more nutrients are made available to the organisms within the lake, and the productivity increases.

Based on Liebig's law of the minimum, it is possible to control excess productivity by eliminating one of the essential nutrients. It is not necessary to eliminate all of them. Since phosphorus is most commonly the limiting nutrient, most efforts have been expended toward controlling phosphorus inputs to lakes in order to control excess productivity. Whereas little can be done to control the natural process of eutrophication, there can be control of the anthropogenic sources of the nutrients that will encourage the unwanted higher productivity. In some instances, reduction of the anthropogenic sources of nutrients has been shown to reverse the eutrophication trend. In other instances the lowering of anthropogenic inputs merely slows down or delays the rate of eutrophication. What works in one lake may not necessarily work in another seemingly similar lake.

Since phosphorus is most frequently the limiting nutrient in a lake, more efforts have been directed toward finding means of reducing phosphorus inputs to a lake. This may include diversion of all storm runoff from gaining direct access into the lake, treatment of wastewaters for phosphorus removal prior to discharge, and use of land application wastewater techniques that have been shown to reduce phosphorus content significantly. Another reason phosphorus has been chosen as the nutrient to remove is that it may be fairly readily removed by chemical precipitation with iron, aluminum, or calcium. The next most important nutrient for removal would be nitrogen. Removal of nitrogen is somewhat more difficult; however, it can be achieved by producing a nitrified treatment plant effluent and then creating denitrification (anaerobic or anoxic) conditions. This will convert the nitrate nitrogen to atmospheric nitrogen, which will then escape to the atmosphere. However, there are certain organisms, particularly blue-green algae, that are able to utilize or "fix" nitrogen directly from the atmosphere into plant material. Since elementary nitrogen makes up approximately 80% of the air, it can be expected that there will be sufficient nitrogen available in the water body to provide for all of the nitrogen

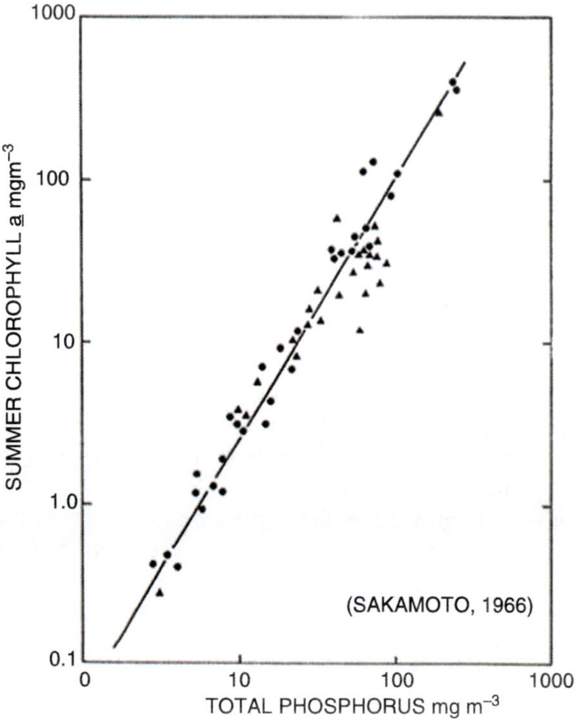

Fig. 6.23 Total phosphorus concentration at spring turnover vs. average chlorophyll-*a* concentration in summer for a number of lakes. Circles represent data from Sakamoto (1966); triangles are for other lakes in the literature. Line is regression line for Sakamoto's points. Correlation coefficient, *r*, is 0.97

fixation that may be needed in most lakes. Thus phosphorus has received the greatest attention in terms of a method for controlling excessive productivity. It has been shown that controlling phosphorus to concentrations of less than 10 μg/L at the time of spring overturn, and 5 μg/L under summer growing conditions, will result in the control of excess productivity in most lakes [18].

Sakamoto [19] has shown a direct correlation between the phosphorus concentration in a lake at the time of spring turnover and the productivity measured as the amount of chlorophyll-*a* present in the summer (Fig. 6.23). It may be seen that there is a good correlation between the total phosphorus and the chlorophyll-*a* concentration. Greater chlorophyll-*a* content, which indicates the presence of algae, would be expected to increase the turbidity of the water and therefore lower the clarity of the water as measured by the Secchi disk depth. Whereas there is good correlation between phosphorus content and chlorophyll-*a*, there is poor correlation between chlorophyll-*a* content and the clarity of the water. Other substances in the water, such as zooplankton that feed on the algae, and particulate matter, such as fine clay or silt, which is carried into the body of water from allochthonous sources, can reduce its clarity. This is especially true near the mouth of a stream after a heavy rain that produces significant runoff. Thus, there is poor correlation between the clarity of the water and the phosphorus or the algae content as measured by chlorophyll-*a*.

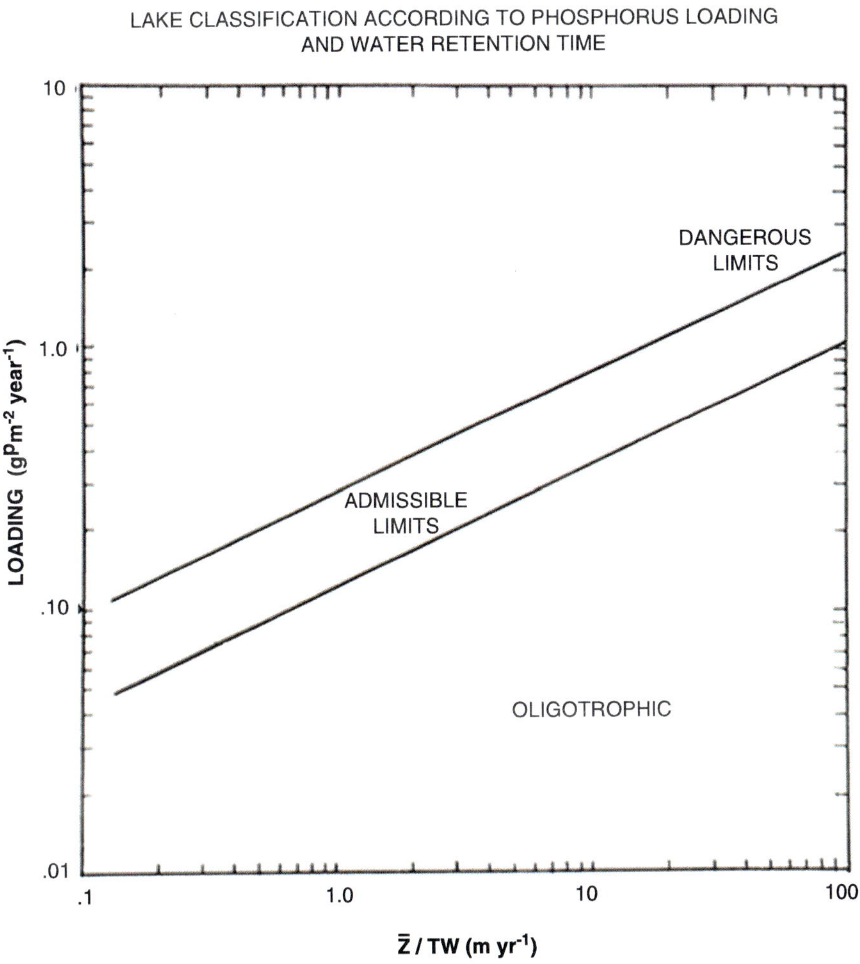

Fig. 6.24 Trophic state of a lake based upon its annual phosphorus loading vs. its mean depth and hydraulic residence time

Many models have been derived to correlate certain specific parameters with the trophic state of a lake. However, two stand out as being quite reliable and simple. These both relate total phosphorus loading to the trophic state of the lake as a function of the body of water. In the original work of Vollenweider [20], a correlation was shown between the total phosphorus loading and the mean depth of a lake, based on observations from many lakes around the world. In most cases there was a good agreement with the general observations of the trophic state of the lakes and these parameters. Vollenweider worked with Dillon [21] to improve this model by comparing phosphorus loadings with the mean depth and the hydraulic detention time of the lake (Fig. 6.24). It was conceded that, in lakes where phosphorus was not the limiting nutrient, this correlation was poor. However, in lakes in which phosphorus is the limiting nutrient, which is the most common occurrence, this correlation has been shown to be extremely satisfactory.

Information such as this is being used to indicate means of controlling productivity in lakes. One of the largest efforts in this direction is the study of controlling nutrients in the Great Lakes of the USA and Canada. Knowing the phosphorus inputs to a lake and the morphology of the lake, it can be determined what amount of phosphorus control would be required to limit the productivity to a certain desired level. This is presently being done by varied means, including the removal of phosphorus from all wastewaters being discharged into the Great Lakes Basin, by controlling urban runoff through the use of holding tanks and subsequent treatment, and by placing buffer zones between farmland and streams into which fertilizers would otherwise be leached. Based on these calculations, the most efficient and/or economical system for control of phosphorus discharges into lakes can be established.

In addition to the overall limitation to biological growth by the major nutrients, special conditions are frequently controlled by minor nutrients. Two of these are iron and silicon. Iron is essential to photosynthesis and can be limiting under certain conditions. Silicon is essential to the growth of diatoms that produce a frustule that is composed mainly of silicate.

Diatom growth is important in relation to control of blue-green algae growth. Diatoms and green algae are considered the base of the food chain in that they can convert inorganic matter to cell growth and oxygen in the presence of sunlight. In addition, they are consumed by higher organisms, thereby controlling their population growth. The blue-green algae, although they also produce cell growth and oxygen by photosynthesis, are much less desirable as a food source by higher organisms. Therefore, they can grow uninhibited to massive numbers, or blooms. Then when they die, their organic matter consumes the oxygen available in the water, creating an undesirable condition.

Studies made on Saratoga Lake [22] indicated that when sufficient silicon is available for diatom growth, they can out-compete the blue-green algae for the available nutrients. However, after all the silicon is depleted by a large diatom growth, their growth diminishes, and the blue-green algae can take over. Since the blue-green algae are only slightly consumed, they proliferate until more soluble silicate is made available, after which they decrease in numbers as the diatoms again increase.

Rabalais et al. [23] studied iron and silicon concentrations along with nitrogen and phosphorus in coastal plankton food webs. They particularly related their results with the hypoxic conditions that exist in the Gulf of Mexico. Babbitt [24] reported that as much as 7000 mi^2 of the Gulf are presently hypoxic. This has become a concern since this restricts the production of shrimp, a major economic crop of the Gulf. The hypoxic conditions correlate with the proliferation of blue-green algae in the affected area. The depletion of silica in the Gulf correlates with the excess blue-green productivity. Rabalais et al. [23] attributed the low levels of silicon reaching the Gulf to the numerous dams along the Mississippi River that trap the silt and sand which normally contribute to the silicate levels in the Gulf. The lack of this essential silicon for the diatoms has allowed the blue-greens to proliferate resulting in less competition to use the large amounts of nitrogen and phosphorus contributed from the Mississippi River.

11 Thermal Impacts on the Aquatic Environment

With increasing human activities, and the accompanying increasing demand for energy, more and more waste heat is being produced. Where possible, means are being considered for the utilization of this waste heat. In other instances, the heat is merely wasted to the environment. The two main categories of the environment receiving this waste heat are the atmosphere and the water. Discharge of heat into the atmosphere may even impact lakes or streams by changing precipitation patterns. If dry cooling towers are used, the atmosphere is heated and the relative humidity thereby decreases. On the other hand, if wet cooling towers are used, significant amounts of water vapor are passed into the atmosphere, thereby increasing the relative humidity and increasing the possibility of precipitation downwind. In terms of discharge into bodies of water, this can directly increase the temperature of the body of water, resulting in increased evaporation and loss of water in the surface lakes and streams.

Excess heat in the aquatic environment may be considered to be another form of pollution: a resource out of place. Every human action produces heat in some form or another. It is the main source of energy and energy itself is usually expended in the form of heat: friction produces heat. The problem is compounded by the fact that industries are expanding and concentrating. When there were many small mills spread around the country, there was no serious problem. However, because of the economy of scale, there are now fewer but much larger mills located at a few choice locations. Thus, the same amount of heat may be produced, but it is discharged into a much more limited local area where it has a greater impact. One of the greatest sources of heat is from electrical power generation. A major problem in a thermal power generation system is the efficiency of utilizing the heat. For fossil fuels, the efficiency is only about 40%, whereas for a nuclear fuel the efficiency is only 33%. Thus, over half of the heat that is generated to produce the electricity must be discharged in some satisfactory manner. Much effort has been expended in finding uses for this waste heat. However, particularly with nuclear power reactors, the heat is frequently produced in a location remote from where it can be readily used. The losses in transmitting such heat long distances preclude its use at distant locations. Thus large amounts of waste heat frequently reach the aquatic environment.

There are many problems that occur as the temperature of a lake or stream is increased. Some of these are obvious, but some represent problems that are difficult to assess. Probably the simplest problem that can be seen is the reduction in the saturation value of dissolved oxygen in a stream at an elevated temperature. At 20 °C, approximately 9.2 mg/L of oxygen is the saturation value. At 30 °C this is 7.6 mg/L; at 40 °C it is 6.6 mg/L. Thus it may be readily seen that there is less

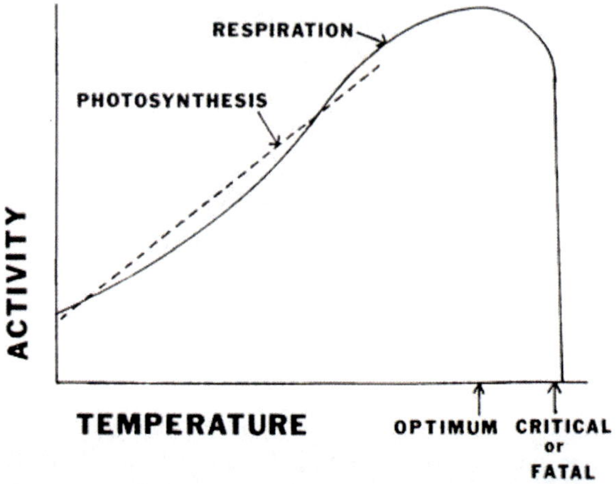

Fig. 6.25 Relationship between biological activity and temperature

oxygen available for biological respiration at the higher temperature. Another physical impact is the reduction of viscosity at a higher temperature. This is a benefit in waste treatment plants in that it allows a more rapid settling of the particulate matter. However, certain plankton require a specific viscosity of water in order to remain in suspension. At a lower viscosity they will sink and may not be able to remain in the euphotic zone. Obviously, greater evaporation occurs at a higher temperature, thereby removing water from the liquid to the vapor phase. This does result in an increased rate of cooling; however, the net effect is a greater loss of water. Rapid heating of water or pressure reduction, such as in passing through a power turbine, causes supersaturation of the nitrogen in the water, which has been shown to kill fish by the formation of nitrogen bubbles in their blood systems.

In terms of chemical effects, most rates of reaction are increased at a higher temperature. Unfortunately, this reduces the biological tolerance to change. This is similar to the reaction with enzymes or catalysts that normally are more effective at higher temperatures. Tastes and odors are also increased at higher temperatures. The BOD rate, which is both chemical and biological, also increases, thus satisfying the BOD in a shorter period of time. This can benefit an area farther downstream, but can result in serious oxygen depletion in the area immediately below a discharge containing a high BOD. Since there is less oxygen available at the higher temperature, it may be seen that the depletion is considerably more rapid and oxygen deficiency occurs in a shorter period of time.

From the standpoint of the biological system, there are many interrelated factors that result from the impact of higher temperatures. In general there is an increased rate of reaction at the higher temperature. Photosynthesis is somewhat increased at the higher temperature. However, the rate of BOD satisfaction, which is a function of respiration, may exceed the increase in oxygen production from photosynthesis. The combined impact of photosynthesis and respiration is shown in Fig. 6.25. The actual ordinates of the graph will vary for each individual organism or species, but the concept is valid. It may be seen that at a lower range of temperature, the increase in photosynthesis may exceed the increase in respiration; however, at higher temperatures, the respiration exceeds the increase of photosynthesis. The respiration increases to a maximum rate at the optimum temperature and at an only slightly higher temperature reaches a critical or fatal temperature at which the organism is destroyed. Since individuals and different species have different optimum and critical temperatures, it may be seen that a change in temperature may seriously change the character of the biological systems present at different temperatures. As a rule of thumb, in the growth range, respiration doubles for each 10 °C rise. The increase in the rate of respiration is correlated with the increased rate of metabolism. Also, the rate of reproduction is increased. This normally is recorded as a shortening of the time for the reproduction cycle to be completed. Thus, more organisms are produced that respire and metabolize at a higher rate, thereby utilizing the diminished oxygen reservoir in a stream at a higher rate. This, of course, will reduce the BOD rapidly, but with a commensurate reduction in DO. As all these factors increase, the total activity may be limited by the availability of food.

Some organisms have a temperature sensitivity in which they prefer warm or cold water. In general in cold streams, fish swim toward warm water, whereas in warm streams, fish swim toward cold water. However, there are also some fish that are quite temperature insensitive. This can result in an accumulation of fish and other swimming aquatic organisms at either a warm spot or cold spot in a stream. Here they may utilize all the food available, ultimately resulting in their demise.

The increase in temperature also adds to the synergistic effects of other harmful substances, such as toxicity, disease, and so on. In general, organisms have a lower tolerance to these impacts at a higher temperature. In some instances, for example, disease may be controlled by a lower temperature. Thus, the higher temperature merely adds to other antagonisms. Another important factor is the rate of change of temperature. Many aquatic organisms can tolerate a slow change in temperature, but not a rapid one. If there is a discharge of hot water into a stream, there can be a narrow zone of mixing resulting in a rapid change in temperature within a short distance.

Another problem regarding a rapid change in temperature is that power plants must shut down at certain times for maintenance and/or accidents. If the thermal discharge from a power plant has been keeping a stream warm through the winter, a sudden stoppage of the power generation would allow the stream to cool very rapidly. This would impact a significant thermal shock on the aquatic organisms. Therefore, most regulations call for prohibiting routine shutdowns during the winter months. Obviously, accidental shutdowns may still occur. Thus there are two main concerns: (a) the impact of the higher temperature and (b) the impact of sudden changes of temperatures because of operation of the thermal generating system.

There are many combinations of impacts related to increasing aquatic temperature. For example, certain organisms will reproduce only at or above a certain temperature. If the body of water is maintained above this temperature, they will reproduce for a longer period of time or constantly. However, their development is dependent upon the food available to them. If the food on which they rely is not available because of other temperature impacts, then they cannot survive. If they

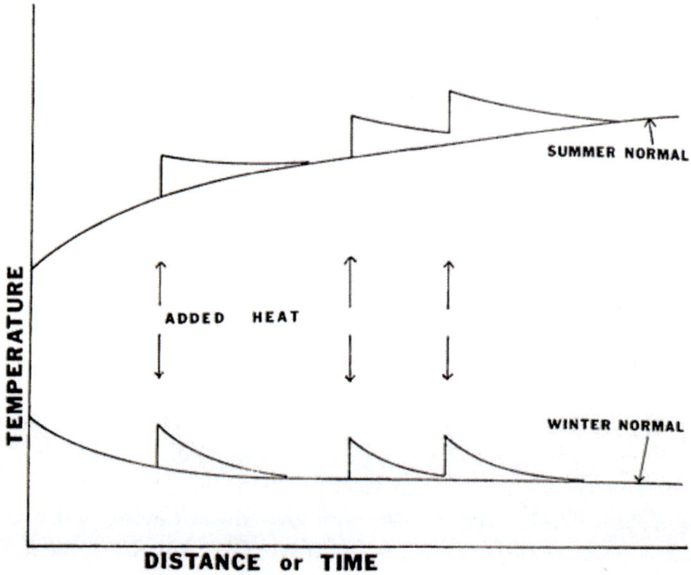

Fig. 6.26 Fate of heat added to a stream

cannot survive, their predators may in turn disappear, because they are not available at the proper time for use by their predators. Thus, it may be seen that a change in temperature at one location can impact the entire food chain throughout a stream or ecosystem. Blue-green algae are more tolerant of higher temperatures. Some have been shown to be present at up to 77 °C in hot springs such as at Yellowstone. Most of the colors in the hot springs are produced by different species of blue-green algae. The blue-green algae, however, are less desirable as food for other organisms. Therefore, they will consume the nutrients available and produce organic matter that is not readily used by higher organisms in the food chain. This obviously distorts the entire system.

The impact of a thermal discharge in a body of water will become dissipated with time and the water body will reach equilibrium after sufficient time. In a stream this implies sufficient flow time from even a constant discharge of heat. However, in a lake, since there is no flow time, the only time would be that from the batch discharge of a slug of heat. The temperature relationship in a stream during summer and winter is depicted in Fig. 6.26.

Although there is no space here for detailed description of control methods, some basic concepts should be mentioned. The first is the use of cooling ponds in which the water is recirculated through the cooling pond and reused for cooling. Crescent-shaped ponds have been shown to be quite economical. The heated water is discharged into one end and, as evaporation causes cooling, the cooler water is taken from the opposite end. Various forms of cooling towers are also used for dissipating heat. There are various combinations of wet and dry towers using natural or forced draft. Detailed descriptions of these may be found elsewhere. Reference section of this chapter has the authors' suggested reading [25–44].

Regulations on thermal discharges are the responsibility of the various states. It is impossible to cite regulations common to all states; however, some of the criteria for evaluation should be useful.

The first criterion is the definition of a thermal discharge, usually as one that is at a temperature above a set value such as 70 °F (21 °C) or if it causes a natural body of water to rise above a recommended limit, such as 3–5 °F (1.5–3 °C). It should be defined separately in terms of lakes, streams, and tidal waters. There is usually a definition for trout and nontrout streams in which the requirements are more stringent for trout streams. There may be greater temperature tolerances during the summer. Summer and winter may be defined as periods when the temperatures reach certain limits. Many criteria involve the degree of mixing in a stream with a description such as at least 50% of the cross-sectional area and/or volume of the stream, including a minimum of 1/2 of the surface, shall not be raised above a certain minimum temperature. This is to allow fish to pass a heated section of the stream. In a lake, frequently the regulations are to prevent a certain temperature rise in a circle or a plume from the point of discharge into the lake. Here it is recommended that a surface area be specified rather than a radius of a circle, which is frequently specified. This will compensate for any distortions in the shape of the plume other than circular. In a summer stratified lake, there should be no discharge of heated effluent into the hypolimnion. Withdrawal of water for cooling may be from the hypolimnion or the epilimnion, depending upon regulations. Discharging the heat to the epilimnion will help maintain the cool water in the bottom and maintain the thermocline relatively constant. These are just some of the generalities to be observed in regulations on thermal discharges.

It is most useful to find alternate uses for this excess heat. When homes or factories may be located close to a source of heat, it may be used for heating or industrial processes. However, during the summer there is less need for this heat unless it can be converted to operating air-cooling systems. The heat has been used for agricultural purposes. Here, again, this has benefit only during certain times of the year. Some aquaculture systems have been put into operation in which the waste heat is placed in ponds, lagoons, or bays to encourage the growth of fish or shellfish. This has been shown to be very beneficial in reducing the reproduction time and the time to reach maturity. However, again this has limited use and will not solve all the problems of the utilization of excess heat. As energy becomes more and more expensive, it is certain that more systems will be devised to utilize this excess heat.

12 Toxics in Water Resources

The concern for toxic materials in water supplies has created one of the greatest controversies of recent time. The US Environmental Protection Agency issued a list of 129 priority pollutants that are of concern and should be eliminated from water supplies. However, there is uncertainty about what concentrations of these substances represent hazardous levels. One problem is that individuals and different age groups may have varying tolerances to these substances; the other problem is our present analytical methods. In some cases we do not have sufficient precision to measure concentration in ranges low enough to detect their presence, whereas in other cases we are improving our techniques by finding even lower concentrations that may or may not prove to be harmful. To add to this dilemma are the present requirements that in some instances tend to add to the problem of toxic materials. A typical example is the requirement for chlorination of wastewaters before discharge and of drinking waters prior to treatment. Chlorination of certain organic materials produces the trihalomethanes, which are on the list of priority pollutants. If chlorination of wastewater is not practiced, there will be fewer trihalomethanes produced. However, the downstream area will have a higher disease potential because the pathogenic organisms have not been reduced sufficiently. Similarly, in the water intake, where poor-quality water is used as the drinking water supply, prechlorination has been practiced to reduce the number of pathogenic organisms. This can result in an increase in the trihalomethane content. In order to reduce trihalomethane production, it is recommended that prechlorination not be applied, but that secondary chlorination, after the removal of the organics, be the only point for chlorination. This of course will reduce the production of the trihalomethanes, but may increase the potential for the carryover of pathogenic organisms in the water supply system. Similarly, several heavy metals are on the list of priority pollutants. Some of these metals have been shown to be essential growth factors in very low concentrations. However, at only slightly higher concentrations, they have been found to be detrimental. Thus, very close tolerance limits must be set, and it is difficult to establish safe levels and still prevent the occurrence of unsafe levels.

Another problem comes in the determination of the priority pollutants. The regulations call for the banning of any substance suspected of being carcinogenic or mutagenic. This usually is based upon feeding massive doses to a susceptible species of mice or rats. If any carcinogenic or mutagenic effects are found, that substance is placed on the priority pollutant list. There is, however, no guarantee that what is harmful to rats and mice will also be harmful to humans. Also it is difficult to extrapolate to humans the concentration limits that were harmful to the animals.

A regulatory agency must of necessity be extremely conservative. Therefore, any substance even suspected of being harmful must be included on this list. Environmental conditions are to be maintained so that even the most susceptible individual is protected from any harm. Thus, the most conservative values and the lowest levels are considered in listing the priority pollutants and their concentration limits.

One of the major problems in the evaluation of priority pollutants is the impact of long-term use of any substance. There may be no immediate symptoms, but symptoms may occur after long periods of time, in another generation, or in a very indirect manner. For example, the use of certain drugs (thalidomide) for increased fertility has been shown to have an impact on the offspring of the second generation of persons who used these drugs. Findings such as these take 20–25 yr to evaluate.

Another is the use of DDT, which has not been shown to be harmful to humans, but indirectly has particularly affected the eggs of certain fish and birds. In society's total concern for the environment, these impacts are also valid concerns. The problem then arises about how one evaluates a new substance that appears to have no deleterious impact at the moment, but may in the future. We certainly cannot ban every new substance. The answer probably lies in continued evaluation and monitoring of any new substances, particularly those that are or are suspected of being related to other substances that have been shown to be deleterious. However, there must also be provision to remove a substance from the priority pollutant list if it has been proven over a long period of time not to be harmful to humans or the environment.

Another condition that has not been evaluated is the risk-benefit ratio. As an example, DDT has been shown to have saved on the order of 500 million lives, particularly in preventing malaria. It is now on the priority pollutant list primarily because of its impact on organisms other than humans. There must be some means of evaluating a substance to determine whether its adverse effects may be more than offset by its benefits.

Of some concern must also be cost; however, cost related to human health must be the final evaluation. Costs for analysis of trace quantities of certain toxics may represent a significant cost to a small water supply system. A question may arise about the need for such a complete analysis for every water supply system. Obviously some evaluation must be made of the potential harm or pollution of water supply systems. A water supply not subject to contamination may not need a complete analysis; however, where potentially toxic chemicals are in the area or in the ground, it is advisable to monitor water supply systems to assure that such contaminants do not reach the public through the water supply system.

There is no simple solution to the concern for toxic materials in the aquatic environment. What is essential is continued monitoring for the presence of toxic materials and a rational evaluation of the impact of these potentially toxic materials upon humans.

13 Goals of Water Pollution Control

Goals should not be the conclusion of any discussion of water pollution control. They should be the reason for establishing some procedure for pollution control. Once proper goals are established, one can make some reasonable decisions in terms of what is necessary and how they may be accomplished. However, some goals have already been established, and it is assumed that the previous information will aid in reaching the prescribed goals.

The goals of water pollution control have been summarized briefly: to provide fishable, swimmable, and drinkable water. Public Law 92-500 passed by the US

Congress and commonly known as the Amendments to the Federal Water Pollution Control Act starts off with a "declaration of goals and policy." A summary of this section is in order. The main objective of this Act is stated to be "to restore and maintain the chemical, physical and biological integrity of the Nation's waters." In order to achieve this, stated goals include the following: (a) discharge of pollutants into the navigable waters should be eliminated by 1995; (b) an interim goal of water quality providing for the protection and propagation of fish, shellfish, and wildlife and providing for recreation should be achieved by July 1, 1983; (c) the discharge of toxic pollutants in toxic amounts should be prohibited; (d) there should be Federal financial assistance to construct publicly owned treatment works (POTW); (e) to assure adequate pollution control areawide, waste treatment management processes should be developed and implemented; and (f) there should be major research and development to develop the technology necessary to eliminate the discharge of pollutants.

The successor to the above law was Public Law 95-217, commonly known as the Clean Water Act. Its main goals were similar to the previous legislation with some minor modifications in the timetable. Greater emphasis was placed on innovative and alternative technology, individual wastewater treatment systems, sludge disposal, and energy conservation.

These basic laws have been renewed, frequently postponing deadlines in recognition of the difficulty in obtaining the information needed to carry them out or the difficulty in enforcing them. Litigation has ensued in both directions: those who want immediate action and those who want to delay enforcement.

Some additional laws have been passed which impact the aquatic environment. The Comprehensive Environmental Response, Compensation, and Liability Act (CERCLA), commonly known as Superfund, was designed to control cleanup of hazardous waste sites, many of which created water pollution problems primarily through leachate. The Endangered Species Act indirectly affects water resources by protecting listed species, some of which require both adequate water supplies and high-quality water. The Federal Insecticide, Fungicide, and Rodenticide Act controls the discharge of these materials that enter water bodies primarily from agricultural runoff. The Oil Pollution Act of 1990 is designed to prevent discharge, either as a waste or from an accident, of oils to the aquatic environment. The Resource Conservation and Recovery Act (RCRA) supports both waste reduction and waste elimination. The Safe Drinking Water Act (SDWA) regulates the quality of drinking water. Details on these and other environmentally related regulations may be obtained from the US Environmental Protection Agency (US EPA) at www.epa.gov/epahome/laws.htm.

All of the laws and regulations that have been passed and promulgated by the US EPA and various other agencies have as their goal the protection of the quality of the water within the USA. To this end, much money and time have been expended. Some improvement has been seen; however, there are some glaring cases where the efforts have not been successful. Some of these most likely will be taken care of in a matter of time; in other specific cases, there may be other problems that will have to be resolved.

Over the years the approach to pollution control enforcement has changed, although both approaches were included in the first act. These approaches were (a) effluent discharge limits and (b) stream quality limits. Initially discharge limits were set for each industry and permits issued to control the amount of a pollutant that may be discharged over a given period of time. This simplified enforcement and evened out treatment costs within an industry. It also lessened the moving of an industry from a state where standards were high to a state with lower standards, since standards were similar throughout the USA. Recently the total maximum daily load (TMDL) concept has gained popularity in which each stream or watershed is evaluated for its ability to absorb a given amount of a pollutant while remaining within its desired stream quality standards. Whereas this is more equitable from the standpoint of the stream, it does require considerable more information, both to set the standards and to enforce them. In addition, standards may have to be set for any potential pollutant that may be discharged into a specific stream. This is more difficult than previous standards based on only BOD loading or dissolved oxygen levels in the stream. Also, it may encourage an industry to move to a less polluted river, thereby allowing it to utilize a larger share of the allowable load, while simultaneously decreasing the quality of the river. Both approaches have their valid points.

One of the major concerns in any effort or program, whether it be that of the Federal or local government or of any other organization, is who is to be the beneficiary of the energy expended. There are two major directions to which pollution control efforts can be directed. One is to improve the quality of the environment for the sake of the environment and the other is to improve the quality of the environment for the sake of society. There is no clear-cut distinction between the two because in general what benefits the environment also benefits humankind. However, there are specific instances where benefits to the environment may not have any obvious benefit to humans; thus, certain individuals may question the expenditure of public monies that will not directly benefit the public.

It is the author's belief that there is room for both points of view. There are cases in which the environment is and should be the sole direct beneficiary of any control practice. For example, there are areas of the Adirondack Park in New York that are to be preserved as "forever wild." Even airplanes are not allowed into these areas to land on the lakes to provide for hiking, hunting, or fishing. Large areas of Alaska have been set aside as wilderness area in which humans are viewed as unwanted interlopers. Many areas of Africa are presently wildlife areas that will have to be preserved if we hope to salvage some of the rarer animals. In some instances it has been questioned whether we really need certain of these rare animals. They may have no direct benefit to humans and in some cases, aside from hunting expeditions that benefit only a very small number of individuals, they could probably be allowed to become extinct without any serious impact upon the environment. On the other hand, we must consider that humankind is also a part of the environment and not the only factor within that environment. Other species have just as much right to exist on this earth as do humans. Another concept that must be considered is the fact that the earth has maintained a balance of biological systems that house various species including humans. As the latter increase excessively in number, they crowd out the other species. This may result in a biological imbalance in which humans may ultimately destroy themselves by overpopulation. It may be necessary to control population so that the earth can continue to support a limited number of human beings. This concept of course rankles many people. Each person feels that he or she has a right to propagate as much as seems desirable and to fill the earth with his or her progeny. Certainly we do not want to have a dictator or government body dictating who may have how many children, when, and where. On the other hand, human beings must take responsible positions and realize that in the total environment, population control is essential. To accomplish this will take a considerable amount of education. It is made more difficult by the fact that traditions and religious beliefs have fostered the concept that humankind is the center of the universe. Thus, we can do whatever we want and everything was made for us. Even though everything may indeed have been made for us, if we utilize everything made for us there will be no more left and of course we will decline. It would be far better for us to control our own numbers than to have these numbers controlled by a large famine, pestilence, or other noxious factors.

In terms of the concept that the environment should be controlled for the benefit of society, there arises the concept that complete pollution elimination may not be a desirable situation. For example, an oligotrophic lake is very unproductive, particularly in terms of producing the large amounts of fish that could provide food for many. Just as adding fertilizer to the land increases the productivity of the terrestrial crop, adding nutrients to a lake will increase the total productivity of that lake. This will result in increased plant growth and ultimately increased animal (fish and shellfish) production. Obviously, limits of productivity are controlled in the aquatic environment by the amount of dissolved oxygen that is available. Here again, in terms of benefits to humans, we could install aerators in all lakes and use them to their maximum productivity. This may even be necessary 1 d if the population continues to increase as it has. Thus, a different set of goals would have to be established if all of the environment were to serve humankind directly.

There is probably some middle ground that offers the best solution. It is certainly desirable to maintain some wilderness areas undisturbed. It is desirable to have some bodies of water that are highly oligotrophic. The problem seems to be that, if these are preserved in their pristine purity, how will humans get to appreciate them. Possibly here limited access will have to be provided to some oligotrophic lakes. Other lakes however may be maintained at higher productivity levels. This will provide greater production of fish and similar related business and recreational activities. Thus, there could be a balanced, diverse environment that would satisfy both phases of the pollution control concept. A problem that arises here, of course, is how does one establish whether a particular area or lake or stream is to be preserved for the benefit of the environment and which other area is to be maintained for the benefit of society?

These are not simple questions to answer. Each individual has a personal concept that has some validity. This will require a considerable amount of discussion and cooperation among individuals and obviously will result in various pressure groups trying to get a little more for themselves or a special cause. However, we must look at the total picture, not just at what affects a few individuals here and now. Considerations must be made for the future so that humankind can coexist with the environment on earth for many years to come. With an educated society, logical thinking, and good information, human beings can make good decisions about what efforts must be expended both now and in the future in order to improve and preserve the environment for the best use of all for now and for time to come.

14 Dissolved Air Flotation Boat Plant for Lake Restoration

One of the primary causes of the death of a lake is excessive biological growth, called eutrophication. Biological growth is limited primarily by the availability of the nutrients necessary for growth. It has been shown that phosphorus is most frequently the limiting nutrient to control biological growth in a lake, but nitrogen is also commonly limiting. Phosphorus may be permanently removed from a lake by various processes, whereas nitrogen is difficult to remove permanently due to the fact that certain blue-green algae can fix atmospheric nitrogen as a nitrogen source. Thus, emphasis has been placed on the removal of phosphorus. There are various methods for the treatment of wastewaters to remove the nutrients before being discharged to a body of water. Once in a lake, phosphorus removal is most frequently achieved by producing an insoluble aluminum salt of the phosphorus, but iron salts are effective under aerobic conditions. Calcium salts are effective in removing phosphorus, but they generally adversely increase the pH of the lake. Precipitated aluminum phosphate salts may be allowed to settle to the bottom of the lake, or they may be removed from the water column. A study showed that removing the phosphate-rich hypolimnetic waters from a summer stratified temperate climate lake, precipitating the phosphorus as either aluminum or iron salts, separating the precipitate by dissolved air flotation (DAF), and returning the phosphate reduced water to the lake were very effective in controlling the phosphorus nutrient content in Devil's Lake, WI, USA.

Figure 6.27 shows a dissolved air flotation boat plant designed by Dr. Lawrence K. Wang of the Lenox Institute of Water Technology (LIWT) and Dr. Milos Krofta of Krofta Engineering Corporation (KEC) [44]. The readers are referred to the authors' other article [30] for the detailed technical information of lake restoration using dissolved air flotation process.

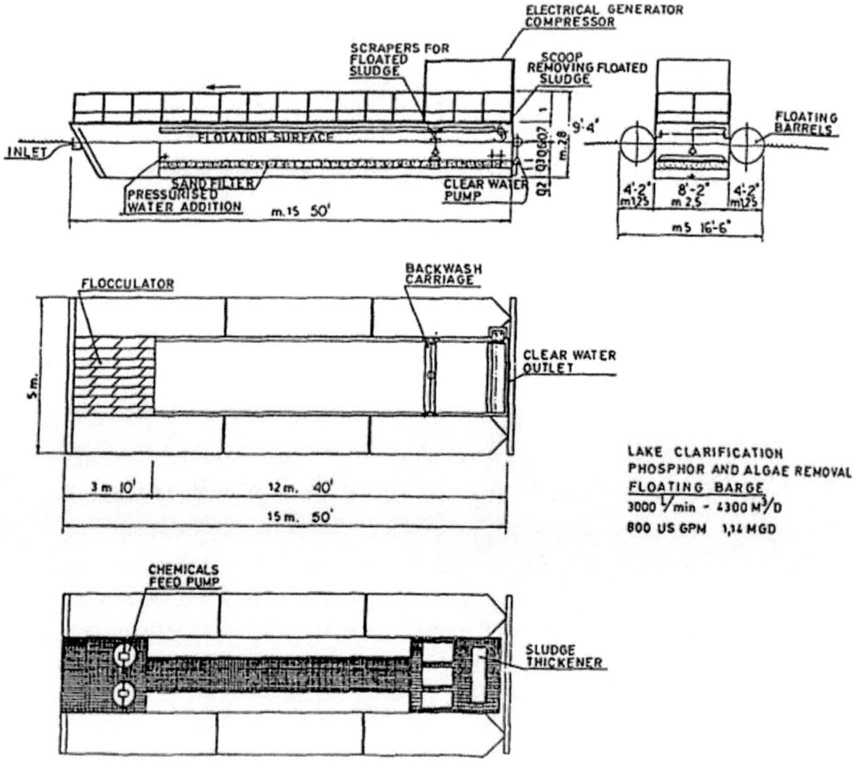

Fig. 6.27 A dissolved air flotation boat plant designed by the Lenox Institute of Water Technology (LIWT) and Krofta Engineering Corporation (KEC) [44]

Dedication This technical paper is published in memory of our coauthors, Dr. Donald B. Aulenbach and Dr. Nazih K. Shammas, who worked closely with Professors Lawrence K. Wang, Mu-Hao Sung Wang, William A. Selke, Milos Krofta, and Daniel Guss from 1981 to 2019 for developing a humanitarian engineering program at the Lenox Institute of Water Technology (LIWT, formerly Lenox Institute for Research), teaching/researching there as an Adjunct Professor for almost two decades, and publishing many research papers and textbooks for academic contribution to the humanity. We salute and will forever remember Don and Nazih.

Glossary

Dissolved air flotation (DAF) One of dissolved gas flotation (DGF) processes when air is used for generation of gas bubbles. A typical example is Krofta Engineering Corporation's Supracell clarifier; see dissolved gas flotation (DGF).

Dissolved gas flotation (DGF) It is a process involving pressurization of gas at 25–95 psig for dissolving gas into water and subsequent release of pressure (to 1 atm) under laminar flow hydraulic conditions for generating extremely fine gas bubbles (20–80 microns) which become attached to the impurities to be removed and rise to the water surface together. The impurities or pollutants to be removed that are on the water surface are called float or scum which are scooped off by sludge collection means. The clarified water is discharged from the flotation clarifier's bottom. The gas flow rate is about 1% of influent liquid flow rate. The attachment of gas bubbles to the impurities can be a result of physical entrapment, electrochemical attraction, surface adsorption, and/or gas stripping. The specific gravity of the bubble-impurity agglomerate is less than one, resulting in buoyancy or non-selective flotation (i.e., Save-All).

Eutrophication Over-enrichment of a water body with nutrients, resulting in excessive growth of organisms and depletion of oxygen concentration.

Ground water The supply of fresh water found beneath the earth's surface (usually in aquifers) which is often accessed through wells and springs.

Hydrologic cycle The cycle of water through its natural process of evaporation and precipitation: from the sea, through the atmosphere, to the land and back to the sea.

Lake An inland body of water, fresh or salt, of considerable size usually greater than 50 acres (200,000 square meters) and occupying a basin or hollow on the earth's surface.

pH An expression of both acidity and alkalinity on a scale of 0–14, with 7 representing neutrality; numbers less than 7 indicate increasing acidity and numbers greater than 7 indicate increasing alkalinity. Acid rain can increase the pH level of the water in a lake, thereby killing all life.

Photosynthesis The manufacture by plants of carbohydrates and oxygen from carbon dioxide and water in the presence of chlorophyll, using sunlight as an energy source. Plants thereby absorb carbon from the atmosphere with the process and store it in their forms.

Phytoplankton Usually microscopic aquatic plants, sometimes consisting of only one cell.

Plankton Those organisms that are unable to maintain their position or distribution independent of the movement of water or air masses.

Pollution The contamination of a natural ecosystem, especially with reference to the activity of humans.

Precipitation Any and all forms of water, whether liquid or solid, that fall from the atmosphere and reach the earth's surface. A day with measurable precipitation is a day when the water equivalent of the precipitation is equal to or greater than 0.2 mm.

Urban runoff Storm water from city streets and adjacent domestic or commercial properties that may carry pollutants of various kinds into the sewer systems and from there to rivers, lakes, or oceans.

Water resources Water in various forms, such as rain, snow, ice, clouds, surface water, ground water, reclaimed water, or reused water – that is potentially useful for domestic, industrial, commercial, or recreational applications.

References

1. Henderson LJ (1913) The fitness of the environment: an inquiry into the geological significance of the properties of matter, vol XV. Macmillan, New York, USA, p 317
2. Osborn B, Harrison PO'N (1965) Water and the land. US Department of Agriculture, Soil Conservation Service, SCS-TP-147
3. Lvovich MI (1972) Fresh water resources of world. Vestnik, Akademii Nauk, (USSR) 70:72
4. Thiessen AH (1911) Monthly Weather Review 39:1082. https://doi.org/10.1175/1520-0493(1911) 39<1082b:PAFLA>2.0 July 1911
5. Meyer AF (1915) Tans American Society of Civil Engineers, Reston, VA 20191, USA 79:1
6. Blaney-Criddle HF, Criddle WD (1957) Trans American Society of Civil Engineers, Reston, VA 20191, USA 122:810
7. Thornthwaite CW (1948) An approach toward a rational classification of climate. Geogr Review 38:55–94
8. Horner WW, Flint FL (1936) Tans American Society of Civil Engineers, Reston, VA 20191, USA 101:140
9. Shammas NK, Wang LK, Queiroz Faria LC, Chaves Ferro MA (2013) Abastecimento De Agua e Remocao De Residuos. GEN (Grupo Editorial Nacional) LTC (Livros Teenicos e Cientificos Editora Ltda), Portugal, p 751, ISBN: 978-85-216-2283-3, www.grupogen.com.br, rotaplanrio@gmail.com, ltc@grupogen.com.br
10. Gray DM (1970) Handbook on the principles of hydrology. Water Information Center, US Geological Survey, Menlo Park, CA and Reston, VA, USA, pp 3.6–8.10
11. Sherman LK (1932) Eng News Record, Mc Graw-Hill, New York, USA, 108:501
12. Viessman W Jr, Lewis G (1966) Introduction to hydrology, 4th edn. Harper & Row, New York, USA
13. Creager WP, Justin JD, Hinds J (1945) Engineering for dams. In: General design, vol 1. Wiley, New York, USA
14. Pickels GW (1941) Drainage and flood control engineering. McGraw-Hill, New York
15. Rippl W (1883) Proc Institute of Civil Engineers, London, UK, 71:270
16. Aulenbach DB, Clesceri NL (1980) Monitoring for land application of wastewater. Water Air Soil Pollut 14:81
17. Linsley RK, Kohler MA, Paulhus JLH (1982) Hydrology for engineers, 3rd edn. McGraw-Hill, New York, USA
18. Sawyer CN, McCarty PL (1978) Chemistry for environmental engineering, 3rd edn. McGraw-Hill, New York, USA
19. Matsumoto GI (1966) Biochemical study of organic substances in Antartica lakes. Hydrobiologia 172:265–299
20. Vollenweider RA (1968) The scientific basis of lake and stream eutrophication, with particular reference to phosphorus and nitrogen as factors in eutrophication, Technical Report to OECD. Paris, DAS/CSI/68.27, pp. 1–182.
21. Vollenweider RA, Dillon PJ (1974) The application of the phosphorus loading concept to eutrophication research. Canada Centre for Inland Waters, NRCC No. 13690, Burlington, Ontario, FL, USA 1974
22. Aulenbach DB (2000) Silicon limitation linked to estuarine pollution problems. In: 32nd Mid-Atlantic Industrial & Hazardous Waste Conference. Rensselaer Polytech. Inst., Troy
23. Rabalais N et al (1996) Nutrient changes in the Mississippi River and system responses on the adjacent Continental Shelf. Estuaries 19:386
24. Babbitt B (2001, October 15) Opening remarks at WEFTEC
25. Industrial waste guide on thermal pollution (1968) Federal Water Pollution Admin.Washington, DC, USA
26. Aulenbach DB, Clesceri NL (2009) Treatment by application onto land. In: Wang LK, Pereira NC, Hung YT (eds) Biological treatment processes. Humana Press/Springer, Totowa, NJ, USA, pp 29–74

27. Clesceri NL, Aulenbach DB, Roetzer JF (2009) Treatment by subsurface application. In: Wang LK, Pereira NC, Hung YT (eds) Biological treatment processes. Humana Press/Springer, Totowa, NJ, USA, pp 75–112
28. Wang LK, Shammas NK, Selke WA, Aulenbach DB (2010) Gas dissolution, release, and bubble formation in flotation systems. In: Wang LK, Shammas NK, Selke WA, Aulenbach DB (eds) Flotation technology. Humana Press/Springer, Totowa, NJ, USA, pp 49–84
29. Aulenbach DB, Shammas NK, Wang LK, Marvin RC (2010) Algae removal by flotation. In: Wang LK, Shammas NK, Selke WA, Aulenbach DB (eds) Flotation technology. Humana Press/Springer, Totowa, NJ, USA, pp 363–400
30. Aulenbach DB, Shammas NK, Wang LK, Kittler DI (2010) Lake restoration using dissolved air flotation. In: Wang LK, Shammas NK, Selke WA, Aulenbach DB (eds) Flotation technology. Humana Press/Springer, Totowa, NJ, USA, pp 429–456
31. Wang LK, Aulenbach DB, Van Dyke JP (2010) Jimmy Peak, Hancak, Massachusetts, Wastewater Treatment Plant: The first RBC-Flotation-UV wastewater treatment plant in the USA. In: Wang LK, Shammas NK, Selke WA, Aulenbach DB (eds) Flotation technology. Humana Press/Springer, Totowa, NJ, USA, pp 457–484
32. Wang LK, Shammas NK, Aulenbach DB, Selke WA, Guss DB (2010) Pittsfield water treatment plant: once the world's largest flotation-filtration plant. In: Wang LK, Shammas NK, Selke WA, Aulenbach DB (eds) Flotation technology. Humana Press/Springer, Totowa, NJ, USA, pp 485–502
33. Wang LK, Aulenbach DB, Shammas NK (2010) Treatment of sea food processing wastewater. In: Wang LK, Shammas NK, Selke WA, Aulenbach DB (eds) Flotation technology. Humana Press/Springer, Totowa, NJ, USA, pp 567–592
34. Shammas NK, Wang LK, Selke WA, Aulenbach DB (2010) Laboratory simulation and testing of air flotation and associated processes. In: Wang LK, Shammas NK, Selke WA, Aulenbach DB (eds) Flotation technology. Humana Press/Springer, Totowa, NJ, USA, pp 593–617
35. Wang LK, Shammas NK, Selke WA, Aulenbach DB (2006) Flotation thickening. In: Wang LK, Shammas NK, Hung YT (eds) Biosolids treatment processes. Humana Press/Springer, Totowa, NJ, USA, pp 71–100
36. Wang LK, Shammas NK, Evanylko GK, Wang MHS (2014) Engineering management of agricultural land application. In: Wang LK, Yang CT (eds) Modern water resources engineering. Humana Press/Springer, Totowa, NJ, USA, pp 313–414
37. Yang CT, Wang LK (2015) Advances in water resources engineering. Humana Press/Springer, Totowa, NJ, USA, p 556
38. Wang LK, Yang CT, Wang MHS (2016) Advances in water resources management. Springer, New York, USA, p 569
39. Wang LK, Wang MHS, Hung YT, Shammas NK (2016) Natural resources and control processes. Springer, New York, USA, p 633
40. Shammas NK, Wang LK (2010) Fair, Geyer and Okung's water and wastewater engineering: water supply and wastewater removal. Wiley, Hoboken, NJ, USA, p 824
41. Shammas NK, Wang LK (2016) Water engineering: hydraulics, distribution and treatment. Wiley, Hoboken, NJ, USA, p 806
42. Wang LK, Wang MHS, Fahey E (2021) Innovative Lee, Massachusetts USA dissolved air flotation potable water filtration plant. In: Wang LK, Wang MHS, Shammas NK (eds) Environmental flotation engineering. Springer Nature Switzerland, pp. 73–94
43. Wang LK (2020) Humanitarian engineering education of the Lenox Institute of Water Technology and its new potable water flotation processes. In: Wang LK, Wang MHS, Shammas NK (eds) Environmental flotation engineering. Springer Nature Switzerland, pp. 1–72
44. Krofta M, Wang LK (2000) Flotation engineering, 1st edn. Lenox Institute of Water Technology, Lenox, MA, USA. Lenox 1-06-2000/368

Chapter 7
Advances in Cadmium Detoxification/Stabilization by Sintering with Ceramic Matrices

Minhua Su, Kaimin Shih, and Diyun Chen

Contents

1 Introduction 301
1.1 Cadmium Pollution and Control Technologies 302
1.2 Thermal Stabilization Technique 304
1.3 Ceramic Matrices for Cadmium Detoxification 306
1.4 XRD-Based Qualitative and Quantitative Analysis 308
1.5 Leaching Tests for Evaluating Metal Stabilization Effects 309
2 Experimental Methods 310
2.1 Materials and Chemicals 310
2.2 Cadmium Incorporation 310
2.3 Qualitative and Quantitative XRD Analysis 310
2.4 Stabilization Effect Evaluation 311
3 Results and Discussion 311
3.1 Crystalline Product Formation for Cadmium Detoxification 311
3.2 Evaluation of Metal Stabilization Effect 314
4 Conclusion 316
Glossary 317
References 317

Abstract Cadmium (Cd) is often found in industrial sludge, fly ash, slag, and flue gas, and it presents serious risks to the environment and biota. Stabilizing and detoxifying Cd in waste streams is thus of great importance. The use of various low-cost and attainable ceramic matrices (amorphous SiO_2, γ-Al_2O_3, α-Fe_2O_3, and Fe_3O_4) to interact with Cd-containing waste is a promising method of Cd stabilization. Heating mixtures of cadmium oxide (CdO) and ceramic matrices at various molar ratios and temperatures (600–1000 °C) for 3 h could achieve the goal of Cd

M. Su (✉) · D. Chen
School of Environmental Science and Engineering, Guangzhou University, Guangzhou, China
e-mail: mhsu@gzhu.edu.cn; cdy@gzhu.edu.cn

K. Shih
Department of Civil Engineering, The University of Hong Kong, Hong Kong, China
e-mail: kshih@hku.hk

L. K. Wang, M. -H. S. Wang, Y. -T. Hung, N. K. Shammas (eds.),
Environmental and Natural Resources Engineering, Handbook of Environmental Engineering 19, https://doi.org/10.1007/978-3-030-54626-7_7

incorporation. Phase transformation was assessed using X-ray diffraction (XRD), and the efficiency of Cd incorporation was quantified through Rietveld analysis of the obtained XRD patterns. The XRD results show that Cd can be crystallochemically incorporated into $CdSiO_3$, Cd_2SiO_4, Cd_3SiO_5, $CdAl_4O_7$, and $CdFe_2O_4$ phases. The treatment temperature greatly affected the Cd incorporation reactions. The Cd incorporation efficiency was quantified and expressed as a transformation ratio according to the weight fractions of crystalline phases in the sintering products. To evaluate the metal stabilization effect of the Cd detoxification process, a series of constant-pH leaching tests was conducted. A remarkable reduction in Cd leachability was achieved by forming different Cd-hosting crystalline products, particularly spinel phase $CdFe_2O_4$. Overall, the efforts to stabilize Cd by sintering with ceramic matrices suggest a promising strategy for the detoxification of Cd in wastes.

Keywords Cadmium pollution control · Ceramic matrices · Detoxification · Leaching tests · Sintering · Thermal stabilization · XRD analysis

Abbreviations and Nomenclature

Al_2O_3	Alumina
AWWA	American Water Works Association
CPLT	Constant-pH leaching test
$CdAl_4O_7$	Cadmium aluminate
$CdFe_2O_4$	Cadmium ferrite
CdO	Cadmium oxide
$CdSiO_3$	Mono-cadmium silicate
Cd_2SiO_4	Di-cadmium silicate
Cd_3SiO_5	Tri-cadmium silicate
d	The spacing between the lattice planes of the corresponding phase
EPA	US Environmental Protection Agency
G	Gibbs free energy
ICDD	International Centre for Diffraction Data
ICP-OES	Inductively coupled plasma optical emission spectrometry
MSW	Municipal solid waste
MSWI	Municipal solid waste incineration
MW	Molecular weight
PDF	Powder Diffraction File
QXRD	Quantitative X-ray diffraction
S	Entropy
SiO_2	Silicon oxide
SPLP	Synthetic precipitation leaching procedure
T	Temperature
TCLP	Toxicity characteristic leaching procedure
TR	Transformation ratio

V	Volume
WHO	World Health Organization
XRD	X-ray diffraction
λ	The applied X-ray wavelength
θ	The angle between the incident beam and the scattering plane(s)

1 Introduction

Contamination of air, water, and soil by heavy metals is a pressing global issue that poses huge threats to the environment and public health, due to long- and short-term toxicological effects [1–3]. Cadmium (Cd) is a heavy metal that is commonly found in zinc/lead mining waste, alkaline batteries, electroplating products, PVC stabilizers, semiconductors, and solar cells [4, 5]. The release of Cd-containing pollutants from solid waste into the environment causes serious contamination [2, 6, 7]. Cadmium accumulates in food chains and is easily absorbed by the human body, where it exerts toxic effects on health and results in a great number of acute and chronic diseases (including kidney failure, skeletal deformation, and lung cancer) [6]. In an attempt to eliminate Cd pollution in solid waste and from combustion processes, many techniques (such as solidification/stabilization by cement or sorption by minerals) have been developed [4, 8–12]. However, most of them have many inconvenient drawbacks, such as their low treatment efficiency, consumption of large amounts of raw materials, and formation of undesired products [10, 13, 14].

Recently, a technology inspired by ceramic sintering processes has proved to be effective and reliable for the stabilization of heavy metals (including Ni, Zn, Cu, and Pb) [15–19]. The interactions of metal-containing waste with a series of inexpensive and easily attainable ceramic matrices have been explored, and various stable ceramic (crystalline) products with high acid resistance have been formed. This novel technique has suggested a possible method for the stabilization of Cd-containing waste. In addition, studies have demonstrated that cadmium oxide (CdO) can thermally react with certain metal oxides (such as Al_2O_3, SiO_2, and Fe_2O_3) or Al/Si-rich materials (e.g., kaolinite and mullite) [20–25]. With thermal treatment, a number of robust crystalline products with particular crystal structures can be harvested. Thermally treating CdO with these ceramic matrices suggests further possible and feasible techniques for Cd incorporation. However, the processing parameters and mechanisms of Cd incorporation should be determined by thorough investigation.

This chapter addresses the sources and toxicity of Cd, methods of Cd detoxification, and techniques to evaluate the Cd stabilization effects. The feasibility of using various Al/Fe/Si-based ceramic matrices to detoxify Cd in wastes via a sintering process is investigated. The detoxification and stabilization of Cd can be achieved by transforming it into crystalline products. The influences of operational parameters (i.e., sintering temperature and the molar ratios of reactants) are considered in detail.

Quantitative X-ray diffraction and constant-pH leaching test (CPLT) results reveal the metal incorporation efficiencies and metal stabilization effects of this method.

1.1 Cadmium Pollution and Control Technologies

The rapid development of industrialization has generated a huge amount of solid waste, including mining waste, electronic devices, industrial and domestic sludge, and incineration residues like bottom and fly ash [2, 26–32]. Such solid waste often contains large amounts of heavy metals and is a main source of environmental pollution. The toxicity of heavy metals has been widely studied. Their effects on human health have been well assessed by international organizations, such as the World Health Organization (WHO). The heavy metals Cd, Pb, Hg, and Cr and most of their compounds have been classified as highly toxic and are known to be carcinogenic. Heavy metals tend to bioaccumulate in food chains and are eventually assimilated in living tissues [33]. As they are non-biodegradable, they are very stable and persistent contaminants. Without adequate treatment, metal-containing waste will release high levels of toxic metals, leading to air pollution and water and soil contamination (Fig. 7.1).

Cadmium can be found in contaminated sites and in solid waste because it is widely used in many fields, including metallurgical alloying, electroplating, alkaline batteries, textile printing industries, metal plating, pigment works, and ceramics [6, 34–37]. In recent years, the use of Cd has been increasing along with its emission to the environment, as Cd-containing products are rarely recycled and are habitually

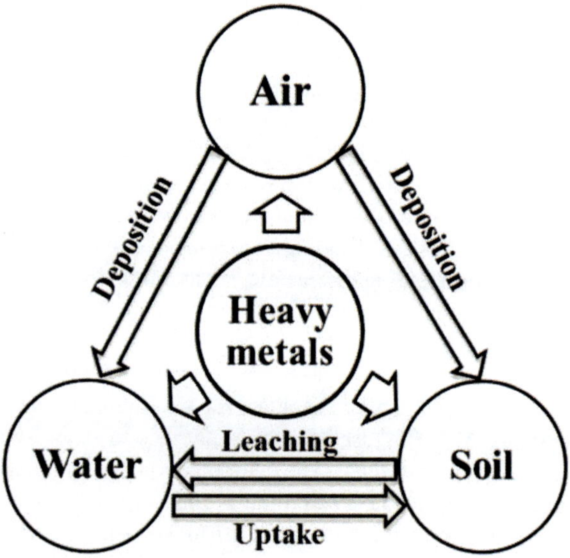

Fig. 7.1 Heavy metals risk to air, water bodies, and soil

abandoned with household waste. The unstable Cd-containing waste contaminates the environment when it is disposed of in landfill or incinerated [2, 35].

Incineration can greatly reduce the mass and volume of municipal solid waste (MSW) [38, 39]. The process creates a considerable amount of incineration residue, mainly composed of bottom ash and fly ash [39–41]. Unlike organic compounds, metal species are not destroyed in high-temperature conditions and remain in the residue. Most of them condense and convert into metallic or metal oxide particles or airborne aerosols. Residues from the incineration of MSW are therefore usually enriched with toxic metals such as Zn, Ni, Cr, Cu, Pb, and Cd [4, 39–41]. Concentrations of Cd from 24 to 1500 $\mu g/m^3$ have been reported for flue gas from incinerating MSW, and most of this Cd eventually accumulates in fly ash [4]. The largest source of Cd pollutants in MSW is Ni-Cd batteries, which contribute 60–70% of the total Cd, and the second largest source is waste plastics [42]. Wan et al. found Cd content of 72 mg/kg in fly ash [43]. In the fly ash from two incineration plants in China, the Cd contents were found to be 37 and 276 mg/kg [44, 45]. Quina et al. reported that the content of Cd in MSW residues ranged from 16 to 1660 mg/kg [46]. Using an adsorption process to remove Cd from aqueous solutions produces a significant amount of waste containing Cd, with 98–600 mg Cd adsorbed per gram of adsorbent [47–52]. Before disposal, the Cd-containing waste must undergo suitable pretreatment or stabilization to reduce its adverse effects on the environment [38, 41, 53–56].

Cadmium can induce a great number of acute and chronic illnesses, including renal damage, hypertension, emphysema, testicular atrophy, and cancers [34, 36, 57, 58]. Due to its high toxicity, Cd has been classified as a group 1 carcinogen [59]. According to a WHO document, the concentration of Cd in drinking water should not exceed 0.005 mg/L [60].

Cadmium-containing waste requires detoxification and metal stabilization before disposal. The development of novel and environmentally friendly technology should be adopted to control and reduce the harm of Cd-containing waste [37, 58, 61]. A solidification/stabilization (S/S) process is commonly used for the treatment of most metal-containing waste, preventing the hazardous substances from migrating into the surroundings by physically fixing toxic contaminants and/or chemically bonding them to binders [22, 62]. Figure 7.2 shows a typical S/S process for toxic metal immobilization. In the S/S process, pollutants can be converted into less mobile, soluble, and toxic forms using various stabilizers, additives, or binders, such as cement, clay, fly ash, zeolite, and red mud [63].

Common S/S technologies using sorption or cementation to immobilize metals may not reliably control metal leaching in a variety of acidic environments because the binding effect is not satisfactory in such environments [64]. It is not possible to fix metals, especially highly mobile metals such as Hg, Cd, and Pb, for long-term disposal because they are easily leached when the products are in acidic conditions [63]. Furthermore, a large amount of cement would be consumed to achieve the goal of stabilization, which may lead to a significant increase in the volume of end products.

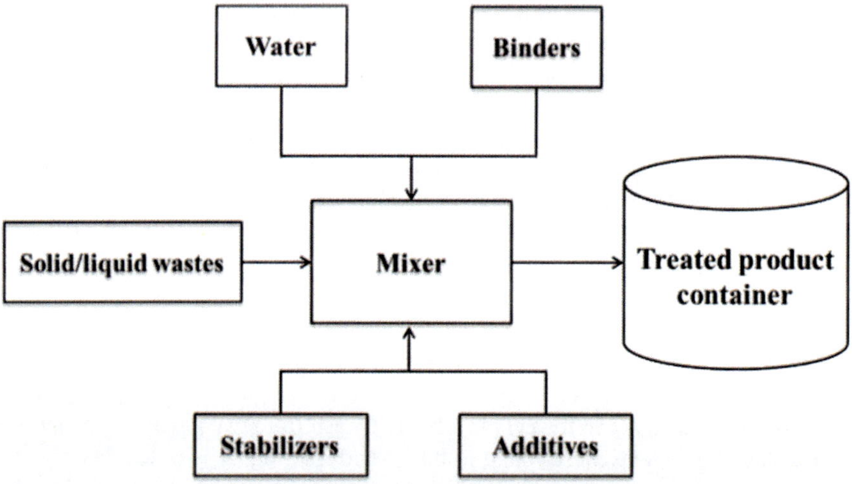

Fig. 7.2 Schematic of a typical S/S process for toxic metal immobilization

Vitrification is an alternative approach to the treatment of metal-containing waste [65–67]. After vitrification, the metals are immobilized in glass matrices [67, 68]. This process can control and reduce the leachability of the harmful constituents of the treated products [69]. However, high temperatures (1600–2000 °C) are required [70], causing a large amount of energy consumption. From the economic and environmental viewpoints, the development of energy-saving and cost-effective metal stabilization is of great importance.

1.2 Thermal Stabilization Technique

Feasible and effective technology used to detoxify metal-containing waste is urgently needed. Using a ceramic sintering process to thermally convert heavy metals into various robust crystal structures via reacting with ceramic matrices is a novel and promising waste management technique [16, 17, 19, 71] and has become a mainstream research interest in the field of waste management [16, 17, 19, 71].

By sintering mixtures of metal-containing waste and ceramic matrices, heavy metals can be incorporated into specific crystalline phases by driving atoms to their most energetically favorable positions. The crystalline products generally have good mechanical properties and high acid resistance, suggesting their potential for heavy metal stabilization [16, 19, 72, 73]. Ceramic sintering is therefore considered to be a good option for the treatment of different types of hazardous waste, including contaminated soil, industrial sludge, and fly ash [61]. It is important to note that comparing to vitrification method ceramic sintering can effectively incorporate toxic components at relatively low temperature, which makes it attractive from economic

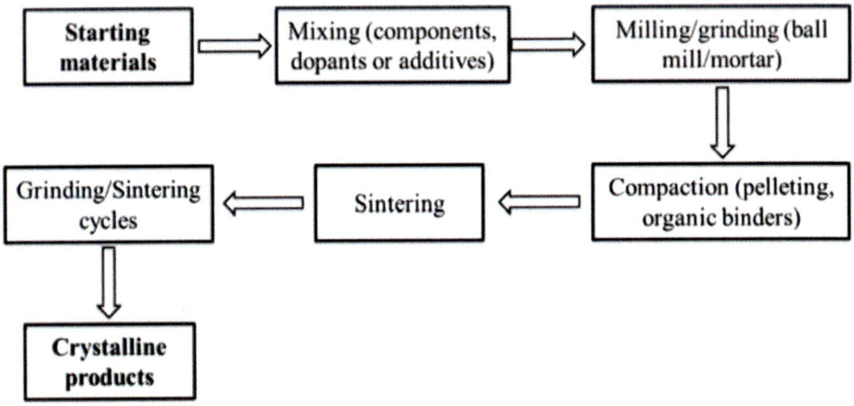

Fig. 7.3 A typical ceramic sintering process

and environmental perspectives because less energy is required [74]. Studies have demonstrated that waste containing relatively high contents of heavy metals, including Pb and Cd, can be successfully incorporated into ceramic products such as bricks and tiles [74, 75]. Figure 7.3 shows a typical ceramic sintering process.

During the ceramic sintering process, a solid-state reaction occurs among the solid particles to transform the raw materials into a dense product upon heating. The consideration of thermal dynamics is one of the most important factors for the incorporation reaction. When sintering takes place, the free energy in the reactive system varies and such variation generates different driving forces. The driving forces generally include (1) particle surface free energy, (2) external pressure, and (3) chemical reaction [76, 77]. Sintering can be achieved by increasing the driving forces via a physical or chemical process [78].

Sintering requires matter transportation among solid materials. The transportation of matter among solid particles can be triggered by various diffusion processes that involve atoms, ions, or molecules [76, 79]. At the atomic level, diffusion is the process of atoms migrating from one lattice site to another. An intermediate stage with higher energy causes the transformation of different phases [77]. In the intermediate stage, the energy barrier must be overcome so that the movement of atoms can occur. The energy required to overcome the energy barrier is the activation energy. The relationship between the energy P for an atom and the energy barrier q is expressed as

$$P = \exp\left(\frac{-q}{kT}\right) \tag{7.1}$$

where k is the Boltzmann constant and T is the absolute temperature.

The Gibbs free energy (G) is an important parameter for the driving force of a reaction and can be used to examine the thermodynamic stability of the reaction. For an irreversible reaction, the change in G is expressed as

$$dG = \left(\frac{\partial G}{\partial T}\right)_p dT + \left(\frac{\partial G}{\partial P}\right)_T dp = -SdT + Vdp \tag{7.2}$$

where S is the entropy and V is the volume of the reactive system [76, 79].

In recent years, the synthesis, structure, properties, and applications of versatile functional materials have become popular research topics in materials science and engineering. To understand the intrinsic characteristic of a new material, knowledge of its crystal structure at different levels is essential. The fundamental level is the electron configuration, which explains characteristics such as the material color, optical properties, electrical conductivity, and magnetic behavior [80]. The distribution of electrons in atoms determines the bonds that can form, which result in differences in crystal structures. Furthermore, knowing the arrangement of atoms in the crystal structure is crucial to understanding the mechanical and electronic properties of the material. During sintering, crystalline and non-crystalline (amorphous) ceramic products can be formed. In crystalline ceramics the atoms show a periodic arrangement, and those in the non-crystalline ceramics such as glass do not have long-range order. Crystals are constructed by regular atomic arrangements in three dimensions, and such arrangements are repeated by the unit cell. The unit cell is the smallest repeating unit in the crystal structure that shows the symmetry of the crystal structure. The crystal systems can be classified as triclinic, monoclinic, orthorhombic, tetragonal, hexagonal, rhombohedral (also called trigonal), or cubic [81].

1.3 *Ceramic Matrices for Cadmium Detoxification*

It has been reported that some oxides (including SiO_2, γ-Al_2O_3, and α-Fe_2O_3) and mineral compounds (including kaolinite and montmorillonite) can thermally react with CdO to produce different Cd-hosting crystalline product phases. For the reaction between CdO and SiO_2, an equilibrium diagram of a Cd-B-Si-O system was reported [23]. It can be seen from the phase diagram that three types of cadmium silicate can be obtained after sintering mixtures of CdO and SiO_2 with different molar ratios of Cd/Si. The phase diagram of a Cd-Al-O system shows that cadmium aluminates (e.g., $CdAl_2O_4$, $CdAl_4O_7$, and $CdAl_{12}O_{19}$) may be formed when sintering CdO with alumina. This may offer an opportunity for Cd stabilization with alumina. According to Kurihara and Suib [82], it is possible to obtain $CdAl_2O_4$ via a two-step synthetic process (sol-gel and sintering). $CdAl_4O_7$ can be prepared via conventional solid-state reactions by sintering mixtures of CdO and alumina [21]. Few studies have mentioned the formation of $CdAl_{12}O_{19}$, although Colin [21] stated that $CdAl_{12}O_{19}$ may be formed by analogy with products from the reactions between β-Al_2O_3 and oxides (e.g., CaO, BaO, and PbO) under thermal conditions. Studies have indicated that $CdFe_2O_4$ can form a Cd-Fe-O system by sintering CdO with α-Fe_2O_3 [83].

Silicon (Si) is the second most abundant element (after oxygen) in the Earth's crust [84, 85]. Silica (SiO_2), the oxide form of Si, has many applications in science and engineering because it is chemically inert, thermodynamically stable, non-toxic, and low-cost [86]. Silica can be crystalline and non-crystalline. In crystalline silica, the Si and O atoms are arranged in a geometric structure. In non-crystalline (amorphous) silica, no spatial ordering of Si and O atoms is observed. Crystalline forms of silica include quartz, cristobalite, and tridymite [87]. Of these, quartz is the most common and is easily attainable from natural sources, such as rocks and soil, including arable loess and clay soils. The three types of amorphous silica are naturally occurring silica, silica formed under uncontrolled conditions, and synthetic silica [87]. Silica fumes comprising very fine amorphous silica particles are a by-product of the manufacture of silicon and ferrosilicon [88]. Studies have demonstrated that silica fumes can fix heavy metals due to their large specific surface area and high amorphous silica content [89–91]. Amorphous and crystalline silicon compounds are also commonly present in sewage sludge ash and MSW fly ash [92]. Most silicates have good physical and chemical stability and thus have a wide range of industrial applications [93]. Silicates can be prepared by various methods, including sol-gel, sintering, and solid-state diffusion. It has been reported that Cd can react with silicates via solid-state reactions, forming different silicate products [23, 94], and thus the use of silicates to stabilize Cd can be a promising strategy.

Aluminum (Al) is the third most abundant element in the Earth's crust and accounts for 8 wt.% of its solid surface [95]. Alumina (Al_2O_3) is the oxide form of Al. The common forms of alumina are α, γ, η, δ, θ, κ, and χ forms. α-Al_2O_3 is the most stable phase and the other forms can be obtained by the thermal decomposition of aluminum hydroxides or oxyhydroxides [96]. Aluminas are widely applied as raw materials in the manufacturing of various functional and conventional ceramic products. Among these aluminas, γ-Al_2O_3 is the most common oxide form of Al in nature. It has great application potential in many fields because of its unique crystal structure, large surface area, highly reactive nature, low cost, and attainability [97, 98]. γ-Al_2O_3 has a cubic lattice in space group *Fd3*. Its crystal structure is often defective, which may promote the incorporation of metals [99–102].

Iron (Fe) is also one of the most abundant elements on Earth and is common in the raw materials of ceramic manufacturing [17, 103]. Several iron oxides exist, including hematite (α-Fe_2O_3), magnetite (Fe_3O_4), maghemite (γ-Fe_2O_3), β-Fe_2O_3, ε-Fe_2O_3, and wustite (FeO). Both β-Fe_2O_3 and ε-Fe_2O_3 are uncommon in nature [104]. Wustite is an intermediate in the reduction process of iron ores and is not thermodynamically stable [103–105]. Hematite and magnetite are the two most widespread iron oxides in nature [103, 104]. Hematite has a corundum structure and is the most thermodynamically stable of all the iron oxides. The close-packed arrangement of Fe and O atoms in hematite means that it has no charge excess or deficit. Magnetite is an inverse spinel and can be easily oxidized to maghemite and finally turned into hematite [103, 104, 106]. Few studies have revealed the effects of different iron oxides on the formation of $CdFe_2O_4$. Hence, it will be of great interest to investigate the reaction of different iron oxides with Cd to form ceramic products.

1.4 XRD-Based Qualitative and Quantitative Analysis

Since the diffraction of X-rays was discovered in 1912, X-ray diffraction (XRD) has become an effective and reliable technique for phase analysis [107]. XRD analysis can provide both qualitative and quantitative information on the phases in a tested sample [108, 109]. XRD is a reliable, precise, and reproducible method that can identify the crystalline phases in solids (e.g., minerals and inorganic compounds) and reveal structural details of their phases [81, 109, 110]. Consequently, XRD is extensively used to characterize natural and industrial materials with the support of detailed database information.

Every crystalline phase has a distinctive diffraction pattern. By analyzing the peak positions (corresponding to *d* values) and peak intensities in the XRD pattern, the crystalline phases in the sample can be identified. The phase analysis of XRD data is based on Bragg's law [108] (Eq. 7.3) and indicates the peaks of crystal lattice scattering on the basis of two conditions: (i) the angle of the incident beam is equal to the angle of the scattered beam, and (ii) the path length difference is equal to an integer number of wavelengths.

$$d = \frac{n \cdot \lambda}{2 \cdot \sin\theta} \tag{7.3}$$

where n is an integer, λ is the applied X-ray wavelength, θ is the angle between the incident beam and the scattering plane(s), and d is the spacing between the lattice planes of the corresponding phase.

With the fast development of computer science, the quantification of XRD data is now well developed [111, 112]. In addition to qualitative analysis, another very important feature of the XRD technique is that it can quantify the weight fractions of crystalline phases in the samples. Quantitative X-ray diffraction (QXRD) analysis is able to quantify the weight fractions of phases in a mixture by refining the XRD data following particular algorithms [113]. QXRD analysis refines the lattice parameters of a unit cell, including coordinates, factors of temperature, atomic occupancies, profile parameters, 2θ displacements, preferred orientation, background radiation parameters, extinction, and micro-absorption [113]. The orientation factors and scale factors of each phase are included in the refinement process.

The methods of QXRD analysis include the reference intensity ratio (RIR) method, Rietveld method, profile stripping (or pattern subtraction), and full-pattern fitting [114–116]. The Rietveld method is the most widely used for phase quantitative analysis and is used in combination with crystal structure models to calculate the XRD pattern for each individual component in the mixture (Fig. 7.4). With a good refinement, the calculated pattern fits well to the observed diffraction pattern and the weight fraction data of crystalline phases can be obtained. The Rietveld refinement method provides not only the weight fractions of crystalline phases but also information about the refined crystal structure, such as lattice parameters, atomic occupancies, and crystal size. The Rietveld method refines structures by minimizing a quantity with Newton-Raphson algorithms [113].

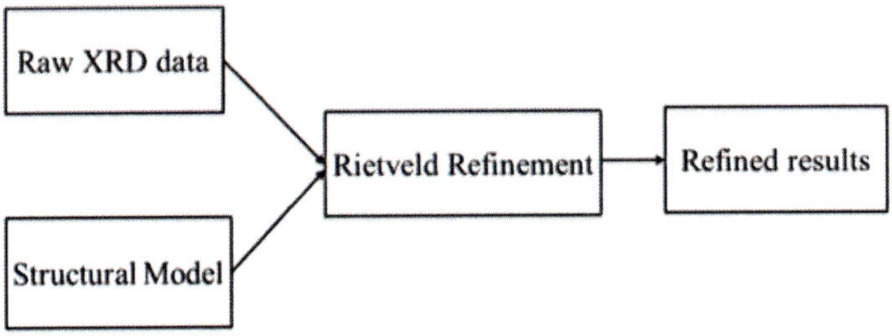

Fig. 7.4 Quantitative X-ray diffraction analysis with Rietveld refinement method for phase composition determination

Ceramic sintering via solid-state reactions is one of the most effective and reliable ways to incorporate metals into crystalline products. The phase transformations that occur via solid-state reactions among CdO and ceramic matrices can be monitored and quantified by a combination of the XRD technique and Rietveld refinement method. The QXRD results can be used to assess the metal incorporation level and efficiency, as well as the stabilization mechanisms.

1.5 Leaching Tests for Evaluating Metal Stabilization Effects

Most pollution caused by metal-containing waste is due to the leaching of metals. The chemical durability of treated products should be examined by a series of leaching tests to assess metal stabilization [117].

Leaching is a relatively complicated reaction [118]. A number of factors, such as chemical composition, pH, redox potential, complexation, liquid-to-solid ratio, and contact time, influence the release rate and leaching behavior of the constituents of waste in different conditions (e.g., acidic or alkaline) [118, 119]. So far, over 50 leaching tests have been developed. Some of them (such as the toxicity characteristic leaching procedure (TCLP), SPLP, and column leaching test) have been frequently used to evaluate the environmental consequences of MSW or industrial waste. The TCLP is one of the most commonly used methods for evaluating the leachability of toxic metals from waste [120]. It is a batch test developed by the United States Environmental Protection Agency for the assessment of the leachability of toxic metals from wastes [63]. The pH significantly influences the substance surface potential, proton competition on surface binding sites, and mineral dissolution [121]. However, during leaching tests, the pH value is not constant, which may impede the leaching process and/or re-precipitation of metal compound(s) [122].

The constant-pH leaching test (CPLT) developed in our recent work can largely overcome the shortcomings of non-constant pH in other leaching tests. The CPLT can effectively examine the chemical durability of the tested materials and products due to its simplicity, improved reproducibility, and shorter time requirements [123, 124]. It can be used to compare the dissolution or leaching behavior of test samples. The test is performed at a constant pH and no buffer solution is involved. Adding buffer solutions is a facile method of controlling the mobility of metals in the leaching system. However, the compounds in buffer solutions may form complexes with the metals, thus influencing the leaching performance of the tested sample [125]. Therefore, the CPLT without buffer solutions can rapidly and precisely assess the leaching performance of the products of a well-controlled thermal treatment scheme.

2 Experimental Methods

2.1 *Materials and Chemicals*

Cadmium oxide (CdO) was used to simulate the Cd source in the waste streams. Silica fumes (amorphous SiO_2), γ-Al_2O_3, α-Fe_2O_3, and Fe_3O_4 were selected as the ceramic matrices for Cd incorporation. γ-Al_2O_3 was prepared via thermally treating alumina powder (Pural SB (Sasol)) at 650 °C for 3 h. Other ceramic matrices were used as received without further purification.

2.2 *Cadmium Incorporation*

To examine the ability of ceramic matrices to incorporate Cd, CdO powder was individually mixed with the ceramic matrices at desired molar ratios of Cd/Si (1/1, 2/1, and 3/1), Cd/Al (1/4), and Cd/Fe (1/2). The mixtures were well mixed in a total weight of 10 g by mortar grinding for 30 min in the presence of absolute ethanol. The mixtures were dried at 105 °C for 24 h in a vacuum oven and then further homogenized by mortar grinding. Subsequently, the homogenous dried powder was pelletized into Φ 20 mm pellets at a pressure of 250 MPa. The pellets were subjected to a well-controlled thermal treatment scheme. The sintering temperature of the scheme was in the range of 600–1000 °C and the dwelling time was set at 3 h.

2.3 *Qualitative and Quantitative XRD Analysis*

The fired pellets were cooled in air and then ground into powder for the XRD test. The step-scanned diffraction data of each powder sample were recorded on a Bruker D8 Advance X-ray powder diffractometer equipped with Cu $K\alpha_{1,2}$ X-ray radiation and a LynxEye detector. The 2θ scanning range was from 10° to 80°. The step size was 0.02° and the scan speed was 0.5 per step. The phase analyses were performed by matching the collected XRD patterns with those derived from the standard powder diffraction database of the International Centre for Diffraction Data (ICDD PDF-2 Release 2008). With Rietveld refinement of the XRD patterns, the weight fractions (in percentages) of the phases in the sample were obtained. The refinement was performed on a TOPAS V4.0 program (Bruker AXS, Karlsruhe). A transformation ratio (TR) index was developed to assess the efficiency of Cd incorporation into Cd-hosting product phase(s). The TR was calculated as follows:

$$\mathrm{TR}(\%) = \frac{\sum \frac{\text{wt\%of Cd-hosting product phase(s)}}{\text{MW of Cd-hosting product phase(s)}}}{\sum \frac{\text{wt\%of Cd-hosting product phase(s)}}{\text{MW of Cd-hosting product phase(s)}} + \frac{\text{wt\%of CdO}}{\text{MW of CdO}}} \quad (7.4)$$

where MW is the molecular weight of the phase, g/mol. A TR of 0% indicates that no Cd was incorporated and a TR of 100% indicates complete transformation of Cd into the Cd-hosting products.

2.4 *Stabilization Effect Evaluation*

The leaching performance of Cd-bearing samples (CdO, $CdSiO_3$, Cd_2SiO_4, Cd_3SiO_5, $CdAl_4O_7$, and $CdFe_2O_4$) was tested by the CPLT so that Cd leachability could be evaluated under acidic conditions. Nitric acid (HNO_3) aqueous solution (pH 4.0) was used as the leaching fluid and the pH of the leaching system was maintained at 4.0 ± 0.2 with the compensation of HNO_3 aqueous solution (1 M) in negligible volume (ca. 20 μL for each adjustment). The CPLT was implemented in a jar with 500 mL of leaching fluid. The weight of tested powders for each CPLT was 0.5 g. During the CPLT, the system was mechanically stirred at 200 rpm and 5 mL of leachate was removed at 10-min intervals. The leachate was filtered with a 0.2-μm syringe filter and then stored in a vial for the determination of metal concentrations by an inductively coupled plasma optical emission spectrometer (ICP-OES 800, PerkinElmer). A standard for Cd was determined before and periodically during each sampling event to generate a satisfactory calibration curve.

3 Results and Discussion

3.1 *Crystalline Product Formation for Cadmium Detoxification*

By sintering CdO with silica fume (amorphous SiO_2), γ-Al_2O_3, α-Fe_2O_3, and Fe_3O_4 at 900 °C for 3 h, cadmium silicate ($CdSiO_3$, Cd_2SiO_4, and Cd_3SiO_5), cadmium aluminate ($CdAl_4O_7$), and cadmium ferrite ($CdFe_2O_4$) phases were formed

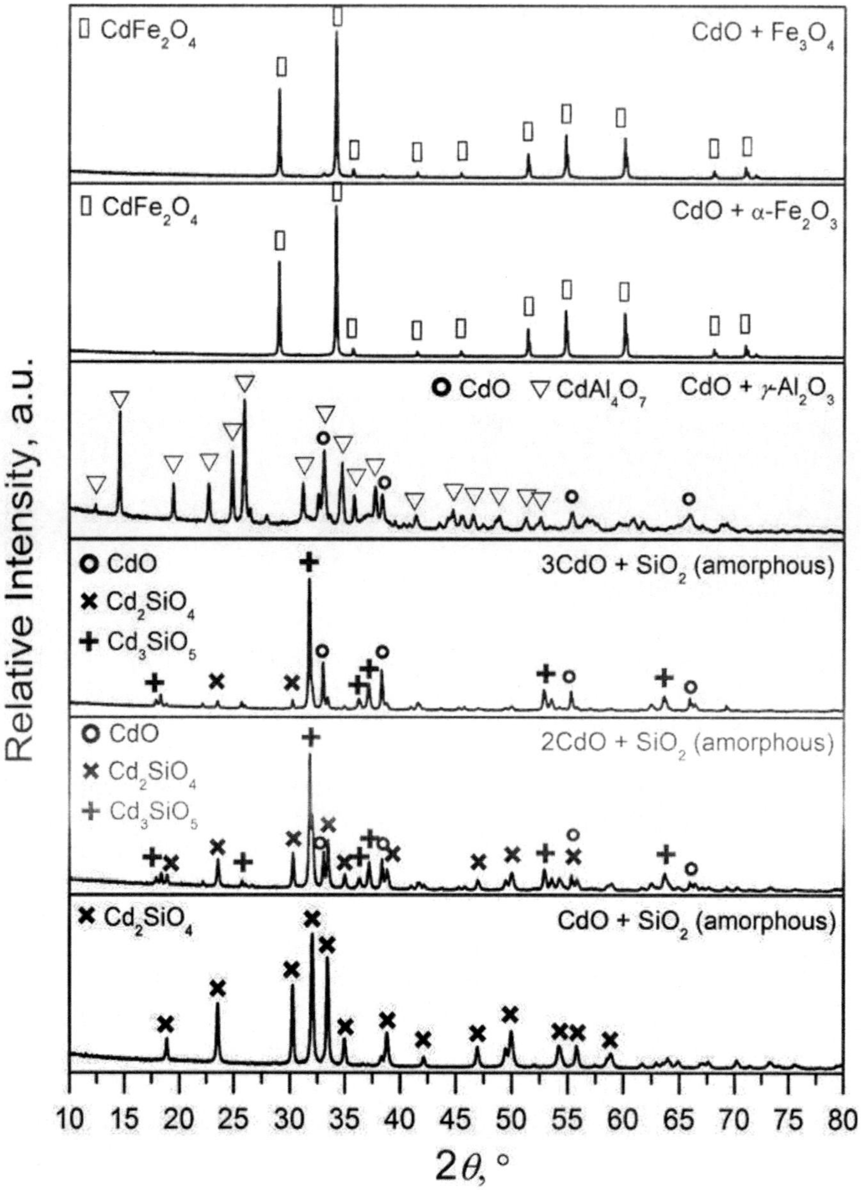

Fig. 7.5 X-ray diffraction patterns of sintered CdO + amorphous SiO_2 (Cd/Si = 1/1, 2/1, 3/1), CdO + γ-Al_2O_3 (Cd/Al = 1/4), and CdO + α-Fe_2O_3 (Cd/Fe = 1/2) at 900 °C for 3 h, showing the formation of Cd silicates, Cd aluminate, and Cd ferrite for Cd incorporation by different ceramic matrices

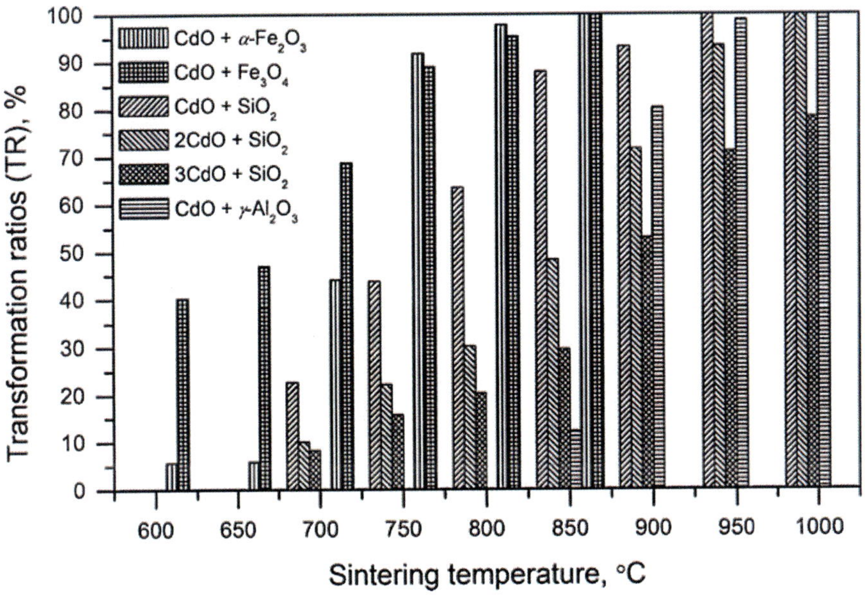

Fig. 7.6 Transformation ratios (TR) of different Cd detoxification systems. The SiO_2 represents silica fumes in amorphous phase

(Fig. 7.5). For the system of $CdO + SiO_2$, the Cd-hosting product phases were identified to be monoclinic $CdSiO_3$ (Eq. 7.5), orthorhombic Cd_2SiO_4 (Eq. 7.6), and tetragonal Cd_3SiO_5 (Eq. 7.7). The Cd incorporation by amorphous SiO_2 was highly influenced by the mixing ratios of the raw materials and by the sintering temperature. In the $CdO + \gamma\text{-}Al_2O_3$ system, the product phase was only $CdAl_4O_7$ with a monoclinic structure (Eq. 7.8), after 900 °C treatment for 3 h [126]. In the $CdO + \alpha\text{-}Fe_2O_3$ and $CdO + Fe_3O_4$ systems, only $CdFe_2O_4$ spinel was formed as a Cd-hosting product (Eq. 7.9) [127].

$$CdO + SiO_2 \rightarrow CdSiO_3 \quad (7.5)$$

$$2CdO + SiO_2 \rightarrow Cd_2SiO_4 \quad (7.6)$$

$$3CdO + SiO_2 \rightarrow Cd_3SiO_5 \quad (7.7)$$

$$CdO + 2\gamma\text{-}Al_2O_3 \rightarrow CdAl_4O_7 \quad (7.8)$$

$$CdO + \alpha\text{-}Fe_2O_3 \rightarrow CdFe_2O_4 \quad (7.9)$$

The transformation ratios for Cd incorporation are summarized in Fig. 7.6 along with the weight fractions of crystalline phases in the sintered sample. The Cd incorporation started at 600 °C and was fully incorporated at 850 °C, when hematite and magnetite were used as ceramic matrices. At low temperatures, magnetite showed the highest Cd incorporation efficiency; the TRs derived from the CdO + Fe_3O_4 system were much higher than those of other incorporation systems. The interactions between CdO and amorphous SiO_2 were triggered at 700 °C. Note that the transformation ratios for the CdO + amorphous SiO_2 system are the sum of the TRs for converting Cd into a specified Cd silicate, as it was found that the systems often contained more than one type of Cd silicate (Fig. 7.5). Amorphous SiO_2 could effectively incorporate Cd at relatively low temperatures, as high TRs were achieved. The Cd-hosting product phase of the CdO + γ-Al_2O_3 system was $CdAl_4O_7$, which starts to form at 850 °C. This temperature is over 200 °C higher than those observed in the CdO + α-Fe_2O_3 and CdO + Fe_3O_4 systems. The ceramic matrices used in this study were able to completely incorporate Cd into various crystalline product phase(s). On the basis of these quantitative results, it can be inferred that although a relatively high incorporation efficiency was achieved by γ-Al_2O_3, more energy may be required to fully incorporate Cd when using γ-Al_2O_3 as matrix, as compared with other ceramic matrices.

For a solid-state reaction, the driving forces for mass transfer are related to the differences in the chemical potentials of the reactants. The standard Gibbs free energies of formation (ΔG_f^0) for amorphous SiO_2, γ-Al_2O_3, α-Fe_2O_3, and Fe_3O_4 are −849.44 kJ/mol [128], −1563.85 kJ/mol [129], −774.4 kJ/mol [130], and −1012.7 kJ/mol [130], respectively. Among these values, the ΔG_f^0 for α-Fe_2O_3 is the highest and thus it can be expected that the reaction between CdO and α-Fe_2O_3 is the most energetically favored [131]. This may explain why α-Fe_2O_3 shows a greater Cd incorporation efficiency than do the systems using amorphous SiO_2 and γ-Al_2O_3 as matrices. Note that Fe_3O_4 will transform into α-Fe_2O_3 at low temperatures in the presence of oxygen and that the crystal size of newly formed α-Fe_2O_3 is generally much smaller than that of Fe_3O_4 [127]. Reactants with small crystal size can stimulate the reactions and allow fast reaction kinetics. Therefore, higher TRs were obtained for Cd incorporation by Fe_3O_4 at temperatures below 750 °C.

3.2 *Evaluation of Metal Stabilization Effect*

Single phases of $CdSiO_3$, Cd_2SiO_4, Cd_3SiO_5, $CdAl_4O_7$, $CdFe_2O_4$, and CdO were prepared by mixing with SiO_2, γ-Al_2O_3, and α-Fe_2O_3 powders at defined stoichiometric molar ratios of Cd/Si (1/1, 2/1, 3/1), Cd/Al (1/4), and Cd/Fe (1/2). The well-mixed materials were pressed into pellets and then sintered in a temperature range of 850–1100 °C for various durations. The homogenization, pelletization, and sintering processes were repeated to further ensure the complete reaction and homogeneity of the crystalline product phases. Their purity and crystallographic features were assessed by XRD (Fig. 7.7). The XRD patterns further confirmed the successful synthesis of $CdSiO_3$, Cd_2SiO_4, Cd_3SiO_5, $CdAl_4O_7$, and $CdFe_2O_4$ phases. No diffraction peaks of the reactants were observed. The CPLT was used to examine the products' ability to resist acid attack.

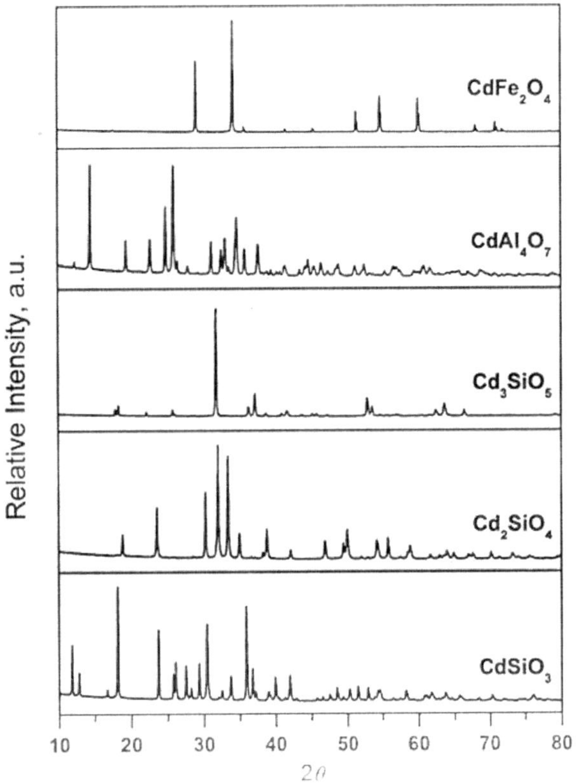

Fig. 7.7 X-ray diffraction patterns of the single phases of $CdSiO_3$, Cd_2SiO_4, Cd_3SiO_5, $CdAl_4O_7$, and $CdFe_2O_4$ for the constant-pH leaching test

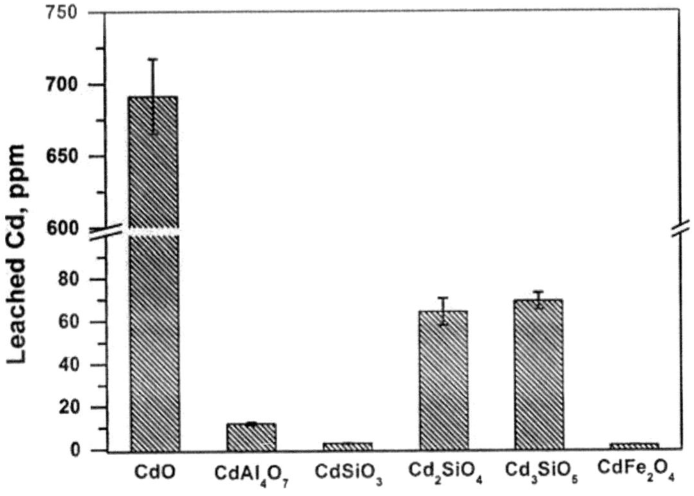

Fig. 7.8 Leached Cd concentrations in the leachates of Cd-hosting products at the end of the constant-pH leaching test (120 min)

The Cd concentrations in the leachates of $CdSiO_3$, Cd_2SiO_4, Cd_3SiO_5, $CdAl_4O_7$, and $CdFe_2O_4$ phases are shown in Fig. 7.8. The CPLT results show that the concentrations of Cd in the leachates of these product phases, particularly for $CdSiO_3$ and $CdFe_2O_4$, were remarkably lower than that in the leachate of CdO after 120 min of leaching by nitric acid at constant pH of 4.0. This indicates that superior stabilization and detoxification could be achieved by the formation of silicates, aluminate, and spinel crystal structures. The leachability of Cd_2SiO_4 and Cd_3SiO_5 phases is more significant than that of other Cd-hosting product phases. The leaching behavior of Cd_2SiO_4 and Cd_3SiO_5 phases tends toward congruent dissolution, whereas the others ($CdSiO_3$, $CdAl_4O_7$, and $CdFe_2O_4$) display incongruent dissolution. The crystal structures of Cd_2SiO_4 and Cd_3SiO_5 together with the high content of Cd (over 70 wt.%) in these two phases suggest that Cd atoms may be more exposed to the leaching fluid, which may lead to a high dissolution of Cd from Cd_2SiO_4 and Cd_3SiO_5 phases. For Cd detoxification, converting Cd into $CdSiO_3$ and $CdFe_2O_4$ can reliably prevent Cd leaching under acidic conditions, compared with the CdO phase and other Cd-hosting product phases.

4 Conclusion

This chapter summarizes the emergence of heavy metal contamination, particularly Cd contamination, the current and new technologies for the control of Cd pollution, the potential ceramic matrices for Cd detoxification, and the quantitative approaches to assessments of Cd incorporation and detoxification. The possibility and feasibility of beneficially using various low-cost and easily attainable ceramic matrices to convert Cd-bearing waste into stable crystal structures via a ceramic sintering process is considered. Ceramic sintering (a type of thermal treatment process) has been demonstrated to be a reliable strategy for transforming metal-contaminated substances into preferable crystalline phases. Four types of common ceramic matrix (SiO_2, γ-Al_2O_3, α-Fe_2O_3, and Fe_3O_4) were thermally reacted with Cd, and their incorporation capabilities were quantified with a combination of the XRD technique and the Rietveld refinement method. The results show that $CdAl_4O_7$ was the only Cd-hosting product of sintering with γ-Al_2O_3. Silicates $CdSiO_3$, Cd_2SiO_4, and Cd_3SiO_5 were the predominant Cd-hosting product phases in the CdO + SiO_2 system. The formation of silicates was greatly affected by the sintering temperature and the Cd/Si molar ratios (1/1, 2/1 or 3/1). $CdFe_2O_4$ spinel was the only Cd-hosting product phase in sintered CdO + α-Fe_2O_3 and CdO + Fe_3O_4 systems. The CPLT results revealed that $CdFe_2O_4$ spinel is the most stable phase, suggesting that the formation of $CdFe_2O_4$ spinel is the most appropriate for Cd detoxification.

Glossary

Cadmium A minor metallic element, one of the naturally occurring components in the Earth's crust and waters and present everywhere in our environment.

Ceramic matrices Materials like clay that can be converted to ceramics after high-temperature treatment.

Detoxification The act of detoxifying.

Heavy metal A general term which applies to the group of metals and metalloids and it has an atomic density more prominent than 4000 kg/m^3.

Pollution control A wide range of techniques, solutions, practices, and services for treating wastewater and waste.

Sintering Using heat or pressure to form a solid mass of material, without melting the material to the point of liquefying.

Solidification/stabilization (S/S) The process utilizes chemically reactive formulations that, together with the water and other components in sludges and other aqueous hazardous wastes, form stable solids.

Thermal stabilization A heating process for metal incorporation into certain stable products.

XRD analysis A technique that can provide both qualitative and quantitative information on the phases in a tested sample.

References

1. Guney M, Zagury GJ (2012) Heavy metals in toys and low-cost jewelry: critical review of US and Canadian legislations and recommendations for testing. Environ Sci Technol 46 (8):4265–4274
2. Järup L (2003) Hazards of heavy metal contamination. Br Med Bull 68(1):167–182
3. Jensen DL, Holm P, Christensen TH (2000) Soil and groundwater contamination with heavy metals at two scrap iron and metal recycling facilities. Waste Manage Res 18(1):52–63
4. Lee M-H, Cho K, Shah AP, Biswas P (2005) Nanostructured sorbents for capture of cadmium species in combustion environments. Environ Sci Technol 39(21):8481–8489
5. Romero-González ME, Williams CJ, Gardiner PH (2001) Study of the mechanisms of cadmium biosorption by dealginated seaweed waste. Environ Sci Technol 35(14):3025–3030
6. Fowler BA (2009) Monitoring of human populations for early markers of cadmium toxicity: a review. Toxicol Appl Pharmacol 238(3):294–300
7. Shiraishi Y (1975) Cytogenetic studies in 12 patients with Itai-Itai disease. Humangenetik 27 (1):31–44
8. Cartledge FK, Butler LG, Chalasani D, Eaton HC, Frey FP, Herrera E, Marty E (1990) Immobilization mechanisms in solidifiction/stabilization of cadmium and lead salts using portland cement fixing agents. Environ Sci Technol 24(6):867–873
9. Gale TK, Wendt JO (2005) In-furnace capture of cadmium and other semi-volatile metals by sorbents. Proc Combust Inst 30(2):2999–3007
10. Kuo J-H, Lin C-L, Wey M-Y (2009) Effects of agglomeration processes on the emission characteristics of heavy metals under different waste compositions and the addition of Al and Ca inhibitors in fluidized bed incineration. Energy Fuel 23(9):4325–4336

11. Scotto MV, Uberoi M, Peterson TW, Shadman F, Wendt JO (1994) Metal capture by sorbents in combustion processes. Fuel Process Technol 39(1–3):357–372
12. Wendt JOL, Lee SJ (2010) High-temperature sorbents for Hg, Cd, Pb, and other trace metals: mechanisms and applications. Fuel 89(4):894–903
13. Ahmaruzzaman M (2010) A review on the utilization of fly ash. Prog Energy Combust Sci 36 (3):327–363
14. Lee EM, Clack HL (2010) In situ detection of altered particle size distributions during simulated powdered sorbent injection for mercury emissions control. Energy Fuel 24 (10):5410–5417
15. Lu X, Shih K, Cheng H (2013) Lead glass-ceramics produced from the beneficial use of waterworks sludge. Water Res 47(3):1353–1360
16. Shih K, White T, Leckie JO (2006) Spinel formation for stabilizing simulated nickel-laden sludge with aluminum-rich ceramic precursors. Environ Sci Technol 40(16):5077–5083
17. Shih K, White T, Leckie JO (2006) Nickel stabilization efficiency of aluminate and ferrite spinels and their leaching behavior. Environ Sci Technol 40(17):5520–5526
18. Tang Y, Chui SS-Y, Shih K, Zhang L (2011) Copper stabilization via spinel formation during the sintering of simulated copper-laden sludge with aluminum-rich ceramic precursors. Environ Sci Technol 45(8):3598–3604
19. Tang Y, Shih K, Wang Y, Chong T-C (2011) Zinc stabilization efficiency of aluminate spinel structure and its leaching behavior. Environ Sci Technol 45(24):10544–10550
20. Alaya MN, Youssef AM, Roumie A, Grge R (2014) Physico-chemical properties of CdO-Al_2O_3 catalysts. I –structural characteristics. Arab J Chem 7(5):722–731
21. Colin F (1968) Des phases formees au cours de la reduction de certains oxydes mixtes $nAl_2O_3{\cdot}MO$. (Phases obtained during the reduction of some nAl_2O_3 MO oxides)Rev. Revue Internationale Des HuatesTemperatures Et Des Refractaires 5(4):267–283
22. Singh TS, Pant KK (2006) Solidification/stabilization of arsenic containing solid wastes using portland cement, fly ash and polymeric materials. J Hazard Mater 131(1–3):29–36
23. Lonsdale TD, Whitaker A (1978) Phase equilibria in the system CdO-B_2O_3-SiO_2 at 800 °C. J Mater Sci 13(7):1503–1508
24. Uberoi M, Shadman F (1991) High-temperature removal of cadmium compounds using solid sorbents. Environ Sci Technol 25(7):1285–1289
25. Panek Z, Fitzner K (1993) Thermodynamic properties and cation distribution of $CdFe_2O_4$ + Fe_3O_4 solid solutions at 1073 K. Thermochim Acta 66(1–3):127–137
26. Zhang Y, Chen Y, Meng A, Li Q, Cheng H (2008) Experimental and thermodynamic investigation on transfer of cadmium influenced by sulfur and chlorine during municipal solid waste (MSW) incineration. J Hazard Mater 153(1–2):309–319
27. Zhuang P, Zou B, Li N, Li Z (2009) Heavy metal contamination in soils and food crops around Dabaoshan mine in Guangdong, China: implication for human health. Environ Geochem Health 31(6):707–715
28. Xu GR, Zou JL, Li GB (2008) Stabilization of heavy metals in ceramsite made with sewage sludge. J Hazard Mater 152(1):56–61
29. Schaider LA, Senn DB, Brabander DJ, McCarthy KD, Shine JP (2007) Characterization of zinc, lead, and cadmium in mine waste: implications for transport, exposure, and bioavailability. Environ Sci Technol 41(11):4164–4171
30. Luna Galiano Y, Fernández Pereira C, Vale J (2011) Stabilization/solidification of a municipal solid waste incineration residue using fly ash-based geopolymers. J Hazard Mater 185 (1):373–381
31. Prabhukumar G, Matsumoto M, Mulchandani A, Chen W (2004) Cadmium removal from contaminated soil by tunable biopolymers. Environ Sci Technol 38(11):3148–3152
32. Winburn RS, Lerach SL, McCarthy GJ, Grier DG, Cathcart JD (2000) Quantification of ferrite spinel and hematite in fly ash magnetically enriched fractions. *JCPDS-International Centre for Diffraction Data 2000.* Adv X-ray Anal 43:350–355

33. Balsamo M, Di Natale F, Erto A, Lancia A, Montagnaro F, Santoro L (2011) Cadmium adsorption by coal combustion ashes-based sorbents-relationship between sorbent properties and adsorption capacity. J Hazard Mater 187(1–3):371–378
34. Nordberg GF (1974) Health hazards of environmental cadmium pollution. Ambio 3(2):55–66
35. Clemens S, Aarts MGM, Thomine S, Verbruggen N (2013) Plant science: the key to preventing slow cadmium poisoning. Trends Plant Sci 18(2):92–99
36. Pan J, Plant J, Voulvoulis N, Oates C, Ihlenfeld C (2010) Cadmium levels in Europe: implications for human health. Environ Geochem Health 32(1):1–12
37. Wang Y, Tang X, Chen Y, Zhan L, Li Z, Tang Q (2009) Adsorption behavior and mechanism of Cd(II) on loess soil from China. J Hazard Mater 172(1):30–37
38. Lam CHK, Ip AWM, Barford JP, McKay G (2010) Use of incineration MSW ash: a review. Sustainability 2(7):1943–1968
39. Li X, Bertos MF, Hills CD, Carey PJ, Simon S (2007) Accelerated carbonation of municipal solid waste incineration fly ashes. Waste Manag 27(9):1200–1206
40. Tian H, Gao J, Lu L, Zhao D, Cheng K, Qiu P (2012) Temporal trends and spatial variation characteristics of hazardous air pollutant emission inventory from municipal solid waste incineration in China. Environ Sci Technol 46(18):10364–10371
41. Ferreira C, Ribeiro A, Ottosen L (2003) Possible applications for municipal solid waste fly ash. J Hazard Mater 96(2):201–216
42. UNEP (2010) Final review of scientific information on cadmium-Version of December 2010
43. Wan X, Wang W, Ye T, Guo Y, Gao X (2006) A study on the chemical and mineralogical characterization of MSWI fly ash using a sequential extraction procedure. J Hazard Mater 134 (1):197–201
44. Xue Y, Hou H, Zhu S, Zha J (2009) Utilization of municipal solid waste incineration ash in stone mastic asphalt mixture: pavement performance and environmental impact. Constr Build Mater 23(2):989–996
45. Shi H-S, Kan L-L (2009) Leaching behavior of heavy metals from municipal solid wastes incineration (MSWI) fly ash used in concrete. J Hazard Mater 164(2):750–754
46. Quina MJ, Bordado JC, Quinta-Ferreira RM (2008) Treatment and use of air pollution control residues from MSW incineration: an overview. Waste Manag 28(11):2097–2121
47. Cetin S, Pehlivan E (2007) The use of fly ash as a low cost, environmentally friendly alternative to activated carbon for the removal of heavy metals from aqueous solutions. Colloids Surf A Physicochem Eng Asp 298(1–2):83–87
48. Shim Y-S, Kim Y-K, Kong S-H, Rhee S-W, Lee W-K (2003) The adsorption characteristics of heavy metals by various particle sizes of MSWI bottom ash. Waste Manag 23(9):851–857
49. Apak R, Tütem E, Hügül M, Hizal J (1998) Heavy metal cation retention by unconventional sorbents (red muds and fly ashes). Water Res 32(2):430–440
50. Apiratikul R, Pavasant P (2008) Sorption of Cu^{2+}, Cd^{2+}, and Pb^{2+} using modified zeolite from coal fly ash. Chem Eng J 144(2):245–258
51. Visa M, Duta A (2013) Methyl-orange and cadmium simultaneous removal using fly ash and photo-Fenton systems. J Hazard Mater 244:773–779
52. Remenárová L, Pipíška M, Florková E, Horník M, Rozložník M, Augustín J (2014) Zeolites from coal fly ash as efficient sorbents for cadmium ions. Clean Techn Environ Policy 16 (8):1551–1564
53. Lewinsky AA (2007) Hazardous materials and wastewater: treatment, removal and analysis. Nova Publishers, New York
54. Jamode A, Sapkal V, Jamode V (2004) Defluoridation of water using inexpensive adsorbents. J Indian Inst Sci 84(5):163
55. Gupta VK, Ali I, Saleh TA, Nayak A, Agarwal S (2012) Chemical treatment technologies for waste-water recycling-an overview. RSC Adv 2(16):6380–6388
56. Ali I, Gupta V (2007) Advances in water treatment by adsorption technology. Nat Protoc 1 (6):2661–2667

57. Anbia M, Davijani AH (2013) Synthesis of L-Cysteine grafted nanoporous carbon (CMK-3) and its use as a new cadmium sorbent. Chem Eng J 223(0):899–907
58. Shin EW, Karthikeyan KG, Tshabalala MA (2007) Adsorption mechanism of cadmium on juniper bark and wood. Bioresour Technol 98(3):588–594
59. NTP (National Toxicology Program) (2014) Report on carcinogens, 13th edn. U.S. Department of Health and Human Services, Public Health Service, Research Triangle Park, USA
60. WHO (2011) Cadmium in drinkingwater–background document for development of WHO guidelines for drinking-water quality. WHO/SDE/WSH/03.04/80/Rev/1
61. Xu GR, Zou JL, Li GB (2010) Stabilization of heavy metals in sludge ceramsite. Water Res 44 (9):2930–2938
62. Voglar GE, Leštan D (2010) Solidification/stabilisation of metals contaminated industrial soil from former Zn smelter in Celje, Slovenia, using cement as a hydraulic binder. J Hazard Mater 178(1):926–933
63. Randall PM, Chattopadhyay S (2009) Bench-scale evaluation of chemically bonded phosphate ceramic technology to stabilize mercury waste mixtures. J Environ Eng 136(3):265–273
64. Shawabkeh RA (2005) Solidification and stabilization of cadmium ions in sand-cement-clay mixture. J Hazard Mater 125(1–3):237–243
65. Yang Y, Xiao Y, Voncken J, Wilson N (2008) Thermal treatment and vitrification of boiler ash from a municipal solid waste incinerator. J Hazard Mater 154(1):871–879
66. Xiao Y, Oorsprong M, Yang Y, Voncken J (2008) Vitrification of bottom ash from a municipal solid waste incinerator. Waste Manag 28(6):1020–1026
67. Li C-T, Lee W-J, Huang K-L, Fu S-F, Lai Y-C (2007) Vitrification of chromium electroplating sludge. Environ Sci Technol 41(8):2950–2956
68. Garcia-Valles M, Avila G, Martinez S, Terradas R, Nogués JM (2007) Heavy metal-rich wastes sequester in mineral phases through a glass-ceramic process. Chemosphere 68 (10):1946–1953
69. Cheng L-C, Wu S-H, Huang K-L, Lin C, Wang C-T, Wang J-W, Kuo Y-M (2013) Evaluation of effect of reducing additives during vitrification via simulation and experiment. J Air Waste Manage Assoc 63(10):1182–1189
70. Bingham P, Hand R (2006) Vitrification of toxic wastes: a brief review. Adv Appl Ceram 105 (1):21–31
71. Tang Y, Lee P-H, Shih K (2013) Copper sludge from printed circuit board production/recycling for ceramic materials: a quantitative analysis of copper transformation and immobilization. Environ Sci Technol 47(15):8609–8615
72. Matsuzawa Y, Mae K, Kitano S, Yoshihara S, Ueno S-I, Nishino J, Kubota N (2006) Leaching behavior of heat-treated waste ash. Fuel 85(3):401–409
73. Jha VK, Kameshima Y, Okada K, Mackenzie KJD (2004) Ni^{2+} uptake by amorphous and crystalline $Ca_2Al_2SiO_7$ synthesized by solid-state reaction of kaolinite. Sep Purif Technol 40 (2):209–215
74. Saikia N, Kojima T (2011) Behaviours of $MeCl_2$ (Me: Pb and Cd) during thermal treatment of kaolin-lime mixture. Afr J Environ Sci Technol 5(10):778–785
75. Saikia N, Sengupta P, Dutta D, Saikia P, Borthakur P (2000) Oil field sludge used to make brick. Am Ceram Soc Bull 79(7):71–74
76. Carter CB, Norton MG (2007) Ceramic materials: science and engineering. Springer Science & Business Media, New York, USA
77. Rahaman MN (2003) Ceramic processing and sintering, 2nd edn. Marcel Dekker,Inc., New York, USA
78. Castro R, van Benthem K (2012) Sintering: mechanisms of convention nanodensification and field assisted processes, vol 35. Springer Science & Business Media, New York, USA
79. Murch GE (1984) Diffusion in crystalline solids. Academic Press, Orlando, FL, USA
80. Ehrenreich H, Spaepen F (2001) Solid state physics. Academic Press, Orlando, FL, USA

81. Jenkins R, Snyder R (2012) Introduction to X-ray powder diffractometry, vol 267. Wiley, Hoboken
82. Kurihara LK, Suib SL (1993) Sol-gel synthesis of ternary metal oxides. 1. Synthesis and characterization of MAl_2O_4 (M= Mg, Ni, Co, Cu, Fe, Zn, Mn, Cd, Ca, Hg, Sr, and Ba) and lead aluminum oxide ($Pb_2Al_2O_5$). Chem Mater 5(5):609–613
83. Bashkirov LA, Kornilova NV (1980) Equilibre de phases dans le systeme MgO-CdO-Fe_2O_3 dans la partie contenant Au moins 50 moles% de Fe_2O_3. Izvestiya AkademiiNauk Sssr – Neorganicheskie Materialy 16(7):1229–1233
84. Chan SH (1989) A review on solubility and polymerization of silica. Geothermics 18 (1–2):49–56
85. McLaughlin JK, Chow WH, Levy LS (1997) Amorphous silica: a review of health effects from inhalation exposure with particular reference to cancer. J Toxicol Environ Health A 50 (6):553–566
86. Nandiyanto ABD, Kim S-G, Iskandar F, Okuyama K (2009) Synthesis of spherical mesoporous silica nanoparticles with nanometer-size controllable pores and outer diameters. Microporous Mesoporous Mater 120(3):447–453
87. Merget R, Bauer T, Küpper H, Philippou S, Bauer H, Breitstadt R, Bruening T (2002) Health hazards due to the inhalation of amorphous silica. Arch Toxicol 75(11–12):625–634
88. Rodella N, Bosio A, Dalipi R, Zacco A, Borgese L, Depero LE, Bontempi E (2014) Waste silica sources as heavy metal stabilizers for municipal solid waste incineration fly ash. Arab J Chem 10:S3676
89. Li X, Chen Q, Zhou Y, Tyrer M, Yu Y (2014) Stabilization of heavy metals in MSWI fly ash using silica fume. Waste Manag 34(12):2494–2504
90. Khan MI, Siddique R (2011) Utilization of silica fume in concrete: review of durability properties. Resour Conserv Recycl 57:30–35
91. Huang W-J, Huang H-S (2008) Using fume silica as heavy metals' stabilizer for high alkali and porous MSWI baghouse ash. J Hazard Mater 152(1):176–182
92. Nowak B, Aschenbrenner P, Winter F (2013) Heavy metal removal from sewage sludge ash and municipal solid waste fly ash-a comparison. Fuel Process Technol 105:195–201
93. Qu X, Cao L, Liu W, Su G, Qu H, Xu C (2009) Synthesis of $CdSiO_3$ powder by sol-gel method. J Alloys Compd 484(1–2):641–644
94. Glasser L (1965) Silicates M_3SiO_5. II. Relationships between Sr_3SiO_5, Cd_3SiO_5, and Ca_3SiO_5. Acta Crystallogr 18(3):455–457
95. Zuckerman B, Koester D, Dufour P, Melis C, Klein B, Jura M (2011) An aluminum/calcium-rich, iron-poor, white dwarf star: evidence for an extrasolar planetary lithosphere? Astrophys J 739(2):101
96. Wolverton C, Hass K (2000) Phase stability and structure of spinel-based transition aluminas. Phys Rev B 63(2):024102
97. Wang S, Lu G (1999) A comprehensive study on carbon dioxide reforming of methane over Ni/γ-Al_2O_3catalysts. Ind Eng Chem Res 38(7):2615–2625
98. Zhang Z, Hicks RW, Pauly TR, Pinnavaia TJ (2002) Mesostructured forms of γ-Al_2O_3. J Am Chem Soc 124(8):1592–1593
99. Gutiérrez G, Taga A, Johansson B (2001) Theoretical structure determination of γ-Al_2O_3. Phys Rev B 65(1):012101
100. Paglia G, Buckley C, Rohl A, Hunter B, Hart R, Hanna J, Byrne L (2003) Tetragonal structure model for boehmite-derived γ-alumina. Phys Rev B 68(14):144110
101. Paglia G, Buckley CE, Udovic TJ, Rohl AL, Jones F, Maitland CF, Connolly J (2004) Boehmite-derived γ-alumina system. 2. Consideration of hydrogen and surface effects. Chem Mater 16(10):1914–1923
102. Alphonse P, Courty M (2005) Structure and thermal behavior of nanocrystalline boehmite. Thermochim Acta 425(1):75–89
103. Schwertmann U, Cornell RM (eds) (2008) Iron oxides in the laboratory. Wiley-VCH Verlag GmbH & Co. KGaA, Weinheim

104. Cornell RM, Schwertmann U (eds) (2006) The iron oxides: structure, properties, reactions, occurrences and uses. Wiley-VCH Verlag GmbH & Co. KGaA, Weinheim, pp 509–551
105. Sakurai S, Namai A, Hashimoto K, Ohkoshi S (2009) First observation of phase transformation of all four Fe_2O_3 phases ($\gamma \rightarrow \varepsilon \rightarrow \beta \rightarrow \alpha$-phase). J Am Chem Soc 131(51):18299–18303
106. Sharma VK, Klingelhofer G, Nishida T (2013) Mossbauer spectroscopy: applications in chemistry, biology, industry, and nanotechnology. Wiley, Hoboken
107. Bragg W, Bragg W (1913) The reflection of X-rays by crystals. Proc R Soc London A 88 (605):428–438
108. Bragg W (1913) The reflection of X-rays by crystals.(II.). Proc R Soc London A 89:246–248
109. Norrish K, Taylor RM (1962) Quantitative analysis by X-ray diffraction. Clay Miner Bull 5 (28):98–109
110. Brindley G (1980) Quantitative X-ray mineral analysis of clays. In: Crystal structures of clay minerals and their x-ray identification, vol 5. Mineralogical Society, London, pp 411–438
111. Rietveld H (1988) The Rietveld method? A historical perspective. Aust J Phys 41(2):113–116
112. Rietveld HM (2014) The Rietveld method. Phys Scr 89(9):098002
113. Young R, Wiles D (1982) Profile shape functions in Rietveld refinements. J Appl Crystallogr 15(4):430–438
114. Bish DL, Howard S (1988) Quantitative phase analysis using the Rietveld method. J Appl Crystallogr 21(2):86–91
115. Hillier S (2000) Accurate quantitative analysis of clay and other minerals in sandstones by XRD: comparison of a Rietveld and a reference intensity ratio (RIR) method and the importance of sample preparation. Clay Miner 35(1):291–302
116. Bish DL, Post JE (1993) Quantitative mineralogical analysis using the Rietveld full-pattern fitting method. Am Mineral 78(9–10):932–940
117. Hu H-Y, Liu H, Shen W-Q, Luo G-Q, Li A-J, Lu Z-L, Yao H (2013) Comparison of CaO's effect on the fate of heavy metals during thermal treatment of two typical types of MSWI fly ashes in China. Chemosphere 93(4):590–596
118. Garrabrants AC, Kosson DS, van der Sloot HA, Sanchez F, Hjelmar O (2011) Background information for the leaching environmental assessment framework (LEAF) test methods. U.S. Environmental Protection Agency, Washington, DC, EPA/600/R-10/170
119. Van der Sloot H, Comans R, Hjelmar O (1996) Similarities in the leaching behaviour of trace contaminants from waste, stabilized waste, construction materials and soils. Sci Total Environ 178(1):111–126
120. U.S. EPA (1997) Test methods for evaluating solid waste, physical chemical methods, SW-846. United States Environmental Protection Agency, Washington, D.C.
121. Fällman A-M (1997) Performance and design of the availability test for measurement of potentially leachable amounts from waste materials. Environ Sci Technol 31(3):735–744
122. Poon CS, Lio KW (1997) The limitation of the toxicity characteristic leaching procedure for evaluating cement-based stabilised/solidified waste forms. Waste Manag 17(1):15–23
123. Al-Abed SR, Jegadeesan G, Purandare J, Allen D (2007) Arsenic release from iron rich mineral processing waste: influence of pH and redox potential. Chemosphere 66(4):775–782
124. Islam MZ, Catalan LJ, Yanful EK (2004) A two-front leach model for cement-stabilized heavy metal waste. Environ Sci Technol 38(5):1522–1528
125. Cappuyns V, Swennen R (2008) The application of pH_{stat} leaching tests to assess the pH-dependent release of trace metals from soils, sediments and waste materials. J Hazard Mater 158(1):185–195
126. Su M, Liao C, Chuang K-H, Wey M-Y, Shih K (2015) Cadmium stabilization efficiency and leachability by $CdAl_4O_7$monoclinic structure. Environ Sci Technol 49(24):14452–14459
127. Su M, Liao C, Lee P-H, Li H, Shih K (2017) Formation and leaching behavior of ferrite spinel for cadmium stabilization. Chem Eng Sci 158:287–293
128. Su C, Harsh JB (1994) Gibbs free energies of formation at 298 K for imogolite and gibbsite from solubility measurements. Geochim Cosmochim Acta 58(6):1667–1677

129. Chase MWJ (ed) (1998) NIST-JANAF thermochemical tables, 4th edn. The American Chemical Society/American Institute of Physics, Washington, D.C./New York, USA
130. Robie RA, Hemingway BS (1995) Thermodynamic properties of minerals and related substances at 298.15 K and 1 bar (10^5 pascals) pressure and at higher temperatures, in Bulletin. United States Government Printing Office, Washington, D.C., USA
131. Shih K, Su M (2018) Stabilization of cadmium in waste incineration residues by aluminum/iron-rich materials. In: Wang LK, Wang MHS, Hung YT, Shammas NK, Chen JP (eds) Handbook of advanced industrial and hazardous wastes management. CRC Press/Taylor & Francis Group, Orlando, FL, USA, pp 239–254

Chapter 8
Advances in Treatment of Vegetable Oil Refining Wastes

Devrim Kaya and Yung-Tse Hung

Contents

1 Introduction 327
2 Refining 328
2.1 Degumming 330
2.2 Neutralization 330
2.3 Bleaching and Filtration 331
2.4 Hardening 332
2.5 Deodorization 332
3 Environmental Impacts 333
3.1 Water and Energy Consumption 334
3.2 Wastewater Sources and Quantities 335
3.3 Wastewater Treatment 338
3.4 Solid Wastes and By-Products 353
3.5 Emissions to Air 355
4 Case Study 358
4.1 Solid Wastes 358
4.2 Factory and Gas Pollutant Emission Sources 359
4.3 Wastewater Characterization 360
4.4 Analysis of Wastewater Treatment Plant Efficiency 360
Glossary 369
References 371

Abstract Vegetable oils are mainly extracted from soybean, sesame, sunflower, corn, canola, and cotton seeds. Their yields, compositions, and physical and chemical properties determine their usefulness in various applications aside from edible uses. Crude oils obtained by pressing of such vegetable seeds are not usually

D. Kaya (✉)
Department of Civil and Environmental Engineering, University of Maryland, College Park, MD, USA

Y.-T. Hung
Department of Civil and Environmental Engineering, Cleveland State University, Cleveland, OH, USA

L. K. Wang, M. -H. S. Wang, Y. -T. Hung, N. K. Shammas (eds.), *Environmental and Natural Resources Engineering*, Handbook of Environmental Engineering 19, https://doi.org/10.1007/978-3-030-54626-7_8

considered to be edible before the removal of various nonglyceride compounds through an operation known as refining. The vegetable oil refinery uses various types of physical and chemical processes to offer a premium quality of oil. The refining processes remove undesirable materials, such as phospholipids, monoacylglycerols, diacylglycerols, free fatty acids, color and pigments, oxidized materials, etc., but may also remove valuable minor components, such as antioxidants and vitamins (carotenes and tocopherols). The major steps involved in chemical refining include degumming, deacidification, deodorization, and neutralization processes. During these processes, a high amount of water is used, and highly polluted effluents are formed. The treatment of vegetable oil refinery wastewaters has been a major problem of environmental concern in developing countries for the last decades due to their complex nature consisting of water and soluble and insoluble substances that contain fats and oil, carbohydrates, phenolic compounds, and suspended solids. Therefore, a suitable wastewater treatment prior to their discharge into the receiving bodies is required. Numerous treatment technologies have been applied to treat oily wastewaters. Coagulation/flocculation, electrocoagulation, reverse osmosis, flocculation/membrane filtration, air flotation, microfiltration, and enzymatic catalysis are the most common ones. However, due to the complex nature and low biodegradability of the oily wastewater, some of these technologies may not be efficient in treating such wastewaters, while others may be of high cost and generally require a pretreatment. As a result, in this chapter, the processes involved in the vegetable oil refining, the environmental impacts of those processes, the characterization of the waste produced during the processes, waste reduction at source, recovery from waste, and treatment technologies are discussed.

Keywords Vegetable oil industry · Vegetable oil · Refining · Industrial wastewater · Waste characterization · Reduction · Removability · Treatment technologies

Acronym

BOD	Biochemical oxygen demand
COD	Chemical oxygen demand
DAF	Dissolved air flotation
DGF	Dissolved gas flotation
FFA	Free fatty acids
IAF	Induced air flotation
MBR	Membrane bioreactors
NO_x	Oxides of nitrogen
TSS	Total suspended solids
UF	Ultrafiltration
VOC	Volatile organic compounds

1 Introduction

Vegetable oils and fats are essential to humans as a source of nutrients, but they are also used in animal feed, for medicinal purposes, and for certain technical applications. There are numerous vegetable oils derived from various sources. Soybean, sesame, sunflower, corn, canola, and cotton seeds are the most commonly used raw materials for the production of edible oil [1]. Their yields, compositions, and physical and chemical properties determine their usefulness in various applications aside from edible uses.

Until the 19th century, vegetable oils were extracted from oil seeds through mechanical or hydraulic means. Today, vegetable oils are obtained by different pressing methods, solvent extraction, or a combination of these [2]. The selection of the extraction process depends primarily on the oil content of the source material, the amount of residual oil in the meal allowed, the amount of protein denaturation allowed, the amount of investment capital available, and local environmental laws concerning emissions of volatile organic compounds (VOCs).

A basic process chart for vegetable oils and related products is given in Fig. 8.1. Oilseeds are dehulled, cracked and flaked, and prepared for oil extraction. Hexane is

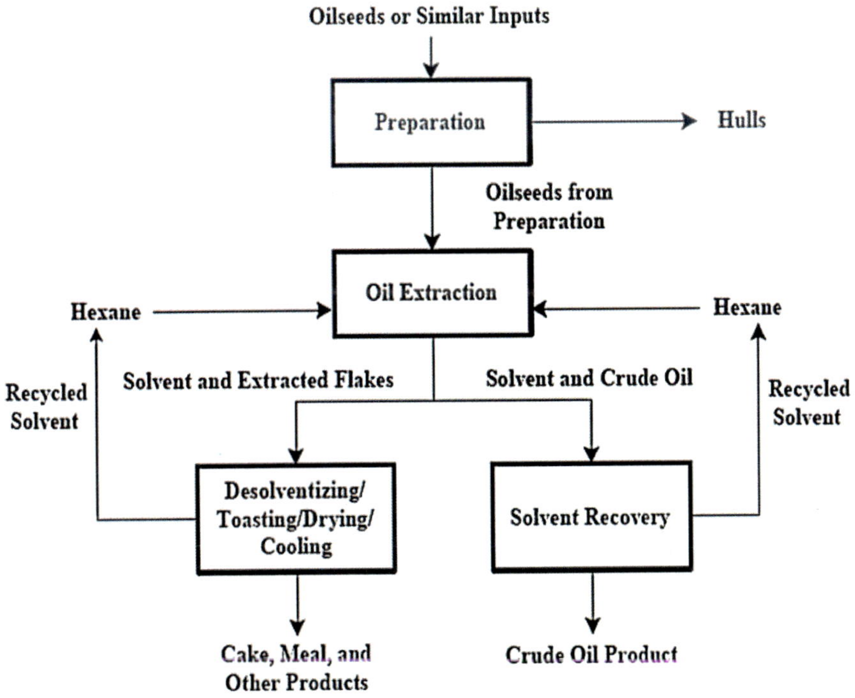

Fig. 8.1 A simple solvent extraction process for vegetable oils [3]

used to dissolve the oil in the prepared oilseed or similar input and then recovered in a desolventizing (evaporation) process and recycled for reuse in the process [3, 4].

Crude vegetable oil obtained at these facilities is then transferred to a refining facility where it is prepared for human consumption or prepared for use in animal feeds. Refining of edible oil is a process where free fatty acids are volatilized, condensed, and recovered simultaneously with vacuum decoloring operation.

However, processes in vegetable oil refining, such as filter backwash, acidification of soap-stock, and washing of equipment, produce big amounts of wastewaters [5]. These wastewaters have high chemical oxygen demand (COD), oil and grease levels as well as elevated phosphorus and dissolved solids concentrations [6].

2 Refining

Crude oils are not usually considered to be edible before the removal of numerous nonglyceride compounds through operations collectively known as refining. The refining processes remove undesirable materials, such as phospholipids, monoacylglycerols, diacylglycerols, free fatty acids, color and pigments, oxidized materials, flavor components, trace metals, sulfur compounds, trace solvent residue, and water, but may also remove valuable minor components which are antioxidants and vitamins, such as carotenes and tocopherols. Unfortunately, some of the refining operations are not very selective and remove some beneficial compounds along with the targeted undesirable ones. However, some vegetable oils (e.g., olive, tallow, and lard) have been consumed without refining.

Figure 8.2 shows a general diagram of the processing steps involved in the refining of vegetable oils. There are two major types of refining: chemical and physical. The major steps involved in chemical refining include degumming, neutralizing, bleaching, and deodorizing. Physical refining removes free fatty acids (FFA) and flavors by distillation, to combine the steps of neutralization and deodorization into one operation.

Conventional chemical refining involves degumming for the removal of phospholipids, neutralization for the removal of FFA, and bleaching for decolorization and deodorization. Water is added during degumming to hydrate any gums present and the mixture is then centrifuged for separation. Non-hydratable gums are removed using phosphoric or citric acid before water is added and separation takes place in a centrifuge.

In physical refining, the crude oil is degummed and bleached and then steam stripped to remove FFA, odor, and VOCs all in one step. This replaces chemical reaction mode using an alkali (caustic soda) in chemical refining. A physical pretreatment is used to achieve low phospholipid content by degumming and using bleaching earth. Following this, FFA can be stripped from the physically pretreated oil using steam distillation in a vacuum at temperatures of around 250 °C and refined by the oil flowing over a series of trays countercurrent to the flow of the stripping steam [7]. Previous neutralization stages are eliminated because the

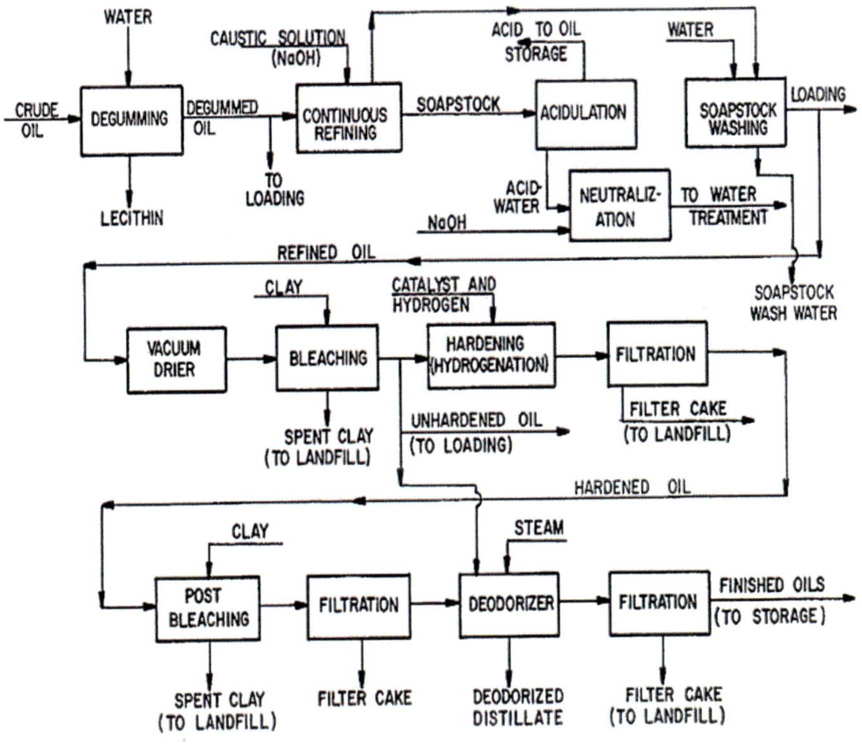

Fig. 8.2 Oil refining flow diagram [4]

neutralization and deodorization are combined. A scrubber is then used to condense the greater part of the fat from the vapors as a water-free product.

The advantages of physical refining method over the chemical refining are listed as follows:

- Better yields
- High-quality fatty acids as by-products
- Good oil stability
- Simultaneous distillation of fatty acids and deodorization
- Lower cost of equipment
- Simplicity of operation

Ultimately, the marketing of a refined vegetable oil depends on the specifications met at the end of the processing [8]. The typical steps followed by the industry in the refining of vegetable oils are discussed below.

2.1 *Degumming*

Degumming is a water-washing process to remove phosphatides. At this step, free fatty acids are converted to water-soluble soaps, soap-stock. Phosphatides can hydrate from moisture in the air during storage or in the headspace; thus, phosphatides become insoluble in the oil and precipitate, resulting in an oil that is unattractive because of unpleasant sludge or gums [7]. It is vital to remove the phosphatide content in the crude oil because the presence of this component will result in undesirable flavor and color and shorten the shelf life of oil. Phosphatides, also known as lecithin, are important food emulsifiers and have an economical value. Therefore, they should be recovered where possible.

During degumming, the phosphatide portion of oil is either removed or conditioned by the addition of phosphoric acid or citric acid. Hot water is added to hydrate any gums present in crude oil prior to centrifugal separation. Non-hydratable gums are removed by converting them first to a hydratable form using phosphoric or citric acid, followed by the addition of water. This causes the separation of phosphatides from oil and they are removed after certain settling time. Occasionally, segregated gums are not directly removed but with soap-stock formed during neutralization [9]. During degumming, caustic soda preheated to between 90 °C and 110 °C is added to the oil to saponify the FFA. The dosage of phosphoric acid is normally within the range of 0.05–0.1% of oil weight and the acid concentration is about 80–85%. Excessive use of phosphoric acid should be avoided since it increases phosphorus acid concentration that may cause further refining problems. However, degumming is not as reliable for bleached oils. Degumming may be conducted either as a separate process or concurrently with neutralization.

2.2 *Neutralization*

FFA naturally present in oil causes acidity which should be neutralized. Neutralization is the most important operation in refining edible oils. The neutralization process consists of three steps such as neutralization, washing, and drying. Neutralization is achieved by accomplishing a reaction between the FFA and caustic soda (sodium hydroxide, NaOH) to form soap referred to as soap-stock. The chemical reaction [9] involved in this operation is given in Eq. 8.1.

$$\frac{\text{R-COOH}}{\text{Fatty acid}} + \text{NaOH} \rightarrow \frac{\text{R-COONa}^{+}}{\text{Soap-stock}} + \text{H}_2\text{O} \tag{8.1}$$

The soap-stock is then removed by precipitation followed by sedimentation or centrifugation and may be further processed into acid oils by splitting. Soap-stock is a valuable by-product of refineries in that it can be acidulated with sulfuric acid to produce a marketable product. When re-acidified, the fatty acids (95% fatty acids) will separate in settling basins as 35–40% free liquid, called acid oil, from an emulsified layer (high in phosphatides) and a water layer. To regain fatty acid from soap-stock, the soap-stock is heated to between 70 °C and 100 °C and reacts with sulfuric acid to reform the fatty acids. The fatty acids are insoluble in water and float to the surface where they are skimmed. The mineral salts and any water-soluble material remain in the water layer that is neutralized and discharged to the waste treatment facilities. The chemical reaction [9] taking place in this step is shown in Eq. 8.2.

$$\underset{\text{Soap-stock}}{\underline{\text{R-COONa}^+}} + H_2SO_4 \rightarrow \underset{\text{Fatty acid}}{\underline{\text{R-COOH}}} + Na_2SO_4 \tag{8.2}$$

The oil from caustic refining contains trace amounts of soaps and sodium hydroxide; hot water is added to the oil and agitated to remove them. The oil and water from this process form a weak (1–3%) soap solution with a relatively high pH. The oil from water washing contains trace amounts of water that can clog the filter during the subsequent filtration operations.

In vacuum drying, the oil is heated under a vacuum to remove this trace moisture. There is no significant waste stream from this process. Citric acid may be used instead of phosphoric acid, which brings a range of advantages, including reduced phosphorus load in the wastewater and a slight reduction in the amount of sludge. Enzymatic degumming uses enzymatic hydrolysis of the phosphatides. Environmental benefits from this approach include reduced consumption of phosphoric and sulfuric acid, as well as caustic soda, water, and energy. Efficiency of neutralization depends on using the proper amount of sodium hydroxide, proper mixing, proper temperature, sufficient contact time, and efficient separation.

2.3 *Bleaching and Filtration*

Vegetable oils are commonly treated with bleaching earth after alkali or physical refining and before hydrogenation and final deodorization. However, bleaching can be applied at different stages of the edible oil refining operation, for example, after water or acid degumming. The aim of bleaching is to remove red pigments (carotenoids, xanthophyll, and gossypol), or green colors (chlorophyll), metals (nickel or iron), residual soaps, and phospholipids from the oil or fat as well as the breakdown of peroxides prior to deodorization [7].

In this stage, the oil is taken into cylindrical tank and mixed with bleaching material called bleaching earth and, then, heated up to 90 °C with steam under vacuum. Thus, the moisture from oil is evaporated and oil becomes dry. The dried oil is treated with bleaching earths or clays such as bentonite, diatomaceous earth, attapulgite, montmorillonite, and carbon [9] at about 1–2%. During this stage, trace metal complexes such as iron and copper, pigments, phosphatides, and oxidation products are removed by adsorptive effect of the bleaching earth. Then, the mixture of oil and bleaching agent is filtered through a standard plate and frame press for separation. The clear oil obtained is bleached oil and is very much lighter in color than the neutralized oil. The oil charge is dehydrated under vacuum to avoid any further deterioration due to oxygen. In this operation, the colored pigments in oil are adsorbed by certain bleaching agents under specific conditions. Used bleaching material is called spent earth. Spent bleaching earth contains some entrained oil, as much as 30–50%. The spent bleaching earth is removed in pressure filters and blown with steam to reduce its oil content to about 20%. It is possible to reduce the oil content to about 1% if the cake removed from the filter is extracted with a solvent.

2.4 *Hardening*

The aim of hardening is to raise the melting point of the oil for use in shortening or margarine manufacture by the addition of hydrogen to unsaturated oil molecules over a nickel catalyst. The degree of hardening is determined by the temperature, pressure, and quantity of hydrogen gas added to the reaction. The catalyst is filtered out of the oil after the reaction is complete. No significant wastewater is generated by hydrogenation.

2.5 *Deodorization*

The final step in refining fats and oils is deodorization. The oil after bleaching is practically pure, but it contains small amounts of original odoriferous matter and also the chemicals used during neutralization process. The filtered oil is channeled into the deodorizer for the deacidification and deodorization treatment. Deodorization is mainly aimed to remove compounds responsible for undesirable odors and flavors, such as residual free fatty acids (especially low-molecular-weight fatty acids), aldehydes, ketones, and alcohols from crude degummed and/or alkali refined edible oils and fats after bleaching. This process utilizes a combination of high-temperature heating about 200 °C, under vacuum (2–4 mmHg), and direct steam injection of about 2.5–4.0% by weight of oil [10]. The volatile materials are evaporated off with some carrier (commonly direct steam). This oil is then cooled and clarified through a filter press to get sparkling oil. The purpose of deodorization is to make oil bland and tasteless. During this process, FFA in the form of fatty acid distillate is removed as refining waste, at the upper section of deodorizer. Apart from FFA, carotenoid pigments and primary and secondary oxidation products are also being removed as they may contribute to off-flavors.

3 Environmental Impacts

The major problem met by the vegetable oil industry is wastewater, both quantitatively and qualitatively. Wastewater generation in an edible oil industry can be divided into two categories:

- Wastewater generated directly from processes, e.g., neutralization, washings, etc.
- Wastewater generated from auxiliary systems, e.g., cooling and vacuum systems, etc.

Wastewater generated from both sources varies significantly in pollution load and concentration. Process wastewater contains high amounts of BOD, COD, oil and grease, TSS, TDS, and nickel while wastewater generated from the auxiliary systems is huge in quantity and relatively higher in temperature. It sometimes contains traces of VOCs. Boiler condensate recovery system is not efficient in some and practically non-existent in most.

Most of the water used for neutralization and deodorization in vegetable oil processing facilities ends up as wastewater with a high organic load. The complexity of wastewater sources makes such wastewater hard to treat. The sources of wastewater can be listed as cleaning of greasy floors and equipment, leakages, heat exchangers, gas scrubbers, barometric condensers, process water from the refining plant, and soap-stock splitting effluents. Typically, vegetable oil refinery wastewater contains used chemicals including alkalis such as caustic soda and sodium carbonate; heavy metals from bleaching earth; acids such as phosphoric acid, citric acid, and sulfuric acid; Ni catalysts; and methylates. Also, pesticides used during the growth of vegetable oilseed have been found in vegetable oil processing industrial wastes. Additionally, vegetable oil refinery wastewater contains chemical compounds like phenol, heavy metals from bleaching earth, catalysts used in the hydrogenation process, oxidizable substances, and fats and oils to lesser degree. Because of quantity and characteristic variations and complexity, wastewater treatment to meet the desired effluent standards is complicated, and the selection of wastewater treatment method depends on many local conditions and, therefore, cannot be standardized.

Solvents such as acetone, ethanol, and methanol are sometimes used instead of, or to supplement, hexane in the extraction process. Hexane can cause health problems in relatively low concentrations and other dangerous chemicals, including strong acids and bases, present significant health and safety hazards.

Besides liquid waste, solid waste and air emissions are also generated. Solid wastes include spent earth, filter cloth, and spent catalyst. Spent earth and spent catalyst are in slurry form and are combined together to extract before their final disposal. Also, during production of vegetable oil, some by-products, such as oils for animal feed or pharmaceutical products, are often produced by further processing of residues. This processing can reduce solid waste production, including fractions like spent bleaching earth that can be reused for energy production through direct incineration or biogas production either on-site or at another location. Citric acid and phosphoric acid can generally be used interchangeably in degumming operations.

3.1 Water and Energy Consumption

An average 1,000 metric tons per day vegetable oil refinery can consume as much as 75,000 m^3 of fresh water annually. Water comes in contact directly with the oil during a number of processes used within the typical refinery. The processes with the most significant water consumption are crude oil production, chemical neutralization, and subsequent oil washing and deodorization [11]. Table 8.1 gives information on water consumption and wastewater generation for some processes in the vegetable oil processing industry. For example, water is used to remove phosphatides in the degumming process, to dilute the caustic soda in the chemical refining process, and to further reduce unwanted constituents such as soaps, gums, and metal

Table 8.1 Water consumption and wastewater production [9]

Inputs per unit of product	Water	Wastewater
Crude oil production (m^3/t raw material)	0.2–14	0.2–14
Chemical neutralization (m^3/t product)	1–1.5	1–1.5
Deodorization (m^3/t product)	10–30	10–30
Hardening (m^3/t product)	2.2–7	N.D.
Chemical refining (m^3/t product)	0.25–0.8	14–35

N.D. No data

Table 8.2 Energy consumption in crude vegetable oil refining [9]

Processing step	Total energy consumption (MJ/t final product)	Steam consumption[a] (MJ/t final product)	Electricity consumption (MJ/t final product)
Neutralization	145–330	112–280	22–44
Soap splitting	620–2,850[b]	560–2,800[b]	11-36[b]
Deodorization	510-1,350	420-1120	60–150
Hardening	400–1,000	N.D.	N.D.
Bleaching	N.D.	N.D.	N.D.
	(kWh/t final product)	(kWh/t final product)	(kWh/t final product)
Neutralization	40–92	31–78	6–12
Soap splitting	172–792[b]	156–778[b]	3–10[b]
Deodorization	142–375	117–311	17–42
Hardening	111–278	N.D.	N.D.
Bleaching	N.D.	N.D.	N.D.

Final product = refined vegetable oil
N.D. No data
[a]MJ/t = 2.8 × kg steam/t
[b]MJ/t soap or kWh/t soap

Table 8.3 Example of semi-continuous deodorization [12]

Parameter	Industry benchmark
Latent heat steam (kJ/g)	2,000
Amount of feedstock (kg)	1,000
FFA removal (kg steam)	0.35
Stripping steam (kg steam)	5
Motive steam (kg steam)	35
Total (kg steam)	65
Heating (kg steam)	24.2
Electricity (kWh)	5

Note: Semi-continuous deodorization, 0.2% FFA, 0.5% stripping steam, and average electricity consumption
FFA free fatty acid

ions in the water-washing step. Also, some water is consumed in other indirect applications, including barometric condensing water in deodorization vacuum systems, boiler feed water, cooling tower make-up water, fresh water in minor cooling applications, and centrifuge service water.

The energy consumption of vegetable oil processing facilities depends on the type of raw material, the equipment, and the manufacturing processes. Heating to produce steam for process applications (especially for soap splitting and deodorization), cooling, drying, milling, pressing, evaporation, and distillation are the major energy-consuming steps. Some amount of energy is lost either during the cooling of the product for the subsequent process or in terms of heat contained by the wastewater.

About 200–500 kg of steam is consumed per ton of processed seed (155–390 kWh/t) and the electricity consumption is in the range 25–50 kWh/t processed seed (90–180 MJ/t). Tables 8.2 and 8.3 show energy and steam consumption for some processes in crude vegetable oil refining in German installations [11], which can be considered as indicators of this sector's efficiency and may be used to track performance changes over time.

3.2 *Wastewater Sources and Quantities*

In a vegetable oil industry, wastewaters are mainly generated from the degumming, deacidification, deodorization, and neutralization steps [12]. Table 8.2 shows wastewater generation of some processes in the vegetable oil processing industry. Boiler blowdown and wash water from the de-oiling of the bleaching earth also contribute to the effluent in small amounts [13]. Characteristics of the wastewater depend mostly on the type of oil processed and the process implemented [14, 15]. On the other hand, the volume of wastewater produced changes highly with the raw material processed and the technology applied; for example, wastewater volumes produced during the oil production from palm fruit can often be limited to 3–5 m^3/t of feedstock [14].

Table 8.4 Characteristics of vegetable oil refinery wastewaters [18]

Parameter	Soap-stock shredding	Deodorization
pH	2	4–7
Temperature (°C)	80–90	20–30
BOD (mg/L)	13,400	600
COD (mg/L)	22,300	2,000
Total sulfate (mg/L)	40,000	–
Total phosphorus (mg/L)	2,500	–

Wastewater generated during oil washing and neutralization may have a high content of organic material and, therefore, a high biochemical oxygen demand (BOD) and COD [13]. Wastewater may also have a high content of suspended solids, organic nitrogen, and oil and fat and may contain pesticide residues from the treatment of the raw materials; therefore, wastewaters from this industry need special attention [16]. The characteristics of vegetable oil refinery wastewaters produced at soap-stock shredding and deodorization stages are given in Table 8.4.

The extraction plant wastewater consists of condensate. About 500–600 m^3/d of condensate water with 300–400 mg/L of BOD is generated during heating oils and jacket steam throughout the refinery. This stream contributes about 150–200 kg of BOD per day to the wastewater load. It is usually hot (80–90 °C), clean water and suitable to be recycled as wash water in the refining step. The pH of this water is near neutral between 6 and 8.

Wash water generated in the oil washing step is high in BOD and fat due to dissolved soaps. The stream will contribute 450–2,200 kg/day of fat or BOD_5 to the sewer load. If the oil is not degummed before caustic refining, some water-soluble phosphatides will be removed in this step and will cause higher loads. This is generally the case when corn oil is being processed. Wash water and barometric condensing water offer nearly ideal conditions for water recovery using evaporation. Both of these applications give an oil-water effluent that is greater than 97% water with only a small amount of organics such as neutral oil, free fatty acids, soaps, light ends, and phosphatides.

Greasy cooling water can be a major source of BOD_5 depending on the operation of the deodorizer scrubber and atmospheric conditions. Any organic material that passes through the scrubber becomes emulsified in the greasy water in the vacuum system. The greasy cooling tower dissipates the heat of condensation of the steam primarily by evaporation of the greasy water. In cold weather the water loses more heat by convection which decreases the amount of evaporation necessary. When the amount of water entering the tower via steam condensation is greater than the wind drift plus evaporation, it becomes necessary to direct the tower blowdown to the sewer. The organic material in the water is primarily short-chain free fatty acids. Typical COD values range up to 3,000 mg/L. For an overflow of 110 L/day, 340 kg of COD would be discharged to the sewer based on 3,000 mg/L of fat.

Table 8.5 Effluent levels for vegetable oil processing [12]

Parameter	Guideline value
pH	6–9
BOD_5 (mg/L)	50
COD (mg/L)	250
Total nitrogen (mg/L)	10
Total phosphorus (mg/L)	2
Total suspended solids (mg/L)	10
Oil and grease (mg/L)	50
Temperature increase (°C)	<3[a]
Total coliform bacteria (MPN[b]/100 mL)	400
Active ingredients/antibiotics	To be determined on a case-specific basis

[a]At the edge of a scientifically established mixing zone which takes into account ambient water quality, receiving water use, potential receptors, and assimilative capacity
[b]MPN = Most probable number

Tank car washings vary from one to four cars per day. The effluent is characterized by high temperature, 70 °C, and a variable amount of emulsified oil. The BOD_5 can be anywhere from 2,000 to 60,000 mg/L.

The refining of crude vegetable oils generates large amounts of wastewater. The neutralization step, in particular, produces sodium salts of free fatty acids (soapstocks) which breaks down through the use of H_2SO_4 and generates highly acidic and oily wastewater [13]. Its characteristics depended largely on the type of oil processed and on the process implemented. This type of wastewaters typically has high COD, oil and grease, sulfate, and phosphate contents resulting in both high inorganic and organic loading of the relevant process.

The neutralized acidulation water flow and COD are highly variable based on the amount of flush water being used on the centrifuges and whether or not the oil is being degummed prior to the refining. The phosphatides which would be removed in degumming significantly increase the strength of the acidulation water due to their water solubility. Fresh water can increase the moisture content of the soap from 50% to 95% which does not affect the total BOD_5 load, but it does reduce the residence time in treatment facilities and the processing time which is necessary to achieve good oil separation. BOD_5 of this stream is very high (25,000 mg/L) and the fat is relatively low (300–500 mg/L). This is the only waste stream from the refinery that is not composed primarily of hexane-soluble material.

A typical day in vegetable oil processing facility would yield loadings of 1,800 kg/day BOD and 30 kg of fat based on 75,000 L at a 25,000 mg/L BOD_5 and 400 mg/L fat. Table 8.5 presents effluent guidelines for vegetable oil processing [14]. These guidelines can be reached under normal operating conditions in properly planned and operated facilities through the application of pollution prevention and control techniques. Effluent guidelines are applicable for direct discharges of treated effluents to surface waters for general use. Site-specific discharge levels may be established based on the availability and conditions in use of publicly operated sewage collection and treatment systems or, if discharged directly to surface waters, on the receiving water use classification.

There are some precautions for the prevention of process wastewater which are listed below:

- Emulsion breaking techniques (e.g., dissolved air flotation (DAF)) should be used to remove high BOD and COD oils from wastewater.
- Condensates should be recycled.
- Grids are needed to be used to protect drains in the production area and to prevent solid wastes and concentrated liquids from entering the wastewater stream.
- The dose and application of cleaning chemicals should be carefully determined and applied.
- Cleaning-in-place procedures should be carefully prepared and applied to help reduce chemical, water, and energy consumption in cleaning operations.
- Cleaning solutions (e.g., through a soap-splitting process) should be properly treated and discharged to separate oil and fatty acids from the water phase and then through a fat trap.
- Phosphoric acid may be replaced with citric acid in degumming operations if feasible; thus, the phosphorus load in the wastewater is reduced and a small reduction in sludge quantities occurs.

3.3 Wastewater Treatment

Treatment of industrial wastewaters is difficult and generally requires chemical or physicochemical treatment prior to biological treatment or discharge to a sewer system. Wastewaters from the vegetable oil refining processes fall in this category. The operating conditions and processes carried out influence the amount and characteristics of the by-products and wastes produced. The wastewater varies both in quantity and characteristics from one oil industry to another. The composition of wastewater from the same industry also varies widely from day to day [17]. These fluctuations may also be attributed to different types of oils processed. Because of quantity and characteristic variations and complexity, wastewater treatment to meet the desired effluent standards is complicated, and the selection of wastewater treatment method depends on various local conditions and, therefore, cannot be standardized. Treatability studies become very important at this point and must be performed to describe wastewater characteristics before selecting any pretreatment process to treat wastewater containing emulsified oil and grease wastewater. Thus, an oil-grease removal process can be designed efficiently and effectively [18].

Oily wastewater treatment in many respects is similar to the treatment domestic of wastewater. Firstly, settleable solids are removed in domestic wastewater treatment plant. Similarly, for the treatment of vegetable oil refining wastewaters, the separation of free-floating oil from disperse emulsified and soluble fractions is performed at the first stage of wastewater treatment. At the second stage of treatment, oil-water emulsions are broken and dispersed oils are removed. Typical treatment technology is chemical treatment. In some applications, tertiary treatment is also used to remove finely dispersed and soluble oil fractions. Ultrafiltration, biological treatment, or carbon adsorption may be used as a tertiary treatment method.

Physicochemical (skimming of oil, air flotation, flocculation, coagulation) treatment to remove colloidal pollutants followed by biological treatment to degrade dissolved organics is the most commonly used technique applied to vegetable oil refining wastewater effluent [18].

Primary treatment used in the vegetable oil sector generally includes grease traps, skimmers or oil-water separators for separation of floating solids, flow and load equalization, and sedimentation for suspended solids reduction using clarifiers [11]. The purpose of primary treatment is to remove free-floating oil and grease and suspended solids. This will also reduce the BOD and COD concentration. High correlation in terms of correlation coefficient (R^2) has been determined between COD with O&G on a large scale by Saatci et al. [19] as 0.92 and by Aslan [20] as 0.90 for sunflower oil refining wastewaters and 0.81 for corn oil refining wastewaters indicating that a decrease in O&G may highly reduce the COD values. This can be achieved either by gravity settling or by dissolved air flotation (DAF).

Further, secondary treatment is applied and either chemical or biological processes are used. In general, the wastewater is well suited to biological treatment. Biological treatment, typically anaerobic followed by aerobic treatment (activated sludge, trickling filters, and rotating biological contactors) for reduction of soluble organic matter (BOD), biological nutrient removal for reduction in nitrogen and phosphorus, chlorination of effluent when disinfection is required, dewatering and disposal of residuals, and, in some instances, composting or land application of wastewater treatment residuals of acceptable quality may be possible [11].

The basic process of chemical treatment is coagulation [21, 22]. The efficiency of treatment depends on the ratio of free oil to emulsified oil. The free oil can be easily removed from wastewater by physical processes using coagulating agents, e.g., alum and ferric chloride for phase separation [23]. Additional engineering controls may be required to contain and neutralize nuisance odors [24, 25].

3.3.1 Unit Operations in Wastewater Treatment of Vegetable Oil Refining

The wastewater to be treated should also be analyzed for total oil and grease, including determination of the free, emulsified, and dissolved oil fractions. Although oil-water separators are designed to remove free oil, they also remove solids. Therefore, the solids content of the wastewater is important in overall system design. Analyses should include TSS, volatile suspended solids (VSS), and settleable solids. These analyses will help determine the amount and frequency of settled materials that will need to be removed from the bottom of the oil-water separator and the advisability of providing grit removal upstream of the oil-water separator.

Oil traps are used in oily wastewater treatment to remove free-floating oils [24]. Oleophilic (oil-attractive) plates attract and retain small oil droplets under laminar flow conditions. Stokes' law, the physical law governing the rise and fall rate of a droplet or particle in a fluid stream, can predict the settling time of solid particles and the rise time of oil droplets, which have specific gravities higher or lower than that of the main fluid. The plates may be made of an oleophilic material, such as polyethylene, fiberglass, or nylon, to promote coalescence of oil droplets. For this reason, the units are sometimes referred to as coalescing plate separators. Parallel or coalescing plate separators are usually recommended only for light oil loadings when a higher level of oil removal is required and when the wastewater stream contains low solids concentrations.

Gravity grease traps, centrifugal separators, and rarely electroflotation methods are also used as the other removal methods for free fats present in the oily wastewater. The majority of oil-water separators used is conventional gravity separators that rely on the different densities of oil, water, and solids for successful operation. The wastewater is fed to a vessel sized to provide a quiescent zone of sufficient retention time to allow the oil to float to the top and the solids to settle to the bottom. For the construction of gravity grease traps, durable construction materials must be used because the wastewater of vegetable oil refining industry has high sulfate content due to the use of acids at some steps of the production. In a recent study, it was demonstrated that gravity separation procedure, in which the wastewater was kept in quiescence in a separation column and the aqueous phase was collected from the bottom, yielded oil and grease removals over 90% and COD removals between 58 and 90% after 24 h [26]. This is a proper procedure as a primary treatment for vegetable oil refinery wastewaters, and it would improve overall efficiency of wastewater treatment plant.

3.3.2 Flotation

Flotation processes are designed to remove all suspended particles, colloids, emulsions, and even some ions or soluble organics that can be precipitated or adsorbed on suspended solids (SS) [27]. It is widely used in the treatment of oily wastewaters. In practice, the natural flotation process differs from the flotation with air. In the first case, the material particles lighter than the water (oils, greases) eventually associated with gas bubbles have the tendency to rise to the stationary liquid surface. In the flotation process with air, the material particles heavier than the water are transported to the liquid surface by attaching them with air bubbles [28]. In the second case, the air could be introduced in water in very fine bubbles by the mechanical agitation flotation with dispersed air [29]. The separation of materials lighter than water, e.g., edible oil/fat, can be enhanced by using flotation. This technique does not facilitate the separation of emulsified fat-oil-grease (FOG) from wastewater and, hence, is used in the vegetable oil refining sector for the removal of free FOG [11].

Flotation requires the generation of small bubbles which can be produced by dispersing air into the wastewater, applying a vacuum to the wastewater, dissolving air into pressurized wastewater, and then releasing the pressure. Oil is usually removed from wastewater mainly by using the dissolved air flotation (DAF) technique for bubble generation. The other flotation techniques are namely dispersed or induced air and electrolytic flotation, which are less frequently used [30]. Brief information on each technique is provided below.

Dispersed (Induced) Air Flotation (IAF)

This method is widely used for oil-water separation (oily sewage). In this technique, air bubbles are formed by mechanically dispersing air injected under rotating impellers or sparged by diffuser. The technology makes use of the centrifugal force developed [31, 32].

Vacuum Flotation

In this system, air is first dispersed into the wastewater to achieve saturation conditions. Partial vacuum is then applied to the wastewater. This results in the generation of small air bubbles that attach themselves to the solid particles and make them rise. Typically, a cylindrical tank maintained under vacuum and continuously fed with wastewater is used for this purpose [29].

Dissolved Air Flotation (DAF)

DAF is very common method used for removal of suspended solids, oils, and other contaminants by introducing small air bubbles into the wastewater containing the suspended solids/free oil to be floated [33]. The fine air bubbles attach themselves to the chemically conditioned particles and, as they rise to the surface, the solids float to the surface with them, where they are mechanically skimmed and removed from the tank. The air is dissolved under pressure of 3–6 bar. The air is introduced into a recycle stream of treated wastewater after being passed through a DAF unit. This supersaturated mixture of air and wastewater flows to a large flotation tank in which the pressure release and generation of numerous small air bubbles occur. Mechanical skimming or suction withdrawal is used to remove the thickened solids/oil particles attached to the air bubbles. Chemicals such as polymers, aluminum sulfate, or ferric chloride can be used to enhance flocculation and, therefore, the adhesion of bubbles [11]. Flotation units can be provided for recycling to prevent the incoming solids to be subjected to the shearing action of the pressurizing pump. By using this technique, free FOG, BOD/COD, SS, nitrogen, and phosphorus levels are reduced.

Electroflotation

Electrochemical processes are investigated in several studies to treat oily wastewater [34–37]. Briefly, in this process, bubbles are generated to carry the pollutants such as sludge and fine particles to the top of the solution where they can be more easily concentrated, collected, and removed. An electrochemical reactor consists of two sets of iron electrodes, anodes and cathodes, in which treatment is done in 2–20 min with intense iron dissolution. The electrodes are placed a few millimeters apart from each other and submerged in the wastewater that is to be treated for oil contamination. The cathode and anode are connected to the negative and positive outlets of a DC power supply, respectively. When direct current is applied to water through the electrodes, the energy barrier is overcome, and water molecules are broken down into hydrogen and oxygen gases. The reaction is observed through the hydrogen bubbles that evolve from the cathode and oxygen bubbles that originate from the anode. On the other hand, when the anode is made of metals having lower oxidation potentials than water, the anode is dissolved to produce metal ions instead of generating oxygen. These ions then react with hydroxyl ions, the by-products of hydrogen generation, to produce metal hydroxides [37]. The reactions involved are given in Eqs. 8.3, 8.4, and 8.5 [38].

$$Fe_{(s)} \rightarrow Fe^{2+}{}_{(aq)} + \frac{2e^{-}O_2}{Air} \rightarrow Fe^{3+}{}_{(aq)} + e^{-} \tag{8.3}$$

$$2H_2O + 2e^{-} \rightarrow H_{2(g)} + 2OH^{-}{}_{(aq)} \tag{8.4}$$

$$Fe^{3+}{}_{(aq)} + 3OH^{-}{}_{(aq)} \rightarrow Fe(OH)_{3(s)} \tag{8.5}$$

As a result of the reactions, the emulsion is destabilized, and the colloidal oil particles begin to coalesce. Ultimately, the destabilized oil droplets absorb into the highly dispersed ferric hydroxide colloid formed by the reaction between the electrogenerated Fe(III) and hydroxyl ions. Finally, the oil-rich sludge floats to the top where it is removed by skimming. It also contributes to coagulation by neutralizing any negatively charged colloidal particles, which have been reported to be more compact than sludge obtained by chemical methods [39]. Ibanez et al. [40] revealed that the oil-water emulsion was actually broken down by the charge neutralization of ferric ions (Fe(III)) generated by the anodic dissolution of the iron electrode. The sunflower seed oil-water emulsion was clarified within minutes of electrochemical treatment. It is mostly preferred due to its fast rate of pollutant removal with little sludge generation, compact size of equipment, simplicity in operation, and low capital and operating costs [36].

Coagulation-Flocculation

The choice of removal method for FOG from wastewater depends on oil form such as free or emulsified present. If oil is present in free form, the methods mentioned above can be used for its removal. However, if oil in wastewater is in emulsified form, these emulsions must be broken down by physicochemical methods [41]. Chemical treatment of an emulsion is usually directed toward the destabilization of the dispersed oil droplets or the destruction of emulsifying agents present in a first stage followed by the removal of the separated oil [37]. Coagulation and flocculation are used for this purpose.

Coagulation and flocculation are often used interchangeably, but they refer to two distinct processes. Coagulation refers the process by which colloidal particles and very fine solid suspensions are destabilized so that they can begin to agglomerate if the conditions are appropriate. On the other hand, flocculation is the process by which destabilized particles actually conglomerate into larger aggregates so that they can be separated from the wastewater.

Most solids suspended in water possess a negative charge and, since they have the same type of surface charge, repel each other when they come close together. Therefore, they will remain in suspension rather than clump together and settle out of the water. Coagulation and flocculation processes are used to remove the suspended solids portion from the water. The suspended particles vary considerably in source, composition charge, particle size, shape, and density. Correct application of coagulation and flocculation processes and selection of the coagulants depend upon understanding the interaction between these factors. The small particles are stabilized (kept in suspension) by the action of physical forces on the particles themselves [28].

Coagulation is commonly achieved by adding different types of chemicals (coagulants) to the wastewater to promote destabilization of the colloid dispersion and agglomeration of the resulting individual colloidal particles. The addition of some common coagulants to a wastewater not only produces coagulation of colloids but also typically results in the precipitation of soluble compounds, such as phosphates, that can be present in the wastewater. In addition, coagulation can also produce the removal of particles larger than colloidal particles due to the entrapment of such particles in the flocs formed during coagulation.

The choice of chemicals used for coagulation and flocculation depends upon the nature of the suspended solid to be removed, the raw water conditions, the facility design, and the cost of the amount of chemical necessary to produce the required effluent quality, as well as the intended disposal route for the sludges. Final selection of the coagulant should be made by jar testing and plant-scale evaluation. Common coagulant chemicals used in wastewater treatment are aluminum salts, ferric sulfate, ferric chloride, ferrous sulfate, sodium aluminate, and lime ($Ca(OH)_2$) or quicklime (CaO). The first four will lower the alkalinity and pH of the solution while the sodium aluminate and lime will add alkalinity and raise the pH [42, 43]. Also, polymers (cationic polymers, anionic and non-ionic polymers) which are long-chained, high-molecular-weight, organic chemicals are widely used, especially as coagulant aids together with the regular inorganic coagulants. Polymers are effective over a wider pH range than inorganic coagulants. They can be applied at lower doses, and they do not consume alkalinity. However, they are generally much more expensive than inorganic coagulants.

Previously, Louhichi et al. [44] utilized coagulation and flocculation process to treat real vegetable oily wastewater and investigated the relationship between coagulant concentration, flocculent dosage, and pH besides two important process parameters (COD and turbidity removals). Response surface methodology was used in order to estimate the optimal conditions for maximum real vegetable oily wastewater treatment. $Al_2(SO_4)_3$ was found as the best coagulant aiding maximum COD and turbidity removals when used directly at a concentration of 2.4 g/L followed by a CHT flocculant (an industrial flocculant) at a dosage of 60 mg/L under initial pH of real effluent of 9.23 [44].

3.3.3 Biological Treatment

Despite the high efficiency of COD removal by physicochemical treatment, at the beginning of the WWTP, biological treatment is needed due to presence of lipids in the form of colloidal suspension [45]. Biological treatment is mainly used to remove biodegradable organics and SS through biological methods. Adsorption of pollutants to the organic sludge produced will also remove non-biodegradable materials, e.g., heavy metals. Organic nitrogen and phosphorus can also be partially removed from the wastewater.

Biological treatment options are aerobic processes using dissolved oxygen, anaerobic processes without oxygen supply, and anoxic processes using biological reduction of oxygen donors. These systems can be used alone or in combination, based on the characteristics of the wastewater and the requirements before discharge. Mostly, aerobic and anaerobic biological treatment systems are used in the oil industry effluents.

Aerobic and anaerobic fat and oil removal by microorganisms have been reported in several studies [46–48]. The conventional biological processes such as aerobic and anaerobic digestion require a BOD/COD ratio to be 0.6 [49]. However, vegetable oil industry effluents usually have BOD/COD ratios between 0.15 and 0.2 indicating the requirement of a proper pretreatment before the biological step [50–53].

Aerobic Treatment

The success of the aerobic treatment depends upon sufficient oxygen levels required by the organisms to sustain their life. For an efficient operation of aerobic reactors to degrade oil and grease, it is very important to maintain good oxygen transfer [52]. Activated sludge systems and biofilm processes such as fixed-bed or trickling filter processes are examples of the aerobic treatment [54].

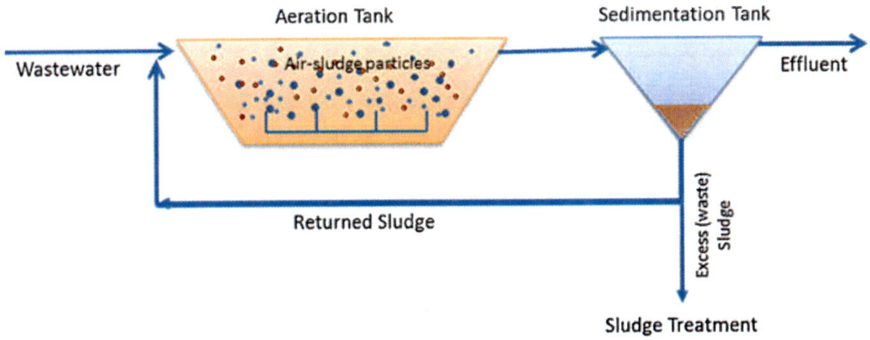

Fig. 8.3 Activated sludge diagram

Activated sludge systems consist of two tanks as shown in Fig. 8.3. In the first tank, air is pumped by means of diffusers from the bottom of the tank into a mixture of wastewater and biological sludge (microorganisms). Air rises to the top in the form of bubbles, creating good agitation and also meeting the desired oxygen demand of the microorganisms. Microorganisms convert wastewater organics into end products such as carbon dioxide, water, and an active mass of microbes (biologic floc) called activated sludge or biomass. This biomass is transferred into the next tank, where the absence of agitation allows the biomass to settle. The biological solids are subsequently separated from the treated wastewater and returned to the aeration process as needed [54, 55].

Activated sludge and wastewater mixture in the aeration tank is called mixed liquor. After a predetermined hydraulic retention time, which varies from several hours to an excess of 10 days, based on an organic loading rate or food/microorganisms (F/M) ratio of about 0.1–0.15 kg BOD/kg MLSS per day, the mixed liquor is directed from the aeration tank to a secondary clarifier where the activated sludge is settled out (Fig. 8.3). The hydraulic retention time (HRT) or sludge age and F/M ratio can all vary as a function of the influent (raw wastewater) characteristics, e.g., composition, availability and degradability of organic substances, and the required final wastewater quality. Most of the settled sludge is returned to the aeration tank (return sludge) to maintain a high population of microbes to allow efficient and rapid degradation of the organics. The excess sludge is diverted or wasted to the sludge handling system for treatment and disposal to maintain mixed liquor suspended solids (MLSS) at a required level, e.g., 3,000 mg/L [11]. This system can be very effective if the level of total fatty matter is kept low after the pretreatment stage. The presence of fatty matter in the activated sludge systems will contribute not only higher BOD but also the formation of foam which reduces the efficiency of the system.

The most common problem of activated sludge is the bulking of sludge indicating the poor settling characteristics of biological sludge. The presence of filamentous bacteria and/or excessive water bound within biological floc is known to be the reason of this problem. A typical solution for bulking problem is to use chemicals, i.e., chlorination or use of other oxidative chemicals or precipitation chemicals, to destroy all filamentous organisms that are not protected by activated sludge flocs. However, these methods are not very selective on microorganisms and can terminate the whole biological activity [11].

The most important parameters of activated sludge system are HRT and the operating temperature. These parameters need to be carefully determined to ensure the breakdown of resistant organic substances. Additionally, separate compartment or selector is commonly used to prevent and control filamentous growth. The selector provides the selective growths of floc-forming organisms by providing high F/M ratios at a controlled dissolved oxygen level. The contact time is generally 10–30 min.

The anoxic selector, which requires the presence of nitrate in the water, is often chosen in activated sludge systems for nitrification purposes. It reduces the oxygen requirements since nitrate-nitrogen is used as a terminal electron acceptor for the oxidation of influent biodegradable organics and reduces the consumption of alkalinity during nitrification, by means of recovered alkalinity in the anoxic zone. Anoxic selectors can also be quite effective in controlling filament growth [11].

In fixed-film aerobic processes such as trickling filters, the biomass grows as a film on the surface of packaging medium and the wastewater is evenly distributed across it. The trickling filter medium typically consists of rocks or various types of filling materials, such as slats, or plastic materials. The wastewater is distributed over the surface of media by a rotating arm. Rock filter diameters may range up to 60 m. As wastewater trickles through the bed, a microbial growth establishes itself on the surface of the stone or packing in a fixed film. The wastewater passes over the stationary microbial population, providing contact between the microorganisms and the organics. Oxygen is provided from the bottom to the biomass by blowing outside air. The treated wastewater is collected under the media and directed to a sedimentation tank where a part of the liquid can be recycled to dilute the strength of the incoming wastewater (Fig. 8.4).

Anaerobic Treatment

Anaerobic processes are commonly used for the treatment of industrial wastewaters. Whereas aerobic processes effectively treat low-strength wastes, anaerobic processes more effectively treat high-strength wastes. These high-strength wastes have COD levels greater than 2,000 mg/L. Anaerobic digestion occurs in a heated reactor with the absence of molecular oxygen that results in methane and carbon dioxide production [29]. In comparing anaerobic processes and aerobic processes, anaerobic processes have distinct advantages including energy and chemical efficiency and low biological sludge yield, in addition to the possibility of treating high-strength wastewaters with a COD of 5,000 up to 40,000 mg/L, with the requirement of only a small reactor volume [56]. The specific biomass loading of typical anaerobic processes treating soluble industrial wastewaters is approximately 1 kg COD utilized per kg biomass-day. There is a wide range of anaerobic reactors with different

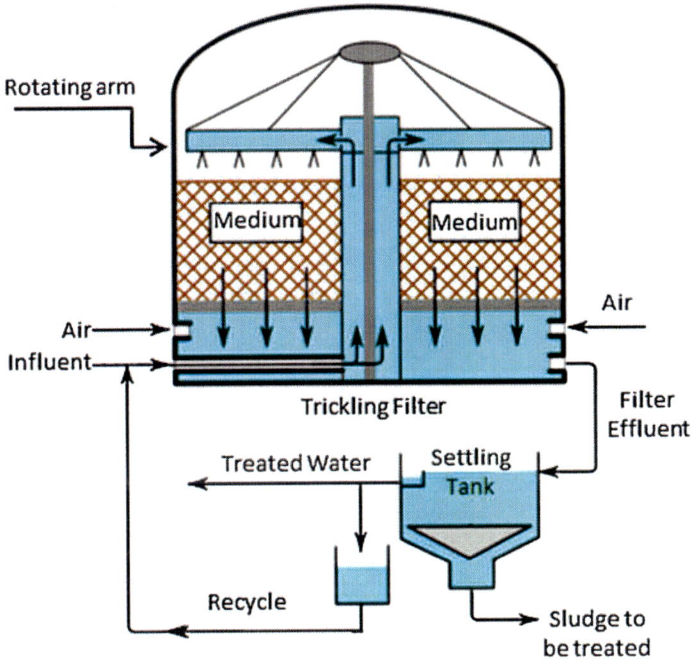

Fig. 8.4 A schematic representation of a trickling filter system [51]

Table 8.6 Typical process and performance data of anaerobic wastewater treatment processes [9]

Process	Input COD (mg/L)	HRT (h)	OLR (kg COD/m^3. day)	COD removal (%)
Anaerobic lagoons	N.D.		0.6–1	
Anaerobic contact process	1,500–5,000	2–14	0.5–5.3	75–90
Fixed-bed reactor	10,000–70,000	24–48	1–15	75–85
UASB	5,000–15,000	4–12	2–12	75–85
Expanded bed reactor	5,000–10,000	5–10	5–30	80–85
Fluidized bed reactor	N.D.	N.D.	40–60	N.D.
Internal circulation (IC) reactor	N.D.	N.D.	31	N.D.

N.D. No data

organic loading rate (OLR), HRT, and influent COD ranges. Typical data of some anaerobic process is presented in Table 8.6.

As shown in Fig. 8.5, anaerobic digestion has four steps which are hydrolysis, acidogenesis, acetogenesis, and methanogenesis [56]. The first step is the solubilization (hydrolysis) step in which hydrolytic microorganisms degrade insoluble polymer-type organic material such as polysaccharides and proteins by extracellular enzymes to monomers. This reduction results in no reduction of COD. Hydrolysis is the rate-limiting step of overall anaerobic digestion because of the semi-rigid

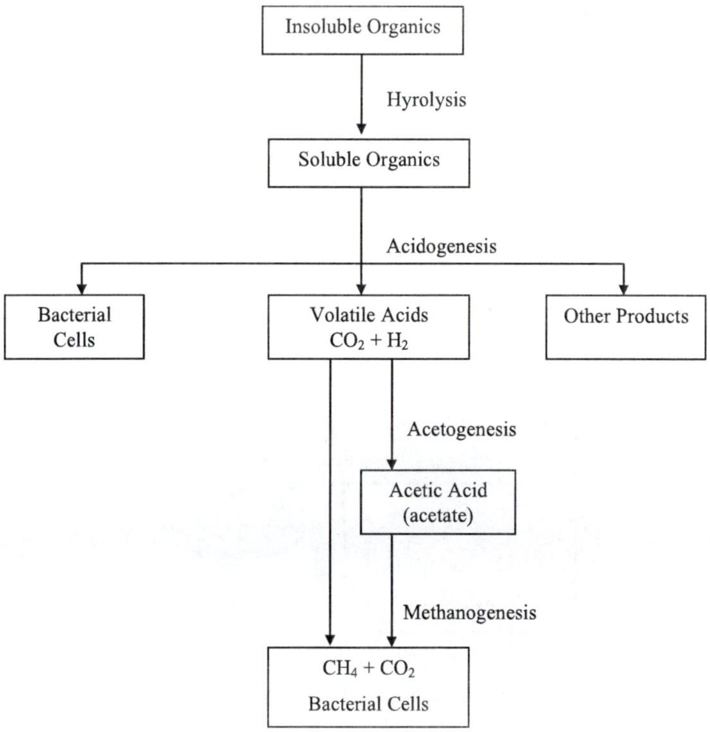

Fig. 8.5 Anaerobic degradation route [58]

structure of the bacterial cells [56]. In the second step, acid formers degrade soluble organics and they mainly form propionic and acetic acid and also alcohols, hydrogen, and carbon dioxide. In the acidification stage, there is minimal reduction of COD. Third step is the acetogenesis in which acetogenic bacteria convert organic acids and alcohols into acetic acid. Finally, methanogenic organisms, methanogens, convert the organic acids into methane and carbon dioxide. At this final step, about 70% of methane is produced from acetic acid [56].

Methane organisms, methanogens, are very sensitive to pH fluctuations and they function over a pH range of 6.6–7.6 with an optimum near pH 7.0 [56]. There are two distinct classes of methanogenes, namely, *Methanothrix* and *Methanosarcina*. Under standard temperature and pressure conditions, the degradation of 1 kg of COD or ultimate BOD results in 0.35 m^3 of methane production [56]. The methane produced has a high calorific value and can be reused as fuel, e.g., elsewhere in the process. The quantity of cells produced during methane fermentation will depend on the strength and character of the waste and the retention of the cells in the system.

Anaerobic digestion can be operated in mesophilic (30–38 °C) or in thermophilic (50–57 °C) conditions. Higher-temperature operation leads to faster reactions and more biogas productions [29]. Effect of temperature on biogas production is given in Fig. 8.6.

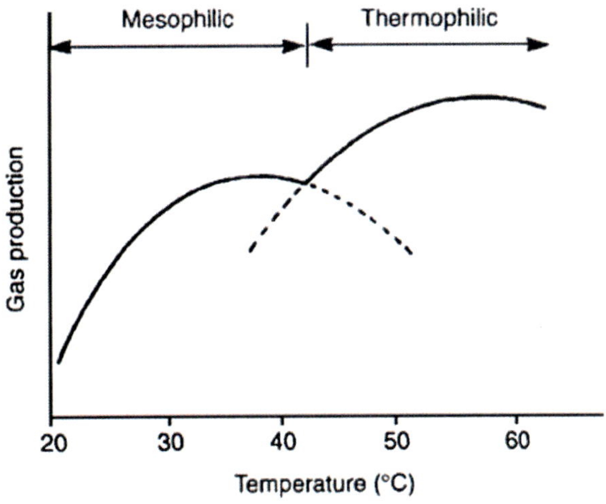

Fig. 8.6 Effect of temperature on biogas production [54]

However, it is well-known that anaerobic digestion of oily wastes is not always easy and straightforward, since anaerobic bacteria are very sensitive to lipid-rich substances along with the intermediate compounds of oily waste degradation process. Therefore, a treatability study together with determination of BOD and COD concentrations, pH, temperatures, and flow rates of the wastewaters is essential, especially for the wastewaters containing low-soluble compounds such as fats, oils, or greases or complex carbohydrates [56]. Still, lipids are attractive substrates for anaerobic digestion and co-digestion due to the higher methane yield obtained when compared to carbohydrates.

In addition, anaerobic digestion has some limitations. The rate-limiting step of hydrolysis is the most important limitation of anaerobic digestion. In some wastewaters most of the organics are located in the microbial cell and the stable semi-rigid structure of the cell walls protects the cell from lysis. Therefore, high resistance to biodegradation leads to long HRT for the biological stabilization in digesters [56]. Thus, anaerobic digestion is a slow process because of the rate-limiting step of hydrolysis. However, the time required to initiate gas production can be reduced by adapting the bacteria to waste. These methods are normally employed strictly as pretreatment measures. Due to the slow microbial growth, there is no phosphorus removal. No nitrification and denitrification occur so nitrogen cannot be removed by anaerobic treatment. Some common operational problems experienced with anaerobic treatment processes are given in Table 8.7.

Foaming and bulking are other problems faced in anaerobic digesters. Some certain filamentous microorganisms lead to foaming and bulking which cause operational problems in anaerobic digesters such as reduced effective volume of the digesters and shortened digestion periods [57].

Anaerobic digestion is sensitive to shock loads and toxic materials [56, 58]. Methane formers can be easily affected by the unfavorable conditions which lead to a poor digestion. An anaerobic system alone would not satisfy a final wastewater quality

Table 8.7 Some common operational problems experienced with anaerobic treatment processes [9]

Problem	Solution
Lack of macro-nutrients	BOD/N/P ratios are normally maintained at 500:5:1
pH	The pH is maintained at 6.8–7.5
Temperature	The optimum temperature for mesophilic bacteria is 35–37 °C
Lack of micro-nutrients	Minimum quantities of micro-nutrients are maintained, especially for Fe, Ca, Mg, and Zn, according to the specific process employed
Physical blockage of the reactor inlet pipework	Effective screening and primary treatment are essential
Overloading	Care needs to be taken to ensure the original hydraulic, solid, and organic loading design rates do not exceed the manufacturer's recommendations

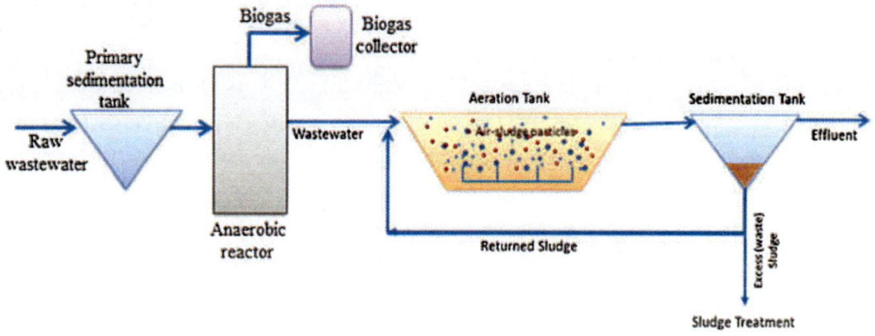

Fig. 8.7 Anaerobic-aerobic treatment method

high enough for discharge to a watercourse. Therefore, an aerobic system is usually installed after anaerobic reactors (Fig. 8.7), so that discharge limit values are satisfied by assisting in the breakdown of the remaining BOD and hydrogen sulfide is removed by well aeration. The energy gained from the anaerobic plant can be used to compensate the energy that is consumed in the aerobic plant.

3.3.4 Advance Treatment

Membrane Filtration

Oil in wastewater is in three forms such as free-floating oil, unstable oil-water emulsions, and highly stable oil-water emulsions. Membrane processes namely microfiltration (MF), ultrafiltration (UF), nanofiltration (NF), and reverse osmosis (RO) are proven promising alternatives for conventional industrial separation methods for the removal of oil forms from such wastewaters [59, 60]. These technologies are considered as the most useful technique in the treatment of stable emulsions particularly with oily wastewater as well as for oilfield produced water treatment, besides being effectively and efficiently used in the treatment of wastewaters from various industries such as agri-food, textile, pharmaceutical, and petrochemical industries [15]. These processes offer various advantages such as high selectivity, easy separation, high oil removal efficiency, easy maintenance, economic and fast operation, and stationary parts with compact modular construction [15, 61, 62].

Membrane systems or membrane bioreactors (MBR) are used in some variations on conventional activated sludge whereby a number of membrane modules, or cartridges, are placed within the body of the reactor vessel. After biological treatment, effluent is pumped under static head pressure to the membrane unit in which the solids and liquids are separated, the clean wastewater is discharged, and the concentrated mixed liquor is pumped back to the bioreactor. The MBR can be operated in either aerobic or anaerobic mode [11].

MBRs are typically used to obtain high-quality effluents free from contaminants that can be reused and recycled in the process itself or in other applications [63]. In another study, Masoudnia et al. [64] used MF consisting of polyvinylidene fluoride membrane for separation of oil-water emulsions. A number of studies have been conducted to treat oily wastewaters by employing various types of inorganic and organic membranes [65–67]. Despite their low cost and various advantages, most organic membranes cannot tolerate harsh conditions, for example, high temperature and pressure, and fly ash that limit the membrane regeneration [68]. In contrast, inorganic membranes such as ceramic membranes can work efficiently under high temperatures and pressures due to their excellent thermal and mechanical durability as well as their chemical resistivity. Therefore, they can be used under extreme process condition for regeneration [69].

An MBR is very efficient in biomass separation, allowing the biomass concentration within the upstream reactor to be up to ten times greater than the concentration normally attainable in a conventional suspended growth system. Secondary sedimentation is not required in MBR process and MLSS levels achieved can vary from 12 to 17,000 mg/L. However, MBR system has very high operational cost especially the energy cost of pumping. This cost may be minimized by applying gravity feed of the wastewater. Another drawback of these systems which is the most common one is fouling. The main forms of fouling are identified as the deposition of solids as a cake layer, pore plugging/clogging by colloidal particles, adsorption of soluble compounds, and biofouling. In order to control this problem, aeration and backwashing have been tried, but these cause an additional production of wastewater. DAF can be used to scour and clean the membrane surfaces to prevent biofouling [11]. Fouling in MBR systems can potentially be controlled by coupling coagulation and flocculation to the process [70].

The space requirement for MBR is very low compared to conventional activated sludge. This system is ideal for higher-strength, acidic, and lower-volume wastewaters. It is more useful in situations where a long solids retention time is required to reach the necessary biological degradation of the pollutants. Likewise, wastewater containing not readily degradable compounds, e.g., phenols, pesticides, herbicides, and chlorinated solvents, and high organic pollution which are present in vegetable oil refining wastes can be treated with MBR.

Hybrid Coagulation/Flocculation-Membrane Filtration

Membrane fouling is a major challenge for the development of membrane technologies due to the high operation and maintenance cost resulting from increased energy and chemical demand for cleaning and the frequency of membrane replacement [60, 71]. Therefore, in order to overcome membrane fouling during oily wastewater treatment, the hybrid processes uniting the advantages of pretreatment methods such as pre-coagulation, pre-adsorption, pre-filtration, and pre-oxidation with membrane processes are often preferred [60, 72, 73]. Among the various pretreatments, coagulation/flocculation is the most applied method since it is easy to operate and relatively inexpensive and can improve UF performance significantly [74]. During coagulation, oil droplets and colloids destabilize and form larger flocs; hence, they can be easily held by UF membranes [75].

In a recent bench-scale experiment, Khouni et al. [76] investigated several technologies to identify the most efficient one for complete removal of oil from synthetic and industrial oily wastewaters. They employed various treatment processes such as coagulation/flocculation and membrane filtrations such as UF directly or in hybrid mode. The treatment processes were compared based on their efficiency in COD, TOC, and turbidity removals. The experiment was done using a commercial ceramic UF membrane (0.245 m^2 and 150 kDa) in cross-flow filtration mode with flow rate of 3000 L/h and a trans-membrane pressure at 1.2 bars. However, UF membrane is susceptible to fouling and not efficient for reusable quality water. Therefore, in that study, the aim was to improve the purification performances of UF by investigating its combination with other treatment options such as optimized coagulation/flocculation as pretreatment and dead-end filtration as post-treatment for high-quality reusable effluent [76]. When coagulation/flocculation was used as pretreatment under optimal conditions (defined as 2.4 g/L of aluminum sulfate, 60 mg/L of CHT flocculant, initial pH of 9.2), the hybrid process uniting coagulation/flocculation with UF membrane provided the highest-quality effluent in terms of turbidity, COD, and TOC removals at about 100%, 98%, and 97%, respectively, compared to those of coagulation/flocculation and UF membrane processes when used separately, with membrane permeability of 135 L/(h.m^2.bar) [76]. The membrane permeability for cases of direct UF and hybrid coagulation/flocculation/UF was in agreement with an industrial application. However, Khouni et al. [76] found that the coagulation/flocculation pretreatment was inefficient in terms of reducing membrane fouling since it did not improve the membrane permeability, 135 L/(h.m^2.bar), compared to that attained with direct UF (150 L/(h.m^2.bar)). The findings of the study in terms of water quality and membrane performance indicated that the hybrid coagulation/flocculation/UF process is a promising alternative to water recovery and pollution removal despite the fact that it does not offer the advantage for valorization and vegetable oil recovery.

Overall, the obtained results can be useful for the development and performance optimization of industrial installation to treat and reuse vegetable oil refinery wastewater using a hybrid CF/UF process as feasible, safer, and more environmentally friendly alternative to other available treatment options.

3.4 Solid Wastes and By-Products

Vegetable oil processing activities also generate significant amounts of organic solid waste and by-products, such as empty fruit bunches (EFBs) and waste palm kernels. The amount of waste generated depends on the quality of the raw materials and the use or reprocessing of the discarded materials into commercially viable by-products. Other solid wastes from the vegetable oil manufacturing process include soap-stock and spent acids from chemical refining of crude oil; spent bleaching earth containing gums, metals, and pigments; deodorizer distillate from the steam distillation of refined edible oils; mucilage from degumming; and spent catalysts and filtering aid from the hardening process. Solid wastes resulting from vegetable edible oil refining plants can be classified as follows:

- General waste
- Tin cans and scrap
- Filter cloth
- Spent bleaching earths
- Soap-stock
- Lecithin
- Spent nickel
- Deodorizer distillate
- Sludge from wastewater treatment plant

General waste includes waste pipe, angle iron, steel, cans, plastic, glass bottles, and papers. A portion of such wastes is sold for recycling; the rest is temporarily stored in the factory area and sent to landfills [77]. Tin cans and scrap waste, used containers, caps, and cans are the wastes resulting from production area. Some factories purchase cans from subcontractors so that these kinds of wastes are not produced in such factories. These wastes are temporarily stored in the factory area and then sold for recycling. The amount of scrap constitutes 2–10% of total amount of tin.

Filter cloths are used in filter presses at winterization and bleaching steps during the refining of oil. When first installed filter cloth can be used 4–6 times. Then, it has to be removed and washed before reuse. In this way before being discarded, a filter cloth can be reused 2–3 times [16].

Spent bleaching earths (SBE) is generated at the vegetable oil refining industry after bleaching of crude oil. The annual world generation of SBE is between 1.5 and 2.0 million tons based on the world edible oil production of 128.2 million metric tons in 2007 [77] and 1.2–1.6 kg of SBE generation per metric ton of edible oil production. SBE contains 20–40% by weight of oil, fat, and colored pigments [78]. This means that the bleaching earths, especially those containing polyunsaturated fats, have self-flammable properties [79]. Therefore, spent bleaching earths are listed as hazardous wastes and their disposal or treatment is difficult in the environment. Also, spent bleaching earths include color pigments like carotene and chlorophyll and oxidation products which are present in the vegetable oil [9]. In order to reduce their fat/oil contents, spent bleaching earth can be extracted with hexane and the oil present in them can be recovered and used as second-quality oil for different purposes. By this way, the oil content of spent bleaching earth can be reduced to below 1%. Thus, bleaching earth becomes a solid waste consisting of inorganic chemicals. This still might be reused as bleaching earth at the second-quality oil production facilities [63].

Soap-stock is accepted as a by-product, which is generated from the alkali-refining of the vegetable oil, otherwise treated as waste. The soap-stock contains 50% total fatty matter and, therefore, it has a commercial value [8]. But it also contains significant amount of phosphatides, pigments, and other contaminants. Therefore, soap-stock cannot be used in good-quality soap production. They chemically break down into acid oil. The breakdown can be accomplished by using a strong concentrated acid. This produced acid oil can be assessed by selling it to soap production factories [9].

Lecithin is formed at the degumming stage of the oil refining. Basically, it is a mixture of phosphatides and small amounts of water-soluble components like glycolipids and oligosaccharides. The gums from the centrifuge are processed by treating with H_2O_2 and then dried. After reduction of moisture, the product is cooled. Thus, lecithin is obtained and used in foods and in cosmetic and pharmaceutical industries. The best-quality lecithin can be produced from fresh oil [8, 9].

Spent nickel catalyst is generated from hydrogenation process of refined vegetable oil during the production of Vanaspati (hard oil). The continuous use of nickel catalyst leads to high content of fat adherence to it [8].

Deodorizer distillate is the volatile organic material which is steam distilled in the deodorization of vegetable oil and hydrogenated vegetable oil. The distillate produced in the industry contains tocopherols and sterols, which are valuable raw materials for production of vitamin E and steroid drugs [8].

Wastewater treatment sludge consists of all chemical or biological sludges gathered from the bottom of sedimentation tanks of physicochemical and/or biological treatment. Biological sludge production is clearly located in the aeration basin and sludge is composed of minerals (10–30%), biomass (20–50%), and non-biomass organics (20–40%). A portion of the sludge is returned to the biological treatment processes while the excess sludge is sold as a fertilizer or stored in landfills. Chemical sludge or lime sludge generated during the process after dewatering of produced wastewater has beneficial reuses [8]. Although its contents vary depending on the chemicals used during the refining of oil, waste sludge from physicochemical treatment contains low-molecular-weight oils produced during refining of oils at the deodorization stage, esters, alcohols, and other semi-volatile organic compounds found in oil, which are organic substances produced during chemical adsorption of long-chain oil. After dewatering processes of this sludge, they can be stored in aboveground storage areas/landfills [80].

The following options can be applied to prevent and control solid wastes and by-products:

(a) Waste kernels should be collected for fuel for steam and power generation at refineries. The waste kernel cannot be incinerated by typical crude oil mills since, unlike refineries, their boilers are not designed to handle the high silica content of the kernel.
(b) Uncontaminated sludge and effluent from on-site wastewater treatment should be used as fertilizer for agricultural purposes.
(c) Contaminated sludge from wastewater treatment should be properly disposed at a sanitary landfill or by incineration. Incineration should only be applied in legalized facilities operating under international recognized standards for pollution prevention and control.
(d) Production processes should be carefully designed and controlled to reduce product losses; for example, air humidity should be monitored and adjusted to prevent product losses caused by the formation of molds on edible materials.
(e) Autoclave condensate should be recycled to remove vegetable oil.
(f) Spent bleaching earth can be disposed of as follows:

- A feedstock for brick, block, and cement manufacturing, and also as fertilizer, if not contaminated with heavy metals such as nickel, pesticide residues, and other contaminants.
- After anaerobic digestion they can be spread on land.
- If contaminated, they should be handled according to the waste management guidance.

(g) Distillates (e.g., free fatty acids and volatile organic compounds), depending on the level of pollutants (pesticides and/or residues) can be used as follows:

- Animal feed if uncontaminated
- A feedstock for chemical industry processes (e.g. antioxidants)
- Fuel for energy production

(h) The nickel catalyst from hydrogenation should be either:

- Recycled and recovered for reuse as a nickel catalyst or as nickel metal, salt, or other uses
- Stored and disposed of according to the waste management guidance

3.5 *Emissions to Air*

Particulate matter (dust) and volatile organic compounds (VOCs) are the main emissions from vegetable oil processing. Dust results from the processing of raw materials, including cleaning, screening, and crushing, whereas VOC emissions are

Table 8.8 Air emission levels for vegetable oil processing [12]

Parameter	Guideline value
Dust (mg/Nm^3)	10 (dry dust) 40 (wet dust)
Hexane/VOCs (mg/Nm^3)	100

caused by the use of oil-extraction solvents, typically hexane [16]. Table 8.8 presents emission guidelines for vegetable oil refining industry although it may change from location to location.

3.5.1 Volatile Organic Compounds

Solvent emissions result from various sources within vegetable oil processing plants, including the solvent-recovery unit, the dryer and cooler, and leaks in piping and vents. Small quantities of solvent may be present in the crude vegetable oil if the oil has been extracted by a solvent and will volatilize during the oil refining process, particularly during deodorization.

The suggested management strategies to prevent and control VOCs are listed below as follows:

(a) The efficient recovery of solvent by distillation of the oil from the extractor should be ensured.
(b) Solvent vapors should be recovered, where feasible, during vegetable oil extraction.
(c) Re-boiler and a gravity separator should be used to treat condensates with high solvent content in order to reduce solvent emissions and reduce the risk of explosions in the sewer.

The following techniques are recommended as management techniques to prevent and control dust and odors:

(a) Proper maintenance of cleaning, screening, and crushing equipment, including any ventilation and air handling systems, should be ensured to reduce emissions of fugitive dust.
(b) Cyclones and/or fabric filters or electrostatic precipitators should be installed on selected vents, including dryers, coolers, and grinders to remove odor emissions.
(c) Odor emissions (e.g., from soap splitting, cookers in the extraction process, vacuum systems, and pressurized systems) should be kept at a minimum by using a caustic, alkaline, or ozone scrubber system or the gas should be incinerated in a boiler plant or in separate incinerator system.

Table 8.9 Gas pollutants resulting from vegetable oil refining plants from different sources [12]

Parameter	Boiler	Generator
CO (mg/Nm^3)	40–130	2,060
CO_2 (mg/Nm^3)	2–205	0.8
NO_x (mg/Nm^3)	60–125	1,300
SO_2 (mg/Nm^3)	–	110
Smoke (Ringelmann scale)	–	5
Particulate matter (mg/Nm^3)	–	275

3.5.2 Exhaust Gases

Vegetable oil processing plants are large energy and steam consumers making use of auxiliary boilers for the generation of steam energy. Emissions related to the operation of these steam energy sources typically consist of combustion by-products such as NO_x, SO_x, PM, volatile organic compounds (VOCs), and greenhouse gases (viz., CO and CO_2). Table 8.9 shows gaseous pollutants resulting from vegetable oil refining plants and their typical quantities together with their source. Recommended management techniques include adoption of a combined strategy which includes a reduction in energy demand, use of cleaner fuels, and application of emissions controls where required.

Oxides of nitrogen (NO_x) are emitted from boiler stack and the generator exhaust. Continuous or intermittent exposure of humans to NO_x may cause certain illnesses, such as irritation in the respiratory tract and abnormal accumulation of fluid in the lungs leading to pulmonary edema. Direct exposure of NO_x to soil causes necrosis, vegetation loss, and inhibition in plant growth. NO_x is undergone various photochemical and chemical reactions in the atmosphere and leads to the formation of photochemical smog and acid rain. Although emissions of NO_x from generator are for a short period of time, still the cumulative effects of NO_x at global scenario should not be ignored. Depletion of ozone at stratosphere level and formation of photochemical smog and acid rain may occur due to this.

Sulfur oxides (SO_x) result from the burning of solid or liquid fuels in the generator. The severity of SO_x effects depends on its concentration and the duration of exposure. Direct exposure to these oxides can be very harmful to human health, plants, and vegetation. The harmful effects are dependent on the concentration and exposure duration. Photochemical smog and acid rain occur due to indirect reactions of SO_x in the atmosphere.

Carbon monoxide (CO) is a colorless non-irritating gas. It is generated due to incomplete combustion. At high concentrations exceeding 5,000 ppm with an exposure of few minutes, this gas can be fatal for human or animal lives, by reacting with hemoglobin to form carboxyhemoglobin. At much lesser concentrations, but with a high duration of exposure, this gas may still be dangerous for human beings, as it may cause damages to visual perception, manual dexterity, and the ability to learn. Concentrations of CO from the methane cracking plants and the generator exhausts of the audited mills are very high. Therefore, its long-term impacts cannot be ignored.

Carbon dioxide (CO_2) is generated in large quantities during natural gas cracking in edible oil mills. In some mills this is collected and sold to beverage industry, while in others this gas is exhausted into the atmosphere. According to laboratory results of gas cracking plant exhaust, the concentration of CO_2 in the exhaust is about 500 times of its concentration in clean air. CO_2 is a greenhouse gas and its higher concentration in the atmosphere is responsible for global warming

Particulate matter covering a large variety of particles varies in size and chemical composition. It has adverse effects on human respiratory system. It lowers the aesthetic value of a place as well as general visibility at high concentrations. The general corrosion reactions on building materials, due to the presence of NO_x and SO_x in the air, may also get catalyzed due to the presence of particulate matter [16].

4 Case Study

In this study, a vegetable oil factory (Kirlangic Oil Factory) was selected as representative of this sector and its effluent was investigated. The production capacity of Kirlangic Oil Factory Yarımca Facilities, Kocaeli, Turkey, is 40,000 tons/year for sunflower-corn oil and 15,000 tons/year for olive oil. Also, fatty acids are produced as a by-product at about 3,000 tons/year. Sunflower, corn, and olive oil undergo the full refining process, except the winterization process. Hydrogenation process is not applied.

4.1 Solid Wastes

Solid wastes are bleaching earth used in the bleaching process (bentonite, Tonsil, diatoms, etc.), perlite used in winterization process produced during the refining of crude oil, chemical and biological wastewater treatment plant sludge, scrap arising during the production of PVC plastics, tins, and paper and cardboard wastes.

Bleaching earth is used at the refining process by 1–2%. Bleaching earth includes color, glue, iron, nickel, and oxidized compounds [80]. According to the analysis carried out in 2002, the amounts of bleaching earth used during the processing of corn, sunflower, olive, and soybean oil were reported as 26,208 tons/year, 26,732 tons/year, 1,399 tons/year, and 1,842 tons/year, respectively. According to the same year data, the amount of oil found in bleaching earth was 14%. These results indicated that the amount of oil in the bleaching earth was at recoverable levels; hence, bleaching earth has been sold to the second-quality oil-producing companies [80].

Perlite is used in the factory during winterization. Perlite contains triglycerides in general. Also, perlite contains silicon oxide and sodium oxide due to its chemical nature. According to the analysis carried out in 2002, the amounts of perlite used during the processing of corn, sunflower, olive, and soybean were reported to as 21,840 tons/year, 26,732 tons/year, 1,105 tons/year, and 1,842 tons/year, respectively. According to the analysis conducted in 2002, the fat content of perlite was found as 25%. These results indicated that the amount of oil in the perlite was at recoverable levels; hence, perlite has been also sold to the second-quality oil-producing companies [80].

Sludge is produced in the WWTP of the factory during chemical and biological wastewater treatment; hence, it has biological and chemical properties. Chemical treatment sludge contains lime, iron(III) chloride and polyelectrolyte and low-molecular-weight oils, esters, alcohols, and other semi-volatile organic compounds found in oil, which are organic materials produced as a result of chemical adsorption of long-chain fats. Sludges from biological treatment contain microbial biomass known as activated sludge and organic matter. According to data obtained in 2002, sludge produced from chemical and biological wastewater treatment was about 140 tons/year and 60 tons/year, respectively, which were sold to the Gubretas Company [80].

Other plastic bottles and can scraps, paper, and cardboard along with other solid wastes are produced at different amounts in a year and they are stored temporarily/separately in the factory area. Then, they are sold to the recycling companies.

4.2 Factory and Gas Pollutant Emission Sources

The sources of emissions in the factory can be listed as boilers used for the production of steam, cooling towers used to cool the equipment, and machines used for the production of PVC. There are three steam boilers in operation. Two of them are located in the boiler room; these boilers are in operation and produce all the steam needed. One of the boilers works continuously while the other is a backup. The other boiler is used in the refining section for the production of heat and high-pressure steam needed for the removal of odor. In two of these four boilers, fuel oil grade 4 is used as fuel while in the boiler diesel fuel is used. Dry flue-gas flow measured at the chimney of the main and auxiliary boiler was 6,000 Nm^3/h, and that of the steam boiler was 300 Nm^3/h [80].

On the other hand, there were cooling tower emissions due to absorption of volatile organic compounds produced as a result of the oil deodorization process to the water used for the cooling of the equipment during refining. The flows of tower fan and dry flue gas were 66,350 m^3/h and 61,400 Nm^3/h, respectively [80].

The other emission sources of the factory are the chimneys of two machines used in the production of PVC bottles, which emit chemicals resulting from the decomposition of the additives in PVC due to the effect of heat and pressure. Emissions from factory chimneys include gaseous pollutants such as CO, CO_2, SO_2, NO_2, dust, oil and unsaturated fatty acids, saturated and unsaturated aldehydes, ketones (such as acetone), phenols, total nitrogen (as NH_3), total sulfur (as H_2S), and HCl [80].

At the factory, emissions such as CO, SO_2, NO_x, and dust burning result from boiler chimneys. Process dust emissions result from the two PVC machines. Organic materials in the form of unsaturated fats, fatty acids, aldehydes, ketones, and phenol emissions are only due to the cooling tower. Inorganic chlorine emissions result from the two PVC bottle production machines. Total nitrogen in the form of NH_3 and total sulfur in the form of H_2S result from the cooling tower [80].

4.3 *Wastewater Characterization*

Factory wastewaters consist of industrial wastewater and municipal wastewater each at about 60 m^3/d. Industrial wastewaters include wash water coming from segregators, wastewaters coming from deodorization, and extraction wastewaters resulting from soap-stock shredding unit. Domestic wastewater results from wastewaters coming from toilet, shower, and dining halls. Since deodorization unit is operated as closed loop, it discharges wastewater at certain times during the cleaning of the system. On the other hand, soap-stock and washing water are continuously produced and discharged to the treatment during the vegetable oil production. Also, a small amount of the wastewater during washing of filters used in production occurs. Filter washing is done at certain times of the year. The amounts of wastewaters produced at this factory from refining unit, soap-stock shredding, and wash water, deodorization unit, and other processes are 36, 10, and 14 m^3/d [80].

Aslan [20] carried out a study to characterize wastewater produced during the refining of corn oil and sunflower oil in Kirlangic Oil Factory. For this purpose, triplicate samples were taken from the inlet of equalization basin 1 at different times of production of both types of oil. Due to the completion of olive oil refining in very short time, wastewaters from olive oil refining are mixed with other wastewaters. For this reason, the characterization study was not carried for olive oil wastewaters. The characterization results of corn and sunflower oil wastewaters are presented in Tables 8.10 and 8.11.

4.4 *Analysis of Wastewater Treatment Plant Efficiency*

Wastewater volume at the treatment plant is 4 m^3/h. Emulsified oils, floating substances, and acid oils present in wastewater are physically separated from wastewater by grease traps and two submersible pumps connected in series with self-leveling pool balancing. During the chemical treatment, suspended and colloidal substances are removed from the wastewater by the addition of coagulants and flocculants. After the chemical treatment, wastewater is treated in a two-stage biological treatment system consisting of a facultative and a completely mixed extended aeration activated sludge system.

Table 8.10 Characterization of wastewater formed during corn oil refining at Kirlangic Oil Factory [80]

Parameter	Minimum	Maximum	Mean	Standard deviation
pH	2.72	2.85	2.79	0.07
COD (mg/L)	11,580	15,450	12,880	2225.7
Soluble COD (SCOD) (mg/L)	5140	6700	5680	884.7
BOD_5 (mg/L)	1250	2278	1932	591.2
Soluble BOD_5 ($SBOD_5$) (mg/L)	1100	1207	1135	61.78
TSS (mg/L)	1058	2990	2850	1077.3
TKN (mg/L)	1125	1458	1261	174.7
NH_3 (mg/L)	38	62	48	12.5
Total P (mg/L)	275	775	583	269.6
Oil-grease (mg/L)	307.8	498.6	375	107.2
Sulfate (mg/L)	11,900	12,300	12,006	257.2
Color (Pt-Co)	868	1700	1161	467.1

Table 8.11 Characterization of wastewater formed during sunflower oil refining at Kirlangic Oil Factory [80]

Parameter	Minimum	Maximum	Mean	Standard deviation
pH	3.5	4.5	4.01	0.5
COD (mg/L)	8,345	9,700	9,215	756.1
SCOD (mg/L)	5,195	5,560	5,385	182.8
BOD_5 (mg/L)	1,500	1,900	1,850	327.9
$SBOD_5$ (mg/L)	1,450	1,800	1,550	217.9
TSS (mg/L)	1,516	1,985	1,733	236.5
TKN (mg/L)	458.3	625	517.6	93.2
NH_3 (mg/L)	18	51	29.6	18.5
Total P (mg/L)	52	420	177.3	210.2
Oil-grease (mg/L)	533	760.2	664.3	117.6
Sulfate (mg/L)	10,800	11,750	11,416	534.1
Color (Pt-Co)	224	272	242	21.6

In order to determine, during the production of sunflower and corn oil which are produced at different times, the efficiency of the treatment, triplicate samples were taken from four different points as indicated in Fig. 8.3 and analyzed. Wastewater samples taken from points 1, 2, 3, and 4 represent the inlet of the wastewater treatment plant, the outlet of chemical treatment, and the outputs of the first stage (facultative treatment) and second stage of biological treatment (completely mixed aeration basin). Wastewater samples except for point 1 were the supernatant of each treatment stage taken at the outlet weir of the treatment unit [20]. Average values of the measured parameters and results of the analysis of wastewater samples are given in Tables 8.12 and 8.13. Sampling points were indicated in accordance with statements given the above. Soluble COD and soluble BOD_5 are shown as SCODs and $SBOD_5$.

Table 8.12 Analysis results of corn oil refining wastewater samples taken from all treatment steps of WWTP at Kirlangic Oil Factory [80]

Sampling point/parameter	1	2	3	4
pH	2.79	5.92	7.23	7.7
COD (mg/L)	12,880	8,015	1,705	225
SCOD (mg/L)	5,680	3,535	1,470	198
BOD_5 (mg/L)	1,932	1,375	620	27
$SBOD_5$ (mg/L)	1,135	783.5	290	19
TSS (mg/L)	2,850	1,074	129	127
TKN (mg/L)	1,261	564.5	250	228
NH_3 (mg/L)	48	84.5	15.5	1.6
Total P (mg/L)	583	3.65	2	1.4
Oil-grease (mg/L)	375	13.4	7.75	6.7
Sulfate (mg/L)	12,006	8,983	7,051	4,585
Color (Pt-Co)	1,161	821	683	151

Table 8.13 Analysis results of sunflower oil refining wastewater samples taken from all treatment steps of WWTP at Kirlangic Oil Factory [80]

Sampling point/parameter	1	2	3	4
pH	4.01	7.49	7.39	8
COD (mg/L)	9,215	5,047	1,770	265
SCOD (mg/L)	5,385	3,170	1,425	190
BOD_5 (mg/L)	1,850	1,095	220	44
$SBOD_5$ (mg/L)	1,550	960	51	18
TSS (mg/L)	1,733	267	261.6	152.6
TKN (mg/L)	517.6	270.8	395.8	215
NH_3 (mg/L)	29.6	10.5	11.5	11.4
Total P (mg/L)	177.3	7.15	4.1	1.7
Oil-grease (mg/L)	664.3	12.8	7.6	3.9
Sulfate (mg/L)	11,416	9,640	7,750	5,050
Color (Pt-Co)	242	186	180	125

As presented in Tables 8.10 and 8.11, pH values of the wastewater indicate that factory wastewaters resulted during the production of corn and sunflower oil refining have an acidic character. The pHs of corn and sunflower oil refining wastewaters were 2.79 and 4.1, respectively. Corn oil refining wastewater is more acidic because it has higher amount of fatty acids in crude oil of corn oil and, accordingly, higher production of soap-stock during the refining, hence, higher amount of acid use for the production of acid oil compared to sunflower oil.

It is clear from COD results of the plants that wastewater of the factory is heavily polluted. In corn oil refining wastewaters, total COD values were ranging from 11,581 to 15,449 mg/L, and in sunflower refining wastewaters, total COD values were ranging from 8,344 to 9,202 mg/L Also, the sulfate content of the wastewater is very high, 11,000–12,000 mg/L (Tables 8.10 and 8.11). The reason for this is due to that during vegetable oil refining effluents at neutralization stage of soap-stock are decomposed with sulfuric acid for the production of acid oil. The chemical equation for this reaction is given in Eq. 8.2. Phosphorus content of the wastewater is also high which is due to that crude oil has phosphorus compounds in its structures and due to the use of phosphoric acid at the degumming stage. It can be seen that wastewaters resulted during sunflower oil as well as corn oil refining are highly colored due to the extraction of oil from vegetables together with color pigments and the presence of natural organics. Vegetable oil refining wastewater has high color content because it contains a lot of dissolved organic matter. However, the color content of such wastewater has not been given much attention in the literature. It is observed that corn oil refining wastewaters have more polluting load and more acidic character when compared to sunflower oil refining wastewater.

When the results of are examined, it can be seen that BOD_5values are quite low when compared to corresponding COD values. This result indicates that there might be some hidden BOD which cannot be completely determined in 5 days and needs longer period for accurate measurement, and therefore, BOD_5 is inadequate in characterization of this kind of wastewaters.

The BOD_5/COD ratios of the factory wastewaters are 0.15 and 0.2 for corn oil and sunflower oil refining wastewaters [20]. Aslan et al. [53] investigated the biodegradability of vegetable oil refinery wastewaters and concluded that they have a low BOD_5/COD ratio and that these wastewaters cannot be treated only by biological treatment. Furthermore, their transport presents high risk of clogging in pumps and piping. In addition, COD/N/P (C/N/P) ratios of corn oil and sunflower oil refining effluents calculated from the average values of the total COD, TKN, and total phosphorus ratio were found as 52/3/1 and 22/3/1, respectively. These ratios indicate that there is no nutrient deficiency for wastewater microorganisms in terms of N and P. However, very low BOD_5/COD ratios suggest that only biological treatment is not enough for treatment of the factory wastewaters.

4.4.1 Caustic Neutralization and Physical Treatment

In soap-stock unit, wastewaters produced during soap-stock chopping in the reaction tanks, containing a strong acid (pH = 0–1), are taken into reaction tank 4 as shown in Fig. 8.8a. This tank is dosed with caustic soda from the top and the air is provided from the bottom of the tank for the reaction of caustic soda with wastewater to adjust desired pH. The wastewater is then sent to the pumping pit.

During the production at oil refining stage, wash water coming from separators reaches to the soap-stock unit after leaving the system. The pH of these waters is around 9–11. As shown in Fig. 8.8b, these waters are taken into tank 3 and sulfuric acid is added and heated by steam in order to facilitate the separation of oil and soap residues from water. The effluent of this tank is then directed to another tank, upgrade pit, and here, pH of the water is automatically adjusted to pH range of 3–5 by the addition of caustic soda. This operation is performed during the

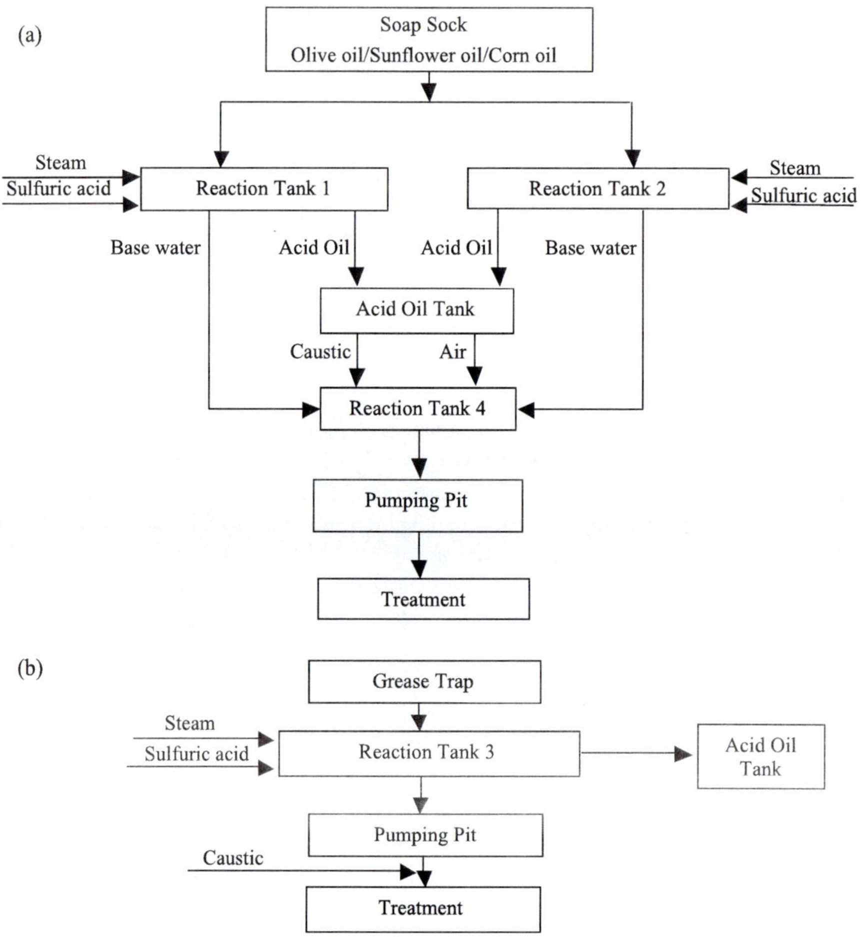

Fig. 8.8 (**a**) Refinery soap-stock processing. (**b**) Refinery wash water processing [80]

continuous stream of water. If the flow rate is appropriate, the pH adjustment is not required. If the pH of the water is not suitable, caustic soda may be added manually by the operators to adjust pH. pH-adjusted wash waters and soap-stock extraction unit wastewaters are taken into equalization basin 1 and 2. Here, oil-water separation from these wastewaters is achieved by screens and by extending hydraulic retention time of the huge volume of the wastewater.

In addition, by considering that the wash waters and rain waters of the factory field may also contain some oil residues, these waters are collected in transfer basins after passing them through grease trap placed at the output of the rain water collection channel. From there, they are taken into either equalization basin or aeration tank. Acid oils and fatty substances are separated from the wastewater by the help of the screens placed in two equalization basins and these substances pumped back to acid oil and reaction tanks as accumulated over time.

4.4.2 Chemical Treatment

Wastewaters that have undergone physical treatment become ready for chemical treatment. In such wastewaters, pollutants are mostly suspended and colloidal organic and inorganic substances and to a lesser degree some oil particles. Most of the suspended particles possess a negative charge and, since they have the same type of surface charge, repel each other when they come close together. Therefore, they will remain in suspension rather than clump together and settle out of the water. The amount of oil present in the wastewater makes this situation even worse.

Chemical treatment methods applied to the system involve the addition of iron (III) chloride ($FeCl_3$), anionic polyelectrolyte chemicals, and limes to the wastewater, so that they stick together and form heavy clusters (flocs) in a polymerized hydrated form. The precipitation of these flocs provides removal of them from wastewater. Thus, the COD of the wastewater is targeted to be at the lowest level before biological treatment.

In WWTP of Kirlangic Oil Factory, wastewaters are taken into the first rapid-mix chamber, wherein $FeCl_3$ purchased as a 40% solution is dosed as 10% solution. Thus, colloidal substances present in wastewater interact with $FeCl_3$ and form positively charged $Fe(OH)_3$. Then, the wastewater is taken to second rapid-mix chamber where neutralization is done by lime dosing to form more stable flocs, wherein the pH of the wastewater is brought to the range of 7–8 by the addition of 5% lime slurry. After rapid mix, wastewater containing stable flocs is taken into slow-mixing tank. In this tank, the microflocs are brought into contact with each other through the process of slow mixing. Collisions of the microfloc particles cause them to bond to produce larger, visible flocs called pinflocs. The floc size continues to build up through additional collisions and interaction with inorganic polymers formed by the coagulant or with organic polymers, 0.002% of the anionic polyelectrolyte, added. Macroflocs are formed. After this process, wastewater is directed to the chemical precipitation basin where formed macroflocs are settled down at the bottom of the basin. Also, foams formed on the surface of the basin during chemical precipitation are skimmed by scrapers and sent to the equalization basin. The effluent free from flocs is taken into CTP tank for final pH adjustment. There, the pH of the wastewater is brought to the suitable range for biological treatment (pH 7–9).

As known, iron(III) chloride has an acidic character. The studies also showed that it drops pH to about two units. Therefore, the necessity of pH adjustment range to ensure the efficiency of treatment requires using of lime in high amounts which causes increased levels of sludge formation. Due to the good quality of the wastewater produced in soap-stock unit and the long retention time of the wastewater in the equalization basin at the first stage of treatment that resulted in a clear wastewater free from most of the colloidal substances, the use of lime in chemical treatment is deemed appropriate both economically and in terms of creating less sludge. Therefore, in the present system, only the lime has been used as a coagulant.

4.4.3 Biological Treatment

Wastewater undergoes biological treatment after chemical treatment. Domestic wastewater of the factory is also taken directly to the biological treatment. Urea and diammonium are added to the wastewater to ensure there is a sufficient amount of nitrogen and phosphorus phosphate which is necessary for microorganisms in biological treatment. Carbon source is the organic matter present in the wastewater. Biological treatment is carried out in two stages: facultative and completely mixed extended aeration tanks consisting of two independent compartments are served for this purpose. In the first stage, some portion of the biological and chemical load of the wastewater is removed in the facultative aeration tank by aerobic microorganisms present at the upper level of the tank to which oxygen is provided with the help of the surface aerators and by anaerobic microorganisms present at the lower/bottom level of the tank. Then, wastewater is taken into the first biological clarification basin and the biomass is removed from wastewater by settling. The supernatant of partially treated wastewater is directed from this tank to the second completely mixed tank. The concentration of biomass settled as the waste sludge is measured frequently and is recycled to the facultative aeration pond. The excess sludge is removed from the system when the biomass concentration increased.

In the second stage, the organic pollution load of wastewater is removed by aerobic microorganisms in a completely mixed aerated tank which is mixed by diffused air provided by the surface and submerged aerators placed at the top and bottom of the tank, respectively. The mixture of wastewater and activated sludge is sent to final sedimentation tank, and the excess waste activated sludge settled to the bottom of the tank is removed while a part of it is recycled back to the aeration tank. After sedimentation pool of clear supernatant is discharged. There is a sand filtration unit at the WWTP before clear supernatant storage tank, but in the current case it is not needed to be used.

4.4.4 Sludge Removal and Dewatering Process

Sludge produced in this WWTP consisted of two different characteristics: chemical and biological sludge. Sludge coming from chemical treatment was stored in the sludge thickener for a while for thickening and settlement purposes. Supernatant above the thickened sludge is recycled back to the chemical treatment while bottom thickened sludge is taken into bags wherein the sludge is more concentrated. The excess sludge coming from biological treatment is taken into sludge drying bed which is a kind of big tank. Sludge dried there is sold to a fertilizer company together with the sludge of chemical treatment at certain times. Treatment plant flow diagram and the sampling points taken for wastewater characterization are shown in Fig. 8.9. Table 8.14 shows the sizes of the tanks and basins in the treatment plant.

According to the results of that study, corn oil refining wastewater treatment efficiency was 98.1%, 98.5%, 98.2%, 95.47%, and 86.8% for total COD, BOD_5,

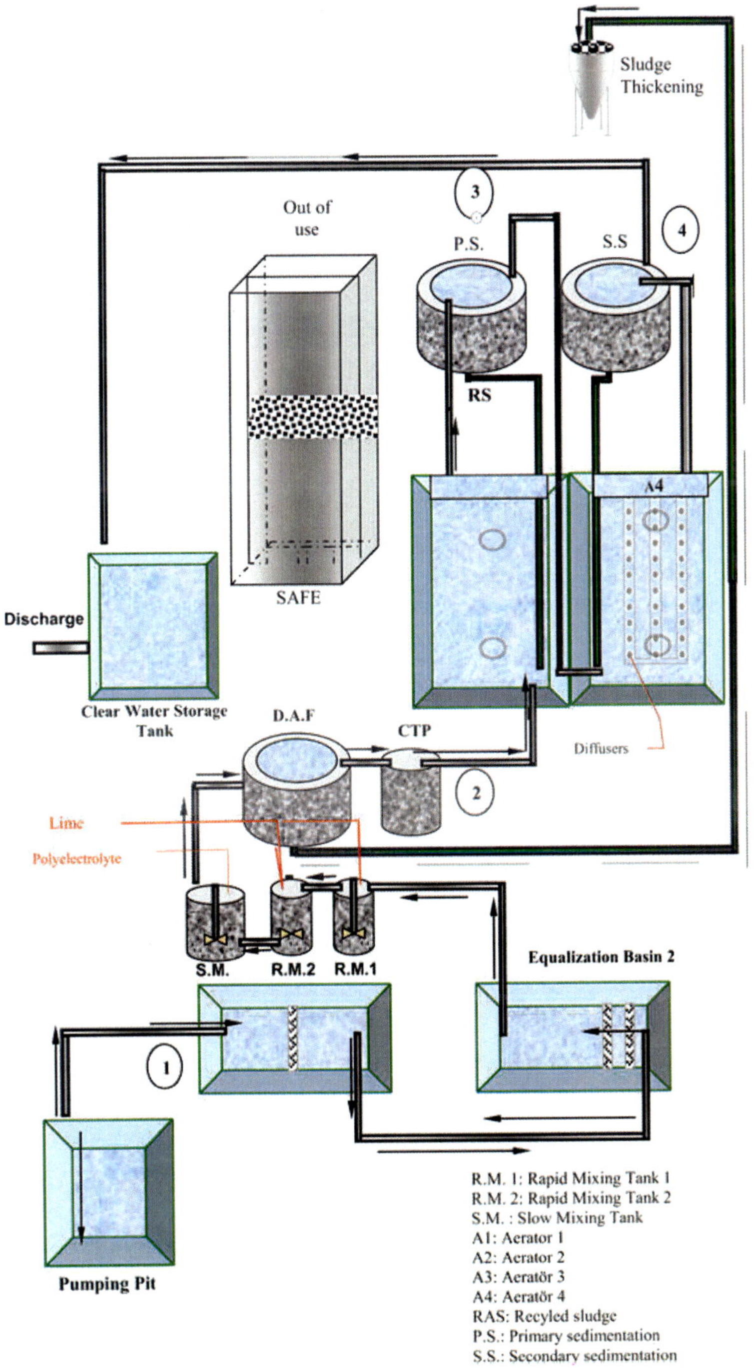

Fig. 8.9 Kirlangic Oil Factory WWTP flow chart [80]

Table 8.14 The sizes of the tanks and basins in WWTP [80]

	Equalization basin							
	(2)			Equalization basin			Aeration tank	
Tanks	A1		A2	(1)			1st compartment	
Depth (m)	3.50		3.25	3.60			4.00	
Width (m)	4.50		4.50	3.00			5.00	
Length (m)	5.23		9.25	7.00			10.50	
Volume (m^3)	82		135	76			210	
1-cm water height volume (*L*)	240		415	210			525	
Tanks	Rapid mixing	Slow mixing	DAF	CTP	Polyelectrolytes	Fe $(Cl)_3$	Lime	Sedimentation basin 1 and 2
Diameter (m)	1.10	1.90	3.0	2.0	1.0	1.00	1.50	4.0
Depth (m)	1.50	1.50	3.0	2.0	1.0	1.20	1.50	4.0
Volume (m^3)	1.40	2.80	21.0	6.0	0.78	0.94	2.50	50.0
1-cm water height volume (*L*)	13.0	15.0	70.0	30.0	7.85	7.80	17.0	125.0

oil-grease, TSS, and color removal, respectively, while treatment efficiency for total COD, BOD_5, oil-grease, TSS, and color removal was 97%, 97.4%, 99.27%, 91%, and 48.2% for sunflower oil refining wastewater. It is clear from the findings that the concentrations of all wastewater pollutants parameters were decreased at all levels of the treatment plant. The study data reveals that the degree of chemical treatment itself is not efficient enough for the removal of such pollutants. For this reason, performing only chemical treatment for edible oil refining wastewater is not enough; the treatment system must be considered in conjunction with biological treatment.

The wastewater treatment plant of the factory operates very efficiently and it is a good model for the treatment of vegetable edible oil refinery wastewaters. However, since the MLSS concentration of wastewaters discharged from the plant is higher than the discharge limits, it is necessary to re-operate sand filter.

Glossary [81–83]

Aerobic An environmental condition in which free and dissolved oxygen is available in an aqueous environment (for instance, nitrification is an aerobic process).

Anaerobic (a) An environmental condition in which free, dissolved, and combined oxygen is unavailable in an aqueous environment (i.e., dissolved oxygen = 0 mg/L). (b) A condition in which atmospheric or dissolved molecular oxygen is *NOT* present in the aquatic (water) environment.

Coagulation (a) The clumping together of very fine particles into large particles (floc) caused by the use of chemicals (coagulants). (b) A process of destabilizing charges of suspended and colloidal particles in water by adding chemicals (coagulants). In coagulation process, positively charged chemicals are added to neutralize or destabilize these negative charges and allow the neutralized particles to accumulate and be removed by clarification (flotation or sedimentation) and/or filtration.

Dissolved air flotation (DAF) (a) A method of solids separation, whereby a side stream is saturated with air at high pressure and then injected into the flotation tank to mix with the incoming water stream. As the air bubbles rise to the surface, they attach to floc particles and create a sludge layer at the surface of the tank, which is then removed for disposal. (b) One of the dissolved gas flotation (DGF) processes when air is used for generation of gas bubbles. A typical example is Krofta Engineering Corporation's Supracell clarifier; see dissolved gas flotation (DGF).

Dissolved gas flotation (DGF) It is a process involving pressurization of gas at 25–95 psig for dissolving gas into water and subsequent release of pressure (to 1 atm) under laminar flow hydraulic conditions for generating extremely fine gas bubbles (20–80 microns) which become attached to the impurities to be removed and rise to the water surface together. The impurities or pollutants to be removed that are on the water surface are called float or scum which are scooped off by sludge collection means. The clarified water is discharged from the flotation clarifier's bottom. The gas flow rate is about 1% of influent liquid flow rate. The attachment of gas bubbles to the impurities can be a result of physical entrapment, electrochemical attraction, surface adsorption, and/or gas stripping. The specific gravity of the bubble-impurity agglomerate is less than one, resulting in buoyancy or non-selective flotation (i.e., Save-All).

Electroflotation It is process involving the generation of hydrogen and oxygen bubbles in a dilute electrolytic aqueous solution by passing a direct current between two electrodes: (a) anode and (b) cathode. Anode reaction generates oxygen bubbles and hydrogen ions, while cathode reaction generates hydrogen bubbles and hydroxide ions. Either aluminum or steel sacrificial electrodes can be employed for generating the gas bubbles as well as coagulants at the same time. Non-sacrificial electrodes are employed for generating the gas bubbles only and can be made of titanium (as the carrier material) and lead dioxide (as the coating material). Electrical power is supplied to the electrodes at a low voltage potential of 5–20 VDC by means of a transformer rectifier. Small bubbles in the range of 20–50 m microns are produced under laminar hydraulic flow conditions feasible for flotation separation of fragile flocs from water in a small system. The floats on the water surface are the impurities/pollutants removed from water. The clarified water is discharged from the flotation clarifier's bottom. There can be unexpected advantages and disadvantages when electroflotation is employed. For instance, chlorine bubbles may be generated as a water disinfectant if the water contains significant amount of chloride ions. Certain unexpected gas bubbles may be generated and may be undesirable.

Filtration It is usually a granular medial filtration process which involves the passage of wastewater or water through a bed of filter media with resulting deposition of suspended solids. Eventually the pressure drop across the bed becomes excessive or the ability of the bed to remove suspended solids is impaired. Cleaning is then necessary to restore operating head and effluent quality. The time in service between cleanings is termed the filter run time or run length. The head loss at which filtration is interrupted for cleaning is called the terminal head loss, and this head loss is maximized by the judicious choice of media sizes. Dual media filtration involves the use of both sand and anthracite as filter media, with anthracite being placed on top of the sand. Gravity filters operate by either using the available head from the previous treatment unit or by pumping to a flow split box after which the wastewater flows by gravity to the filter cells. Pressure filters utilize pumping to increase the available head. A filter unit generally consists of a containing vessel; the filter media; structures to support the media; distribution and collection devices for filter influent, effluent, and backwash water flows; supplemental cleaning devices; and necessary controls for flows, water levels, and backwash sequencing. Backwash sequences can include air scour or surface wash steps. Backwash water can be stored separately or in chambers that are integral parts of the filter unit. Backwash water can be pumped through the unit or can be supplied through gravity head tanks.

Induced air flotation (IAF) It is one of the induced gas flotation processes in which the gas is air.

Refinery A refinery is a production facility composed of a group of chemical engineering unit processes and unit operations refining certain materials or converting raw material into products of value.

Refining waste A refining waste or refinery waste is the waste or wastewater from a refinery.

Vacuum flotation In vacuum flotation, the influent process water to be treated is usually almost saturated with air at atmospheric pressure. There is an air-tight enclosure on the top of the flotation chamber in which partial vacuum is maintained. The fine air bubbles (20–80 microns) are generated under laminar hydraulic flow conditions by applying a vacuum (negative pressure) to the flotation chamber. The theory is that the lower the pressure, the lower the air solubility in water. The soluble air originally in water is partially released out of solution as extremely fine bubbles due to a reduction in air solubility caused by negative vacuum pressure. The bubbles and the attached solid particles rise to the water surface to form a scum blanket, which can be removed by a continuous scooping or skimming mechanism. Grit and other heavy solids that settle to the bottom are raked to a central sludge sump for removal. Auxiliary equipment includes an aeration tank for saturating the water or wastewater with air, vacuum pumps, and sludge pumps.

References

1. Bucas G, Saliot A (2002) Sea transport of animal and vegetable oils and its environmental consequences. Mar Pollut Bull 44(12):1388–1396
2. Bennion M (1995) Introductory foods, 10th edn. Prentice-Hall Inc., Upper Saddle River
3. U.S. EPA (2001) Economic impact analysis for the proposed vegetable oil processing NESHAP. Report #452/R-01-005. U.S. Environmental Protection Agency, Research Triangle Park
4. U.S. EPA (1976) Treatment of effluent waters from vegetable oil refining. Report # 600/2-76-294. U.S. Environmental Protection Agency, Cincinnati
5. Patwardhan AD (2008) Industrial waste water treatment. PHI Learning, New Delhi
6. Pandey RA, Sanyal PB, Chattopadhyay N, Kaul SN (2003) Treatment and reuse of wastes of a vegetable oil refinery. Res Conserv Recycl 37:101–117
7. Wiedermann LH (1981) Degumming, refining and bleaching soybean oil. J Am Oil Chem Soc 58:159–165
8. O'Brien RD (1998) Fats and oils. Formulating and processing applications. Technomic Publishing, Lancaster
9. Wolf H, Hamilton R (2000) Edible oil processing. Sheffield Academic Press, Sheffield, pp 1–275
10. Ceriani R, Meirelles AJA (2006) Simulation of continuous physical refiners for edible oil deacidification. J Food Eng 76:261–271
11. European Commission (2006) Integrated pollution prevention and control. Reference Document on Best Available Techniques Food, Drink and Milk Industries Full text is at: http://eippcb.jrc.es/reference/fdm.html (Visited in 2013)

12. Kale V, Katikaneni SSP, Cheryan M (1999) Deacidification of ricebran oil by solvent extraction and membrane technology. J Am Oil Chem Soc 76(6):723
13. Azbar N, Yonar T (2004) Comparative evaluation of a laboratory and full-scale treatment alternatives for the vegetable oil refining industry wastewater (VORW). Process Biochem 39:869–875
14. IFC (2007) Environmental, health, and safety guidelines for vegetable oil processing. World Bank Group. Full text is at: http://www1.ifc.org/wps/wcm/connect/83b8628048855b84893cdb6a6515bb18/Final%2B-%2BVegetable%2BOil%2BProcessing.pdf?MOD=AJPERES (Visited in 2013)
15. Decloux M, Lameloise ML, Brocard A, Bisson E, Parmentier M, Spiraers A (2007) Treatment of acidic wastewater arising from the refining of vegetable oil by crossflow microfiltration at very low transmembrane pressure. Process Biochem 42(4):693
16. Environmental Technology Program for Industry (1999) The edible oil& ghee sector environmental report. The Federation of Pakistan Chambers of Commerce&Industry, Pakistan, pp 1–24
17. Chipasa KB (2001) Limits of physicochemical treatment of wastewater in the vegetable oil refining industry. Pol J Environ Stud 10(3):141–147
18. Pathe PP, Nandy T, Kaul SN (1999) Wastewater management in vegetable oil industries. Indian J Environ Prot 27:481–492
19. Saatci Y, Hasar H, Cici M (2001) Treatability of vegetable oil industry effluents through physical-chemical methods. Fresenius Environ Bull 10(12):854–858
20. Aslan S (2003) Characterization, COD fractions of an edible oil plant wastewaters and investigations of current wastewater treatment plant: MSc Thesis. Kocaeli University, Kocaeli
21. Basu AK (1967) Treatment of effluents from the manufacture of soap and hydrogenated vegetable oil. J WPCF 39:1653–1657
22. Chin KK, Ng WJ, Ma AN (1987) Palm oil refinery effluent treatment by sequencing batch reactors. Biol Wastes 20:101–109
23. Sengul F, Filibeli A, Karagunduz A (1992) Cotton oil industry wastewater characteristics and treatment. In: 3rd Symposium on Industrial Pollution by Istanbul Technical University (ITU), İstanbul, Turkey, pp 121–137
24. Güvezne E (1984) Industrial wastewater treatment: Turyag case. In: Çevre'84, Turkish-German Environmental Engineering Symposium by TUBITAK, İzmir, pp 29–33
25. Öztürk İ, Şan H, Eroğlu V (1989) Pilot and full scale treatability studies on wastewaters from edible oil refining industry. In: Proceedings of the 44th Industrial Waste Conference'89. Purdue University, West Lafayette, pp 769–777
26. Pintor AMA, Vilar VJP, Botelho CMS, Boaventura RAR (2014) Optimization of a primary gravity separation treatment for vegetable oil refinery wastewaters. Clean Techn Environ Policy 16:1725–1734
27. Zabel T (1992) Flotation in water treatment. In: Mavros P, Matis KA (eds) Innovations in flotation technology. Kluwer Academic Publishers, Dordrecht
28. Reynolds TD (1982) Unit operations and process in environmental engineering. Wadswort Inc., Belmont, pp 512–517
29. Metcalf & Eddy (1991) Wastewater engineering: treatment, disposal, and reuse. McGraw-Hill, New York, pp 243–248, 486, 804–805
30. Matis KA, Zouboulis AI (1995) In: Mathis KA (ed) Flotation science and engineering. Marcel Dekker, Inc., New York, p 385
31. Zheng YY, Zhao CC (1993) A study of kinetics on induced-air flotation for oil-water separation. Separ Purif Technol 28(5):1233–1240
32. Bennett GF (1988) The removal of oil from wastewater by air flotation: a review. CRC Crit Rev Environ Contr 18(3):189–253
33. Kiuri HJ (2001) Development of dissolved air flotation technology from the 1st generation to the newest 3rd one (very thick microbubbles) with high flow-rates (DAF in turbulent flow conditions). Water Sci Technol 8:1–8

34. Sanchez-Calvo L, Tanguy G, Cames MC, Paternotte G, Valentin G, Rostan A, Lapicque P (2003) An electrocoagulation unit for the purification of soluble oil wastes of high COD. Environ Prog 22(1):57–65
35. Mostefa NM, Tri M (2004) Coupling flocculation with electroflotation for waste oil/water emulsion treatment: optimization of the operation conditions. Desalination 161:115–121
36. Xu X, Zhu X (2004) Treatment of refectory oily wastewater by electro-coagulation process. Chemosphere 56:889–894
37. Yang CL (2007) Electrochemical coagulation for oily water demulsification. Separ Purif Technol 54:388–395
38. Chen G (2004) Electrochemical technologies inwastewater treatment. Separ Purif Technol 38 (1):11–41
39. Larue O, Vorobiev E, Vu C, Durand B (2003) Electrocoagulation and coagulation by iron of latex particles in aqueous suspensions. Separ Purif Technol 31(2):177–192
40. Ibanez JG, Takimoto MM, Vasquez RC, Basak S, Myung N, Rajeshwer K (1995) Laboratory experiments on electrochemical remediation of the environment: electrocoagulation of oilywastewater. J Chem Educ 72(11):1050–1058
41. Yilmaz T, Yüceer A (1997) Physico-chemical treatability of vegetable oil industry wastewater. In: 2nd Symposium on Environmental Pollution Priorities in Turkey. Gebze Institute of Advanced Technology, Istanbul, pp 215–221
42. Boisvert JP, To, T.C, Berrak A, Julicocur C (1997) Phosphate adsorption in flocculation processes of aluminium sulphate and poly-aluminium-silicate-sulphate. Water Res 31:1937–1946
43. Okuda T, Baes AU, Nishijima W, Okada M (1999) Improvement of extraction method of coagulation active components from Moringa oleifera seed. Water Res 33(15):3373–3378
44. Louhıchı G, Bousselmı L, Ghrabı A, Khounı I (2019) Process optimization via response surface methodology in the physico-chemical treatment of vegetable oil refinery wastewater. Environ Sci Pollut R 26:18993–19011
45. Ahmed Z (2001) Effect of iron compounds on the anaerobic treatment of fat-containing wastewaters. Cleaner Production Program, Pakistan
46. Nunn WD (1986) A molecular view of fatty acid catabolism in *Eschericia coli*. Microbiol Rev 50:179–192
47. Ratledge C (1992) Microbial oxidations of fatty alcohols and fatty acids. J Chem Technol Biotechnol 55:399–400
48. Lalman JA, Bagley DM (2000) Anaerobic degradation and inhibitory effects of linoleic acid. Water Res 34:4220–4228
49. Tezcan Un U, Koparal AS, Bakir Ogutveren U (2009) Electrocoagulation of vegetable oil refinery wastewater using aluminum electrodes. J Environ Manag 90:428–433
50. Chian ESK, Dewalle FB (1977) Treatment of high strength acidic wastewater with completely mixed anaerobic filter. Water Res 11:295
51. Rao DG, Meyyapan N, Feroz S (2012) Wastewater treatment: advanced processes and technologies. CRC Press, New York, p 264
52. Campo P, Zhao Y, Suidan M, Venosa A (2012) Aerobic fate and impact of canola oil in aquatic media. Clean Techn Environ Policy 14(1):125–132
53. Aslan S, Alyuz B, Bozkurt Z, Bakaoğlu M (2009) Characterization and biological treatability of edible oil wastewaters. Pol J Environ Stud 18(4):533–538
54. Davis M, Cornwell D (1991) Introduction to environmental engineering, 2nd edn. McGraw-Hill International, New York
55. Stolting B, Bolle WF (2000, 13–14, 29–30 April) Treatment processes for liquid and solid waste from olive oil production. In Proceedings of Workshop "Improlive-2000", (Annex A1), Seville, Spain
56. Gray NF (2005) Chapter 19: Anaerobic treatment. In: Water technology: an introduction for environmental scientists and engineers, 2nd edn. Boston Elsevier, Oxford

57. Müller JA (2001) Prospects and problems of sludge pre-treatment processes. Water Sci Technol 44(10):121–128
58. Speece RE (1996) Anaerobic biotechnology for industrial wastewater. Archae Press, Nashville
59. Hemmati M, Rekabdar F, Gheshlaghi A, Salahi A, Mohammadi T (2012) Effects of air sparging, cross flow velocity and pressure on permeation flux enhancement in industrial oily wastewater treatment using microfiltration. Desalin Water Treat 39:33–40
60. Wu C, Li A, Li L, Zhang L, Wang H, Qi X, Zhang Q (2008) Treatment of oily water by a poly (vinyl alcohol) ultrafiltration membrane. Desalination 225:312–321
61. Hafidi A, Pioch D, Teyssier ML, Ajana H (2004) Influence of oil conditioning on the permeate flux and cake properties during microfiltration of lampante olive oil. Eur J Lipid Sci Technol 106:152–159
62. Cheryan M, Rajagopalan N (1998) Membrane processing of oily streams, wastewater treatment and waste reduction. J Membrane Sci 151:13–28
63. Cañizares P, Pérez Á, Camarillo R (2002) Recovery of heavy metals by means of ultrafiltration with water-soluble polymers: calculation of design parameters. Desalination 144:279–285
64. Masoudnia K, Raisi A, Aroujalian A, Fathizadeh M (2015) A hybrid microfiltration/ultrafiltration membrane process for treatment of oily wastewater. Desalin WaterTreat 55:901–912
65. Gryta M, Karakulski K, Morawski AW (2001) Purification of oily wastewater by hybrid UF/MD. Water Res 35:3665–3669
66. Yan L, Li YS, Xiang CB, Xianda S (2006) Effect of nano-sized A12O3-particle addition on PVDF ultrafiltration membrane performance. J Membrane Sci 276:162–167
67. Yan L, Hong S, Li ML, Li YS (2009) Application of the A12O3–PVDF nanocomposite tubular ultrafiltration (UF) membrane for oily wastewater treatment and its antifouling research. Sep Purif Technol 66:347–352
68. Kaur H, Bulasara VK, Gupta RK (2018) Influence of pH and temperature of dip-coating solution on the properties of cellulose acetate-ceramic composite membrane for ultrafiltration. Carbohydr Polym 195:613–621
69. Parma S, Chowdhury P (2014) Preparation and characterization of microfiltration ceramic membrane for oily waste water treatment. Int J ResEng Technol 3:725–730
70. Leiknes T (2009) The effect of coupling coagulation and flocculation with membrane filtration in water treatment: A review. J Environ Sci 21(1):8–12
71. Schwarze M (2017) Micellar-enhanced ultrafiltration (MEUF)–state of the art. Environ Sci Water Res Technol 3:598–624
72. Choo KH, Choi SJ, Hwang ED (2007) Effect of coagulant types on textile wastewater reclamation in a combined coagulation/ultrafiltration system. Desalination 202:262–270
73. Gao WJ, Leung KT, Qin WS, Liao BQ (2011) Effects of temperature and temperature shock on the performance and microbial community structure of a submerged anaerobic membrane bioreactor. Bioresour Technol 102:8733–8740
74. Arhin SG, Banadda N, Komakech AJ, Pronk W, Marks SJ (2017) Optimization of hybrid coagulation-ultrafiltration process for potable water treatment using response surface methodology. Water Sci Tech 18:862–874
75. Xiangli Q, Zhenjia Z, Nongcun W, Wee V, Low M, Loh CS, Hing NT (2008) Coagulation pretreatment for a large-scale ultrafiltration process treating water from the Taihu River. Desalination 230:305–313
76. Khouni I, Louhichi G, Ghrabi A, Moulin P (2020) Efficiency of a coagulation/flocculation–membrane filtration hybrid process for the treatment of vegetable oil refinery wastewater for safe reuse and recovery. Process Saf Environ Prot 135:323–341
77. Soystats (2007) World vegetable oil consumption 2007. American Soybean Association, St. Louis, MO, USA
78. Werner Z (1994) Spent bleaching earth-practical solutions. Inform 5:1375
79. Taylor DR, Jenkins DB (1991) Acid-activated clays, Society of Mining Engineers of AIME. Transactions 282:1901–1910

80. KOU (2002) Kirlangic oil factory solid waste and process analysis report. Environmental Engineering Department of Kocaeli University, Izmit, Turkey
81. Wang LK, Wang MHS, Hung YT, Shammas NK (eds) (2021) Environmental and resources engineering. Springer Nature Switzerland, 525 p
82. Wang LK, Wang MHS, Shammas NK, Aulenbach DB (eds) (2021) Environmental flotation engineering. Springer Nature Switzerland, 438 p
83. Wang LK, Wang MHS, Hung YT, Shammas NK (eds) (2021) Integrated natural resources management. Springer Nature Switzerland, 456 p

Chapter 9
Environmental Engineering Education in the Modern Era

Henry R. Bungay III

Contents

1 Introduction 378
2 Implementation 379
3 Presentation Skills 380
4 Information Retrieval 382
5 Digression 384
6 Conclusion 384
Glossary 385
References 386

Abstract Environmental engineering education is enhanced by the internet with web pages full of information, instructions, photos, and videos of installations and equipment, and with means of communication that facilitate alerts to new developments. Students who have grown up with years of using the internet have different expectations and computer sophistication that are yet to be exploited fully at our colleges and universities. Some reasonable modifications of the undergraduate curriculum can improve matters.

Keywords Engineering education · Environmental engineering · Students as teachers · Environmental curriculum · Technical presentations · Interactive web pages · Personalized books

---

H. R. Bungay III (✉)
H. P. Isermann Department of Chemical and Biological Engineering, Rensselaer Polytechnic Institute, Troy, NY, USA

L. K. Wang, M. -H. S. Wang, Y. -T. Hung, N. K. Shammas (eds.), *Environmental and Natural Resources Engineering*, Handbook of Environmental Engineering 19, https://doi.org/10.1007/978-3-030-54626-7_9

1 Introduction

There are dozens and perhaps hundreds of published articles about the state of engineering education and about how to improve it. The potential for enhancing engineering education with the internet was appreciated long ago [1], and Van Ness claimed that we may never get it right because education evolves [2].

Pollution is a necessary evil. As an index of industrial health and of lifestyle for individuals, creating waste is good, and environmental engineers are at the forefront for cleaning up the messes. Environmentalists, on the other hand, decry pollution and scare the public with images of dead fish from a waterway that was mistreated, with tales of air pollution disasters, with accounts of destruction that may last forever from radioactive leakage, toxic landfills, and the like. Environmental engineers are much more likely to appreciate cost/benefit analyses and compromises for acceptable environmental impact.

With no disrespect to education for environmental awareness, better habits for the general public such as not flushing drugs down toilets, better recycling, less careless damage to our environment, and conservation are not enough. In addition to better citizenship, there must be engineers and technicians for tasks such as planning, designing, constructing, and operating what is needed for managing and disposing of wastes, developing more cost-effective treatments of potable water, and enforcing environmental regulations.

Other engineering disciplines, notably chemical engineering, encroach on environmental engineering and have sometimes outperformed us in areas such as industrial waste treatment. However, the key difference is evident in the term *environmental*. For the chemical engineer, a high percentage of projects are indoors, while much of environmental engineering is outdoors, a vast environment. Other disciplines target profits that can pay for expensive equipment. Very few waste treatment facilities generate any profit, and low technology is mandatory.

To train environmental engineers, we must introduce the proper mindset. Environmental engineers must shun expensive equipment such a vessels of stainless steel, cannot control the temperatures of high-volume waste treatment process, must accept wastes that often come as pulses or have diurnal and seasonal variations, cannot afford highly skilled plant operators, and will not have the financial resources to do it right. Because perfect treatment would be impossibly expensive, we set targets for percentages of pollutants that must be removed only to find that some governmental agency changes the rules to make reaching these goals more expensive.

When the goal is treatment of ordinary water and waste water, polluted air, radioactive or toxic materials, and solid waste, the clear choice is a conventional environmental engineer who understands how to satisfy regulations, because departure from the norm risks lawsuits should the installation fail to operate properly. A method that follows established practices is less likely to face lawsuits or prosecution when it fails to work well. Off-the-shelf technology will perform as expected or maybe a little better or a little worse.

Those trained in the traditional manner have transformed our civilization with advances for drinking water, for air pollution, for sewage treatment, and for disposal of solid wastes. We understand the interactions whereby burning residues of wastes creates air pollution, how particles collected from polluted air can be hazardous waste, and how treatment that was once routine and acceptable can be damned because of new knowledge or for vulnerability to disasters such as floods. Those environmental engineers who address public health challenges have improved many more lives than have other engineering disciplines.

Tinkering with waste treatment for decades has optimized performance of ordinary processes to the point where major new gains are unlikely. Along the way, a few significant advances have become standard practice, for example, collection of solid wastes with lifting and tilting of containers at street curbs. Unfortunately, the challenges become more daunting, while potential improvements are stagnant.

2 Implementation

While a convincing case can be made for major changes in environmental engineering education, the hurdles of obtaining approval and financial recourses from the university administration are formidable and may take years to implement. Immediate improvements can come with little or no cost by simply changing the required courses. Here are some examples:

Freshman chemistry. There are often special sections for students majoring in chemistry or chemical engineering. Some extra rigor and competition with brighter students than those in the other sections of general chemistry would benefit environmental engineers.

Mass and Energy Balances: This tends to be an absolute jewel of a course for sophomores majoring in chemical engineering everywhere. The instructors tend to be excellent, and the problems all have a practical bent. Some problems have environmental context, especially those concerned with humidity and combustion. Students learn to think as an engineer. While difficult, a student is rewarded with new skills and pride in solving interesting problems. In its present state, it would be excellent for environmental engineering students and could be almost perfect when more of the assigned problems were for environmental situations.

Chemical Process Control: This can be taken in the junior or senior year. An obvious way to improve waste treatment is with better process control. For a few years, this author managed to get a course in chemical process control required for our undergraduate environmental engineering students. Chemical process control is a premier course at RPI and is taught in studio format with mini lectures interspersed with teams of students solving problems with computers. With few exceptions, the environmental students loved the course and began to appreciate the importance of environmental dynamics better as contrasted with the traditional emphasis on environmental processing at steady state. They learned how improved control can make a process more stable and why improper control can be worse than no control. Now that course is now longer required, no environmental students have opted to take it for many years.

Laboratory Courses: The environmental engineers in charge of waste treatment are seldom blessed with well-trained technicians. While a technician who was educated at a community college might enjoy working at a water treatment plant, the stink of sewage treatment or of municipal solid waste is discouraging. There are nicer jobs in hospital labs or industrial labs. The usual result is environmental technicians who must be trained on the job and who may have no education after high school. The good ones flourish while the bad ones make mistakes, and an effective boss must understand the methods in order to spot the good technicians and to weed out the bad. Laboratory courses are essential for students in environmental engineering. To save money, one approach is to have more options in existing laboratory courses. With less time spent on stretching asphalt or breaking specimens, there could be more experiments for measuring BOD, COD, chlorine residual, and the like. These tests and more as videos can be found on YouTube or elsewhere.

As education evolves and knowledge expands, there is not enough room in a curriculum to cover everything that is relevant; something has to give. Few civil engineering students now learn surveying. The civil/environmental option has focus on microbiology, appreciation of ground water hydrology, and introductory courses about water treatment and waste treatment. To earn their degrees in civil engineering, there are required courses such as structures and the like that consume the available time. To make room for more courses in the environmental option, it may be wise to compact some of the basic civil engineering into sections only for environmental engineers while providing less depth.

Another way to make room in environmental courses is to devote less time to background topics. Environmental engineering courses tend to attract students from other departments who have taken several courses in chemistry and biology. The civil/environmental students need to catch up. This author used to spend 2 weeks lecturing about these topics until deciding to assign web pages to supplement only 1 week of lectures. Good students who learned and retained this material and paid for it in previous courses breezed through these web pages, while other students struggled but reached a sufficient level of understanding for the environmental engineering.

3 Presentation Skills

This author has been blessed to have had two or three wonderful teachers and to have collaborated with two or three colleagues who were magnificent lecturers. Developing such outstanding presentation skills for students is so unlikely that other measures deserved a trial. One that had a bad start was assigning term projects for which teams of three students would videotape a short technical presentation. All of the students were required to watch the videos of at least three other teams and to submit a paragraph with comments. The bad news was that a friend who lived down the street and who was in charge of the university's studios refused to let my students use his professional equipment. However, he would be pleased to teach each team how to exploit television and then to direct their presentations. They now were learning from a pro instead of from watching commercial television.

This author was walking past one of the rooms in the university library that had a video player and saw a dozen of his students overflowing the room and howling with laughter. They were joking and commiserating in a friendly way with those appearing on the screen. Feedback after graduation told that the term project had had inspired them to excel at job interviews and at their initial technical reporting to their employers or at their postgraduate institutions when that was their choice.

Perhaps it was a mistake to try another idea, and assigning two term projects seemed to be overkill. The new term project was to stop complaining and to do something about improving their education by forming teams with not more than three classmates to author web pages on any topic in the course that piqued their interest. In those days, many of the students asked how to code a web page and were told by this author to learn by themselves by finding a web page on the internet and cutting out its content and pasting in their own. Today there are few students who do not already know how to design and implement a web site.

Being thanked by a number of students for the web page assignment was a unique experience. The instructions were to devote 8 or 10 hours to the term project, but many students spent upward of 30 hours. One student said that her parents were delighted by her sequence of web pages and had commented that they were getting their money's worth for her college expenses. The various projects ranged in quality from good to outstanding, and the best ones were edited (with attribution) for inclusion in a textbook. Hyperlinks to material on the internet enhanced most of the term projects.

A third experiment was based on what many professors were already doing. Students in a laboratory course were required to use PowerPoint and to take turns presenting it. One difference was constructive criticism immediately afterward. Several students said that they had never before been graded on poise, word choice, fluffs, demeanor, and enunciation.

Expectations were high for favorable student evaluations of these courses, but few students were enthusiastic. Perhaps a professor should be content to earn reasonably good course evaluations at a university with far more than its fair share of prominent professors. What really had happened was that students were learning from each other, and using the internet was to them nothing special. Increasing student involvement and developing their teaching skills is in its infancy, and the approaches mentioned here are primitive.

Other computer programs for presentation are not dependent on the internet. Smart spreadsheets should be handy tools for all engineering students, and solving problems with calculations in a spreadsheet can be useful throughout a career as a template for more complicated problems or as a routine solver that accommodates new input data. It would add no extra effort for the instructor and would challenge the students to require that homework with calculations be submitted as the printout of a spreadsheet. Warning: students often are so comfortable with spreadsheets that they prefer a clumsy spreadsheet calculation over writing a good computer program.

After half a century of educational movies and videos, many of which were authored by master teachers or professional teams, most of us still prefer a live presentation. Microsoft PowerPoint dominates for technical lectures and incorporates a variety of excellent visual aids, but the slides without the speaker tend to fall flat. Authoring a PowerPoint presentation takes time and creates valuable intellectual property that another teacher can edit and supplement to suit a different audience. An instructor who gives it away for free to educators at other institutions may hesitate to provide aid and comfort to the competition. Not so for a nonacademic who can find fame and recognition. Nevertheless, education would benefit by sharing presentations and encouraging innovation and improvements while attributing each contributor.

4 Information Retrieval

People who browse the internet know about favorites or bookmarks for instant access to web sites that are consulted frequently. An engineer should realize how seldom a textbook is the primary source when seeking information. Some handbooks and articles may be at your fingertips, but it is more convenient to type search words into your internet browser than to select from your bookcase a text that is probably out of date. When the web pages that you most often consult such as those for rules and regulations are among your favorites in the browser, you instantly have current information.

Educational material for supplementing formal courses is termed “Open Courseware.” This author decided not to collect web pages for complete courses but to feature challenges for students, interactive exercises, short stories, and links to engineering tools [3, 4]. The code for these pages can be templates for others to author their own web pages, but the main goal is providing educational material for anyone to supplement various courses. It is also a way for a retired professor to leave a legacy of best ideas (Figs. 9.1 and 9.2).

A good habit for students is to collect information by topics. With a program such as Microsoft Word, the headings and subheadings can be indexed easily, and search features can locate information by using key words. When class notes are entered into the collection, the student benefits from review and most often will do this while the material is fresh and not a hazy memory the night before an examination. Instructors with useful internet bookmarks and with topical collections of relevant information should provide them to their students as examples. Their students can cut and paste to personalize these templates and can share discoveries with their classmates. The instructor also has the responsibility of integrating lectures with internet sessions, updating the courses, discovering internet resources, and authoring or editing these resources to meet the objectives of the courses.

Rensselaer

Howard P. Isermann Department of Chemical and Biological Engineering

Open Courseware for Biotechnology

- Analysis of Microbial Growth Rate
- Interactive Javascript exercises for Biochemical Engineering
- Javascript exercises for Environmental Unit Operations
- True story introducing unit operation of filling powders
- Challenges for Fermentation Engineering
- Sun Valves
- Interactive story aboout Brewing
- Sample pages for textbooks
- Help for creating web pages
- Introduction to engineering economics

At YouTube:

- Microbial Cellulose With Solid Inclusions
- Affordable Photobioreactors

Fig. 9.1 Example of an index page for Open Courseware (https://www.rpi.edu/dept/chem-eng/Biotech-Environ/OpenCourseware.htm)

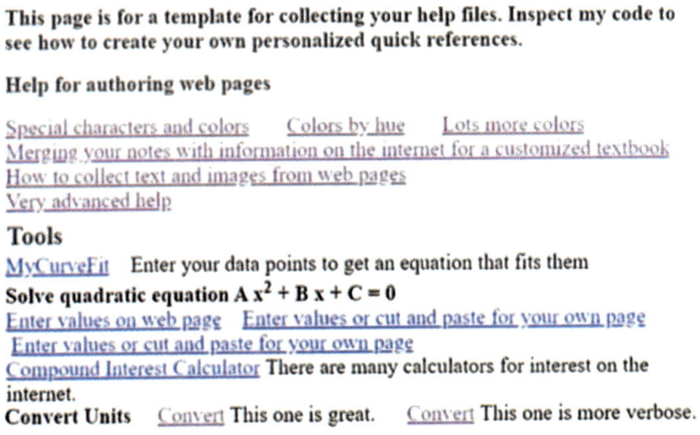

This page is for a template for collecting your help files. Inspect my code to see how to create your own personalized quick references.

Help for authoring web pages

Special characters and colors Colors by hue Lots more colors
Merging your notes with information on the internet for a customized textbook
How to collect text and images from web pages
Very advanced help

Tools

MyCurveFit Enter your data points to get an equation that fits them
Solve quadratic equation $A x^2 + B x + C = 0$
Enter values on web page Enter values or cut and paste for your own page
Enter values or cut and paste for your own page
Compound Interest Calculator There are many calculators for interest on the internet.
Convert Units Convert This one is great. Convert This one is more verbose.

Fig. 9.2 Example of personalized web page for quick help (https://www.rpi.edu/dept/chem-eng/Biotech-Environ/Authoring.htm)

This author has several thumb drives for collecting information from the internet by topic. For reliable web sites such as Wikipedia, only the URL is kept because the information online will be updated and current. More often, the web pages are copied to the flash drive, because they might disappear. Files on a flash drive can be edited or duplicated easily, and content from different files can be combined to integrate and to customize the concepts and explanations.

5 Digression

In a typical lecture, some members of the audience would prefer a faster pace, some are lagging behind, and a scant few consider the pace fully acceptable. A recorded lecture can be replayed to review what needs reinforcing, but doing so is rare. Why not videotape a good lecture and upload it to YouTube?

This author has viewed educational videos of lecture by teachers who are considered excellent but has yet to find one that is very effective. Furthermore, students may skip class thinking that they can learn enough on the computer. This is highly unlikely as evidenced by seminars that would transmit information much less effectively without the speaker.

With web pages or PowerPoint slides, you click to navigate and to select the pace that suits you. When appropriate, sound clips may have been included to reinforce the written text.

6 Conclusion

The internet can make life easier for professors. When information on the internet is very good, it can be assigned to substitute for a lecture. When not quite right for a course, it can be captured and improved to meet the objectives of the course. Wise choices about term projects can improve a course while not infringing much of the professor's time constraints.

Usually, improvements are aimed at classroom or laboratory instruction with perhaps homework assignments that depend on the internet, but a passive approach misses the point. Active participation is better. Information transfer takes several forms, and the internet provides photos, videos, interactive pages, and educational games. Entire pages from the internet can be captured for the document, or text and images can be copied and pasted.

Some logical restructuring of undergraduate education should be sufficient for filling the jobs for pollution control and for jobs with regulatory agencies. However, an old-fashioned Bachelor's degree may not be adequate for skills needed for pursuing an advanced degree. Restructuring the curriculum and sophisticated use of the internet will benefit engineers and society.

Dedication I treasure my warm friendship with Donald Aulenbach that started in 1976 and lasted until his recent death. We were together when the university administration at RPI experimented with merging environmental engineering and chemical engineering. I continued to teach environmental engineering courses for almost two decades after Don and the other environmental professors went to civil engineering as I opted for chemical engineering. His unique perspectives were inspirational, and he was outstanding for research in the field while training students at Lake George and Dyken's Pond. After retirement, we often sat together at events such as monthly luncheon gatherings. I pray for his soul.

Glossary

Chemical engineering It is a branch of engineering that uses principles of chemistry, physics, mathematics, biology, and economics to efficiently use, produce, design, transport, and transform energy and materials. The work of chemical engineers can range from the utilization of nanotechnology and nanomaterials in the laboratory to large-scale industrial processes that convert chemicals, raw materials, living cells, microorganisms, and energy into useful forms and products.

Civil engineering It is a professional engineering discipline that deals with the design, construction, and maintenance of the physical and naturally built environment, including public works such as roads, bridges, canals, dams, airports, sewerage systems, pipelines, structural components of buildings, and railways. Civil engineering is traditionally broken into a number of subdisciplines. It is considered the second-oldest engineering discipline after military engineering, and it is defined to distinguish nonmilitary engineering from military engineering. Civil engineering takes place in the public sector from municipal through to national governments, and in the private sector from individual homeowners through to international companies.

Environmental engineering It is a professional engineering discipline that takes from broad scientific topics like chemistry, biology, ecology, geology, hydraulics, hydrology, microbiology, and mathematics to create solutions that will protect and also improve the health of living organisms and improve the quality of the environment. Environmental engineering is a subdiscipline of civil engineering, chemical engineering, and mechanical engineering.

Environmental resource management It is the management of the interaction and impact of human societies on the environment. It is not, as the phrase might suggest, the management of the environment itself. Environmental resources management aims to ensure that ecosystem services are protected and maintained for future human generations, and also maintain ecosystem integrity through considering ethical, economic, and scientific (ecological) variables. Environmental resource management tries to identify factors affected by conflicts that rise between meeting needs and protecting resources. It is thus linked to environmental protection, sustainability, and integrated landscape management.

Environmental science It is an interdisciplinary academic field that integrates physical, biological, and information sciences (including ecology, biology, chemistry, mathematics, microbiology, geology, information science, etc.)

Environmental studies It is a multidisciplinary academic field, which systematically studies human interaction with the environment in the interests solving complex problems. Environmental studies brings together the principles of the physical sciences, commerce/economics, and social sciences so as to solve contemporary environmental problems. It is a broad field of study that includes the natural environment, the built environment, and the sets of relationships between them. The field encompasses study in basic principles of ecology and environmental science, as well as associated subjects such as ethics, geography, anthropology, policy, politics, urban-planning, law, economics, philosophy, sociology and social justice, planning, pollution control, and natural resource management. There are also many degree programs in Environmental Studies including a Master's degree and a Bachelor's degree.

Environmental technology This field may include (envirotech), green technology (greentech) or clean technology (cleantech) is the application of one or more of environmental science, green chemistry, environmental monitoring and electronic devices to monitor, model and conserve the natural environment and resources, and to curb the negative impacts of human involvement. The term is also used to describe sustainable energy generation technologies such as photovoltaics, wind turbines, bioreactors, etc. Sustainable development is the core of environmental technologies. The term environmental technologies is also used to describe a class of electronic devices that can promote sustainable management of resources.

Internet The Internet (portmanteau of interconnected network) is the global system of interconnected computer networks that uses the Internet protocol suite (TCP/IP) to link devices worldwide. It is a network of networks that consists of private, public, academic, business, and government networks of local to global scope, linked by a broad array of electronic, wireless, and optical networking technologies. The Internet carries a vast range of information resources and services, such as the interlinked hypertext documents and applications of the World Wide Web (WWW), electronic mail, telephony, and file sharing.

References

1. Kuchinski WD (1995) The use of electronic mail, newsgroups, and the world-wide web to improve chemical engineering education. MS Thesis. Rensselaer Polytechnic Institute
2. Van Ness HC (1989) Chemical engineering education: will we ever get it right? Chem Eng Prog 85(1):18
3. https://www.rpi.edu/dept/chem-eng/Biotech-Environ/OpenCourseware.htm
4. https://www.rpi.edu/dept/chem-eng/Biotech-Environ/Authoring.htm

Chapter 10
Environmental Control of Pests and Vectors

Lenore S. Clesceri and Erika J. Clesceri

Contents

1 Introduction 388
2 Pests and Vectors 389
3 Environmental Control of Pests 390
4 Pesticides for Control 393
5 Pesticide Control of Organisms Pathogenic to Humans 396
5.1 Microbes and Worms 397
5.2 Arthropods 397
5.3 Vertebrates 399
6 Pesticide Control of Organisms Destructive to Plants 400
7 Pesticide Control of Nuisance Plants and Other Organisms 401
7.1 Terrestrial Plants 401
7.2 Aquatic Plants 402
7.3 Arthropods 403
7.4 Vertebrates 404
8 Resistance to Chemical Pesticides 405
9 Pollution and Human Health Risks from Chemical Pesticides 406
9.1 Zone of Influence 406
9.2 Fate and Transport of Pesticides in the Environment 407
9.3 Biological Magnification 407
9.4 Toxic Effects on Human Health 408
10 In Summary and Other Considerations 408
Glossary 409
References 410

Abstract Among the plants and animals inhabiting the earth, there are very few taxonomic categories that do not contain some organisms capable of producing human sickness or of causing economic loss, or that simply may be classified as

---

L. S. Clesceri (✉)
Department of Biology, Rensselaer Polytechnic Institute, Troy, NY, USA

E. J. Clesceri
Bureau for Democracy, Conflict and Humanitarian Assistance, US Agency for International Development, Washington, DC, USA

L. K. Wang, M. -H. S. Wang, Y. -T. Hung, N. K. Shammas (eds.),
Environmental and Natural Resources Engineering, Handbook of Environmental Engineering 19, https://doi.org/10.1007/978-3-030-54626-7_10

nuisances, i.e., a "pest." Control of undesirable pests (i.e., certain plants, insects, animals, and microbes) includes both the management of the pest occurrence (e.g., the rodent) as well as any vector organisms that may carry a pathogen (e.g., a malaria-carrying mosquito). Such control or management of the occurrence of pests is addressed by both environmental approaches and that of pesticides. As a necessary precursor to the application of pesticides, successful strategies are employed in the environmental control of undesirable organisms through manipulation of the physical and ecological factors within the pest habitat. Together, this approach is regarded as Integrated Pest Management (IPM). Pesticide approaches encompass a wide arrange of organic and inorganic chemistries, which for many years have been developed to both great success and unintended damage to nontarget organisms, including humans. Specific pests and vectors are to be discussed within this chapter that highlight the economic impact of such agents on human health and agricultural productivity.

Keywords Pest · Vector · Environmental control · Pesticide · Pathogenic organisms · Humans · Plants · Nuisance plants · Pollution · Health risks

Acronym

$Ca(OH)_2$	Slaked lime
CCA treatment	Chromium, copper, and arsenic treatments
$CuSO_4$	Copper (II) sulfate
FFDCA	Federal Food, Drug, and Cosmetic Act
IPM	Integrated Pest Management (IPM)
UC	University of California (UC)
USEPA	US Environmental Protection Agency

1 Introduction

Among the plants and animals inhabiting the earth, there are very few taxonomic categories that do not contain some organisms capable of producing human sickness or of causing economic loss, or that simply may be classified as nuisances, i.e., a "pest". Control of undesirable pests (i.e., certain plants, insects, animals, and microbes) includes both the management of the pest occurrence (e.g., the rodent) as well as any vector organisms that may carry a pathogen (e.g., a malaria-carrying mosquito). Such control or management of the occurrence of pests is addressed by both nonchemical or environmental approaches and chemical pesticides.

As a necessary precursor to the application of chemical pesticides, successful strategies are employed in the environmental control of undesirable organisms through manipulation of the physical and ecological factors within the pest habitat.

Chemical pesticide approaches encompass a wide range of organic and inorganic chemistries, which for many years have been developed—to both great success and unintended damage to nontarget organisms, including humans. Together, these environmental and chemical approaches are regarded as Integrated Pest Management (IPM). Pesticides have been used in the management of pests and vectors for many decades in the United States and elsewhere. The term, poison, is very broad, referring to practically any substance that can produce sickness or death if its concentration is high enough. Thus, even essential nutrients can be poisonous at high concentrations. For example, copper, which is highly toxic to all forms of plant life, and used in wood preservatives, is an essential element for plant growth at the trace level [1].

Pesticides, for the purpose of this chapter, will include only those agents intentionally used by humans for the control of undesirable pest organisms. The pesticides identified herein are by no means an exhaustive listing of available agents. This discussion will omit the very large group of poisonous wastes generated by industry, government, agriculture, and various institutions such as hospitals and laboratories.

Specific pests and vectors are to be discussed within this chapter that highlight the economic impact of such agents on human health and agricultural productivity.

2 Pests and Vectors

Among the plants and animals inhabiting the earth, there are very few taxonomic categories that do not contain some organisms capable of producing human sickness or of causing economic loss, or that simply may be classified as nuisances, i.e., a "pest." The term "pest" is a very broad term that defines organisms that are an annoyance to humans simply because these pests inhabit places where they are unwanted, like our homes (e.g., ants, bed bugs, lice, mosquitoes, and wasps). Whereas, vectors are organisms that carry a pathogen with it that transmits disease to humans or animals such as, malaria- or Zika- carrying mosquitoes, ticks, and fleas. In most circumstances, vectors are also considered pests.

One finds bacteria, fungi, and arthropods responsible for food spoilage and economic losses through decay of natural products. In addition, there are bacteria and fungi that are pathogenic to humans, domestic plants, animals, and wildlife. There are pathogenic worms, as well as poisonous echinoderms, arthropods, and reptiles. For example, the fall armyworm was reported in 2016 for the first time in West and Central Africa, so it now threatens crops in Africa and Europe.

Mollusks harboring poisonous toxins, trash fish, disease-carrying arthropods, rodents and nuisance birds, and creatures of various types are considered pests. There are taste-, odor-, and slime-producing microbes, allergenic plants, plants that are poisonous, and weeds occupying our lands and waters. Finally, pests also include viruses with no known usefulness for humanity.

3 Environmental Control of Pests

Environmental control, via nonchemical means, includes targeted alterations or control of the natural and man-made environment in which the pest occurs. These control measures are a necessary precursor to the application of chemical pesticides, as environmental manipulative methods require an entomological understanding of the pest life cycle and habitat. It is not advised to apply any chemical control measures without first establishing an understanding of the pest. This type of environmental control approach has a wide range of methods that are used in concert with chemical poisons and is considered Integrated Pest Management or "IPM." Indeed, the practice of good hygiene and manipulation of pest habitats can obviate the need for chemical applications under well-managed systems. A longer term view of production and pest prevention is favored over the short-term, reactive view.

A good definition of IPM from University of California (UC)-Davis includes: "Integrated pest management (IPM) is an ecosystem-based strategy that focuses on long-term prevention of pests or their damage through a combination of techniques such as biological control, habitat manipulation, modification of cultural practices, and use of resistant varieties. Pesticides are used only after monitoring indicates they are needed according to established guidelines, and treatments are made with the goal of removing only the target organism. Pest control materials [pesticides] are selected and applied in a manner that minimizes risks to human health, beneficial and non-target organisms, and the environment."

There is no general prescription that can apply to every situation regarding the need for pest control. There are, however, some general rules that can be applied as outlined in a study of pest control strategies made under the aegis of the National Research Council [2]. The approach of Integrated Pest Management is applicable to most situations and includes some of the following principles:

- Practice prevention first, by managing crops or households to prevent pest problems before they occur.
- Monitor and identify pests. Not all insects, weeds, and wild animals around the farm are harmful; some may even help.
- Set an action threshold—the point where a pest problem must be dealt with. Sighting a single pest may not justify a campaign.
- Learn to tolerate a certain level of the pest since most pests are very adaptable to host, environment, and the control scheme, making them very difficult to eradicate completely.
- Take direct action against pests by suppressing reproduction.
- Enhance the destruction of pests by drawing them to natural predators or making them susceptible to parasites.

While environmental control through prevention is the recommended first approach to pest management, as an alternative to the use of chemical poisons, nonchemical means do not, by in large, provide the immediate response produced by chemical control. They are, nonetheless, much more desirable as means for pest control. When used in a preventative program, the effect is a stable, controlled environment without the hidden costs of chemical poisoning. Many of the environmental control measures rest in the realm of biological control of one type or another. Some of these attempts have backfired as the result of insufficient study before employing them. Others, however, are tried and true measures. Illustrative examples of these environmental control measures include the following:

1. Sanitation

   Community and local sanitation to eliminate food, water, and harborage for vertebrate pests (rodents, pigeons, etc.). This is the simplest and most effective means for controlling such animals. A similar policy applies to the control of wildlife, although this is much more complicated since many food alternatives exist outside of the urban area. Thus, removing a single food source or even a class of food sources will in all probability exert some control over the particular wildlife species, but not to the extent that may be desired.
2. Maintenance of Healthy Plants

   Plants that are less stressed are not as *prone* to attack by insects as unhealthy plants. Insects mostly infest unhealthy plants, as strong plants can withstand certain levels of herbivory. Plants that receive appropriate amount of sunlight, that are well-watered, and that are in nutrient-rich and well-drained soils that are protected from wind damage tend to be able to resist certain levels of insect or fungal attack. In nature, weak organisms are typically attacked first. However, the application of too much fertilizer or excess nitrogen is well documented to correlate to increased pest prevalence.
3. Construction and Packaging

   Exclusionary methods can take the form of different modalities. Tight construction prevents the use of the building as harborage for pests. In packaging of perishable foodstuffs in storage bags, there are new applications for hermetic sealing [3], which creates an anoxic environment inside the bag, reducing the growth of insects that can be found in stored agricultural commodities requiring oxygen for survival. Also, avoiding the contact of wooden structural members with the ground and elimination of buried wood will discourage the development of termite colonies.
4. Elimination of Breeding Places

   This is the most desirable way of controlling most arthropod pests. The housefly thrives on any kind of waste (rotting vegetation, manure, sewage sludge, etc.). Proper sanitation, installation of garbage grinders in kitchen sinks, and use of fly traps are all measures that help in controlling flies without insecticides. The elimination of standing water, controlled fluctuation of water levels to alternatively flood and dry breeding areas, and the stocking of ponds with top-feeding fish that consume larvae are effective means for combating the mosquito where it breeds. Burning brush and long grasses that harbor mosquitoes, chiggers, etc., are another alternative to chemical poisoning.

5. Multiple Crops on Agricultural Land

   Insect infestation that frequently occurs with monoculture is controlled by multiple plantings. Alternating rows of susceptible with nonsusceptible crops serves to confine the pests and prevent the widespread infestation that can occur when an unlimited supply of food is contiguous to the initial infection.
6. Natural Predators

   The use of natural predators that consume developing or adult stages of pests is ideal if the predator itself does not become a pest. Ladybugs are among the most effective destroyers of a variety of plant-eating insects. The preying mantis lives exclusively at the expense of other insects. Yellow jackets feed on soft-bodied insects. Adult dragonflies feed on mosquitoes, whereas their young, naiads, prey on the developing stages of the mosquito and other insects in the waters below.
7. Sterile Mating

   Some success has been met in combating undesirable organisms by capturing, sterilizing (usually with radiation), and releasing sterile mates into the environment. The relative effectiveness of this program, of course, depends upon how many fruitful matings can be prevented.
8. Competition

   Enhancement of the growth of a competitor is frequently possible. For example, one can replace weeds with a desirable ground cover by seeding among the weeds. The cutting of weeds, especially when seeded with grass, will accelerate succession to a more stable vegetation.

   Mycotoxin contamination of agricultural commodities is a major threat to stability of foods in storage. Aflasafe is an example of a product of fungal spores used for competing with toxigenic strains to prevent the formation of toxic aflatoxin-producing strains in soils. It is applied to soils to reduce toxicity of the native soil fungal populations.
9. Use of Attractants

   Many insects locate food, mating partners, and sites to lay eggs by being able to respond to chemical stimuli emanating from these sources. Traps permeated with these materials (pheromones) may physically restrain the insect by means of a sticky surface or by drawing it through a small orifice, by killing the insect outright by electric shock, or poisoning it with insecticides. Both natural and synthetic attractants are used. Light is also commonly used as an attractant for certain insects.
10. Physical Removal of Aquatic Weeds

    Preferable to the use of pesticides is the physical removal of aquatic weeds. This is feasible for the rooted macrophytes through weed cutting procedures. Weed cutting and removal of the plants are better than removal by dredging, since it leaves the sediment less disturbed to continue its role in the decomposition of sedimented organic material.
11. Use of Hormones

    Hormonal pesticides can be used to selectively disturb growth, development, and reproduction. When applied in excessive quantities or at abnormal times in a life cycle, they disrupt a wide range of body functions. Dormancy-breaking chemicals produce germination before a suitable environment is available, resulting in seedling death. In a like manner, insect diapause has been terminated prematurely through the use of juvenile hormones.

12. Cold, Heat, Dehydration, and Radiation

    For certain insects and microorganisms, the use of cold, heat, dehydration, or UV radiation is an effective means of protection, sometimes applied simultaneously or in concert.

4 Pesticides for Control

Since the 1930s in the United States, to combat the array of undesirable organisms (i.e., pests and vectors), an arsenal of chemical agents has been developed for control and management. Some of these agents are simple inorganic compounds, others are organic compounds of relatively complex structure, and yet others are biopesticides, such as the *Bacilli* bacteria.

It is generally agreed that materials used to control, destroy, or mitigate undesirable organisms (other than bacteria, rickettsia, and viruses) are called pesticides, whereas disinfectants, antibiotics, and virucidal agents are used to control both cellular and acellular microbes. Yet in many cases, pesticides indirectly control the microbe through control of the vector as, for example, in the battle against the tick that harbors the rickettsia known to induce Rocky Mountain Spotted fever.

"Pesticides" is an overarching term for chemically and biologically based materials that control a wide array of pests. More specifically, it includes pesticides that are designed to control the occurrence of targeted pests: such as rodenticides (for rodents), molluscicides (snails), nematicides (worms), miticides (mites), insecticides (insects), fungicides (fungi), algicides (aquatic algae), bactericides (bacteria), and herbicides (weed and brush killers, defoliants, and desiccants). Plant or insect growth regulators and insect sex attractants are also normally considered among the pesticides. The US EPA registers the disinfectant, sodium hypochlorite or chlorine additive for water treatment, as a pesticide as well.

Federal legislation in the United States strictly controls the use of pesticides through the Federal Insecticide, Fungicide, and Rodenticide Act (FIFRA). This act requires the classification of all pesticides according to whether they may be used generally or are restricted. A pesticide is classified as restricted if it can be hazardous to human health or the environment. These restricted pesticides can be applied only by people certified by the US Environmental Protection Agency (USEPA) as competent in using and handling pesticides. Only registered pesticides can be marketed in the United States.

USEPA has developed a multivolume registry of pesticides [4] containing uses of all registered products. The registry specifies dosage, mode of application, and limitations along with a classification for restricted or general usage. The registry does not recommend one pesticide over another and thus is most useful to persons with some prior knowledge. Chemical and generic names are used, but not trade names. For trade name information, one is referred to the *Pesticide Handbook* [5], which is an annual publication containing information on trade names, generic names, and the composition of commercial pesticide formulations. The series, *Advances in Pest Control Research* [6], is an excellent reference on the fundamental aspects of pest control.

Federal regulations also require US Department of Agriculture registration of any pesticide shipped in interstate commerce. The Department of Health and Human Services has established tolerance levels for residues resulting from pesticide treatment of agricultural products. Pesticides added to processed foods are considered additives and are subject to regulations of the Federal Food, Drug, and Cosmetic Act (FFDCA).

For any pesticide, one must check federal, state, and local regulations regarding its use in chemical poisoning programs. The chemical examples that are given in this chapter do not imply that current usage is permitted. This would be dangerous, since the listing of chemicals that can be used in vector and pest control is changing rapidly as new information becomes available.

When considering global usage of pesticide poisons in international developing world settings, the market readiness of pesticides is far more fragile than in the United State. Given limited literacy and numeracy in structurally remote and impoverished regions in Africa, Latin America, and South Asia, the human capacity to safely and effectively utilize pesticides is more tenuous. It is also important to note that registration and oversight of pesticides is far less controlled by national and local municipal governments than in developed nations, like the United States. As such, assurances that in the United States one can largely take for granted, like the quality of the chemicals purchased in formal markets cannot be assured. This uncertainty in pesticide performance results in even greater challenges to poor farmers who are planting crops on heavily degraded landscapes that are subject to repeated shocks like floods and droughts. It is not unheard of for pesticides with high aquatic toxicities to be applied in, for example, Ramsar protected wetlands in Africa. As such, the application of pesticides by farmers near the wetland could undermine the livelihoods of fishermen, which supply nutrient-rich proteins to larger portions of the lesser developed world.

The chemical insecticides can be most conveniently classified into organic and inorganic types, where an illustrative, but not exhaustive, list is provided below. Some of these have been widely used in the past but are currently only used with narrow usage restrictions.

1. Organic chemical pesticides

   (a) Chlorinated hydrocarbons such as DDT, lindane, methoxychlor, and dieldrin
   (b) Organophosphorus insecticides such as parathion, malathion, diazinon, and glyphosate
   (c) Carbamate insecticides such as carbaryl and propoxur
   (d) Neonicotinoids such as imidacloprid
   (e) Synthetic pyrethroids, like the permethrins

Neonicotinoids were developed in the 1980s as alternatives to organophosphates, carbamates, and pyrethroids, many of which had exhibited pest resistance issues. Neonicotinoid insecticides are very water soluble, allowing them to be applied to soil and taken up by plants. This also makes aquatic ecosystems and associated organisms particularly vulnerable to this class of pesticides. It should be noted that development of neonicotinoids includes a class of insecticides that are registered by US Environmental Protection Agency (USEPA) for a variety of uses, including the control of sucking insects on vegetables, fruits, cotton, ornamentals, turf, and/or as a seed treatment. The five neonicotinoids registered by USEPA are acetamiprid, clothianidin, dinotefuran, imidacloprid, and thiamethoxam.

2. Inorganic chemical pesticides
   (a) Arsenicals such as Paris green, lead arsenate, and arsenic trioxide
   (b) Sulfur and its compounds
   (c) Others such as compounds of fluorine and mercury, and hydrogen cyanide
   (d) Bordeaux mixture: a mixture of copper(II) sulfate ($CuSO_4$) and slaked lime ($Ca(OH)_2$)
   (e) Aluminum or magnesium phosphide(fumigant)
3. Biopesticides
   (a) Natural plant-based pesticides such as nicotine, rotenone, pyrethroids, Neem (tree extract)
   (b) Fungal-based mycopesticides, like Green Muscle [7]
   (c) Bacterial-based like *Bacillus popilliae* and *Bacillus thuringiensis*
   (d) Herbal repellents like those extracted from garlic and red chili pepper

US EPA states that biopesticides "include naturally occurring substances that control pests (biochemical pesticides), microorganisms that control pests (microbial pesticides), and pesticidal substances produced by plants containing added genetic material (plant-incorporated protectants)." Effective control of pests through the use of specific pathogenic microorganisms has been demonstrated for a variety of insects.

A particularly dramatic example is the decimation of the European pine sawfly that produced tremendous damage to pine plantations and nurseries in southeastern Canada and the northeastern United States in 1949. Infection of the insects with a virus imported from Sweden quickly brought the insects under control. Only two pathogens are federally registered for use in insect control programs. These are *Bacillus popilliae* and *Bacillus thuringiensis* for the control of Japanese beetles and various caterpillars, respectively. Several other pathogens await proof of safety and reliability and are currently being tested.

Other natural biopesticides include those containing *Beauveria bassiana* (used against coffee berry borer), *Myrothecium verrucaria* (used against nematodes), and *Trichoderma harzianum* (used against soil fungal diseases). The fungal-based mycopesticide, Green Muscle, has been used to counteract locust plagues in Africa, while soil-dwelling bacterial extracts like Spinosad, an insecticide based on chemical compounds found in the bacterial species *Saccharopolyspora spinosa*, can be used in water treatment [8]. Use of organic botanical insecticides like azadirachtin or Neem (tree extract) can be combined with diatomaceous earth.

One can classify pesticides according to usage, as below.

1. Stomach poisons

   These are absorbed through the insect's alimentary tract and are applied before an insect feeds. Most of the stomach poisons are inorganic chemicals and familiar examples are the arsenicals and sodium fluorosilicate.
2. Contact poisons

   These kill by direct contact with some part of the insect's body.

   (a) The contact may be made by direct application of sprays, dusts, or aerosols. An example of this type is parathion.
   (b) The pesticide may be applied to a surface on which the insect passes (residual treatment). The chlorinated hydrocarbon compounds are insecticides of this type.
   (c) The pesticide may be applied in a closed space as a fumigant. Among these are hydrogen cyanide, methyl bromide, paradichlorobenzene, phosphine, sulfuro chloride, and carbon dioxide.
3. Systemic poisons

   These compounds are taken up by the plant or animal at the point of application whereby they are transported to another part of the body. Some of the organophosphorus insecticides are systemic poisons.

Most of the organic pesticides act as nerve poisons of the pest in question. They induce continuous nerve action through inhibition of cholinesterase, producing tremors, convulsions, and, eventually, death. Since the biochemistry of pest insects is no different from that of beneficial insects (e.g., honeybee or preying mantis), and very similar to that of higher animals, including humans, these chemicals must be used very cautiously.

In contrast to the nerve-poisoning role of nerve poisons, the chlorinated hydrocarbon insecticides interfere with the organism's production of enzymes, resulting in a variety of abnormalities such as the thinning of eggshells and subsequent unsuccessful development to maturity, and the elevation and depression of hormone levels, producing erratic behavior.

While an important and active area of research, this chapter will only include a very brief discussion on human health risk assessment of acute or chronic effects chemical pesticide (Sect. 9), as is largely beyond the scope of this work.

5 Pesticide Control of Organisms Pathogenic to Humans

Organisms that are pathogenic for human health widely vary, and are a critical drain on human productivity and well-being worldwide.

5.1 Microbes and Worms

Some microorganisms are among organisms that are a health problem. These organisms include certain bacteria, protozoa, fungi, and algae. Viruses and rickettsia (acellular microbes) are also in this group since methods used to control them are similar to those used for bacterial control. Recently, however, much concern has been expressed over the inadequacy of detection methods to quantitatively determine infectious viruses. Researchers recognize that water deemed safe by the coliform index may be virologically unsafe. Disinfection of water by chlorination is the most common choice for the control of viruses, rickettsia, and infectious bacteria such as the enterics and the various spirochetes responsible for leptospirosis. Chlorination also is often used to control other water-borne pathogens such as the protozoan *Entameba histolytica* as well.

Certain water-borne flatworms and round worms that infect humans are also controlled by disinfection techniques. More commonly, however, since foods are common vehicles of infection, proper cooking measures can control the spread of these parasitic worms. A summary of food and water-borne diseases, along with their etiological agents, prevention, and control, is given by Salvato [9].

Microorganisms that produce respiratory diseases are usually spread via the air or by direct contact with an infected individual. These organisms are best controlled through personal cleanliness of the infected individual. Other measures include disinfecting aerosols and disinfection of surfaces upon which the pathogens fall.

Disinfection is likewise used to control fungi that produce various forms of skin disease (dermatophytosis). People most commonly contract skin diseases of fungal origin where conditions include moisture, warmth, and other people, such as swimming pools and shower rooms. Dermatophytosis is controlled at the source by measures capable of killing the fungal spores. The simplest of these is the drying of the infected area, which destroys both the vegetative fungi and the fungal spores. Chemical means normally used are either disinfection with sodium hypochlorite (500 mg/L available chlorine) or scrubbing the facility frequently with a hot solution of strong detergent followed by a rinse with a disinfectant.

Health hazards originating from algae arise via their endotoxins, some of which are highly poisonous to humans. Most noteworthy are shellfish contaminated by the alga *Gonyaulax*. The control of algae will be discussed later in this chapter.

5.2 Arthropods

Arthropods affecting human health contain a wide class of organisms, most inclusive of insects for purposes of this section. More people have died from malaria than the combined deaths of World Wars I and II. The protozoan parasite is carried by mosquitoes of various species in the genus *Anopheles*. In addition, mosquitoes transmit dengue fever, encephalitis, filariasis, Rift Valley fever, and yellow fever,

Table 10.1 Illustrative human diseases transmitted by arthropods[a]

Disease	Causative agent	Vector	Reservoir
Cholera[b]	Bacteria, *Vibrio cholera*	Housefly, *Musca domestica*	Humans
Dengue[b]	Virus	Yellow fever mosquito, *Aedes aegypti*	Humans
Dysentery, amebic[b]	Protozoan, *Entameba histolytica*	Housefly, *M. domestica*	Humans
Dysentery, bacillary[b]	Bacteria, *Shigella dysenteriae* and other species	Housefly, *M. domestica*	Humans
Encephalitis (St. Louis, Western, and Eastern)	Virus	Mosquitoes, *Culex tarsalis* and others	Birds and mammals
Filariasis[b]	Worm, *Wuchereria bancrofti* and possible *malayi*	Mosquitoes, Culex, *Aedes, Anopheles*, and *Mansonia*	Humans
Leishmaniasis[b]	Protozoan, *Leischmania*	Sandfly	Dogs, foxes, rodents
Lyme disease	*Borrelia burgdorferi*	Ticks	Small mammals
Malaria	Protozoan, *Plasmodium vivax, P. falciparum, P. malariae*	Mosquitoes, *Anopheles*	Humans
Plague	Bacteria, *Pasteurella pestis*	Oriental rat flea, *Xenopsylla cheopis*, and other fleas	Rats and wild animals
Tularemia	Bacteria, Pasteurella, *tularensis*	Deerfly, *Chrysops*, hard ticks, *Dermacentor*	Rabbits and other wild animals
Yellow fever[b]	Virus	Yellow fever mosquito, *Aedes aegypti*	Humans and monkeys
Zika infection	Virus	Mosquitoes, *Aedes sp.*	Humans

[a]Adapted from Pratt [16]
[b]Not found in the United States

among others of less importance. A summary of mosquito-borne diseases, along with etiologic agent, reservoir, mode of transmission, incubation period, and control, is found in Salvato [9]. A listing of illustrative human diseases transmitted by arthropods is given in Table 10.1.

Historically, the insecticide qualities of DDT were first observed in 1939. In 1942, the United States used the compound to control the massive malaria outbreaks sustained by the military. It became available commercially in 1945. Malaria has largely been controlled in most parts of the world through the use of DDT, which was the first synthetic organic insecticide of significance. Since then, vast numbers of synthetic organic insecticides have been developed. Many such as the chlorinated hydrocarbons have been subsequently banned in the United States due to their effects on human health and the environment.

Many arthropods besides the mosquito are disease vectors for humans. Houseflies carry disease microorganisms and parasitic worms. They spread diseases such as typhoid fever, paratyphoid fever, bacillary and amebic dysentery, cholera, typhus, anthrax, gastroenteritis, conjunctivitis, trachoma, and various viral diseases. Wood ticks and dog ticks infected with ricksettsia carry Rocky Mountain Spotted fever, Q fever, and typhus. They are also vectors for tuleremia caused by the bacterium *Pasteurella tularensis*.

The list of pesticides suitable for these vector arthropods is long. The nonpersistence of the organophosphorus and carbamate insecticides (1–12 weeks) and neonicotinoids (days to years depending on exposure to light) has made these the pesticides of choice in many instances. And yet, when dealing with a disease problem, the use of more persistent insecticides has been justified. Recommendations for the use of specific agents can be found in various sources, but interested persons should consult the USEPA registry [4] for current regulations regarding the use of any of the pesticides since these are rapidly changing.

Some pesticides which were first registered years ago must be re-registered to ensure that they meet today's more stringent standards.

In mosquito control, carbaryl has been used extensively for control of the adult, and malathion both for adult larvae in outdoors spraying. Dichlorvos (chlorinated organophosphorus insecticide) and malathion have been widely used in outdoor spraying for fly control.

5.3 Vertebrates

The rat is one of the most destructive vectors of disease and human parasites. Because of growth and proliferation, rats annually cause millions of dollars of damage to stored foods and materials in the United States alone.

Some of the diseases carried by the rat are bubonic plague, caused by the bacterium *Pasteurella pestis*, transmitted to humans by fleas, and leptospirosis, which is transmitted by direct bite. Leptospiral jaundice also is transmitted to humans and, more frequently, to the domestic dog.

Rat poisons are toxic against other rodents as well, such as mice, rabbits, and squirrels. Rats migrate in the absence of food, water, or shelter; they are also confined to a certain population size by existing conditions. It is obvious then that a rat-poisoning program without accompanying cleanup procedures is a strictly temporary relief, since survivors or newcomers will eventually multiply to again fill the vacant capacity of the previously infested environment.

The placing of poisoned baits is commonly employed in rat-poisoning programs. Care must be taken that the poison selected is rapidly detoxified within the rat or that the poisoned animals are immediately removed to avoid transmission to other animals.

Anticoagulants such as Warfarin were the preferred rat poisons for most baiting purposes, but continuous feedings were required for 3–10 days, causing high labor costs for large programs. These compounds act by causing capillaries to break down, resulting in internal bleeding, and by interfering with normal blood coagulation.

Some other rodenticides are sodium fiuoroacetate, fiuoracetamide, and a-naphthyl thiourea (ANTU). The fiuoro compounds are highly toxic to other vertebrates as well as to rats. A preferred rodenticide is diphacinone, also an anticoagulant and toxic to other vertebrates, is used in tamper-resistant prebaited traps.

A contraceptive bait, called ContraPest, for fertility control of rats, has recently been developed to reduce rat reproductive capacity in the management of populations of Norway rats. This product has been approved for use in protecting agricultural crops and reducing disease and requires further research [10].

Some other diseases transmitted to humans by vertebrate vectors are anthrax, caused by the bacterium *Bacillus anthracis* and carried by livestock; psittacosis, caused by a virus and carried by birds; and rabies also viral caused and carried by dogs, cats, livestock, and many wild animals.

6 Pesticide Control of Organisms Destructive to Plants

Because humans can chemically modify the effects of climate, weather, and natural predators on plant pests, the ability to produce large agricultural yields has been greatly increased.

Of all plant parasites, fungi are the most numerous [11]. The taxonomic groups of fungi with the largest number of plant pathogens are the *fungi imperfecti* and the *Ascomycetes*. These organisms infect plants by (1) killing tissue prior to their spread through the secretion of pectic enzymes producing soft rot and eventual death or (2) feeding directly from the living plant cell without necessarily killing the plant tissue.

Agricultural fungicides are applied to soil, seed, propagating material, and growing plants for combating fungal infections. The method of application and choice of fungicide is specific for the particular plant. Agriculturalists have used dithiocarbamate fungicides and inorganic sulfur compound preparations widely and effectively to control pathogenic fungi.

The guidelines for control of insects that are destructive or pathogenic to specific plants are beyond the scope of this chapter. However, in brief, substantial losses both during the growing season and in storage result from the action of insects on crops and foodstuffs. Early synthetic insecticides were inorganic (Paris green, silicofluoride, thiocyanates). The organic synthetic insecticide era has produced a formidable array of chemicals with insecticidal (and often homicidal) properties. Much of the search has been directed toward better insect selectivity, safety to higher animals, and desirable levels of persistence.

As an example, fumigation with the pesticide phosphine gas is a critical tool to assure protection of agricultural food commodities from spoilage. It is also an intrinsically dangerous process: phosphine is deadly, flammable, and corrosive. Failure to conduct fumigation properly poses serious risks to human health and of ineffective fumigation, posing both an immediate problem for commodity losses and contributing to phosphine pesticide resistance. In the worst case, resistance could lead to the loss of phosphine fumigation as an effective pest control [12].

Other organisms that destroy plants that must be considered are the nematodes and molluscs. For the most part, nematodes, which are small, unsegmented worms, dwell in the soil and feed on plant roots. Some, however, invade bulbs, stems, or leaves. In so doing, they produce damage to the plant themselves and also open the way for further damage from fungi. Methyl isothiocyanate and formaldehyde are commonly used nematicides.

Slugs and snails are especially damaging in a wet spring. They damage many leaf and root vegetables and produce considerable economic loss. The most effective chemical against them has been a polymeric form of acetaldehyde called meta. Upon contact with meta, desiccation occurs from excessive loss of water from slime secretion.

7 Pesticide Control of Nuisance Plants and Other Organisms

7.1 *Terrestrial Plants*

Attention also should be directed to a large category of miscellaneous terrestrial plants that compete with the plants that humans have chosen to cultivate. Chemical weed killers or herbicides for the control of these plants are widely used in agriculture, for lawns, and along roadways and railways.

When applying chemicals in plant control, care should be taken to determine whether the nuisance plant serves any less evident function like a food source for beneficial organisms such as pollinators.

Herbicides are called pre-emergents when they are sprayed before the weed plant appears. The germinating seed absorbs the herbicide until a lethal concentration occurs within the young plant. Foliage application or postemergent application is made directly on the plant.

Herbicides may be selective or nonselective, i.e., they may have a biochemical specificity that permits selective killing of weeds and spraying of cultivated plants, or they may be broad-spectrum nonselective herbicides for total kill programs.

The herbicides can be further classified on the basis of hormonal or nonhormonal action. The phenoxyacetic acids (2,4-dichlorophenoxyacetic acid and related compounds) are the oldest of the hormonal herbicides. They produce their effect by competing with the action of natural hormone, resulting in lethally abnormal growth. The excessive root thickening produced in treated plants is a typical effect of 2,4-dichlorophenoxyacetic acid (2,4-D) poisoning. The phenoxyacetic acids have been the herbicides of choice for broad leafed weed control in grass and cereal crops.

Among other hormone herbicides available, picloram has been found to be extremely effective in killing trees and is probably the most persistent and most active of the herbicides. In contrast with most other pesticides, the hormonal herbicides have a very low toxicity to animals because of their specific effect on plant metabolism. This is also apparently true for the nonhormone substituted ureas and triazines that act by inhibiting the photolysis step in photosynthesis. Of this group atrazine and simazine have been widely used for massive weed killing programs, and are unusually nontoxic toward corn crops and slightly to moderately toxic to humans.

A widely used herbicide is glyphosate. It operates by interfering with key plant enzymes. Since introduction in the 1970s, its use has increased significantly for both home and agriculture.

Among nuisance plants, one must mention again the fungi, but this time as agents that produce tremendous damage to wood and wood products. In general, damage can be lessened through various treatments. Polymerizing chemicals applied to lumber reduce porosity and water adsorption and mimic the lignification process in nature. Heartwood, which is more widely lignified, is more resistant to fungal attack than sapwood. Lumber can also be treated by pressurized or brush application of various fungicides.

Most preservatives have been developed to be applied in an oil base to facilitate penetration and prevent warping. There are various wood preservatives of low volatile organic compounds (VOCs) content and toxicity, and are water based or whey based that have undergone USEPA registration. For outdoor and commercial purposes only, creosote, which is a crude mixture of aromatic hydrocarbons, is permitted for railroad ties away from drinking water sources. Creosote has no registered residential uses. Pentachlorophenol is a restricted use pesticide that is only used for commercial purposes such as utility poles, railroad ties, and wharf piling.

Wood preservatives containing chromated arsenicals include preservatives containing chromium, copper, and arsenic (i.e., CCA treatments). Effective December 31, 2003, chromated arsenical manufacturers voluntarily canceled virtually all residential uses of CCA, and wood products treated with CCA are no longer used in most residential settings, including decks and children's playsets. Newer arsenic-free options rely on copper as their primary biocide.

7.2 *Aquatic Plants*

Aquatic plants are valuable as food and shelter for aquatic animals. Floating or emergent vegetation is rarely desirable, however, when the body of water is used for recreation. Excessive algal growths severely restrict the use of the water for drinking, cooling, or recreation. They compete with the aquatic fauna for oxygen during low light intensity and, upon death, products of their decomposition produce objectionable tastes and odors in the water.

Thus, it is frequently desirable to chemically rid the water of its flora. It is only as a last resort that chemical control of aquatic weeds is employed, however, and then only after approval by the proper local, state, or federal agency. Weed killers, such as 2,4-D, applied as pellets or granules, are effective for rooted plants, whereas spray application is effective for floating weeds. Copper sulfate should not be used for aquatic weeds, since the concentration required to destroy the vegetation will most certainly kill any fish in the vicinity. Copper sulfate is, however, commonly used for the control of algae.

It should be pointed out that a chemical control program is best done with complete knowledge of the system being treated. The chemistry of the water and its physical characteristics will contribute to the success or failure of the program. For example, it is known that $CuSO_4$ is rapidly inactivated in alkaline hard water by precipitation as the basic carbonate, and that fish kill can occur as readily from asphyxiation as from toxicity following a CuS04 treatment. Decomposition of algae with the concomitant consumption of oxygen may exhaust the oxygen supply producing such a massive fish kill, especially when the water temperature is high. Thus, for an effective program, investigation is highly warranted before chemical application.

7.3 Arthropods

The mention of mosquitoes and houseflies both as nuisances and disease vectors is well justified and certainly few will contest this double entry.

More than an annoyance, termites are highly destructive and impose serious restraints in building construction. Redwood and certain species of cedar are quite resistant to termites, whereas pine and fir are readily attacked under certain conditions. There are both above-ground and below-ground types of termites. Below-ground termites are very sensitive to light and moisture and, consequently, protection is best achieved through lighting and ventilating subterranean areas.

Above-ground termites may be of the damp wood or dry wood destroying type. The dry wood type, along with other nuisance insects (such as roaches, fleas, bedbugs, and ants), is quite susceptible to sorptive dusts such as silica gels. These materials remove the lipoid protective layer covering the insect's body and cause a rapid loss of water and death. Both types are controlled with pentachlorophenol, as well as a wide range of other insecticides, including lindane, chlordane, and dieldrin. Several newer termiticides, such as bifrenthin, deltamethrin, fipronil, and others, are undergoing testing.

Ants can be a serious problem, but have the beneficial effect of controlling termites by entering their tunnels and attacking them. Where ants are to be controlled, chlordane has been most effective, but malathion or diazinon has also been useful.

Pediculosis or infestation with lice can be controlled by dusting with insecticidal powders.

Blackflies are blood-sucking insects that breed in flowing streams. Larvae are found attached to rocks and vegetation in the streams. Certain species in Mexico, South America, and Africa carry disease, and they occur as annoying pests in other parts of the world.

Clothes moths and carpet beetles do extensive damage to woolen or wool-containing materials. Chlordane, lindane, and dieldrin are effective but dangerous, especially inside the home. Preferable control is through exposure to sunlight and storage in tight containers with naphthalene, paradichlorobenzene, or camphor.

Killer bees or Africanized honey bees are venomous and very aggressive, capable of killing large animals when swarming. They have spread throughout Central and South America, and have only slowly moved into the southern United States because of climate limitations [13].

7.4 Vertebrates

The use of poisons to regulate pest populations of vertebrates is rarely popular. With the exception of the killing of rats, many people oppose chemical measures (or any measure) to control the higher animals. These animal pests, however, produce considerable damage and can wipe out whole crops of grains or fruits, interfere with planting by feeding on the sown seed, destroy trees through bark stripping, and damage lawns by burrowing.

Easiest to control are the burrowing animals. Gassing the burrows with hydrogen cyanide or phosphine is effective, as is baiting the burrows with poisoned earthworms.

Overpopulation of certain fish species that have no value for angling or fish production may result in a decision to completely remove all fish. This can be done with several fish toxicants such as rotenone, antimycin A, and toxaphene. A certain amount of success results from partial and selective poisoning [14].

A chemical approach that seems acceptable is the use of repellents. The ultimate effect is to force the animal to alternative foods or to starve it. Populations become regulated according to the size of the food supply. In making part of the food supply unavailable through the use of repellents, competition for the available food becomes more intense, resulting in successful and unsuccessful competitors. The effect of repellents is most often too transient for effective control. Since higher animals have a learning capacity, they quickly find out if a repellent is only distasteful and not harmful. Under such conditions, protection is short if the animal is hungry. For a critical period, however, such as seed germination, the use of repellents can be highly effective. Such an example is the coating of corn seed with liquid coal tar to repel crows.

Another chemical approach is the use of nonlethal narcotics, such as chloralose [9]. This can be used to dust grain, which will produce sleepiness in birds shortly after consumption. Selective control of pest birds can be made by dusting large seeds so that small song birds cannot eat them. The undesirable drugged birds can then be destroyed, allowing the desirable ones to recover in a safe place with no ill effect.

The major difficulty with chemical approaches for vertebrate pest control is that enormous testing is necessary for the use of the chemical. Testing with respect to effect on nontarget organisms, biodegradation, lethal dose, and so on is expensive, and many of the vertebrate pest problems, although very serious, are also very limited in scope.

8 Resistance to Chemical Pesticides

Over a period of time of a particular pesticide's use, it is expected that many pesticides will become resistant to the originally targeted pest. Indeed, the problem of insect resistance has stimulated the development of new insecticides. Nearly all of the established insecticides have induced resistance in some insects. The mechanism of insect resistance may be morphological, behavioral, or biochemical. The selective advantage of some adaptation in structure, behavior, or metabolism in members of a population produces resistance in certain insects. Very commonly the successful mutant possesses an enzyme mechanism for detoxifying an organochemical poison, thereby necessitating the use of a different class of compounds. This problem of resistance does not exist as much with the inorganic pesticides, since the possibility for chemical modification by the organism is much less among the inorganic poisons.

A general rule worth observing is that two classes of organic pesticides may not be used simultaneously. Such use encourages the disastrous appearance of a double mutant resistant to both agents. The agents should be used sequentially instead, if needed. Thus, mutants resistant to the first agent will be killed by the second. Further mutation is unlikely to occur without a loss of the initial resistance.

The use of an insecticide like dichlorvos is not ecologically wise according to this rule. Dichlorvos contains the insecticidal moiety of both the organophosphorus and the organochlorine insecticides making it doubly lethal, but also conducive to the selection of a double mutant resistant to both classes of insecticides.

Unfortunately, all the pests that we wish to control are more adaptable than humans to a changing environmental chemistry, and pesticide resistance in insects is common.

Just as the inappropriate use of pharmaceuticals leads to drug-resistant tuberculosis, pest resistance to pesticides is also being accelerated by human error and misuse. For example, when pesticide phosphine gas exposure times are shortened during the fumigation of agriculture or wood commodity to save time, the remaining insects that are not killed, are more resistant to the next application. Pesticide resistance monitoring is a key facet of effective pesticide use and is typically carried out by municipal agricultural extension agencies, research universities, and professional pesticide applicators.

9 Pollution and Human Health Risks from Chemical Pesticides

Increasing concerns regarding the effects of environmental contaminants on human disease or morbidity, as well as, on ecological systems, such as on organisms important to agricultural productivity, like fish, and pollinators, have driven a great deal of community-based "citizen science", research, pollution prevention technologies, and policy outcomes.

9.1 Zone of Influence

The effect of the pesticides on target organisms, as well as on possible fringe or nontarget organisms, is a critical component to be fully examined prior to selection of pesticides as a control measure. There are numerous examples of unintended harm to nontarget organisms from pesticides that were developed for a useful purpose: from the impact of the organochlorine DDT on avian eggs to more recent examples of neonicotinoid pesticides on the navigation ability of insect pollinators.

Tolerance figures are arrived at, for the most part, by studying the influence of the poisoning agent on small samples of organisms under laboratory conditions for relatively short durations. The objective of these studies is the evaluation of the survival or death of the organism as expressed in LC_{50} values. These are a statistical estimate of the concentration of the pesticide required to kill 50% of the population under study. Such values are usually reported for a given length of exposure. Sometimes this information is reported as survival time for a given percentage of organisms at a given concentration of lethal agent. The point is, however, that these data do not tell us much about survival rates in nature since the level of the chemical agent causing death depends on the organism's physiology and total environment. Variation in these factors modifies the laboratory-based LC_{50} value. In addition, to kill or not to kill is only part of the role played by the poisonous agent. The effect on growth and development for nontarget and target organisms that survive is a role of longer duration and far greater complexity that remains to be unraveled for most chemical pesticides.

For aquatic life, toxicity of an agent is commonly reported as a TLm96 value. These aquatic toxicity ratings are defined as the 96-hour static or continuous flow concentration producing a toxic effect. For aquatic life, toxicity of an agent generally decreases with increasing salinity, turbidity, or organic content of the water. These effects appear to be the result of decreasing the concentration of the toxic agent either by decreasing its aqueous solubility or by adsorption to particulate matter. In general, the toxic effect of the pesticide is proportional to temperature, reflecting the increased metabolic activity of organisms to a certain maximum level as temperature is elevated.

A compendium of toxicity values for selected aquatic organisms to various pesticides is the National Institute of Occupational Safety and Health Registry [14].

Chemical pesticides have a wide sphere of influence and thus must be used with caution and concern for the entire influenced zone. Obviously the rule must be to use the least toxic agent and the minimum quantity for the particular job. Consideration should be given to the most sensitive stage of the pest's development in order to permit this rule to be most effective. In addition, the state of the application equipment and weather conditions must be given consideration before commencing operation.

Although a discrete area may be slated for a spraying or other poisoning operation, the nature of the soil, rainfall, and presence of migratory animals, all contribute to the size of the zone of influence.

9.2 Fate and Transport of Pesticides in the Environment

The effect of pesticide use on the environment is of great concern because of possible unintended consequences. These can include impacts on water supply and soil contamination as well as producing injury or death to nontarget species.

Certain pesticides will have more acute and direct lethal and sublethal effects, while others will have indirect habitat damage and destruction, which has longer-term implications on, for example, fish and aquatic invertebrate survival rates.

A variety of factors control the distribution of chemical pesticides on land, air, and water that are both chemically specific, as well as, dependent on the local environmental conditions. Characteristics of individual chemical compounds such as water solubility, vapor pressure, lipophilic nature or octanol-water partition coefficient, and half-life or chemical persistence are important factors to consider when modeling fate and transport. Local environmental conditions such as air and water temperature, rainfall, sun exposure (photodegradation), and redox condition add a dynamic element with both temporal and spatial variability that needs to be considered to fully predict distribution of pesticides in nature. For example, when applying pesticides in and around a warehouse, judicious use of pesticides in the immediate footprint of existing structures and new construction minimizes the potential for run-off that may contain contaminated soils.

For example, neonicotinoid insecticides are very water soluble, which allows them to be applied to soil and taken up by plants. This also makes aquatic ecosystems and associated organisms particularly vulnerable to this class of pesticides. Linkages of neonicotinoid exposure to unintended consequences of honey bee pollinator Colony Collapse Disorder (CCD) have also been observed.

9.3 Biological Magnification

It must also be recognized that immediate influence does not tell the whole story, since biological magnification can eliminate organisms that were not influenced by the initial contact with the agent. Thus, acceptable water concentrations of pesticides such as the organochlorine insecticides have been found to result in concentrations in fish-eating birds that far exceed levels acceptable for human consumption.

This magnification occurs when a substance is not excreted as rapidly as it is ingested. The concentrated dose is then passed on to its predator in the food chain, which in turn concentrates the substance further. These are the effects that are so difficult to access since an actual lethal effect may not appear until several or many years after exposure to the pesticide.

9.4 Toxic Effects on Human Health

Although beyond the scope of this chapter, one must express concern about the possible human health problems that can arise from the application of chemical pesticide poisons to the environment in pest and vector control.

Toxic effects on human health may be acute, chronic, carcinogenic, mutagenic, or teratogenic. A human health risk evaluation or assessment can inform the level risk of a single chemical based on the level of human exposure, toxicity of the chemical, and sensitivity of the target (child vs. adult). Cocktails of chemicals are not typically assessed using this methodology, but is more representative of the real risk, as chemicals do not occur in nature in isolation from each other.

Acute effects imply that the effect is elicited rather quickly (within hours to a few days) following exposure. The acute effect may range from a mild headache and discomfort to death. Chronic effects occur following long exposure to a chemical. Chronic effects are exemplified by chemically induced liver cirrhosis and ulcers. Carcinogenic and mutagenic effects involve the alteration in the DNA of a cell. Teratogenic effects involve the effects a chemical may have on a developing embryo. For example, neurological effects of DDT and its by-products have been suggested to have linkages to Alzheimer disease occurrence [15].

Among the chemicals that humans handle, organophosphorus insecticides rank high among occupational poisonings. High-risk groups include pesticide manufacturers and spray operators. The organophosphates have been largely replaced since the 1990s by the neonicotinoids in the United States.

While an important and active area of research and policy, further discussions on human health risk assessment of acute or chronic effects of chemical pesticides are beyond the scope of this chapter.

10 In Summary and Other Considerations

Coupled with preventative programs for environmental control and consideration of externalities of fate and transport in the environment, as well as, pesticide trace food residues and human health effects, pesticide-based control of pests and vectors is a necessary approach to protect human health and the economy. Within the limitations provided for by state and federal regulation, chemically and biologically based pesticides are used in pest and vector control throughout the world.

Thus, targeted alterations or control of the natural and man-made environment in which the pest occurs along with pesticide use in an Integrated Pest Management program is a reasonable approach for the control of pests and vectors.

The use of any control measure is bound to produce side effects as well as the particularly desired effects. Therefore, the impact of the side effects must be carefully considered by a knowledgeable authority before commencing a control program. State Health Department, State Cooperative extension programs, the Office of Pesticide Programs of the US Environmental Protection Agency, US Department of Agriculture, universities, and consulting firms can provide local guidance to the selection of suitable control measures.

Of increasing presence in the market, though still a small percentage by comparison, is organic (nonpesticide) production of food and animal commodities that relies on an IPM methodology without the use of additional chemical pesticide agents. With greater emphasis on small-scale local and artisanal agriculture and increased research on unintended human and ecological risks, this market share is expected to grow.

The environment belongs to everyone and is, hence, the responsibility of everyone to use wisely. A careful and considered use of any pesticides applied in nature is required to maintain a good stewardship of our local land and water. And, when applied, as needed, we must ensure that the benefits outweigh any longer-term risks.

Glossary

Chemical pesticide It is a chemical substance used to kill harmful insects, small animals, wild plants, and other unwanted organisms: The pesticides that farmers spray on their crops kill pests, but they can also damage people's health.

Environmental control It is the degree to which individuals, groups, or business units can modify and adapt features of their physical workplace or living environment to enhance environmental or public health effectiveness.

Pest It is a destructive insect or other animal that attacks crops, food, livestock, human, etc.

Pollution The presence in or introduction into the environment of a substance or thing that has harmful or poisonous effects.

Risk It is a situation involving exposure to danger.

Vector It is an organism, typically a biting insect or tick, that transmits a disease or parasite from one animal or plant to another.

References

1. Levitt J (1972) Responses of plants to environmental stresses. Academic Press, New York, USA, p 542
2. McNew GL Pest control strategies for the future. National Academy of Sciences, Wash
3. https://cite-archive.mit.edu/reports/food-aid-packaging-evaluation, 2018
4. Compendium of Registered Pesticides, Vol. 1–5, Technical Services Division, Office of Pesticides Programs (updated continually)
5. Frear DEH Pesticide handbook. College Science Publishers, State College, Pennsylvania (annual)
6. R. L. Metcalf, Advances in Pest Control Research, Interscience, New York, USA. I 957- present
7. Lomar CJ et al (1997) Development of Metarhizium spp.for the control of grasshoppers and locusts. Mem Entomol Soc Can 171:265–286
8. WHO, 2010. Spinosad DT in Drinking-water: Use for vector control in drinking-water sources and containers
9. Salvato JA Jr (1972) Environmental engineering and sanitation, 2nd edn. Wiley-Interscience, New York, USA
10. Elsen RJ et al (2013) Evidence that rodent control strategies ought to be improved to enhance food security and reduce the risk of rodent-borne illnesses within subsistence farming villages in the plague-endemic West Nile region, Uganda. Int J Pest Manag 59(4):259–270
11. Moore-Landecker E Fundamentals of the Fungi. Prentice-Hall, Englewood Cliffs
12. USAID, Programmatic Environmental Assessment for Phosphine, Washington, DC, USA. 2013
13. V. Lazaneo, Africanized Honey Facts, University of California, USA, ANR, publication 8068, 2002
14. Registry of Toxic Effects of Chemical Substances, National Institute of Occupational Safety and Health, 1979
15. Richardson JR et al (2014) Elevated serum pesticide levels and risk for Alzheimer disease. JAMA Neurol 71(3):284–290
16. H.D. Pratt, Mosquitoes of Public Health Importance, Center for Disease Control, US Public Health Service, Department of Health, Education and Welfare, USA, 1960

Chapter 11
Book Reviews: Water Environment

Lawrence K. Wang and Mu-Hao Sung Wang

Contents

1 *Wastewater Biology: The Microlife*, Third Edition (2017). Water Environment Federation (WEF), WEF Press, 601 Wythe Street, Alexandria, VA 22314, USA, 445 pp., $95, softcover, ISBN: 978-1-57278-337-9 ... 412
2 *Guidelines for Grit Sampling and Characterization* (2016). Water Environment Federation (WEF) Task Force. WEF Press, 601 Wythe Street, Alexandria, VA 22314-1994, USA, www.wef.org, 105 pp., $125, softcover, ISBN: 978-1-57278-330-0 ... 413
3 *Activated Sludge and Nutrient Removal, Manual of Practice No. OM-9*, Third Edition (2018). Water Environment Federation (WEF), 601 Wythe Street, Alexandria, VA 22314-1994, USA, 450 pp., $125, softcover, ISBN: 978-1-57278-340-9 ... 414
4 *The Nutrient Roadmap*. Water Environment Federation (2015), WEF Press, 601 Wythe Street, Alexandria, VA 22314-1994, USA, 184 pp., softcover, $105., ISBN: 978-1-57278-314-0 ... 416
5 *Wet Weather Design and Operation in Water Resource Recovery Facilities*. Water Environment Federation (2014), 601 Wythe Street, Alexandria, VA 22314, USA, 306 pp., $130, softcover, ISBN: 978-1-57278-304-1 ... 417
6 *Operation of Water Resource Recovery Facilities Study Guide* (2018), Water Environment Federation (WEF), 601 Wythe Street, Alexandria, VA 22314-1994, USA, www.wef.org, 300 pp., $85, softcover, ISBN: 978-1-57278-338-6 ... 417
References ... 419

Abstract The authors have reviewed six technical books published by the Water Environment Federation (WEF), Alexandria, VA, USA, in 2014–2018. These six books are (1) *Wastewater Biology: The Microlife*, Third Edition, 2017; (2) *Guidelines for Grit Sampling and Characterization*, 2016; (3) *Activated Sludge and Nutrient Removal, Manual of Practice No. OM-9*, Third Edition, 2018; (4) *The Nutrient Roadmap*, 2015; (5) *Wet Weather Design and Operation in Water Resource Recovery Facilities*, 2014; and (6) *Operation of Water Resource Recovery Facilities Study Guide*. (2018). All six reviewed books are in the professional area of

L. K. Wang (✉) · M.-H. S. Wang
Lenox Institute of Water Technology, Newtonville, NY, USA

L. K. Wang, M. -H. S. Wang, Y. -T. Hung, N. K. Shammas (eds.),
Environmental and Natural Resources Engineering, Handbook of Environmental Engineering 19, https://doi.org/10.1007/978-3-030-54626-7_11

wastewater treatment and resources recovery. In each book review, the authors introduce the publisher, authors, editors, and previous old editions of the book if they exist. Each book is reviewed and discussed in terms of its technical coverage, professional level, area of applications, affordability to readers, advantages, disadvantages, suitability for international distribution, recommended readership, possible improvements, etc.

Keywords Book review · Water Environment Federation · Environmental engineering · Environmental science · Natural resources · Wastewater treatment · Water resources recovery · Wastewater biology · Grit sampling · Grit characterization · Activated sludge · Nutrient removal · Wet weather design · Water resource recovery · Facilities · Study guide

1 *Wastewater Biology: The Microlife*, Third Edition (2017). Water Environment Federation (WEF), WEF Press, 601 Wythe Street, Alexandria, VA 22314, USA, 445 pp., $95, softcover, ISBN: 978-1-57278-337-9

This book is about the unique microbial world encountered within a biological water resource recovery facility (WRRF). With more than 12 chapters, the authors provide extensive information regarding different types of microlife, measurement techniques, and the operating conditions that affect the microorganisms [1].

> *Wastewater Biology: The Microlife*, Third Edition is an essential reference for wastewater operators and engineers.

For more than 20 years, *Wastewater Biology: The Microlife* has served as an essential resource for operator training and engineers. This new edition explores the organisms that are considered the most important in the treatment of wastewater and disease transmission and provides the readers with the knowledge needed to regulate and control treatment processes properly.

The third edition offers detailed information on bacteria, protozoa, rotifers, nematodes and other metazoans, filamentous organisms, pathogens, and parasites. This new edition also includes a discussion of the microscope and photomicroscopy as well as other tools, equipment, and procedures that are necessary for examining, identifying, and enumerating these life forms.

Key features of the third new edition:

- New! Generously illustrated with more realistic visuals
- New! Now includes 13 color plates to assist in identifying the microlife
- New! Extensive discussion of the tools, equipment, and procedures necessary for proper treatment

- New! Discussion of the microscope and photomicroscopy
- Expanded! The most up-to-date material on the microorganisms critical to disease transmission and the treatment of wastewater
- New! Available as an e-book

Table of Contents:

- Chapter 1: Introduction
- Chapter 2: Microorganisms
- Chapter 3: The Microscope
- Chapter 4: Bacteria
- Chapter 5: Protozoa
- Chapter 6: Rotifers
- Chapter 7: Nematodes and Other Metazoa
- Chapter 8: Filamentous Organisms
- Chapter 9: Indicator Bacteria
- Chapter 10: Wastewater Pathogens
- Chapter 11: Wastewater Parasites
- Chapter 23: Photomicroscopy
- Appendix A Test Procedures and Methods
- Glossary

Both Appendix A (Test Procedures and Methods) and Glossary sections are very useful.

The book has something for all stakeholders involved in wastewater treatment, including design engineers, microbiologists, facility managers, and operators.

As the world grapples with the challenges of population growth, urbanization, and water scarcity, the effects will be felt on WRRFs and mostly the microorganisms that are at the thick of the action. This book will be an ideal reference companion for WRRF managers and technicians who deal with various operational challenges.

2 *Guidelines for Grit Sampling and Characterization* (2016). Water Environment Federation (WEF) Task Force. WEF Press, 601 Wythe Street, Alexandria, VA 22314-1994, USA, www.wef.org, 105 pp., $125, softcover, ISBN: 978-1-57278-330-0.

This special publication is a typical example of how a group of water professionals unselfishly compile their knowledge and experience together for dissemination to their colleagues [2].

Thirty-five environmental engineers and scientists formed a Grit Sampling and Characterization Task Force and jointly prepared this special publication. It was conducted under the leadership of Task Force Chair Kendra Sveum and Vice Chair

Lucas Botero under the direction of the Municipal Design Subcommittee of the Technical Practice Committee in which 30 more water professionals helped review the manuscript before its publication. The work of these volunteers is especially noteworthy because grit sampling and characterization are small, narrow subjects, which are normally ignored by engineers and managers.

In this publication, the currently used grit sampling and characterization methods are introduced and reviewed, and modeling and expectations are discussed.

The book consists of four chapters. Chapter 1 introduces the grit characteristics, historical and current definitions of wastewater solids, and the roles of service utilities, consulting engineers, testing companies, and grit removal equipment manufacturers. Chapter 2 describes the various methods and equipment used for obtaining a grit sample, such as bucket sampling, single-point pumped sampling, multiple-port vertical sampling, vertically integrated slotted sampling, manifold pumped sampling, siphon cross-channel sampling, and complete influent mixing/sampling. Chapter 3 covers various methods for pretreatment, solids determination, settling velocity determination, grit particle size and shape analysis, and calculations. Chapter 4 focuses on international practices, design/operational expectations, grit characterization modeling, and future expectations.

This book is meant for utility owners, consulting engineers, and grit equipment manufacturers who want to understand the implications and selection of a grit sampling and characterization method for proper grit chamber design and grit removal optimization. The book is not intended to address various types of grit removal equipment and grit slurry processing equipment, evaluate the equipment performance, or discuss the grit removal and slurry processing costs.

For further improvement of this excellent publication, practical examples should be added in Chapter 3, Section 5 "Calculations," so licensed wastewater operators with a high school education can be included as intended readers. Normally, operators at a resource recovery facility will operate the grit chambers and take grit samples for analysis.

3 *Activated Sludge and Nutrient Removal, Manual of Practice No. OM-9*, Third Edition (2018). Water Environment Federation (WEF), 601 Wythe Street, Alexandria, VA 22314-1994, USA, 450 pp., $125, softcover, ISBN: 978-1-57278-340-9.

This extensively revised third edition of *Activated Sludge and Nutrient Removal, Manual of Practice No. OM-9*, reflects best practices and the latest advances. This is one of the best books ever published by WEF. The book is not only a primary reference for the design and operation of the activated sludge process but also an excellent supplemental college textbook for environmental biotechnology courses [3].

This book was published by the Water Environment Federation (Alexandria, VA) and prepared by a WEF task force under the direction of the Plant Operations and Maintenance Subcommittee of the Technical Practice Committee (TPC). Congratulations are deserved for task force chair Barton Jones, the 26 task force members, TPC chair Eric Rothstein, the 28 TPC control group members, and the 15 outstanding biological wastewater treatment experts for a job well done.

The table of contents of this book introduces the knowledge to be learned by readers and gives credits to the dedicated scholars who have contributed to the book chapters. Some of the chapter topics are activated sludge process fundamentals, process control methods for the secondary clarifier, advanced concepts in nutrient removal, and process simulators.

All topics in the book are important for design, operation, and learning of the activated sludge process and nutrient removal. Contributing authors are all experienced and qualified biotechnologists, professional engineers, or educators. The task force chair/editor did an excellent job; all chapters are very smooth in terms of presentation although each chapter is contributed by different authors.

This five-star book includes an updated process control section, with step-by-step examples for calculations; a new laboratory chapter, with detailed directions for common process control tests; and an introduction to using modeling for process control. Design engineers, facility managers, operators, educators, students, and researchers may find this book very useful because they may be guided through selecting an appropriate sludge age, calculating wasting rates, optimizing return activated sludge flow, managing clarifier blankets, and setting dissolved oxygen and oxidation-reduction potential set points.

This book has many other useful features and benefits. For instance, both US customary and international metric units are adopted. There are ample figures, tables, equations, and practical examples for illustration and explanation of theories, principles, biochemistry, microbiology, and engineering applications of activated sludge process and nutrient removal. Practice questions with answers have been added to the end of each chapter to help the reader enhance his or her understanding of the material and retain information vital to solving performance problems and improving operations. The book also discussed such modern activated sludge processes as membrane bioreactors and sequencing batch bioreactors. Updated references are provided in each chapter for further study by researchers.

The reviewers cannot find any shortcomings. However, future items to include to further enhance these resources are a list of symbols and acronyms, conversion factors, a glossary, and examples of other modern secondary clarification, such as secondary dissolved air flotation clarification.

4 *The Nutrient Roadmap*. Water Environment Federation (2015), WEF Press, 601 Wythe Street, Alexandria, VA 22314-1994, USA, 184 pp., softcover, $105., ISBN: 978-1-57278-314-0

This is a reference book prepared by the Nutrient Roadmap Task Force of the Water Environment Federation (WEF) for guiding decision-makers to retool their wastewater treatment plants (WWTPs) into water resource recovery facilities (WRRFs) that not only produce clean plant effluent but also recover phosphorus and nitrogen for reuse in an environmentally conserving manner. The reviewers give credits to Dr. Samuel S. Jeyanayagam, Task Force Chair, and his 83 task force members for publication of this nutrient removal/recovery roadmap [4].

The book has seven chapters covering the important subjects of nutrient roadmap purpose, nutrient removal/recovery drivers and trends, staff education, financing, planning, the US regulatory requirements, receiving water environment, WRRF existing performance, nutrient sources and characteristics, process models and modifications/additions, effluent limits, permit conditions, water-nutrient-energy nexus, nutrient removal/recovery systems, nonpoint source control technologies, point source control technologies, mainstream treatment, sidestream treatment, biosolids management and recovery options, water recycle, other resources recovery, alternatives evaluation steps, risks identification and management, and case studies. Although all seven chapters are equally important, Chapter 4 ("Resource Recovery"), Chapter 5 ("Alternatives Evaluation"), and Chapter 7 ("Case Studies") actually show the readers what the nutrient removal/recovery roadmap is. An experienced environmental engineer may adopt the roadmap for determination of a feasible nutrient removal/recovery system for his/her clients.

It should be noted, however, the book is a timely, high-level overview document that provides useful guidance to only experienced decision-makers and other knowable parties, such as regulatory managers, public health professionals, professors, and researchers. For an in-depth technological discourse, the readers are directed to other publications, including the USEPA manuals and the WEF Manuals of Practice and special publications listed at the end of each chapter.

For international adoption and distribution, the book has adopted both the US customary units and the international metric units. Further improvements of the book may include (1) addition of a glossary of terms section and (2) inclusion of some international regulatory requirements for nutrient removal/recovery. Overall it is a five-star, excellent reference book.

5 *Wet Weather Design and Operation in Water Resource Recovery Facilities*. Water Environment Federation (2014), 601 Wythe Street, Alexandria, VA 22314, USA, 306 pp., $130, softcover, ISBN: 978-1-57278-304-1.

This special publication was prepared by the Wet Weather Design and Operation in Water Resource Recovery Facilities Task Force of the Water Environment Federation (WEF) and the WEF Technical Practice Committee Control Group [5].

Forty-three experts on wet weather facility design and operation contributed 16 chapters covering design considerations, integrated systems approach for design and operation, site selection and facility arrangement, facility hydraulics and pumping, support systems, occupational health and safety, modeling for wet weather, on-site storage and flow equalization, preliminary treatment, primary treatment, biological treatment, high-rate treatment, disinfection, and residuals. All chapters are grouped around planning and configuration for wet weather events and facility processes.

Both US customary units and metric units are used side-by-side—except in the figures. The text presentation style is consistent throughout the entire book. Every chapter has a list of updated references and/or suggested readings. The book also presents many useful planning examples or case studies for wet weather facility design and operation.

Overall, this is a five-star book, which provides professionals involved in the design and operation of water resource recovery facilities (WRRFs) with a comprehensive reference of modern practices for dealing with the unique challenges associated with wet weather conditions. For detailed design calculations, the book refers readers to another book, *Design of Municipal Wastewater Treatment Plants* (WEF MOP No. 8). For additional planning reference, readers are referred to a guidebook, *Guide for Municipal Wet Weather Strategies*.

6 *Operation of Water Resource Recovery Facilities Study Guide* (2018), Water Environment Federation (WEF), 601 Wythe Street, Alexandria, VA 22314-1994, USA, www.wef.org, 300 pp., $85, softcover, ISBN: 978-1-57278-338-6

This study guide was prepared by a WEF Task Force of the Technical Practice Committee Control Group (Chair Eric Rothstein and Vice Chair D. Medina) and authored by B. Ahmed, J. A. Brown, V. deSilva, G. Grissop, N. Jensen, M. Kasi, I. LaRue, D.M. Mason, and N. Naureen [6].

The study guide is a companion to two other WEF publications: (1) the seventh edition of *Operation of Water Resource Recovery Facilities*, Manual of Practice

(MOP) 11, and (2) a WEF video, *Water Resource Recovery Facility 3D Virtual Tour*. The video, manual, and study guide should be used together for training classes, studying for certification exams, and improving the quality of operations within the facility or firm.

The video can be downloaded from the Internet free of charge (news.wef.org/wef-teaches-about-operation-of-water-resource-recovery-facility-2/). It is a 10-minute, narrated educational video providing an excellent virtual tour of a typical water resource recovery facility about wastewater treatment, water reuse, biosolids generation, land application, energy generation, anaerobic digestion, facility infrastructure, and wastewater treatment operation. The readers should download and view the WEF video first.

The next step for the readers is reading the official training material MOP 11, *Operation of Water Resource Recovery Facilities*, and learning the state-of-the-art development in water resource recovery facility management and operation.

Finally the questions and solutions in this study guide can be used by the readers for a comprehensive review and examination preparation. The video, training manual MOP11, and this study guide all emphasize principles of treatment, facility management, troubleshooting, and preventive maintenance.

Specifically this study guide contains challenging questions and detailed solutions to help readers with continuing education as well as on the job demands. These questions can be used to help develop advanced knowledge and ensure that your water resource recovery facility is fulfilling its mission of environmental protection. The following is this study guide's table of contents: Chapter 1, "Introduction"; Chapter 2, "Permit Compliance and Wastewater Treatment Systems" (20 problems and solutions); Chapter 3, "Fundamentals of Management" (20 problems and solutions); Chapter 4, "Industrial Wastes and Pretreatment" (15 problems and solutions); Chapter 5, "Safety" (22 problems and solutions); Chapter 6, "Management Information Systems—Reports and Records" (13 problems and solutions); Chapter 7, "Process Instrumentation" (19 problems and solutions); Chapter 8, "Pumping of Wastewater and Sludge" (16 problems and solutions); Chapter 9, "Chemical Storage, Handling, and Feeding" (17 problems and solutions); Chapter 10, "Electrical" (12 problems and solutions); Chapter 11, "Utilities" (12 problems and solutions); Chapter 12, "Maintenance" (25 problems and solutions); Chapter 13, "Odor Control" (12 problems and solutions); Chapter 14, "Integrated Process Management" (20 problems and solutions); Chapter 15, "Outsourced Operations Services and Public/Private Partnerships" (13 problems and solutions); Chapter 16, "Training" (16 problems and solutions); Chapter 17, "Wastewater Sampling" (14 problems and solutions); Chapter 18, "Preliminary Treatment" (10 problems and solutions); Chapter 19, "Primary Treatment" (14 problems and solutions); Chapter 20, "Activated Sludge" (24 problems and solutions); Chapter 21, "Trickling Filters, RBC, Combined Processes" (15 problems and solutions); Chapter 22, "Biological Nutrient Removal" (20 problems and solutions); Chapter 23, "Natural Biological Processes" (20 problems and solutions); Chapter 24, "Physical-Chemical Treatment" (16 problems and solutions); Chapter 25, "Process Performance" (12 problems and solutions); Chapter 26, "Effluent Disinfection" (15 problems and solutions); Chapter 27,

"Solids Management" (12 problems and solutions); Chapter 28, "Sludges and Residuals" (12 problems and solutions); Chapter 29, "Thickening" (13 problems and solutions); Chapter 30, "Anaerobic Digestion" (10 problems and solutions); Chapter 31, "Aerobic Digestion" (22 problems and solutions); Chapter 32, "Stabilization Methods" (12 problems and solutions); and Chapter 33, "Dewatering" (14 problems and solutions).

Both the international standard units and the US customary units are adopted in this study guide for international distribution. The study guide's listing cost of $85 is reasonable. Many subjects, such as permit compliance, industrial wastewater pretreatment, physical-chemical treatment, UV disinfection, DAF thickening, nutrient removal, energy recovery, instrumentation, safety, etc., are either first time included or further emphasized. The WEF video, training material MOP 11, and this study guide are all excellent self-study references for facility managers, superintendents, wastewater operators, college students, and consulting engineers.

The already excellent WEF training materials set (video, MOP 11, and this study guide) may still be further improved by adding important resource recovery processes and technologies, such as membrane filtration (microfiltration, ultrafiltration, nanofiltration, reverse osmosis), membrane bioreactor, sequencing bioreactor, primary DAF clarification, secondary DAF clarification, powdered activated carbon adsorption, oil-water separation, wet land treatment, living machines, nutrient recovery, heavy metal recovery, carbon dioxide recovery, VOC reduction, activated carbon regeneration, chemical sludge treatment/recovery, etc.

References

1. Wastewater Biology: The Microlife, Third Edition (2017). Water Environment Federation (WEF), WEF Press 601 Wythe Street, Alexandria, VA 22314, USA; 445 pages. ISBN 978–1–57278-337-9
2. Guidelines for Grit Sampling and Characterization (2016). Water Environment Federation (WEF) Task Force. WEF Press, 601 Wythe Street, Alexandria, VA 22314–1994, USA. www.wef.org, 105 pp. ISBN: 978–1–57278-330-0
3. Activated Sludge and Nutrient Removal, Manual of Practice No. OM-9, Third Edition (2018). Water Environment Federation (WEF), 601 Wythe Street, Alexandria, VA 22314–1994, USA; 450 pp., softcover, ISBN: 978–1–57278-340-9
4. The Nutrient Roadmap (2015). Water Environment Federation WEF Press, 601 Wythe Street, Alexandria, VA 22314–1994, USA, 184 pp., softcover, ISBN 978–1–57278-314-0
5. Wet Weather Design and Operation in Water Resource Recovery Facilities (2014). Water Environment Federation, 601 Wythe Street, Alexandria, VA 22314, USA, 306 pp., softcover, ISBN: 978–1–57278-304-1
6. Operation of Water Resource Recovery Facilities Study Guide (2018). Water Environment Federation (WEF), 601 Wythe Street, Alexandria, VA 22314–1994, USA; www.wef.org; 300 pp., softcover, ISBN :978–1–57278-338-6

Chapter 12
Glossary of Natural Resources and Environmental Pollution Control

Mu-Hao Sung Wang and Lawrence K. Wang

Contents

Glossary of Natural Resources and Environmental Pollution Control 423
References 494

Abstract This chapter presents the technical terms and acronym terms in the wide academic areas of environmental engineering and natural resources management. The authors have tried their best to selectively cover the subject areas in the Handbook of Environmental Engineering (HEE), Volume 19 (Environmental and Natural Resources Engineering), Volume 20 (Integrated Natural Resources Management), Volume 21 (Environmental Flotation Engineering), and Volume 22 (Integrated Natural Resources Research).

Keywords Glossary · Acronym · Environmental engineering · Natural resources · Bees · Waste reclamation · Water reuse · Biological processes · Water resources protection · Endocrine disruptors removal · Environmental protection · Cooling · Thermal pollution control · Thermal energy reuse · Hydrology · Dissolved air flotation boat plant · Cadmium detoxification · Sintering · Ceramic matrices · Vegetable oil · Refining waste treatment · Education · Pests and vectors control · Books review · Global warming · Climate change · Glacier protection · Salmon protection · Community-based latrine development · Engineers Without Borders-USA · Analysis for surface water quality management · Water quality control of tidal rivers and estuaries · Geographic information systems · Remote sensing applications, · Water loss investigation and management · Wet infrastructure · Mathematical energy models and cost models · Water pollution control · Activated carbon treatment · Laundry wastes treatment · Dissolved protein · Hazardous waste management · Good laboratory practice · Surfactant analysis · Flotation Engineering; Lenox Institute of Water Technology · Pittsfield Water Treatment Plant · Potable

---

M.-H. S. Wang · L. K. Wang (✉)
Lenox Institute of Water Technology, Newtonville, NY, USA

L. K. Wang, M. -H. S. Wang, Y. -T. Hung, N. K. Shammas (eds.),
Environmental and Natural Resources Engineering, Handbook of Environmental Engineering 19, https://doi.org/10.1007/978-3-030-54626-7_12

water dissolved air flotation plants · Independent physicochemical wastewater treatment · Primary flotation clarification · Secondary flotation clarification · Acid rain · Ozone layer protection · Sludge treatment

Acronym

BC	Black carbon
BOD	Biochemical oxygen demand
CBOD	Carbonaceous biochemical oxygen demand
CCS	Carbon capture and sequestration
CDE	Carbon dioxide equivalent
CERT	Community Emergency Response Team
COD	Chemical oxygen demand
CPG	Comprehensive Preparedness Guide
CWA	Clean Water Act
DAF	Dissolved air flotation
DAFF	Dissolved air flotation and filtration.
DEM	Digital Elevation Model
DGF	Dissolved gas flotation
DO	Dissolved oxygen
ED	Endocrine disruptor
EDC	Endocrine-disrupting compounds
FEMA	Federal Emergency Management Agency
F-Cell	Dissolved air flotation cell, or DAF
FF-Cell	Dissolved air flotation and filtration cell or DAFF
GAC	Granular activated carbon
GAO	Glycogen-accumulating organisms
GIS	Geographic Information System
GW	Global warming
GWP	Global warming potential
HC	Hydrocarbons
HCFC	Hydrochlorofluorocarbons
HRT	Hydraulic retention time
IE	Industrial ecology
IPCC	Intergovernmental Panel on Climate Change
KAMWT	Krofta Advanced Municipal Water Treatment
MBR	Membrane bioreactor
MGD	Million gallons per day
ML	Mixed liquor
MLSSs	Mixed liquor suspended solids
$MMTCO_2Eq$	Million metric tons of carbon dioxide equivalents
MPN	Most probable number
N	Nitrogen
N_2	Nitrogen gas

N_2O	Nitrous oxide
NMVOC	Nonmethane volatile organic compound
NOx	Nitrogen oxides
NPDES	National Pollutant Discharge Elimination System
O_3	Ozone
ORP	Oxidation reduction potential
PAC	Powdered activated carbon
PAO	Phosphorus-accumulating organism
PFC	Perfluorocarbons
PM	Particulate matter
PM2.5	Fine particles
ppb	Parts per billion
ppmv	Parts per million by volume
ppt	Parts per trillion
RAS	Returned activated sludge
SBOD	Soluble biochemical oxygen demand
SPDES	State Pollutant Discharge Elimination System
SVI	Sludge volume index
TKN	Total Kjeldahl nitrogen
TN	Total nitrogen
TP	Total phosphorus
VFAs	Volatile fatty acids
VIP	Ventilated improved pit latrine
WAS	Waste activated sludge
WASH	Water, Sanitation, and Hygiene

Glossary of Natural Resources and Environmental Pollution Control [1–18]

100-Year flood levels Severe flood levels with a one-in-100 likelihood of occurring in any given year.

Abrupt climate change Sudden (on the order of decades), large changes in some major component of the climate system, with rapid, widespread effects.

Absconding swarm It is an entire colony of bees, which abandons the hive, often because of disease, wax moth, excessive heat or water, lack of resources, or other reasons.

Absconding swarm It is an action when an entire colony of bees abandons the hive, often because of disease, wax moth, excessive heat or water, lack of resources, or other reasons

Acidity It is the capacity of water or wastewater to neutralize bases. Acidity is expressed in milligrams per liter of equivalent calcium carbonate.

Acinetobacter A common heterotrophic bacteria that functions as a phosphorus-accumulating organism (PAO) in biological phosphorus removal process.

Activated carbon adsorption, granular Granular activated carbon (GAC) is used in both water and wastewater treatment. GAC systems generally consist of vessels in which the carbon is placed, forming a "filter" bed. These systems can also include carbon storage vessels and thermal regeneration facilities. Vessels are usually circular for pressure systems or rectangular for gravity flow systems. Once the carbon adsorptive capacity has been fully utilized, it must be disposed of or regenerated. Usually multiple carbon vessels are used to allow continuous operation. Columns can be operated in series or parallel modes. All vessels must be equipped with carbon removal and loading mechanisms to allow for the removal of spent carbon and the addition of new material. Flow can be either upward or downward through the carbon bed. Vessels are backwashed periodically. Surface wash and air scour systems can also be used as part of the backwash cycle. Small systems usually dispose of spent carbon or regenerate it offsite. Systems above about 3 to 5 MGD (million gallons per day) usually provide on-site regeneration of carbon for economic reasons.

Activated carbon adsorption, powdered Powdered activated carbon (PAC) is used in both water and wastewater treatment. For water treatment, PAC systems generally involve dosing powdered activated carbons in slurry form to rapid mixers and flocculators (for mixing and flocculation, and adsorption) before reaching clarifiers for PAC-floc separation and clarifier effluent discharge. For wastewater treatment, powdered activated carbon in slurry form is dosed to either mixing/flocculation chambers (if they exist), or to an aeration basin directly to be mixed with mixed liquor for adsorption reactions. The spent PAC again is recovered or removed by a subsequent clarifier (either sedimentation clarifier or flotation clarifier). When PAC is used in conjunction with a flotation clarifier, it is also called "adsorption flotation."

Activated carbon It is a porous carbon adsorbent with many micropores, microchannels, and high surface area for adsorption (removal) of mainly soluble organic substances, color, odor, and some inorganic heavy metals from water or air. There are three kinds of activated carbons: (a) powdered activated carbon; (b) granular activated carbon; and (c) cloth or fiber-type activated carbon.

Activated sludge They are sludge particles produced in raw or settled wastewater (primarily effluent) by the growth of organisms (including zoogleal bacteria) in aeration tanks in the presence of dissolved oxygen. The term "activated" comes from the fact that the particles are teeming with bacteria, fungi, and protozoa. Activated sludge is different from primary sludge in that the sludge particles contain many living organisms, which can feed on the incoming wastewater.

Activated sludge process (a) It is a continuous flow, biological treatment process characterized by a suspension of aerobic microorganisms, maintained in a relatively homogeneous state by the mixing and turbulence induced by aeration. The microorganisms are used to oxidize soluble and colloidal organics to carbon dioxide and water in the presence of molecular oxygen. The process is generally, but not always, preceded by primary clarification (either primary sedimentation clarification or primary flotation clarification). The mixture of microorganisms and wastewater formed in the aeration basins, called mixed liquor (ML), is transferred to secondary clarification (either secondary sedimentation clarification or secondary flotation clarification) for liquid-solid separation. The major portion of the microorganisms separated from the secondary clarifiers (secondary sedimentation clarifiers or secondary flotation clarifiers) is recycled to the aeration basins to be mixed with incoming wastewater, while the excess, which constitutes the waste activated sludge (WAS), is sent to the sludge-handling facilities. The rate and concentration of the returned activated sludge (RAS) returned to the aeration basins determines the mixed liquor suspended solids (MLSSs) level developed and maintained in the aeration basins. During the oxidation process, a certain amount of the organic material is synthesized into new cells, some of which then undergoes auto-oxidation (self-oxidation, or endogenous respiration) in the aeration basins, the remainder forming net growth or excess sludge. Oxygen is required in the process to support the oxidation and synthesis reactions. In the conventional activated sludge plant, the wastewater is commonly aerated for a period of 4 to 8 hours (based on average daily flow) in a plug flow hydraulic mode. Either surface or submerged aeration systems can be employed to transfer oxygen from air to wastewater. (b) It is a suspended-growth biological wastewater treatment process, which speeds up the decomposition of wastes in the wastewater being treated. Activated sludge is added to wastewater, and the mixture (mixed liquor) is aerated and agitated. After some time in the aeration tank, the activated sludge is allowed to settle out by sedimentation and is disposed of (wasted) or refused (returned to the aeration tank) as needed. The remaining wastewater then undergoes more treatment.

Adaptation Adjustment or preparation of natural or human systems to a new or changing environment, which moderates harm or exploits beneficial opportunities.

Adaptive capacity The ability of a system to adjust to climate change (including climate variability and extremes) to moderate potential damages, to take advantage of opportunities, or to cope with the consequences.

Adsorption flotation When powdered activated carbon (PAC) is used in conjunction with a flotation clarifier for water or wastewater treatment, the unit process is also called "adsorption flotation."

Advanced waste treatment It is any process of water renovation that upgrades treated wastewater beyond secondary treatment to meet specific reuse requirements. Typical processes include chemical treatment, granular activated carbon, and pressure filtration. Also called tertiary treatment.

Advisory A notice that highlights special weather conditions that are less serious than a warning. They are for events that may cause significant inconvenience, and if caution is not exercised, the conditions could lead to situations that may threaten life and/or property.

Aeration The process of adding air to water. In wastewater treatment, air is added to refresh wastewater and to keep solids in suspension. With mixtures of wastewater and activated sludge, adding air provides mixing and oxygen for the microorganisms treating the wastewater.

Aerobes Bacteria that must have molecular (dissolved) oxygen (DO) to survive.

Aerobic bacteria These bacteria live, remove BOD, and reproduce themselves under aerobic environmental conditions containing sufficient dissolved oxygen (DO).

Aerobic digestion The process equipment of aerobic digestion is similar to that of the activated sludge process. Aerobic digestion is not a suspended-growth biological process in principle, and is for sludge treatment (not for wastewater treatment). It is a suspended-destruction biological process. In the presence of oxygen, the biodegradable matter in the activated sludge (biosolids) is oxidized, destroyed, and converted to carbon dioxide and water. Oxygen must be supplied in direct proportion to the mass of volatile solids (VSs) destroyed. Sludge stabilization is not complete until there has been an extended period of primarily endogenous respiration (15 to 20 days).

Aerobic An environmental condition in which free and dissolved oxygen is available in an aqueous environment (For instance, nitrification is an aerobic process).

Aerosols Small particles or liquid droplets in the atmosphere that can absorb or reflect sunlight depending on their composition.

Afforestation Planting of new forests on lands that historically have not contained forests.

After-action findings Documented results of what in your emergency plan worked and did not work after an incident happens. These findings are used to improve your plan and procedures.

Air lift It is a type of pump. This device consists of a vertical riser pipe in the wastewater or sludge to be pumped. Compressed air is injected into a tall piece at the bottom of the pipe. Fine air bubbles mix with the wastewater or sludge to form a mixture lighter than the surrounding water, which causes the mixture to rise in the discharge pipe to the outlet. An air lift pump works like the center of a stand in a percolator coffee pot.

Albedo The amount of solar radiation reflected from an object or surface, often expressed as a percentage.

Algae They are microscopic plants, which contain chlorophyll and live floating or are suspended in water. They also may be attached to structures, rocks, or other similar substances. Algae produce oxygen during sunlight hours and use oxygen during night hours. Their biological activities appreciably affect the pH and dissolve oxygen of the water.

Aliquot It is a portion of a sample, which is often an equally divided portion of a sample.

Alkali bee (*Nomia melanderi*) These bees are important pollinators, especially of alfalfa. Reared in large numbers, the bees are released in areas near this crop, resulting in a greatly increased yield of alfalfa. Each female takes about 30 days to build a single nest of about 20 cells. The bees hibernate in the prepupa stage and complete their life cycle the second year.

Alkalinity The capacity of water to neutralize acids by the water's content of carbonates, bicarbonates, hydroxide, and other compounds, also known as buffering capacity.

Alternative energy Energy derived from nontraditional sources (e.g., compressed natural gas, solar, hydroelectric, wind).

Ammonia nitrogen Elemental nitrogen present in the form of ammonia (NH_3).

Ammonia stripping It is a physicochemical desorption process for removing ammonia content from a wastewater by a gas-stripping operation. In the process, wastewater at elevated pH is pumped to the top of a packed tower with a countercurrent flow of air drawn through the bottom openings. Free ammonia (NH_3) is stripped from the falling water droplets into the air stream, which is then discharged to the atmosphere. Lime or caustic soda is added prior to the stripping to raise the pH of the wastewater to the range of 10.8 to 11.5, converting essentially all ammonium ions to ammonia gas, which can then be stripped by air. Process controls required for the operation are the proper pH adjustment of the influent wastewater, and maintenance of proper air and water flows. Ammonia removal efficiency is highly dependent on air temperature and air/water ratios. As the air temperature decreases, the efficiency drops significantly. For instance, the ammonia removal efficiency can be 75% at air temperature of 10 ° C, while the ammonia removal efficiency can be 90–95% at air temperature of 20 ° C. Very poor ammonia removal efficiency will be experienced in cold weather location (0 to 10 ° C). The influent wastewater should always be clarified before stripping. Residual ammonia in the ammonia stripping effluent can be removed by breakpoint chlorination. The countercurrent flow ammonia stripping process is technically and economically feasible for a wastewater containing about 10–100 mg/L of ammonia.

Ammonification A biochemical or chemical process in which soluble organic nitrogen is converted to ammonia.

Anaerobic (a) An environmental condition in which free, dissolved, and combined oxygen is unavailable in an aqueous environment (i.e., dissolved oxygen = 0 mg/L); (b) A condition in which atmospheric or dissolved molecular oxygen is NOT present in the aquatic (water) environment.

Anaerobic bacteria They are bacteria that live and reproduce in an environment containing no "free" or dissolved oxygen. Anaerobic bacteria obtain their oxygen supply by breaking down chemical compounds, which contain oxygen, such as sulfate ($SO_4{}^{2-}$).

Anaerobic digestion This is a biological destruction process for destroying biodegradable organic sludge (biosolids) in the absence of molecular oxygen, or under anaerobic conditions. It is a three-stage sludge stabilization process for sludge volume reduction: (a) hydrolysis stage: breakdown of particulate matter and large macromolecules; (b) acid formation stage: fermentation of the soluble organic matter formed in the first reaction to volatile acids; (c) methanogenesis stage: conversion of the volatile acids to the stable end produces, such as methane gas, carbon dioxide gas, and water. Four anaerobic digestion operational modes are available: (a) standard rate digestion mode, or one stage, unheated, and unmixed;

(b) high rate digestion mode: one stage, heated, and mixed; (c) two-stage digestion mode; and (d) anaerobic contact process. Anaerobic digestion can occur in two temperature ranges: mesophilic (80–110 ° F) and thermophilic (113–149 ° F). Mesophilic range operation is more stable, while thermophilic range operation offers improved sludge dewatering. The process reduces volatile solids and thereby reduces the overall mass of the sludge, lowering subsequent land disposal costs. Wastewater solids and water (about 5% solids, 95% water) are placed in a large tank where bacteria decompose the solids in the absence of dissolved oxygen.

Analysis (chemical) The determination of the qualitative and/or quantitative composition of a substance.

Analytical reagent (AR) The American Chemical Society's designation for the highest purity of certain chemical reagents and solvents.

Anchor booms They are parts fitted on some dredges for handling and placing the swing anchors from the dredge without the need of a support vessel, or mounted at the bow and suspended from a gantry or gallows frame.

Anionic polymer (or anionic polyelectrolyte) A negatively charged electrolyte used as a coagulant to aid in separating (settling or floating) solids in water or wastewater.

Anionic surfactant A negatively charged surface active agent, such as ABS and LAS, that lowers the surface tension in water, and usually used as a cleaning agent, or as a collector in a flotation process system.

Annual plant It is a plant that completes its life cycle from germination to the production of seed within one year and then dies.

Anoxic (a) An environmental condition in which oxygen is only available in a combined form, such as nitrate (NO_3), nitrite (NO_2), or sulfate (SO_4) in an aqueous environment; (b) oxygen deficient or lacking sufficient oxygen, and dissolved oxygen = 0.0 to 0.1 mg/L.

Anthropogenic Made by people or resulting from human activities. Usually used in the context of emissions that are produced as a result of human activities.

Apiary It is a bee year or area when/where colonies, hives, and other equipment assembled in one location for beekeeping operations, also known as a bee yard.

Apiculture It is a career, or business, or hobby of beekeeping; the keeping of bees, particularly on a commercial scale.

Approach The difference in temperatures of water temperature exiting the tower versus the dry bulb temperature of the air entering the tower.

AquaDAF It is a rectangular dissolved air flotation (DAF) clarifier commercially available from SUEZ Water Technologies and Solutions, Richmond, VA 23229, USA.

Assimilation The biological or microbiological process in which nitrogen is taken in by cells to be used for growth and reproduction.

Atmosphere The gaseous envelope surrounding the Earth. The dry atmosphere consists almost entirely of nitrogen (78.1% volume mixing ratio) and oxygen (20.9% volume mixing ratio), together with a number of trace gases, such as argon (0.93% volume mixing ratio), helium, radiatively active greenhouse gases such as carbon dioxide (0.035% volume mixing ratio), and ozone. In addition the atmosphere contains water vapor, whose amount is highly variable but typically 1% volume mixing ratio. The atmosphere also contains clouds and aerosols.

Atmospheric lifetime Atmospheric lifetime is the average time that a molecule resides in the atmosphere before it is removed by chemical reaction or deposition. This can also be thought of as the time that it takes after the human-caused emission of a gas for the concentrations of that gas in the atmosphere to return to natural levels. Greenhouse gas lifetimes can range from a few years to a few thousand years.

Attached-growth biological processes In an attached-growth biological process system, the biomass responsible for biochemical reactions is attached on the surface of rocks, sands, woods, disks, screens, etc., where the wastewater passes through, and the organic pollutants and nutrients are removed. They include: trickling filters, denitrification filter using coarse media; denitrification filter using fine media, slow sand filters, rotating biological contactors, etc. The effluent from an attached-growth bioreactor discharges to a water-solid separator (sedimentation clarifier, or flotation clarifier, or membrane filter) for further processing.

Attribute It is a characteristic of a geographic feature, typically stored in tabular format and linked to the feature in a relational database. The attributes of a well-represented point might include an identification number, address, and type.

Autotrophic organisms Organisms that use carbon dioxide, sunlight, or other inorganic substances for their cell growth (Nitrifying bacteria are autotrophic).

Auxiliary engine It is used to suitably power the hydraulic system and other auxiliary functions on certain type of dredges.

Bacteria They are living organisms, microscopic in size, which usually consist of a single cell. Most bacteria use organic matter for their food and produce waste products as the result of their life processes.

Baffle It is a flat board or plate, deflector, guide, or similar device constructed or placed in flowing water, wastewater, or slurry systems to cause more uniform flow velocities, to absorb energy, and to divert, guide, or agitate liquids (water, chemical solutions, slurry).

Bar screens There are two types of bar screens (or racks). The most commonly used, and oldest technology, consists of hand-cleaned bar racks. These are generally used in smaller wastewater treatment plants (WWTPs). The second type of bar screen is the type that is mechanically cleaned, which is commonly used in larger facilities.

Base layer It is a primary layer for spatial reference, upon which other layers are built. Examples of a base layer typically used are either the parcels, or street centerlines.

Base, alkalinity (a) A substance, which takes up or accepts protons; (b) A substance, which dissociates (separates) in aqueous solution to yield hydroxyl ions (OH^-); (c) A substance containing hydroxyl ions, which reacts with an acid to form a salt or which may react with metals to form precipitates; (d) A substance that may raise pH.

Basic plan A component of an emergency operations plan that describes expected hazards, outlines roles and responsibilities, and explains how you keep the plan current.

Basin The bottom of the tower, water is collected here after passing the tower.

Beach restoration It is the act of replenishing the sand loss that occurs naturally over time from tidal cycles, currents, waves, and inclement weather. Beaches that have been restored by dredging are designed to protect local communities from future storm damage and to preserve recreational beach space. Dredges are often used as a primary tool for rebuilding beaches.

Bee It is an insect with four wings that is related to the wasps, gathers pollen and nectar from flowers from which it makes beebread and honey for food, and usually lives in large colonies. There many kinds of bees, such as, honey bees, bumble bees, cuckoo bees, carpenter bees, leaf-cutting bees, digger bees, alkali bees, and mason bees.

Belt filter press, combined sludge thickening and dewatering Conditioned sludge is first placed on a gravity drainage section where it is allowed to thicken. In this section, the bulk of the free water is removed from the sludge by gravity. On some units, this section is equipped with a vacuum assist, which enhances drainage and may help to reduce odors. Following gravity drainage, pressure is applied in a low-pressure section, where sludge is squeezed between opposing porous cloth belts. On some units, the low-pressure section is followed by a high-pressure section where the sludge is subjected to shearing forces that induce the release of additional quantities of water from the sludge. The final dewatered sludge cake is removed from the belts by scrapper blades.

Biochemical oxygen demand (BOD) A biochemical quantitative measure of the amount of oxygen used in the biochemical oxidation of organic matter in a specified time period (usually 5 days), at a specified temperature (usually 20 ° C), and under specified conditions (in the dark).

Biofuels Gas or liquid fuel made from plant material (biomass); includes wood, wood waste, wood liquors, peat, railroad ties, wood sludge, spent sulfite liquors, agricultural waste, straw, tires, fish oils, tall oil, sludge waste, waste alcohol, municipal solid waste, landfill gases, other waste, and ethanol blended into motor gasoline.

Biogeochemical cycle Movements through the Earth system of key chemical constituents essential to life, such as carbon, nitrogen, oxygen, and phosphorus.

Biological nutrient removal (BNR) The removal of nitrogen and/or phosphorus by the use of proliferation and selection of certain microbiological populations.

Biological phosphorus removal (BPR) The biological process of phosphorus removal in wastewater that relies on the proliferation and selection of a microbiological population capable of storing phosphorus in excess of their normal growth requirements. Also called biological nutrient removal (BNR) or enhanced biological phosphorus removal (EBPR).

Biomass (a) Materials that are biological in origin, including organic material (both living and dead) from above and below ground, for example, trees, crops, grasses, tree litter, roots, and animals and animal waste. (b) It is a mass or clump of organic material consisting of living organisms feeding on the wastes in wastewater, dead organisms, and other debris.

Biosolids (a) It is general term for any organic solids containing carbon, such as activated sludge, septic tank sludge, manure, etc. (b) It is a primarily organic solid product, produced by wastewater treatment processes or agricultural installations, that can be beneficially recycled. The word biosolids is replacing the word biological sludge, or organic sludge.

Biosphere The part of the Earth system comprising all ecosystems and living organisms, in the atmosphere, on land (terrestrial biosphere) or in the oceans (marine biosphere), including derived dead organic matter, such as litter, soil organic matter, and oceanic detritus.

Biotower It is one type of trickling filters. A biotower is an attached growth biological process that is used to remove organic matter found in wastewater.

Black carbon aerosol Black carbon (BC) is the most strongly light-absorbing component of particulate matter (PM), and is formed by the incomplete combustion of fossil fuels, biofuels, and biomass. It is emitted directly into the atmosphere in the form of fine particles (PM2.5).

Blank sample A clean sample or a sample of matrix processed so as to measure artifacts in the measurement (sampling and analysis) process.

Blank It is a sample bottle containing only dilution water or distilled water, but the sample being tested is not added. Tests are frequently run on a sample and a blank, and the differences are compared.

Blow-down Some of the water collected in the basin is exited from the tower rather than recirculated within the tower. This prevents buildup of solids in the water.

Blowoff It is turbulence in aeration tanks caused by damaged diffusers.

Boat plant A complete small water or wastewater treatment plant installed on a boat, which can navigate in a lake or river for necessary water or wastewater treatment.

BOD **B**iochemical **o**xygen **d**emand. The rate at which organisms use the oxygen in water or wastewater while stabilizing decomposable organic matter under aerobic conditions. In decomposition, organic matter serves as food for the bacteria and energy results from its oxidation. BOD measurements are used as a measure of the organic strength of wastes in water.

Booster pump It is a supplemental dredge type pump used in a dredging operation to increase the distance the slurry can be pumped. A booster is often required when a deposit area is greater than 3500 ft. (about 1 km) from the dredge.

Borehole Any exploratory hole drilled into the Earth or ice to gather geophysical data. Climate researchers often take ice core samples, a type of borehole, to predict atmospheric composition in earlier years.

Breeding It is a business or hobby in the context of managing domesticated animals, to cause an animal to produce offspring in a controlled and organized manner.

Briefing A type of training that is a short meeting to provide information about a specific topic.

Brood They are immature bees that have not yet emerged from their cells; brood of different ages can be in the form of eggs, larvae, or pupae.

Buffer (a) It is a zone of a specified distance around a feature; (b) it is a solution or liquid whose chemical makeup neutralizes acids or bases without a great change in pH.

Buffer solution It is either called pH buffer or hydrogen ion buffer, which is an aqueous solution consisting of a mixture of a weak acid and its conjugate base, or vice versa. Its pH changes very little when a small amount of strong acid or base is added to it.

Bulking They are clouds of billowing sludge that occur throughout secondary clarifiers and sludge thickeners when the sludge does not settle properly. In the activated sludge process, bulking is usually caused by filamentous bacteria or bound water.

Bumble bee (*Bombus occidentalis*) They are natural bees in black and yellow color. A bumble bee's body is densely pilose, and its wings are transparent. They are natural important pollinators. They nest in the ground. Although bumble bees are closely related to honey bees, their life cycle is different, because only the young queen overwinters. In the fall, the female will mate with a male and this young queen will overwinter alone, the male dying shortly after mating. As spring approaches, the queen starts a new colony by herself— finding or building a cavity in the ground, laying eggs, developing larvae, constructing the waxen honey pot as initial foods, and waiting for development of worker bee, drone bees, etc.

Business recovery The identification of the systems in place to continue business and administrative operations after an incident.

Cadmium A minor metallic element, one of the naturally occurring components in the Earth's crust and waters, and present everywhere in our environment.

Calibrate To determine, by measurement or comparison with a standard, the correct value of each scale reading on a meter or other device, or the correct value for each setting of a control knob. The levels of the calibration standards should bracket the range of planned measurements.

Calibration The checking, adjusting, or systematic standardizing of the graduations of a quantitative measuring instrument.

Calibration curve The graphical relationship between the known values for a series of calibration standards and instrument responses.

Calibration standard A substance or reference material used to calibrate an instrument.

Carbon capture and sequestration (CCS) Carbon capture and sequestration (CCS) is a set of technologies that can greatly reduce carbon dioxide emissions from new and existing coal- and gas-fired power plants, industrial processes, and other stationary sources of carbon dioxide. It is a three-step process that includes capture of carbon dioxide from power plants or industrial sources; transport of the captured and compressed carbon dioxide (usually in pipelines); and underground injection and geologic sequestration, or permanent storage, of that carbon dioxide in rock formations that contain tiny openings or pores that trap and hold the carbon dioxide.

Carbon cycle All parts (reservoirs) and fluxes of carbon. The cycle is usually thought of as four main reservoirs of carbon interconnected by pathways of exchange. The reservoirs are the atmosphere, terrestrial biosphere (usually includes freshwater systems), oceans, and sediments (includes fossil fuels). The annual movements of carbon, the carbon exchanges between reservoirs, occur because of various chemical, physical, geological, and biological processes. The ocean contains the largest pool of carbon near the surface of the Earth, but most of that pool is not involved with rapid exchange with the atmosphere.

Carbon dioxide A naturally occurring gas, and also a by-product of burning fossil fuels and biomass, as well as land-use changes and other industrial processes. It is the principal human-caused greenhouse gas that affects the Earth's radiation balance. It is the reference gas against which other greenhouse gases are measured and therefore has a global warming potential of 1.

Carbon dioxide equivalent (CDE) A metric measure used to compare the emissions from various greenhouse gases based upon their global warming potential (GWP). Carbon dioxide equivalents are commonly expressed as "million metric tons of carbon dioxide equivalents ($MMTCO_2Eq$)." The carbon dioxide equivalent for a gas is derived by multiplying the tons of the gas by the associated GWP. $MMTCO_2Eq$ = (million metric tons of a gas) * (GWP of the gas)

Carbon dioxide fertilization The enhancement of the growth of plants as a result of increased atmospheric CO_2 concentration. Depending on their mechanism of photosynthesis, certain types of plants are more sensitive to changes in atmospheric CO_2 concentration.

Carbon footprint The total amount of greenhouse gases that are emitted into the atmosphere each year by a person, family, building, organization, or company. A person's carbon footprint includes greenhouse gas emissions from fuel that an individual burns directly, such as by heating at home or riding in a car. It also includes greenhouse gases that come from producing the goods or services that the individual uses, including emissions from power plants that make electricity, factories that make products, and landfills where trash gets sent.

Carbon sequestration Terrestrial, or biologic, carbon sequestration is the process by which trees and plants absorb carbon dioxide, release the oxygen, and store the carbon. Geologic sequestration is one step in the process of carbon capture and sequestration (CCS) and involves injecting carbon dioxide deep underground where it stays permanently.

Carbonaceous biochemical oxygen demand (CBOD) A chemical quantitative measure of the amount of dissolved oxygen required for the biological oxidation of carbon-containing compounds in a sample.

Carpenter bee (*Xylocopa micans*) They are considered to be household pests in porch posts, barn beams, and other wooden structures. Both males and females overwinter as adults in nests constructed in wooden materials.

Cationic polymer (or cationic polyelectrolyte) A positively charged electrolyte used as a coagulant to aid in separating (settling or floating) solids in water or wastewater.

Cationic surfactant (a) A positively charged surface active agent that lowers the surface tension in water, and can be used as a collector in a flotation process, or as an organic disinfectant in personal cleaners, swimming pools, house roofs, restaurants, etc. (b) an organic quaternary ammonia compound with positively charged surface active moieties.

Cavitation The formation and collapse of a gas pocket or bubble on the blade of an impeller or the gate of a valve. The collapse of this gas pocket or bubble drives water into the impeller or gate with a terrific force that can cause pitting on the impeller or gate surface. Cavitation is accompanied by loud noises that sound like someone is pounding on the impeller or gate with a hammer.

Centrate Dilute effluent stream from a centrifuge.

Centrifuge (a) It is a water-solid separation device that separates the solids from liquid by centrifugal force. Centrifuges are used for separating liquids of different density, thickening slurries, or dewatering solids. (b) It is a mechanical device that uses centrifugal or rotational forces to separate solids from liquids.

Centrifuge thickening It is sludge preliminary concentration process before sludge watering process involving the use of centrifugal force. Water in a wet dilute sludge is forced out of the wet sludge placed in the spinning drum of a centrifuge.

Centrifuge, basket centrifuge for sludge thickening/dewatering They are suitable for small plants. It is a semibatch type operation. Sludge cake is collected on the sides of the spinning bowl while the centrate (dilute effluent stream from the centrifuge) overflows the bowl rim. Once solids have built up to a maximum thickness, the feed sludge is stopped and scrapper blades peel the sludge from the walls. The process is then resumed.

Centrifuge, solids bowl centrifuge for sludge thickening/dewatering Sludge is fed at a constant flowrate into the rotating bowl where it is separated into a dense cake and a dilute stream called centrate. The centrifuge units can be used with no prior chemical conditioning, but solids caputer and centrate quality are improved considerably when the sludge is conditioned with polymer.

Ceramic matrices Materials like clay that can be converted to ceramics after high temperature treatment.

Characteristic Wastes In the USA, characteristic wastes may be regulated for exhibiting one of the four characteristics (ignitability, corrosivity, reactivity, and toxicity) of the waste as determined by the US Environmental Protection Agency (USEPA).

Chemical engineering It is a branch of engineering that uses principles of chemistry, physics, mathematics, biology, and economics to efficiently use, produce, design, transport, and transform energy and materials. The work of chemical engineers can range from the utilization of nanotechnology and nanomaterials in the laboratory to large-scale industrial processes that convert chemicals, raw materials, living cells, microorganisms, and energy into useful forms and products.

Chemical analysis The use of a standard chemical analytical procedures to determine the concentration of a specific analyte in a sample, or qualitatively or quantitatively measure a specific parameter of a sample.

Chemical oxygen demand (COD) An environmental chemical test used to measure indirectly organic compounds in water or wastewater.

Childproofing Making the physical environment safe for the children in your care to reduce the risks of injury.

Chlorination (a) The application of chlorine to water or wastewater, generally for the purpose of disinfection, but frequently for accomplishing other biological or chemical results, and (b) one of disinfection processes involving the use of chlorine gas. Disinfection with chlorine is accomplished by adding chlorine gas to the water or wastewater. Sometimes application of sodium hypochlorite (liquid) for disinfection is also called chlorination. Academically disinfection using sodium hypochlorite should be called hypochlorination.

Chlorine demand Chlorine demand is the difference between the amount of chorine added to wastewater and the amount of residual chlorine remaining after a given contact time. Chlorine demand may change with dosage, time, temperature, pH, or nature or amount of the impurities in the water. Chlorine demand, mg/L = chlorine applied, mg/L – chlorine residual, mg/L

Chlorine requirement The amount of chlorine that is needed for a particular purpose. Some reasons for adding chlorine are reducing the number of coliform bacteria (Most Probable Number), obtaining a particular chlorine residual, or oxidizing some substance in the water. In each case, a definite dosage of chlorine will be necessary. This dosage is the chlorine requirement.

Chlorine sponge Incomplete nitrification resulting in excess nitrite levels that react with chlorine. Also known as "Nitrite Lock."

Chlorofluorocarbons Gases covered under the 1987 Montreal Protocol and used for refrigeration, air conditioning, packaging, insulation, solvents, or aerosol propellants. Since they are not destroyed in the lower atmosphere, CFCs drift into the upper atmosphere where, given suitable conditions, they break down ozone. These gases are being replaced by other compounds: hydrochlorofluorocarbons, an interim replacement for CFCs that are also covered under the Montreal Protocol, and hydrofluorocarbons, which are covered under the Kyoto Protocol. All these substances are also greenhouse gases.

Chloroform It is a colorless, volatile, sweet-smelling liquid ($CHCl_3$) used as an organic solvent in an analytical procedure and formerly as a general anesthetic in a medical procedure.

Civil engineering It is a professional engineering discipline that deals with the design, construction, and maintenance of the physical and naturally built environment, including public works such as roads, bridges, canals, dams, airports, sewerage systems, pipelines, structural components of buildings, and railways. Civil engineering is traditionally broken into a number of subdisciplines. It is considered the second-oldest engineering discipline after military engineering, and it is defined to distinguish nonmilitary engineering from military engineering.

Civil engineering takes place in the public sector from municipal through to national governments, and in the private sector from individual homeowners through to international companies.

Clari-DAF It is rectangular dissolved air flotation (DAF) clarifier commercially available from Xylem Water and Wastewater, Zelienople, PA 16063, USA.

Clarifier, settling tank, sedimentation basin A tank or basin in which water or wastewater is held for a period of time during which most of the heavier solids settle to the bottom and small amount of the lighter material will float to the water surface.

Clarifier, flotation tank A tank in which water or wastewater is held for a period of time during which most of the lighter air-TSS flocs float to the water surface and small amount of the heavier solids will settle to the tank bottom.

Clean sample A sample of a natural or synthetic matrix containing no detectable amount of the analyte of interest and no interfering material.

Climate Climate in a narrow sense is usually defined as the "average weather," or more rigorously, as the statistical description in terms of the mean and variability of relevant quantities over a period of time ranging from months to thousands of years. The classical period is 3 decades, as defined by the World Meteorological Organization (WMO). These quantities are most often surface variables such as temperature, precipitation, and wind. Climate in a wider sense is the state, including a statistical description, of the climate system.

Climate change Climate change refers to any significant change in the measures of climate lasting for an extended period of time. In other words, climate change includes major changes in temperature, precipitation, or wind patterns, among others, that occur over several decades or longer.

Climate feedback A process that acts to amplify or reduce direct warming or cooling effects.

Climate lag The delay that occurs in climate change as a result of some factor that changes only very slowly. For example, the effects of releasing more carbon dioxide into the atmosphere occur gradually over time because the ocean takes a long time to warm up in response to a change in radiation.

Climate model A quantitative way of representing the interactions of the atmosphere, oceans, land surface, and ice. Models can range from relatively simple to quite comprehensive.

Climate sensitivity In Intergovernmental Panel on Climate Change (IPCC) reports, equilibrium climate sensitivity refers to the equilibrium change in global mean surface temperature following a doubling of the atmospheric (equivalent) CO_2 concentration. More generally, equilibrium climate sensitivity refers to the equilibrium change in surface air temperature following a unit change in radiative forcing (degrees Celsius, per watts per square meter). One method of evaluating the equilibrium climate sensitivity requires very long simulations with Coupled General Circulation Models (Climate model). The effective climate sensitivity is a related measure that circumvents this requirement. It is evaluated from model output for evolving nonequilibrium conditions. It is a measure of the strengths of the feedbacks at a particular time and may vary with forcing history and climate state.

Climate system (or Earth system) The five physical components (atmosphere, hydrosphere, cryosphere, lithosphere, and biosphere) that are responsible for the climate and its variations.

Coagulant (a) A chemical (alum or iron salts) added to water to destabilize particles, allowing subsequent floc formation and removal by clarification (flotation or sedimentation) and/or filtration. (b) Chemicals that cause very fine particles to clump (floc) together into larger particles. This makes it easier to separate the solids from the water by settling, skimming, draining or filtering.

Coagulation (a) The clumping together of very fine particles into large particles (floc) caused by the use of chemicals (coagulants); (b) A process of destabilizing charges of suspended and colloidal particles in water by adding chemicals (coagulants). In coagulation process, positively charged chemicals are added to neutralize or destabilize these negative charges and allow the neutralized particles to accumulate and be removed by clarification (flotation or sedimentation) and/or filtration.

Coal mine methane Coal mine methane is the subset of coalbed methane that is released from the coal seams during the process of coal mining.

Coalbed methane Coalbed methane is methane contained in coal seams, and is often referred to as virgin coalbed methane, or coal seam gas.

Cobenefit The benefits of policies that are implemented for various reasons at the same time including climate change mitigation acknowledging that most policies designed to address greenhouse gas mitigation also have other, often at least equally important, rationales (e.g., related to objectives of development, sustainability, and equity).

Coliform A type of bacteria. The presence of coliform-group bacteria is an indication of possible pathogenic bacterial contamination. The human intestinal tract is one of the main habitats of coliform bacteria. They may also be found in the intestinal tracts of warm-blooded animals, and in plants, soil, air, and the aquatic environment. Fecal coliforms are those coliforms found in the feces of various warm-blooded animals, whereas the term "coliform" also includes various other environmental sources.

Collector A device or system designed to collect filter backwash water, or other treatment unit.

Colony It is a gathering of all the worker bees, drones, queen, and developing brood living together in one hive or other dwelling.

Colorimetric measurement A means of measuring unknown chemical concentrations in water by measuring a sample's color intensity. The specific color of the sample, developed by addition of chemical reagents, is measured with a photoelectric colorimeter or is compared with "color standards" using, or corresponding with, known concentrations of the chemical.

Comminuting It is a grinding or shredding operation for reducing the particle size of objects or debris in the influent wastewater. They are installed with a screen directly in the influent wastewater flow's channel, with the shredded particles returned to the flow downstream of the screen. The influent flow is channeled to and through these units. The debris is collected against the screen, or outside drum, and the teeth, which penetrate this screen cut up the solids. When the solids are reduced to the size of the screen or drum openings, they pass through and on for additional downstream wastewater treatment. The barminutor is a comminuting device that incorporates revolving cutters that move up and down the upstream face of a bar screen, shredding and cutting whatever debris has accumulated against the screen. The screenings are transported to the cutting device, shredded, and then allowed to fall back into the influent channel downstream of the bar screen.

Comminutor A device used to reduce the size of the solid chunks in wastewater by shredding (comminuting). The shredding action is like many scissors cutting or chopping to shreds all the large influent solids material in the wastewater.

Community Emergency Response Team (CERT) A program that educates people about disaster preparedness and provides training in basic disaster response skills.

Complete mix activated sludge process It is an activated process, in which aeration tank (with about 3–5 hours DT) is usually circular and the influent wastewater and the mixed liquor are completely mixed during aeration operation. It is most useful for treating wastewater with fluctuating organic strength, because it is least susceptible to shock loads among all activated sludge processes. A mechanically stirred reactor best simulates a complete mix condition. When diffused aeration in a rectangular aeration tank is used, the mixture of wastewater influent and the return sludge should enter the aeration tank at several points in a central location with the effluent going into channels on the sides of the aeration tank.

Complex A substance consisting of many different and connected parts.

Composite A composite sample is a collection of individual samples obtained at regular intervals, usually every one or two hours during a 24-hour time span. Each individual sample is combined with the others in proportion to the rate of flow when the sample was collected. The resulting mixture (composite sample) forms a representative sample and is analyzed to determine the average conditions during the sample period.

Compost It is decayed organic material used as a plant fertilizer.

Composting It is an aerobic biological process designed to reduce organic concentrations, reclaim nutrients, and eliminate pathogenic organisms in the waste sludge or solid wastes. There are at least three major types of composting systems: (a) aerated static pile composting system; (b) windrow composting system; and (c) the in-vessel composting system. Sludge, wood chips leaves, etc., that have been composted properly is a sanitary, nuisance-free, humus-like material. The end product can be used as a soil additive or landfill cover component.

Composting latrine It is a type of dry (not using water) latrine, which treats human excreta using a biological process called composting, which decomposes excreta and organic matter into a compost-like material but does not destroy all pathogens.

Composting, aerated static pile system The aerated stack pile consists of a grid of aeration or exhaust piping underlying a mixture of dewatered sludge or small solid wastes (such as leaves, wood chips, waste foods) and bulking agent. Material is composted for 21–28 days and then cured for another 30 days or longer.

Composting, in-vessel system In this in-vessel system, composting of sludge, solid wastes, and bulking agent is accomplished inside an enclosed container or vessel. Mechanical devices are designed to minimize odors and process time by controlling environmental conditions such as air flow, temperature, and oxygen concentration. There are two types: (a) the plug flow system and (b) the dynamic system.

Composting, windrow system This composting system is similar to the aerated static pile system. Piles of dewatered sludge, small solids wastes, and bulking agent are periodically turned for aeration and temperature control. The major drawback to this system is that turning the windrows is often accompanied by the release of offensive odors.

Comprehensive Preparedness Guide (CPG) 101 A document designed to provide guidance for developing emergency operations plans. It promotes a common understanding of the fundamentals of risk-informed planning and decision-making to help planners examine a hazard or threat and produce integrated, coordinated, and synchronized plans. The goal of CPG 101 is to assist in making the planning process routine across all phases of emergency management and for all homeland security mission areas. It helps planners at all levels of government in their efforts to develop and maintain viable, all-hazards, all-threats emergency plans.

Computer-aided design (CAD) It is an automated system for the design, drafting, and display of graphically oriented information.

Concentration (a) Amount of a chemical in a particular volume or weight of air, water, soil, or other medium; (b) in solutions, the mass, volume, or number of moles of solute present in proportion to the amount of solvent or total solution. Common measures are: molarity, normality, percent, molality, and by specific gravity scales.

Concept of operations (CONOPS) A component of the basic plan that clarifies the childcare site's or school's overall approach to an emergency (i.e., what should happen, when, and at whose direction) and identifies specialized response teams and/or unique resources needed to respond to an incident.

Confined space Confined space means a space that (a) is large enough and so configured that an employee can bodily enter and perform assigned work; and (b) has limited or restricted means for entry or exit; and (c) is not designed for continuous employee occupancy.

Contact stabilization process It is a modification of the activated sludge process. In this modification, the adsorptive capacity of the floc is utilized in the contact tank to adsorb suspended, colloidal, and some dissolved organics. The hydraulic detention time in the contact tank is only about 30–60 minutes (based on average daily flow). After the biological sludge is separated from the wastewater in the secondary clarifier (either secondary sedimentation clarifier, or secondary flotation clarifier), the concentrated sludge is separately aerated in the stabilization tank with a detention time of 2 to 6 hours (based on sludge recycle flow). The adsorbed organics undergo oxidation in the stabilization tank and are synthesized into microbial cells. If the detention time is long enough in the stabilization tank, endogenous respiration will occur, along with a concomitant decrease in excess biological sludge production. Following stabilization, the reaerated sludge is mixed with incoming wastewater in the contact tank and the cycle starts anew. Volatile compounds are driven off to a certain extent by aeration in the contact and stabilization tanks. Metals will also be partially removed, with accumulation in the sludge. This process requires smaller total aeration volume than the conventional activated sludge process. It also can handle greater organic shock and toxic loadings because of the biological buffering capacity of the stabilization tank and the fact that at any given time. The majority of the activated sludge is isolated from the main stream of the plant flow. Generally, the total aeration basin volume (contact basin plus stabilization basin) is only 50–75 percent of that required in the conventional activated sludge system. An equalization basin is generally recommended for equalizing the wastewater influent flow.

Contamination A general term signifying the introduction into water of microorganisms, chemicals, wastes, or sewage, which renders the water unfit for its intended use.

Conventional biological wastewater treatment system It normally includes (a) preliminary treatment units (i.e., screen, comminutor, grit chamber, etc., for removal of sand, gravel, cinders, coffee grounds, small stones, cigarette filter tips, logs, cans, and other large-sized unwanted materials from raw wastewater), (b) primary sedimentation clarification for removing mainly total suspended solids from preliminary effluent, (c) secondary biological treatment units (such as activated sludge aeration or equivalent plus secondary sedimentation clarification) for removing dissolved organic/inorganic pollutants from primary effluent, and (d) tertiary treatment plant units (i.e., filtration, granular activated carbon adsorption, ion exchange, oxidation, nitrification, denitrification, and/or disinfection) for final polishing the secondary effluent in order to meet the effluent discharge standards.

Conventional physicochemical wastewater treatment system It normally includes (a) preliminary treatment units (i.e., screen, comminutor, grit chamber, etc., for removal of sand, gravel, cinders, coffee grounds, small stones, cigarette filter tips, logs, cans, and other large-sized unwanted materials from raw wastewater), (b) primary sedimentation clarification for removing mainly total suspended solids from preliminary effluent, (c) secondary physicochemical treatment units (such as chemical precipitation/coagulation or equivalent plus secondary sedimentation clarification) for removing dissolved organic/inorganic pollutants from primary effluent, and (d) tertiary treatment plant units (i.e., filtration, granular activated carbon adsorption, ion exchange, oxidation, nitrification, denitrification, and/or disinfection) for final polishing the secondary effluent in order to meet the effluent discharge standards. In the nitrification and denitrification steps, only tertiary sedimentation clarification will be used for solid-water separation.

Cooling It is a thermodynamic process to lower the temperature.

Cooling range The temperature difference between hot water entering the tower and cooled water exiting the tower.

Coordinate It is an x,y location in a Cartesian coordinate system or an x,y,z coordinate in a three-dimensional system. Coordinates represent locations on the Earth's surface relative to other locations.

Coral bleaching The process in which a coral colony, under environmental stress, expels the microscopic algae (zooxanthellae) that live in symbiosis with their host organisms (polyps). The affected coral colony appears whitened.

Corrosivity Any substance, which exhibits any of the following properties, is considered a hazardous waste due to corrosivity: (a) an aqueous material with pH less than or equal to 2 or greater than or equal to 12.5; and (b) a liquid that corrodes steel at a rate greater than 1/4 inch per year at a temperature of 55 ° C.

Cross connection A connection between a drinking (potable) water system and an unapproved water supply. For example, if you have a pump moving nonpotable water and hook into the drinking water system to supply water for the pump seal, a cross connection, or mixing, between the two water systems can occur. This mixing may lead to contamination of the drinking water.

Cryosphere One of the interrelated components of the Earth's system, the cryosphere, is frozen water in the form of snow, permanently frozen ground (permafrost), floating ice, and glaciers. Fluctuations in the volume of the cryosphere cause changes in ocean sea level, which directly impact the atmosphere and biosphere.

CSHO It is an abbreviation for an OSHA Compliance Safety and Health Officer.

Cuckoo bee (*Epeolus compactus*) They are parasitic bees of wild species, not affecting humans. They lack pollen-carrying baskets or brushes. So they are parasitic, usually on species of bees of the genus *Colletes*.

Cutter It describes the type of excavator head such as a basket cutterhead, plain-edge cutter, or bucketwheel, which houses the suction intake and is used for cutting or agitating the materials being dredged.

DAF Dissolved air flotation process.

DAFF Dissolved air flotation and filtration.

Data Facts or figures from which conclusions can be inferred.

Database It is a logical collection of interrelated information, managed and stored as a unit. A GIS database includes data about the spatial location and shape of geographic features recorded as points, lines, and polygons as well as their attributes.

Debris flow Another term for a landslide or mudslide. See landslide definition.

Dechlorination It is a process for removal of excess amount of residual chlorine in a chlorination disinfection process. Dechlorination is accomplished by using a reducing chemical such as sulfur dioxide (gas), sodium bisulfate (liquid), or sodium metabisulfite (liquid).

Dechlorination The removal of chlorine from the effluent of a treatment plant.

Deforestation Those practices or processes that result in the conversion of forested lands for nonforest uses. Deforestation contributes to increasing carbon dioxide concentrations for two reasons: 1) the burning or decomposition of the wood releases carbon dioxide; and 2) trees that once removed carbon dioxide from the atmosphere in the process of photosynthesis are no longer present.

Denitrification filter It is an attached-growth biological process filter using wood, plastic, rock, granular activated carbon, or sand as filtration media for denitrification (conversion of nitrate to nitrogen gas) of wastewater under anoxic/anaerobic condition.

Denitrification followed by flotation clarification The process involves the reduction of nitrates and nitrites to nitrogen gas through the action of facultative heterotrophic bacteria. In suspended-growth separate stage denitrification processes, nitrified wastewater containing primarily nitrates is passed through a mixed anaerobic vessel containing denitrifying bacteria. Since the nitrified feedwater contains very little carbonaceous materials, a supplemental source of carbon (such as methanol, sugar, acetic acid, ethanol, or other compounds) is required to maintain the denitrifying biomass. This supplemental energy is provided by feeding methanol, for instance, to the biological reactor along with the nitrified wastewater. Mixing in the anaerobic denitrification reaction vessel may be accomplished using low-speed paddles analogous to standard flocculation equipment. Flotation clarification follows the denitrification step with the floated sludge being either returned to the head end of the denitrification system or wasted. An intermediate aeration step for stabilization (much less than 40 minutes) may be needed to guard against carryover of carbonaceous materials. The denitrification reactor may be covered but not airtight to assure anaerobic conditions.

Denitrification followed by membrane clarification Same as "denitrification followed by flotation clarification" except that the flotation clarification is replaced by a membrane filtration reactor.

Denitrification followed by sedimentation clarification The process involves the reduction of nitrates and nitrites to nitrogen gas through the action of facultative heterotrophic bacteria. In suspended-growth separate stage denitrification processes, nitrified wastewater containing primarily nitrates is passed through a mixed anaerobic vessel containing denitrifying bacteria. Since the nitrified feedwater contains very little carbonaceous materials, a supplemental source of carbon (such as methanol, sugar, acetic acid, ethanol, or other compounds) is required to maintain the denitrifying biomass. This supplemental energy is provided by feeding methanol, for instance, to the biological reactor along with the nitrified wastewater. Mixing in the anaerobic denitrification reaction vessel may be accomplished using low-speed paddles analogous to standard flocculation equipment. Following the reactor, the denitrified effluent is aerated for a short period (5 to 10 minutes) to strip out gaseous nitrogen formed in the previous step, which might otherwise inhibit sludge settling. Sedimentation clarification follows the gas-stripping step with the collected sludge being either returned to the head end of the denitrification system, or wasted. An intermediate aeration step for stabilization (about 50 minutes) between the denitrification reactor and the stripping step may be used to guard against carryover of carbonaceous materials. The denitrification reactor may be covered but not airtight to assure anaerobic conditions.

Denitrification (a) The anoxic biological reduction of nitrate-nitrogen to nitrogen gas. (b) The removal of some nitrogen from a system. (c) An anoxic process that occurs when nitrite or nitrate ions are reduced to nitrogen gas and nitrogen bubbles are formed as a result of this process. (d) The biochemical reduction reaction of reducing nitrate-nitrogen to nitrogen gas in an anoxic environment.

Derived-From Wastes USEPA's "derived-from" rule regulates any solid waste generated from the treatment, storage, or disposal of a hazardous waste, including any sludge, spill residue, ash, emission control dust, or leachate (but not including precipitation run-off) is a hazardous waste under 40 CFR 261.4(c)(2)(i), with some specified exclusions.

Desertification Land degradation in arid, semiarid, and dry subhumid areas resulting from various factors, including climatic variations and human activities. Further, the UNCCD (The United Nations Convention to Combat Desertification) defines land degradation as a reduction or loss, in arid, semiarid, and dry subhumid areas, of the biological or economic productivity and complexity of rain-fed cropland, irrigated cropland, or range, pasture, forest, and woodlands resulting from land uses or from a process or combination of processes, including processes arising from human activities and habitation patterns, such as (i) soil erosion caused by wind and/or water; (ii) deterioration of the physical, chemical, and biological or economic properties of soil; and (iii) long-term loss of natural vegetation. Conversion of forest to nonforest.

Detention time The time required to fill a tank at a given flow or the theoretical time required for a given flow of wastewater to pass through a tank.

Detoxification The act of detoxifying.

Detritus The heavy, coarse mixture of grit and organic material carried by wastewater (also called grit).

Diffused-air aeration A diffused air activated sludge plant takes air, compresses it, and then discharges the air below the water surface of the aerator through some type of air diffusion device.

Diffuser A device used to break the air stream from the blower system into fine bubbles in an aeration tank or reactor.

Digester A tank in which sludge is placed to allow decomposition by microorganisms. Digestion may occur under anaerobic (more common) or aerobic conditions.

Digger bee (*Ptilothrix bombiformis*) They are natural bees of wild species, not affecting humans. Adults burrow into the ground where they store a mixture of honey and pollen to be used as larval food. Adults often are seen visiting the flowers of hibiscus.

Digging depth, dredging It is the maximum design depth from which the dredge can excavate material, which is usually stated as a measurement from water level and typically expressed in feet or meters.

Digital Elevation Model (DEM) It is a terrain elevation data provided in digital form.

Digitize It is an action to encode map features as x,y coordinates in digital form. Lines are traced to define their shapes. This can be accomplished either manually or by use of a scanner.

Disinfection Potable water disinfection is used in an attempt to destroy all pathogenic agents in drinking water for public health protection. Wastewater disinfection is used in an attempt to destroy pathogenic agents in a wastewater stream and protect the best uses of the receiving stream or groundwater. Environmental disinfection is used in an attempt to destroy pathogenic agents in rooms, houses, swimming pools, buildings, restaurants, streets, parks, etc., to prevent the spread of pathogenic agents, such as pathogenic bacteria and Coronavirus in the environment. Typical disinfection techniques used include chlorine (gas), sodium hypochlorite (liquid), bromine (gas), bromine dioxide (gas), UV (light), ozone (gas), and cationic surfactant (water-soluble solid).

Disinfection The process designed to kill or inactivate most microorganisms in wastewater, including essentially all pathogenic (disease-causing) bacteria. There are several ways to disinfect, with chlorination being the most frequently used in water and wastewater treatment plants.

Dispersion A uniform and maximum separation of extremely fine particles, often of colloidal size.

Dissolved air flotation (DAF) One of dissolved gas flotation (DGF) processes when air is used for generation of gas bubbles. See dissolved gas flotation (DGF).

Dissolved gas flotation (DGF) It is a process involving pressurization of gas at 25 to 95 psig for dissolving gas into water, and subsequent release of pressure (to one atm) under laminar flow hydraulic conditions for generating extremely fine gas bubbles (20–80 microns), which become attached to the impurities to be removed and rise to the water surface together. The impurities or pollutants to be removed are on the water surface and are called float or scum, which are scooped off by sludge collection means. The clarified water is discharged from the flotation clarifier's bottom. The gas flow rate is about one percent of influent liquid flow rate. The attachment of gas bubbles to the impurities can be a result of physical entrapment, electrochemical attraction, surface adsorption, and/or gas stripping. The specific gravity of the bubble-impurity agglomerate is less than 1, resulting in buoyancy or nonselective flotation (i.e., Save-All).

Dissolved oxygen (DO) Molecular (atmospheric) oxygen dissolved in water or wastewater.

Distilled water Water that has been purified by distillation (boiling the water off as steam and condensing it back to a liquid, leaving the impurities behind). Having been boiled, it is also sterile.

Dredge mining It is a practical means for extracting submerged sand, gravel, sand, iron ore, coal, fine tailings, and other minerals. Mining with a dredge is an efficient way to obtain materials and hydrotransport them to on-shore processing facilities, whether it be sand, gravel, or hard salt deposits.

Dredge pump It is a horizontal centrifugal pump designed to handle abrasive granular materials and solids of limited size in suspension. They are distinct from water and slurry pumps, which carry liquid and homogeneous mixtures, respectively.

Dredge It is a piece of machinery used for removing materials that have been partially or fully immersed under water, with the intent of collecting the sediment and redistributing the content at a separate location.

Dredger It is the European term to describe a dredge.

Dredging It describes the digging or mining action used to collect sediment from below the water surface. The gathered sediment is then either transported elsewhere or repurposed for other uses.

Drift eliminators Baffles, or other specialized collection products, placed in the air exit path to capture water droplets. (Note: as per the previous mention of scrubbing, again, mist elimination in scrubbing is a similar application.)

Drift Also known as windage, small droplets of water, which are carried out of the tower by the airflow, normally expressed as a % of total water flow rate.

Drill A type of exercise that is a coordinated, supervised activity usually employed to test a single specific operation or function in a single agency. Drills are commonly used to provide training on new equipment, develop or test new policies or procedures, or practice and maintain current skills.

Drone bee It is the sexually active male honey bee. Drones are most numerous during the summer, but as food becomes scarce, the worker bees kill the drones and remove them from the hive. A drone bee can only mate the female queen bee once, and then die.

Dry bulb temperature Air temperature, measured by standard means.

Dryland farming A technique that uses soil moisture conservation and seed selection to optimize production under dry conditions.

Dual Wheel Excavator (DWE) It is an Ellicott's term for its patented underwater bucket wheel excavator, also known as a Wheel Dragon™ excavator.

Earthquake The sudden movement of the Earth caused by the breaking and shifting of rock beneath the Earth's surface.

Eccentricity The extent to which the Earth's orbit around the Sun departs from a perfect circle.

Ecology It is a branch of biology that deals with the mutual relations between organisms and their environment. Ecology implies more the webs of natural forces and organisms, their competition and cooperation, and how they live off one another.

Ecosystem Any natural unit or entity including living and nonliving parts that interact to produce a stable system through cyclic exchange of materials.

Effluent Wastewater or other liquid—raw (untreated), partially or completely treated—flowing from a reservoir, basin, treatment process, or treatment plant.

Elutriation The washing of digested sludge with fresh water, plant effluent, or other wastewater. The goal is to remove fine particles and/or the alkalinity in the sludge. This process reduces the demand for conditioning chemicals and improves settling or filtering characteristics of the sludge.

Emergency action plan A simple set of emergency procedures that includes, at a minimum, parent/guardian contact information; medical information for each child; emergency services contact information; how to accommodate the needs of each child; and procedures for shelter-in-place, evacuation, and reunification.

Emergency kit Items to help you survive during and after an emergency, including food, water, and other supplies.

Emergency operations plan A comprehensive formal plan based on the Federal Emergency Management Agency's (FEMA's) Comprehensive Preparedness Guide (CPG) steps. It describes how people and property will be protected; details who is responsible for carrying out specific actions; identifies the personnel, equipment, facilities, supplies, and other resources available; and outlines how all actions will be coordinated.

Emissions The release of a substance (usually a gas when referring to the subject of climate change) into the atmosphere.

Emissions factor A unique value for scaling emissions to activity data in terms of a standard rate of emissions per unit of activity (e.g., grams of carbon dioxide emitted per barrel of fossil fuel consumed, or per pound of product produced).

Emotional and psychological recovery Identification of strategies to address disruption of services, psychological injury, and external pressure.

Endocrine-disrupting compounds (EDCs) They are chemicals that mimic endogenous hormones, interfere with pharmacokinetics, or act by other mechanisms. Such adverse effects as compromised reproductive fitness, functional or morphological birth defects, cancer, and altered immune functions, among others, have been reported.

Endocrine disruptors (a) They are substances that are not produced in the body but act by mimicking or antagonizing natural hormones and are mostly man-made found in various materials. By interfering with the human body's endocrine system, endocrine disruptors produce adverse developmental, reproductive, neurological, and immune effects in humans, abnormal growth patterns and neurodevelopmental delays in children. (b) They are chemicals that mimic endogenous hormones, interfere with pharmacokinetics, or act by other mechanisms resulting in adverse effects such as compromised reproductive fitness, functional or morphological birth defects, cancer, and altered immune functions. (c) They are environmental pollutants with adverse hormonal effects on wildlife and humans, such as pesticides (methoxychlor), surfactants (nonylphenol), plasticizers (diethylphthalate), and organohalogens (polychlorinated biphenyls or PCBs and dioxin).

Endocrine systems They are also referred to as hormone systems and are found in all mammals, birds, fish, and many other types. The endocrine system regulates all biological processes in the body from conception through adulthood and into old age, including the development of the brain and nervous system, the growth and function of the reproductive system, as well as the metabolism and blood sugar levels.

Energy efficiency Using less energy to provide the same service.

ENERGY STAR A US Environmental Protection Agency voluntary program that helps businesses and individuals save money and protect our climate through superior energy efficiency.

Energy (a) In engineering physics, energy is the quantitative property that must be transferred to an object in order to perform work on, or to heat, the object. (b) Energy is a conserved quantity; the law of conservation of energy states that energy can be converted in form, but not created or destroyed. (c) Energy is the power derived from the utilization of physical or chemical resources, especially to provide light and heat or to work machines.

Engineering It is a professional field involving the use of science and mathematics to invent, create, design, develop, or improve technologies.

Engineers Without Borders: USA It is a nonprofit organization established to partner with developing communities in order to improve their quality of life.

Enhanced biological phosphorus removal (EBPR) The process of phosphorus removal in wastewater that relies on the proliferation and selection of a microbiological population capable of storing phosphorus in excess of their normal growth requirements. Also called biological nutrient removal (BNR).

Enhanced greenhouse effect The concept that the natural greenhouse effect has been enhanced by increased atmospheric concentrations of greenhouse gases (such as CO_2 and methane) emitted as a result of human activities. These added greenhouse gases cause the Earth to warm.

Enteric fermentation Livestock, especially cattle, produce methane as part of their digestion. This process is called enteric fermentation, and it represents one-third of the emissions from the agriculture sector.

Environmental engineering It is a professional engineering discipline that takes from broad scientific topics like chemistry, biology, ecology, geology, hydraulics, hydrology, microbiology, and mathematics to create solutions that will protect and also improve the health of living organisms and improve the quality of the environment. Environmental engineering is a subdiscipline of civil engineering, chemical engineering, and mechanical engineering.

Environmental remediation (a) It is a means of dredging that is beneficial to both recreational users and natural wildlife habitats. The natural flow of sediments and industrial runoff can often choke off channels that feed fresh water to inland wetlands or ponds. Dredging these channels allows water to circulate in and out of these inland sites, bringing in nutrients and flushing out waste. (b) It is an action of pollution control that is beneficial to both people and ecological system.

Environmental resource management It is the management of the interaction and impact of human societies on the environment. It is not, as the phrase might suggest, the management of the environment itself. Environmental resources management aims to ensure that ecosystem services are protected and maintained for future human generations, and also maintain ecosystem integrity through considering ethical, economic, and scientific (ecological) variables. Environmental resource management tries to identify factors affected by conflicts that arise between meeting needs and protecting resources. It is thus linked to environmental protection, sustainability, and integrated landscape management.

Environmental risk management The conscious and coordinated effort in appraising the potential and/or existing impact of various productive activities on their environment and people.

Environmental science It is an interdisciplinary academic field that integrates physical, biological, and information sciences (including ecology, biology, chemistry, mathematics, microbiology, geology, information science, etc.)

Environmental studies It is a multidisciplinary academic field, which systematically studies human interaction with the environment in the interests solving complex problems. Environmental studies brings together the principles of the physical sciences, commerce/economics, and social sciences so as to solve contemporary environmental problems. It is a broad field of study that includes the natural environment, the built environment, and the sets of relationships between them. The field encompasses study in basic principles of ecology and environmental science, as well as associated subjects such as ethics, geography, anthropology, policy, politics, urban-planning, law, economics, philosophy, sociology and social justice, planning, pollution control, and natural resource management. There are also many degree programs in Environmental Studies including a Master's degree and a Bachelor's degree.

Environmental technology It may include (envirotech), green technology (greentech), or clean technology (cleantech). It is the application of one or more of environmental science, green chemistry, environmental monitoring, and electronic devices to monitor, model, and conserve the natural environment and resources, and to curb the negative impacts of human involvement. The term is also used to describe sustainable energy generation technologies such as photovoltaics, wind turbines, bioreactors, etc. Sustainable development is the core of environmental technologies. The term environmental technologies is also used to describe a class of electronic devices that can promote sustainable management of resources.

Equalization basin A holding basin in which variations in flow and composition of a liquid are averaged. Such basins are used to provide a flow of reasonably uniform volume and composition to a treatment unit. Also called a balancing reservoir.

Estuary Body of water that is located at the lower end of a river and is subject to tidal fluctuations.

Eutrophication (a) Nutrient overenrichment of a body of water, causing increased growth of algae and rooted aquatic plants. (b) Overenrichment of a water body with nutrients, resulting in excessive growth of organisms and depletion of oxygen concentration. (c) The increase of nutrient levels of a lake or other body of water; this usually causes an increase in the growth of aquatic animal and plant life.

Evacuation The organized, phased, and supervised withdrawal, dispersal, or removal of children, personnel, and visitors from dangerous or potentially dangerous areas.

Evaporation The process by which water changes from a liquid to a gas or vapor.

Evapotranspiration (a) The combined process of evaporation from the Earth's surface and transpiration from vegetation. (b) The process by which water vapor passes into the atmosphere from living plants, also called transpiration. (c) The total water removed from an area by transpiration (living plants) and by evaporation from soil, snow, and water surfaces.

Excessive heat When heat index values meet or exceed daytime highs of 105–110 ° F (depending on local climate).

Excreta It is a waste matter discharged from the body, especially feces and urine.

Exercise A way to train for, assess, practice, and improve performance in prevention, protection, response, and recovery capabilities in a risk-free environment. Exercises can be used for testing and validating policies, plans, procedures, training, equipment, and interagency agreements; clarifying and training personnel in roles and responsibilities; improving interagency coordination and communications; identifying gaps in resources; improving individual performance; and identifying opportunities for improvement.

Extended aeration activated sludge process A long aeration time (over 18 hours DT) is employed in an extended aeration system. It is applicable only to small wastewater treatment plants of less than 3785 m^3/day or 1.0 MGD. Normally, no primary clarifier is provided and no exercise of optimum return sludge control is attempted although the return activated sludge (RAS) flow is about 75% to 100% of plant flow. The objective is to simplify the process both in construction and in operation. Sludge production is very small because of prolonged endogenous oxidation, which minimizes the problem of sludge treatment and disposal. The aeration effluent's sludge normally has an inferior settling characteristic. Therefore, a secondary flotation clarifier is better for an extended aeration plant.

Fallow It is an area (of farmland) plowed and harrowed but left unsown for a period in order to restore its fertility as part of a crop rotation or to avoid surplus production

F-Cell Dissolved air flotation cell or DAF.

Feedback mechanisms Factors that increase or amplify (positive feedback) or decrease (negative feedback) the rate of a process. An example of positive climatic feedback is the ice-albedo feedback.

FEMA The Federal Emergency Management Agency is the USA's lead emergency management and preparedness agency.

Fermentation The process in which bacteria degrade organic matter under anaerobic conditions, such as in a collection system, primary clarifier, anaerobic selector, or fermenter tank

FF-Cell Dissolved air flotation and filtration cell or DAFF.

Field margins It is generally the least productive areas of a field and just a 1-metre grass strip between the outer edge of the hedges, and the crop edge can benefit wildlife in many ways.

Filamentous organism (a) Microorganisms, such as bacterial, algal, or fungal species, that grow in thread-like colonies, or a biological mass that may interfere with settling in clarifiers, or may interfere with drainage through a filter. (b) Organisms that grow in a thread or filamentous form. Common types are Thiothrix and Actinomycetes. A common cause of sludge bulking in the activated sludge process.

Filtration It is usually a granular medial filtration process, which involves the passage of wastewater or water through a bed of filter media with resulting deposition of suspended solids. Eventually the pressure drop across the bed becomes excessive or the ability of the bed to remove suspended solids is impaired. Cleaning is then necessary to restore operating head and effluent quality. The time in service between cleanings is termed the filter run time or run length. The head loss at which filtration is interrupted for cleaning is called the terminal head loss, and this head loss is maximized by the judicious choice of media sizes. Dual media filtration involves the use of both sand and anthracite as filter media, with anthracite being placed on top of the sand. Gravity filters operate by either using the available head from the previous treatment unit, or by pumping to a flow split box after which the wastewater flows by gravity to the filter cells. Pressure filters utilize pumping to increase the available head. A filter unit generally consists of a containing vessel, the filter media, structures to support the media, distribution and collection devices for filter influent, effluent, and backwash water flows, supplemental cleaning devices, ad necessary controls for flows, water levels, and backwash sequencing. Backwash sequences can include air scour or surface wash steps. Backwash water can be stored separately or in chambers that are integral parts of the filter unit. Backwash water can be pumped through the unit or can be supplied through gravity head tanks.

Fire-resistant material A material designed to withstand heat and resist burning to aid in saving lives and protecting property.

First responder Includes organizations and individuals who assume an emergency management role. Also known as emergency management or response personnel.

Floc (a) Collections of smaller particles that have agglomerated together into larger, more separable, floatable or settleable particles as result of the coagulation process; (b) clumps of bacteria and particles or coagulants and impurities that have come together and formed a cluster. Found in aeration tanks, secondary clarifiers, and chemical precipitation processes.

Flocculation (a) A water treatment unit process following coagulation that uses gentle stirring to bring suspended particles together so that they will form larger, more separable (floatable or settleable) floc. (b) The gathering together of fine particles after coagulation to form larger particles by a process of gentle mixing.

Flocculator A process device to enhance the formation of floc in a water. Mixing energy can be provided by slow turning mechanical means or head loss.

Flood A general and temporary condition of partial or complete inundation of normally dry land areas from overflow of inland or tidal waters, unusual or rapid accumulation or runoff of surface waters, or mudslides/mudflows caused by accumulation of water.

Flotation cell A dissolved air flotation cell or unit process.

Flotation thickening In a dissolved air flotation (DAF) sludge thickening process, air is introduced into liquid sludge that is being held at an elevated pressure. When the sludge is depressurized, the dissolved air is released as finely divided air bubbles carrying the solids to the top, where they are compacted and removed.

Flotation-filtration cell A package treatment plant which consists of both dissolved air flotation and filtration unit processes.

Fluidized bed incinerator In the fluidized bed system, dewatered sludge or small solid waste is fed into a bed of hot sand fluidized by circulating air. There is rapid drying and combustion of the sludge or solid waste. Ash is carried out of the incinerator by the combustion gases and is separated by a wet scrubber system or equivalent.

Fluorinated gases Powerful synthetic greenhouse gases such as hydrofluorocarbons, perfluorocarbons, and sulfur hexafluoride that are emitted from a variety of industrial processes. Fluorinated gases are sometimes used as substitutes for stratospheric ozone-depleting substances (e.g., chlorofluorocarbons, hydrochlorofluorocarbons, and halons) and are often used in coolants, foaming agents, fire extinguishers, solvents, pesticides, and aerosol propellants. These gases are emitted in small quantities compared to carbon dioxide (CO_2), methane (CH_4), or nitrous oxide (N_2O), but because they are potent greenhouse gases, they are sometimes referred to as high global warming potential gases (high GWP gases).

Fluorocarbons Carbon-fluorine compounds that often contain other elements such as hydrogen, chlorine, or bromine. Common fluorocarbons include chlorofluorocarbons (CFCs), hydrochlorofluorocarbons (HCFCs), hydrofluorocarbons (HFCs), and perfluorocarbons (PFCs).

Forage (a) The food supply of an animal; (b) an action (of a person or animal) searching widely for food or provisions.

Force main A pipe that carries wastewater under pressure from the discharge side of a pump to a point of gravity flow downstream.

Forcing mechanism A process that alters the energy balance of the climate system, i.e., changes the relative balance between incoming solar radiation and outgoing infrared radiation from Earth. Such mechanisms include changes in solar irradiance, volcanic eruptions, and enhancement of the natural greenhouse effect by emissions of greenhouse gases.

Fossil fuel A general term for organic materials formed from decayed plants and animals that have been converted to crude oil, coal, natural gas, or heavy oils by exposure to heat and pressure in the Earth's crust over hundreds of millions of years.

Freeboard The vertical distance from the normal water surface to the top of the confining wall.

Fuel switching In general, this is substituting one type of fuel for another. In the climate-change discussion, it is implicit that the substituted fuel produces lower carbon emissions per unit energy produced than the original fuel, e.g., natural gas for coal.

Full-scale exercise A multiagency, multijurisdictional operations-based exercise involving actual deployment of resources in a coordinated response as if a real incident had occurred. A full-scale exercise tests many components of one or more capabilities within emergency response and recovery, and is typically used to assess plans and procedures under crisis conditions, and assess coordinated response under crisis conditions. Characteristics of a full-scale exercise include mobilized units, personnel, and equipment; a stressful, realistic environment; and scripted exercise scenarios.

Functional annex A component of an emergency operations plan that describes procedures and missions for many hazards.

Functional exercise A single-agency or multiagency operations-based exercise designed to evaluate capabilities and multiple functions using a simulated response. Characteristics of a functional exercise include simulated deployment of resources and personnel, rapid problem solving, and a highly stressful environment.

Fungicide It is a chemical that destroys fungus.

General circulation model (GCM) A global, three-dimensional computer model of the climate system, which can be used to simulate human-induced climate change. GCMs are highly complex and they represent the effects of such factors as reflective and absorptive properties of atmospheric water vapor, greenhouse gas concentrations, clouds, annual and daily solar heating, ocean temperatures and ice boundaries. The most recent GCMs include global representations of the atmosphere, oceans, and land surface.

Geocode It is a process of identifying a location by one or more attributes from a base layer.

Geographic Information System (GIS) It is an organized collection of computer hardware, software, geographic data, and personnel designed to efficiently capture, store, update, manipulate, analyze, and display all forms of geographically referenced information.

Geosphere The soils, sediments, and rock layers of the Earth's crust, both continental and beneath the ocean floors.

Glacier A multiyear surplus accumulation of snowfall in excess of snowmelt on land and resulting in a mass of ice at least 0.1 km2 in area that shows some evidence of movement in response to gravity. A glacier may terminate on land or in water. Glacier ice is the largest reservoir of fresh water on Earth, and second only to the oceans as the largest reservoir of total water. Glaciers are found on every continent except Australia.

Global average temperature An estimate of Earth's mean surface air temperature averaged over the entire planet.

Global Positioning System (GPS) It is a satellite-based device that records x,y,z coordinates and other data. Ground locations are calculated by signals from satellites orbiting the Earth. GPS devices can be taken into the field to record data while walking, driving, or flying.

Global warming potential (GWP) A measure of the total energy that a gas absorbs over a particular period of time (usually 100 years), compared to carbon dioxide.

Global warming The recent and ongoing global average increase in temperature near the Earth's surface.

Glycogen-accumulating organisms (GAOs) They are organisms that take up glycogen and volatile fatty acids during metabolism, sometimes competing with phosphorus-accumulating organisms for food sources.

Glycogen It is a polysaccharide of glucose, which is energy storage inside a cell. It is present in all cells; however, microorganisms that accumulate glycogen in cells during the anaerobic stage of a bio-P process are not able to perform biological nutrient removal.

Good laboratory practices (GLPs) General guidelines or formal regulations for performing basic laboratory operations or activities that are known or believed to influence the quality and integrity of the results. The purpose of GLP is for personal protection, laboratory protection, proper chemical storage, and proper waste management.

Grab sample A single sample of water collected at a particular time and place, which represents the composition of the water only at that time and place.

Grab sample A single sample which is collected at one point in time and place.

Granular activated carbon adsorption Granular activated carbon (GAC) is used in wastewater substances. GAC systems generally consist of vessels in which the carbon is placed, forming a "filter" bed. These systems can also include carbon storage vessels and thermal regeneration facilities. Vessels are usually circular for pressure systems or rectangular for gravity flow systems. Once the carbon adsorptive capacity has been fully utilized, it must be disposed of or regenerated. Usually, multiple carbon vessels are used to allow continuous operation. Columns can be operated in series or parallel modes. All vessels must be equipped with carbon removal and loading mechanisms to allow for the removal of spent carbon and the addition of new material. Flow can be either upward or downward through the carbon bed. Vessels are backwashed periodically. Surface wash and air scour systems can also be used as part of the backwash cycle. Small systems usually dispose of spent carbon or regenerate it offsite. Systems above about 3 to 5 MGD (million gallons per day) usually provide on-site regeneration of carbon for economic reasons.

Gravity belt thickening It is a process for removing water from a wet dilute sludge using a moving belt. Dilute sludge is conditioned with a polymer and fed into a feed/distribution box of a belt thickening device. The sludge distributes evenly on the moving belt as water drains through the porous belt and the thickened sludge is discharged.

Gravity thickening Its tank design is similar to a conventional sedimentation tank. Dilute sludge is fed to a center feed well. The feed sludge is allowed to settle and compact, and the thickened sludge is withdrawn.

Green solvent An environmentally friendly solvent, or biosolvents, which may be derived from the processing of agricultural crops, such as corn.

Greenhouse effect Trapping and build-up of heat in the atmosphere (troposphere) near the Earth's surface. Some of the heat flowing back toward space from the

Earth's surface is absorbed by water vapor, carbon dioxide, ozone, and several other gases in the atmosphere and then reradiated back toward the Earth's surface. If the atmospheric concentrations of these greenhouse gases rise, the average temperature of the lower atmosphere will gradually increase.

Greenhouse gas (GHG) Any gas that absorbs infrared radiation in the atmosphere. Greenhouse gases include, carbon dioxide, methane, nitrous oxide, ozone, chlorofluorocarbons, hydrochlorofluorocarbons, hydrofluorocarbons, perfluorocarbons, sulfur hexafluoride.

Grinding It is a unit operation for reducing the particle size of objects or debris in the influent wastewater, also termed shredding or comminuting. These devices may be installed with a screen directly in the wastewater flow or separately out of the wastewater flow, with the shredded particles returned to the flow downstream of the screen. Only those shredding and grinding devices that are installed directly in the influent channel are termed comminuting devices.

Grit chamber It is a grit removal device that is designed to allow the settling out of this material. Grit removal is an important process for several reasons: (1) to prevent cementing effects at the bottom of sludge digesters and primary clarification tanks; (2) to reduce the potential for clogging of pipes and sludge hoppers; (3) to protect moving mechanical equipment and pumps from unnecessary wear and abrasion; (4) to reduce accumulations of materials in aeration tanks and sludge digesters, which would result in a loss of usable volume; and (5) to reduce accumulations at the bases of mechanical screens. There are two types of grit chambers. The velocity controlled grit chambers limit the velocity in the rectangular channels to a maximum of 1 foot per second (fps). This velocity is low enough to allow the grit to settle but fast enough to maintain a majority of the organic material in suspension. The aerated grit chambers are normally sized on the basis of both detention time and volume of air. Typically, the detention time is in the range of 2 to 5 minutes, and the air flow is in the range of 0.04 to 0.06 cu ft./gallon of wastewater. The constant head type of system is normally designed using an overflow rate of 15,000 gallons per day per square foot and a 1 minute detention time at peak day flows.

Grit removal Grit removal is accomplished by providing an enlarged channel or chamber which causes the flow velocity to be reduced and allows the heavier grit to settle to the bottom of the channel where it can be removed.

Grit (a) It includes sand, gravel, cinders, coffee grounds, small stones, cigarette filter tips, and other large-sized unwanted materials in wastewater. (b) The heavy material present in wastewater, such as sand, coffee grounds, eggshells, gravel, and cinders.

Groundwater The supply of fresh water found beneath the Earth's surface (usually in aquifers) which is often accessed through wells and springs.

Habitat fragmentation A process during which larger areas of habitat are broken into a number of smaller patches of smaller total area, isolated from each other by a matrix of habitats unlike the original habitat.

Halocarbons Compounds containing either chlorine, bromine, or fluorine and carbon. Such compounds can act as powerful greenhouse gases in the atmosphere. The chlorine- and bromine-containing halocarbons are also involved in the depletion of the ozone layer.

Hazard A natural, technological, or human-caused source or cause of harm or difficulty.

Hazardous wastes A hazardous waste is a material that is subject to special consideration by the US Environmental Protection Agency (USEPA), under 40CFR261. State or local authorities may also designate additional materials as hazardous waste in their areas. The definition given by 40 CFR 261 defines a hazardous waste as a solid waste that is not excluded from regulation and meets one or more of the following criteria: (a) it is a discarded commercial chemical product, off-specification species, container residue, or spill residue of materials specifically listed in 40CFR261.33 (P- and U-codes); (b) it is a waste from a specific source listed in 40CFR261.32 (K-code); (c) it is a waste from a non-specific source listed in 40CFR261.31 (F-code); and/or (d) it displays any of the following characteristics of hazardous wastes: ignitability (such as flash point is below 60 ° C or 140 ° F, it is classified by the US Department of Transportation,DOT, as an oxidizer D001), corrosivity (such as the pH of the waste material is less than or equal to 2, or greater than or equal to 12.5, or classified by DOT as D002), reactivity (such as the waste material is unstable, reacts violently with water, may generate toxic gases when mixed with water, or classified by DOT as D003), or toxicity (such as it is classified by DOT ad D004-D043).

Hazard-specific annex A component of an emergency operations plan that describes strategies for managing specific hazards.

Headloss It is an indirect measure of loss of energy or pressure of flowing water. It is measured as the difference in elevation between the upstream water surface and the downstream water surface.

Headworks The facilities where wastewater enters a wastewater treatment plant. The headworks may consist of bar screens, comminutors, a wet well, and pumps.

Heat island An urban area characterized by temperatures higher than those of the surrounding non-urban area. As urban areas develop, buildings, roads, and other infrastructure replace open land and vegetation. These surfaces absorb more solar energy, which can create higher temperatures in urban areas.

Heat load Amount of heat exchanged within the tower between water and air, expressed per unit of time. It can also be expressed as the product of the quantity of water circulated within the tower per unit time and the cooling range.

Heat waves A prolonged period of excessive heat, often combined with excessive humidity.

Heavy metal A general term, which applies to the group of metals and metalloids, and it has an atomic density more prominent than 4000 kg/m^3.

Herbicide It is a substance that is toxic to plants and is used to destroy unwanted vegetation.

Heterotrophic Microorganisms that use organic matter (carbon) for energy and growth and can grow in both aerobic and anoxic environments using both dissolved and chemically bound oxygen (nitrates)

Hive It is the structure used by bees for a home.

Honey bees (*Apis mellifera*) They are completely domesticated and live in hives provided by humans for the purpose of production of honey, however, they will leave these hives and form "wild" colonies in hollow trees, but there are no wild, native bees known. The castes of bees are: drones (sexually active males), a single queen (the egg-laying female), and workers (sexless females). The worker honey bees visit flowers to obtain nectar and pollen as their food, and construct the waxy combs made from secretions of the wax glands. The worker bees tend the eggs and larvae as well as feed the queen and carry away the eggs produced by the queen.

Honey crop It is the amount of honey produced per bee colony per season.

Hormone A hormone is any member of a class of signaling molecules, produced by glands in multicellular organisms, that are transported by the circulatory system to target distant organs to regulate physiology and behavior.

Horsepower continuous duty rate It is as the term implies, the horsepower which can be supplied continuously (24 hours per day) by an engine without overheating.

Humanitarian engineering A new engineering branch focuses on programs that are affordable, sustainable, and based on local resources, and it involves the creation and application of the innovative methods, processes, systems, and infrastructures to improving the well-being of marginalized people and disadvantaged communities, usually in the developing world. The word of "humanitarian" has been defined as "concerned with or seeking to promote human welfare."

Humanitarian technology A new professional field focuses on programs that are affordable, sustainable, and based on local resources, and it involves mainly application of the already developed simple methods, processes, systems, and infrastructures to improving the well-being of marginalized people and disadvantaged communities, usually in the developing world. Examples of humanitarian technologies for developing communities include affordable: (a) Technologies for clean water and sanitation; (b) Nonpolluting lighting, heating, and cooking methods; (c) Off-grid electricity generation (e.g., solar or bicycle); (d) Agricultural technologies and methods (e.g., aquaponics or irrigation); (e) Healthcare technologies; (f) Shelters; and (g) international STEAM education of Science, Technology, Engineering, Arts and Mathematics.

Hurricane A tropical storm with winds of 74 miles per hour or more. Hurricanes form in the southern Atlantic Ocean, Caribbean Sea, Gulf of Mexico, and eastern Pacific Ocean.

Hydraulic dredging It is a floating barge type machine known as a cutter suction dredge used to remove underwater materials and transfer those sediments mixed with water (referred to as a slurry) through a pressurized pipeline to a deposit location.

Hydraulic retention time (HRT) The given time it takes wastewater, including any return flows, to pass through a certain area

Hydrocarbons Substances containing only hydrogen and carbon. Fossil fuels are made up of hydrocarbons.

Hydrochlorofluorocarbons (HCFCs) Compounds containing hydrogen, fluorine, chlorine, and carbon atoms. Although ozone-depleting substances, they are less potent at destroying stratospheric ozone than chlorofluorocarbons (CFCs). They have been introduced as temporary replacements for CFCs and are also greenhouse gases.

Hydrofluorocarbons (HFCs) Compounds containing only hydrogen, fluorine, and carbon atoms. They were introduced as alternatives to ozone-depleting substances in serving many industrial, commercial, and personal needs. HFCs are emitted as by-products of industrial processes and are also used in manufacturing. They do not significantly deplete the stratospheric ozone layer, but they are powerful greenhouse gases with global warming potentials ranging from 140 (HFC-152a) to 11,700 (HFC-23).

Hydrogen sulfide gas (H_2S) A gas with a rotten egg odor. This gas is produced under anaerobic conditions. Hydrogen sulfide is particularly dangerous because it dulls the sense of smell so that it is unnoticeable after a prolonged period of time and because the odor is not noticeable in high concentrations. The gas is colorless, explosive, flammable, and poisonous to the respiratory system.

Hydrologic cycle The process of evaporation, vertical and horizontal transport of vapor, condensation, precipitation, and the flow of water from continents to oceans. It is a major factor in determining climate through its influence on surface vegetation, the clouds, snow and ice, and soil moisture. The hydrologic cycle is responsible for 25 to 30 percent of the mid-latitudes' heat transport from the equatorial to polar regions.

Hydrology It is a branch of science concerned with the properties of the Earth's water, and especially its movement in relation to land.

Hydrosphere The component of the climate system comprising liquid surface and subterranean water, such as: oceans, seas, rivers, fresh water lakes, underground water etc.

Hypochlorination Application of sodium hypochlorite (liquid) for disinfection is also called hypochlorination. It achieves the same result as that of chlorination using chlorine gas. Its use is generally limited to smaller water or wastewater treatment facilities or where there are significant health and safety concerns regarding the handling and use of chlorine gas.

Ice core A cylindrical section of ice removed from a glacier or an ice sheet in order to study climate patterns of the past. By performing chemical analyses on the air trapped in the ice, scientists can estimate the percentage of carbon dioxide and other trace gases in the atmosphere at a given time. Analysis of the ice itself can give some indication of historic temperatures.

Ignitability Any substance that exhibits any of the following properties is considered a hazardous substance due to ignitability: (a) a liquid that has a flash point less than 60 ° C; (b) a nonliquid capable, under normal conditions, of spontaneous and sustained combustion; (c) an ignitable compressed gas under the US Department of Transportation (DOT) regulations; and (d) an oxidizer under DOT regulations.

Impeller, pump It is a spinning component, the heart of the dredge pump, that transmits energy from the motor which drives the pump to the fluid being pumped by advancing the fluid externally from the center of rotation.

Incident Command System (ICS) A standardized on-scene emergency management construct specifically designed to provide an integrated organizational structure that reflects the complexity and demands of single or multiple incidents, without being hindered by jurisdictional boundaries. The Incident Command System is the combination of facilities, equipment, personnel, procedures, and communications operating within a common organizational structure, designed to aid in the management of resources during incidents. ICS is used for all kinds of emergencies and is applicable to small as well as large and complex incidents. ICS is used by various jurisdictions and functional agencies, both public and private, to organize field-level incident management operations.

Incineration It is dry combustion of sludge or solid wastes to produce an inert ash. This ash can then be beneficially used or disposed of in a sanitary landfill. The process must dry the sludge cake or small solid wastes, destroy the volatile content by burning, and finally produce a sterile ash. A variety of incinerator configurations exist. The most common include: (a) multiple hearth furnace incinerator; and (b) fluidized bed incinerator.

Incinerator, fluidized bed In the fluidized bed system, dewatered sludge or small solid waste is fed into a bed of hot sand fluidized by circulating air. There is rapid drying and combustion of the sludge or solid waste. Ash is carried out of the incinerator by the combustion gases and is separated by a wet scrubber system or equivalent.

Incinerator, multiple hearth furnace Sludge passes downward through a series of hearths. Dewatered sludge cake or small solid wastes are fed to the top hearths. In the upper hearths, the water content is vaporized and the sludge or solid wastes are dried. In the middle hearths, the sludge or solid wastes are ignited and burned. In the lower hearths, the slow-burning materials is continuously burned, and the ash undergoes cooling.

Independent physicochemical wastewater treatment system (IPCWWTS) An Independent Physicochemical Wastewater Treatment System (IPCWWTS) utilizes physicochemical (PC) process technology other than biological process technology to obtain combined primary and secondary treatment efficiency for removals of mainly biochemical oxidation demand (BOD), chemical oxidation demand (COD), total suspended solids (TSS), and phosphate. Typically, an IPCWWTS uses combinations of preliminary treatment (flow equalization, bar screening, comminution, grit chamber, ammonia stripping), chemical precipitation/coagulation, primary clarification (primary sedimentation clarification or primary flotation clarification), secondary clarification (secondary sedimentation clarification, or secondary flotation clarification, without biological treatment), tertiary wastewater treatment (filtration, and/or granular activated carbon adsorption, ion exchange, PC oxidation, etc.), and disinfection. An innovative efficient primary flotation clarifier or a secondary flotation clarifier can be in any shape, circular or rectangular. In general this IPCWWTS requires much less land area than conventional biological secondary treatment systems. Phosphors removal is inherent in this physicochemical process system.

Indirect emissions Indirect emissions from a building, home or business are those emissions of greenhouse gases that occur as a result of the generation of electricity used in that building. These emissions are called "indirect" because the actual emissions occur at the power plant which generates the electricity, not at the building using the electricity.

Industrial Ecology (IE) Industry, according to the Oxford English dictionary, is "intelligent or clever working" as well as the particular branches of productive labor. Ecology is the branch of biology that deals with the mutual relations between organisms and their environment. Ecology implies more the webs of natural forces and organisms, their competition and cooperation, and how they live off one another. Industrial ecology (IE) is now a branch of systems science for sustainability, or a framework for designing and operating industrial systems as sustainable and interdependent with natural systems. It seeks to balance Industrial production and economic performance with an emerging understanding of local and global ecological constraints.

Industrial revolution A period of rapid industrial growth with far-reaching social and economic consequences, beginning in England during the second half of the eighteenth century and spreading to Europe and later to other countries including the United States. The industrial revolution marks the beginning of a strong increase in combustion of fossil fuels and related emissions of carbon dioxide [8].

Industry It is intelligent or clever working as well as the particular branches of productive labor.

Inflow Water discharged into a sewer system and service connections from sources other than regular connections. This includes flow from yard drains, foundation drains and around manhole covers. Inflow differs from infiltration in that it is a direct discharge into the sewer rather than a leak in the sewer itself.

Influent Wastewater or other liquid—raw (untreated) or partially treated—flowing into a reservoir, basin, treatment process, or treatment plant.

Infrared radiation Infrared radiation consists of light whose wavelength is longer than the red color in the visible part of the spectrum, but shorter than microwave radiation. Infrared radiation can be perceived as heat. The Earth's surface, the atmosphere, and clouds all emit infrared radiation, which is also known as terrestrial or long-wave radiation. In contrast, solar radiation is mainly short-wave radiation because of the temperature of the Sun.

Infrared sludge dryer It is a sludge dewatering process using regular direct radiant energy from infrared heating elements. Dewatered sludge is transferred from the dewatering equipment to the dryer by a conveyor belt. The sludge then drops into the augers, where the auger blades agitate the sludge while infrared heating elements heat the sludge. This process then repeats once the sludge is dropped into the secondary drying zone.

Innovative biological wastewater treatment system It normally includes (a) preliminary treatment units (i.e., screen, comminutor, grit chamber etc., for removal of sand, gravel, cinders, coffee grounds, small stones, cigarette filter tips, logs, cans, and other large-sized unwanted materials from raw wastewater), (b) primary flotation clarification for removing mainly total suspended solids from preliminary effluent, (c) secondary biological treatment units (such as activated sludge aeration or equivalent plus secondary flotation clarification) for removing dissolved organic/inorganic pollutants from primary effluent, and (d) tertiary treatment plant units (i.e., filtration, granular activated carbon adsorption, ion exchange, oxidation, nitrification, denitrification, and/or disinfection) for final polishing the secondary effluent in order to meet the effluent discharge standards.

Innovative physicochemical flotation wastewater treatment system It includes (a) preliminary treatment units (i.e., screen, comminutor, grit chamber, etc., for removal of sand, gravel, cinders, coffee grounds, small stones, cigarette filter tips, logs, cans, and other large-sized unwanted materials from raw wastewater), (b) primary flotation clarification for removing mainly total suspended solids from preliminary effluent, (c) secondary physicochemical treatment units (such as chemical precipitation/coagulation or equivalent plus secondary flotation clarification) for removing dissolved organic/inorganic pollutants from primary effluent, and (d) tertiary treatment plant units (i.e., filtration, granular activated carbon adsorption, ion exchange, oxidation, nitrification, denitrification, and/or disinfection) for final polishing the secondary effluent in order to meet the effluent discharge standards. In the nitrification and denitrification steps, only tertiary flotation clarification will be used for solid-water separation.

Innovative physicochemical flotation-membrane wastewater treatment system It includes (a) preliminary treatment units (i.e., screen, comminutor, grit chamber etc. for removal of sand, gravel, cinders, coffee grounds, small stones, cigarette filter tips, logs, cans, and other large-sized unwanted materials from raw wastewater), (b) primary flotation clarification for removing mainly total suspended solids from preliminary effluent, (c) secondary physicochemical treatment units (such as chemical precipitation/coagulation or equivalent plus secondary membrane clarification) for removing dissolved organic/inorganic pollutants from primary effluent, and (d) tertiary treatment plant units (i.e., filtration, granular activated carbon adsorption, ion exchange, oxidation, nitrification, denitrification, and/or disinfection) for final polishing the secondary effluent in order to meet the effluent discharge standards. In the nitrification and denitrification steps, only tertiary membrane clarification will be used for solid-water separation.

Insecticide It is a substance used for killing insects.

Intergovernmental Panel on climate Change (IPCC) The IPCC was established jointly by the United Nations Environment Programme and the World Meteorological Organization in 1988. The purpose of the IPCC is to assess information in the scientific and technical literature related to all significant components of the issue of climate change. The IPCC draws upon hundreds of the world's expert scientists as authors and thousands as expert reviewers. Leading experts on climate change and environmental, social, and economic sciences from some 60 nations have helped the IPCC to prepare periodic assessments of the scientific underpinnings for understanding global climate change and its consequences. With its capacity for reporting on climate change, its consequences, and the viability of adaptation and mitigation measures, the IPCC is also looked to as the official advisory body to the world's governments on the state of the science of the climate change issue.

Internet The Internet (portmanteau of interconnected network) is the global system of interconnected computer networks that uses the Internet protocol suite (TCP/IP) to link devices worldwide. It is a network of networks that consists of private, public, academic, business, and government networks of local to global scope, linked by a broad array of electronic, wireless, and optical networking technologies. The Internet carries a vast range of information resources and services, such as the inter-linked hypertext documents and applications of the World Wide Web (WWW), electronic mail, telephony, and file sharing.

Inundation The submergence of land by water, particularly in a coastal setting.

Ion A charged atom, molecule, or radical that affects the transport of electricity through an electrolyte or gas. An atom or molecule that has lost or gained one or more electrons

KAMET It is the abbreviation of Krofta Advanced Municipal Effluent Treatment (KAMET), and is a circular package plant consisting of both Supracell and Sandfloat with Supracell on the top. KAMET system can be used for either water treatment or wastewater treatment, but is advertised for wastewater treatment.

Kraus activated sludge process When nitrogen deficiency occurs in biological wastewater treatment and exogenous supply of nitrogen is expensive, the Kraus process can be useful in that an internal or endogenous supply of nitrogen is used to maintain the growth of active biomass. The Kraus process has a regular aeration tank (conventional, step aeration or complete mix) plus an extra reaeration tank. The reaeration tank receives (a) return activated sludge from a secondary clarifier, (b) digester supernatant, (c) digested sludge, and (d) possible other nitrogen source, if needed, and aerates the mixture. When the reaeration tank effluent is introduced into the aeration tank with the wastewater influent, the nitrate released from the reaeration tank serves as the supplemental nitrogen source for synthesis.

Ladder, dredge It is a ladder that extends from the front position of the cutter suction dredge (CSD)and is attached to the pontoon by a hinge coupling. It is lowered by the dredge operator via controls located inside the control cab. The cutter head attached at the end of ladder excavates the material being dredged.

Lagoon, aerated An aerated lagoon is a holding basin (3–10 days DT; 6–20 ft. depth) in which air is mechanically introduced to speed up aerobic decomposition of organic pollutants in wastewater. It is essentially the same as the extended aeration activated sludge biological process, except that an earthen basin is used for the aerobic suspended-growth bioreactor. The downstream of an aerated lagoon can be a facultative stabilization pond (or facultative lagoon), a sedimentation clarifier, or a dissolved air flotation clarifier, each incorporating the recycle of biological solids (or biosolids). The oxygen required by the aerated lagoon process is supplied by surface or diffused aerators. As with other suspended-growth systems, the turbulence created is used to maintain the contents of the basin in suspension.

Lagoon, anaerobic Anaerobic lagoons are capable of treating high strength wastewater. They are relatively deep (8–20 ft. depth; 20–50 days DT) ponds with steep sidewalls in which anaerobic conditions are maintained by keeping wastewater organic loading so high that complete deoxygenation is prevalent. Although some oxygenation is possible in a shallow surface zone, once grease form an impervious surface layer, complete anaerobic conditions develop. Waste treatment or stabilization results from thermophilic anaerobic digestion of organic wastes. The treatment process is analogous to that occurring in single stage untreated anaerobic digestion of sludge in which acid forming bacteria break down organics. The resultant acids are then converted to carbon dioxide methane, cells and other products. In a typical anaerobic lagoon, raw wastewater enters near the bottom of the pond (often at the center) and mixes with the active microbial mass in the sludge blanket which is usually about 6 ft. deep. The effluent discharge is located near one of the sides of the pond, submerged below the liquid surface. Excess undigested grease floats to the top, forming a heat retaining and relatively airtight cover. Wastewater flow equalization and heating are generally not practiced. Excess sludge is washed out with the lagoon effluent. Recirculation of waste sludge is not required.

Lagoon, facultative Facultative lagoons are intermediate depth (3–8 ft. depth; 20–180 days DT) ponds in which the wastewater is stratified into three zones. These zones consist of an anaerobic bottom layer, an aerobic surface layer, and an intermediate zone. Stratification is a result of solids settling and temperature-water density variations. Oxygen in the surface stabilization zone is provided by reaeration and photosynthesis. This is in contrast to aerated lagoons in which mechanical aeration is used to create aerobic surface conditions. In general the aerobic surface layer serves to reduce odors while providing treatment of soluble organic by-products of the anaerobic processes operating at the bottom. Sludge at the bottom of facultative lagoons will undergo anaerobic digestion producing carbon dioxide, methane, and cells. The photosynthetic activity at the lagoon surface produces oxygen diurnally, increasing the dissolved oxygen during daylight hours, while surface oxygen is depleted at night. It is used for treating raw, screened wastewater, or primary effluent, secondary effluent, or weak biodegradable industrial wastewater.

Lake dredging Lake dredging helps to maintain a healthy environment for both aquatic life and recreational activities because both natural and man-made lakes eroding sediment, silt accumulation, and algae blooms can gradually build up causing both depth and water quality issues.

Lake An inland body of water, fresh or salt, of considerable size usually greater than 50 acres (200,000 square meters) and occupying a basin or hollow on the Earth's surface.

Land reclamation It is a process consisting of dredging sand, clay, or rock from offshore or inland lakes, ponds, and quarries. The dredged material is repurposed as fill material forming new land along coastal waterways or inland water bodies, such as swamps.

Landfill Land waste disposal site in which waste is generally spread in thin layers, compacted, and covered with a fresh layer of soil each day.

Landfill, sanitary Burying sludge or solid wastes to minimize nuisance conditions or public health problem is called sanitary landfill. Sludge and solid wastes are buried in layers which are covered with fill material excavated at the site. Landfills must be located where nearby wells and groundwater supplies will not be contaminated by leachate from the landfill operation. Sludge landfills are generally separate from refuse landfills.

Landslide Processes that result in the downward movement of falling or flowing rock, soil, organic materials, or a combination of these.

Latitude The location north or south in reference to the equator, which is designated at zero (0) degrees. Lines of latitude are parallel to the equator and circle the globe. The North and South poles are at 90° North and South latitude.

Latrine It is a receptacle (like a toilet or something as simple as a pit in the ground) for use as a sanitation system which functions to receive human excreta.

Laundromat An establishment with coin-operated washing machines and dryers for public use.

Layer, data It is a logical set of thematic data described and stored in a map library. Layers act as digital transparencies that can be laid atop one another for viewing or spatial analysis.

Leaf-cutting bee (*Megachile pascoensis*) They prepare a tunnel in the ground or in rotten wood. The female flies to a suitable flower and cuts parts out of the petals, or the leaves. These round disks are brought back to the tunnel and used construct a small cradle. The female also collect foods for other bees. They are natural bees of wild species, not affecting humans.

Least developed country A country with low indicators of socioeconomic development and human resources, as well as economic vulnerability, as determined by the United Nations.

Line data Lines represent geographic features too narrow to be displayed as an area at a given scale, such as contours, street centerlines, or streams.

Listed wastes In the USA, the government listed wastes are wastes specifically regulated by the US Environmental Protection Agency (USEPA) and appear on one of four lists found in 40CFR 261.3. The four USEPA lists, however, fall into three categories of wastes: (a) nonspecific source wastes, such as generic wastes, commonly produced wastes by CFR 261.32; (c) commercial chemical products, such as wastes including specific commercial chemical products or manufacturing chemical intermediates—creosote and kepone, etc., under 40 CFR 261.33 (e) and (f).

Long-wave radiation Radiation emitted in the spectral wavelength greater than about 4 micrometers, corresponding to the radiation emitted from the Earth and atmosphere. It is sometimes referred to as "terrestrial radiation" or "infrared radiation," although somewhat imprecisely.

Make-up water Water that is added to the tower to replenish water losses due to evaporation, drift, blow-down, and other miscellaneous water losses.

Marina dredging It is an operation for maintaining the depth of a marina, whether private or public, is vital to both marina owners and the customers they serve. Adequate channel depths allow boaters to safely move in and out while protecting vessels from damage.

Masking agents Substances used to cover up or disguise unpleasant odors. Liquid masking agents are dripped into the wastewater, sprayed into the air, or evaporated (using heat) with the unpleasant fumes or odors and then discharged into the air by blowers to make an undesirable odor less noticeable.

Mason bee (*Osmia cobaltina*) This bee nests in the ground or in natural cavities in wood. Mason bees lack the pollen baskets on their tibiae. They carry their pollen in a "pollen brush" on the underside of the abdomen. They are natural bees of wild species, not affecting humans.

Mathematical modeling The process of developing a *mathematical model* is termed *mathematical modeling*.

Mathematical models A *mathematical model* is a description of a system using *mathematical* concepts and language. A *model* may help to explain a system and to study the effects of different components, and to make predictions about costs, impact, population, behavior, frequency, etc.

Mechanical aeration The use of machinery to mix air and water so that oxygen can be absorbed into the water.

Mechanical dredging It is an operation for removing debris or compressed hardened elements from below the water using a clamshell type bucket or heavy-duty excavator. This method is preferred when both dredging and placing elements nearby onshore or onto a barge for further transport to a designated disposal location.

Megacities Cities with populations over ten million.

Melliferous It is a situation of yielding or having to do with honey.

Membrane bioreactor (MBR) Currently it is one of suspended growth biological processes (such as activated sludge process) involving the use of a bioreactor (such as an aeration tank) for carbonaceous oxidation, nitrification and/or denitrification, and the use of membrane filtration for subsequent bioreactor effluent treatment. The membrane filter discharges the membrane filter effluent (as the product water), recycles certain amount of activated sludge to the bioreactor, and wastes the excess amount of activated sludge. It the future, an MBR may also be a combination of an attached growth bioreactor and a membrane filter.

Metadata It is the information about a data set. It may include the source of the data; its creation date and format; its projection, scale, resolution, and accuracy; and its reliability with regard to some standard.

Metazoan It is a group of small animals that may be present in wastewater having bodies composed of cells that are differentiated into tissues and organs, such as rotifers, water bears, and daphnia.

Methane (CH_4) A hydrocarbon that is a greenhouse gas with a global warming potential most recently estimated at 25 times that of carbon dioxide (CO_2). Methane is produced through anaerobic (without oxygen) decomposition of waste in landfills, animal digestion, decomposition of animal wastes, production and distribution of natural gas and petroleum, coal production, and incomplete fossil fuel combustion. The GWP is from the IPCC's Fourth Assessment Report (AR4).

Metric ton Common international measurement for the quantity of greenhouse gas emissions. A metric ton is equal to 2205 lbs. or 1.1 short tons.

Microorganisms Very small organisms that can be seen only through a microscope. Some microorganisms use the wastes in wastewater for food and thus remove or alter much of the undesired matter.

Miticide It is a substance that kill mites.

Mitigation (a) A human intervention to reduce the human impact on the climate system; it includes strategies to reduce greenhouse gas sources and emissions and enhancing greenhouse gas sinks; (b) Activities taken to reduce the loss of life and lessen the impact to property from disasters.

Mixed liquor suspended solids (MLSSs) (a) The concentration of suspended solids present in activated sludge mixed liquor, usually expressed as milligrams per liter (mg/L); (b) Suspended solids in the mixed liquor of an aeration tank.

Mixed liquor volatile suspended solids (MLVSS) (a) The fraction of the suspended solids in biological reactor's mixed liquor that can be burned off by combustion at 550 ° C., or the organic portion of the solids (an estimation of the microorganisms and food). Usually expressed as milligrams per liter (mg/L). (b) The organic or volatile suspended solids in the mixed liquor of an aeration tank. This volatile portion is used as a measure or indication of the microorganisms present.

Mixed liquor (ML) (a) The water containing biosolids in a biological reactor. (b) When the activated sludge in an aeration tank is mixed with primary effluent or the raw wastewater and return sludge, this mixture is then referred to as mixed liquor as long as it is in the aeration tank. Mixed liquor may also refer to the contents of mixed aerobic or anaerobic digesters.

Mixed wastes The entire volume of any waste mixture containing a listed hazardous waste, regardless of concentration, is considered a hazardous waste with the following exceptions: (a) the listed waste in the mixture was listed solely because it exhibits a hazardous waste characteristic and the mixture no longer exhibits that characteristic; (b) a wastewater discharge subject to regulation by the Clean Water Act (CWA) and the hazardous waste mixed with the mixtures of nonhazardous wastes and characteristic hazardous wastes if the mixture no longer exhibits any of the characteristics (ignitability, corrosivity, reactivity, and toxicity); (c) a wastewater discharge subject to regulation by the Clean Water Act (CWA) and the hazardous waste mixed with the waste water which concentration of carcinogenic and non-carcinogenic spent solvents listed in 40 CFR Part 261.31 in the wastewater, provided the concentration does not exceed 1 ppm and 25 ppm, respectively; (d) a wastewater discharge subject to regulation by the Clean Water Act (CWA) and the hazardous waste mixed with the small loss of discarded commercial chemical products or intermediaries used as raw materials in manufacturing or produced as by-products; (e) a wastewater discharge subject to regulation by the Clean Water Act (CWA) and the hazardous waste mixed with a laboratory wastewater containing small amounts of listed toxic wastes; (f) a wastewater discharge subject to regulation by the Clean Water Act (CWA) and the hazardous waste mixed with the heat exchanger bundle cleaning sludge (petroleum refining industry).

Multiple hearth furnace incinerator Sludge passes downward through a series of hearths. Dewatered sludge cake or small solid wastes are fed to the top hearths. In the upper hearths, the water content is vaporized and the sludge or solid wastes are dried. In the middle hearths, the sludge or solid wastes are ignited and burned. In the lower hearths, the slow-burning materials is continuously burned, and the ash undergoes cooling.

Municipal solid waste (MSW) Residential solid waste and some non-hazardous commercial, institutional, and industrial wastes. This material is generally sent to municipal landfills for disposal.

Natural gas Underground deposits of gases consisting of 50 to 90 percent methane (CH_4) and small amounts of heavier gaseous hydrocarbon compounds such as propane (C_3H_8) and butane (C_4H_{10}).

Natural resources These resources naturally exist, such as surface water, groundwater, ice, snow, air, plants, animals, minerals, metals, soils, rocks, oils, underground gases, light, winds, radiation, etc. that are useful or potentially useful. Uses of natural resources include agricultural, industrial, household, recreational and environmental activities.

Natural variability Variations in the mean state and other statistics (such as standard deviations or statistics of extremes) of the climate on all time and space scales beyond that of individual weather events. Natural variations in climate over time are caused by internal processes of the climate system as well as changes in external influences, such as volcanic activity and variations in the output of the sun.

Nectar flow (a) It is a time of harvest when nectar is plentiful and bees produce and store surplus honey; also called honey flow. (b) It is called honey flow which is a time when nectar is plentiful and bees produce and store surplus honey.

NGO It is a Nongovernmental Organization. Any nonprofit, voluntary citizens' organization which can be organized at a local, national or international level.

Nitrate (NO_3) An oxygenated form of nitrogen.

Nitrification It is a biological process by which ammonia in wastewater is converted by *Nitrosomonas* and *Nitrobacter* to nitrite, then to nitrate in the presence of oxygen. The biological reactions involved in these conversions may take place during activated sludge treatment, or a separate stage following removal of carbonaceous materials. Separate stage nitrification may be accomplished via suspended growth or attached growth unit processes. In either case, the nitrification step is preceded by a pretreatment sequence (i.e., preliminary treatment, primary treatment, and secondary treatment). Possible secondary treatment may be (a) biological secondary treatment, such as activated sludge, trickling filter, roughing filter, plus secondary clarification; or (b) physicochemical secondary treatment, such as chemical precipitation/coagulation plus secondary clarification. Low BOD (i.e., BOD_5/TKN ratio of less than 3) in the secondary effluent will assure a high concentration of nitrifiers (*Nitrosomonas* and *Nitrobacter*) in the nitrification biomass. The most common separate stage nitrification process is the plug flow suspended growth configuration with clarification. In this process, pretreatment effluent (i.e., nitrification influent) is pH adjusted as required, and aerated, in a plug flow mode. Because the carbonaceous demand is low, nitrifiers predominate. A clarifier (either sedimentation clarifier or flotation clarifier) follows aeration, and nitrification sludge is returned to the aeration tank. A possible modification is the use of pure oxygen in place of conventional aeration during the plug flow operation.

Nitrifier (nitrifying bacteria) Bacteria that are capable of oxidizing nitrogenous material, such as *Nitrobacter* and *Nitrosomonas.*

Nitrifying bacteria Bacteria that change the ammonia and organic nitrogen in wastewater into oxidized nitrogen (usually nitrate).

Nitrite (NO_2) It is an intermediate oxygenated form of nitrogen

Nitrite Lock Incomplete nitrification resulting in excess nitrite levels that react with chlorine. Also known as "chlorine sponge."

Nitrobacter It is a group of nitrifying bacteria that oxidize nitrite to nitrate.

Nitrogen (N) N is an essential nutrient that is often present in wastewater as ammonia, nitrite, nitrate, and organic nitrogen. The sum of these is expressed as total nitrogen.

Nitrogen cycle (a) It is a chemical transformation cycle of nitrogen through various stages of decomposition and assimilation. (b) The natural circulation of nitrogen among the atmosphere, plants, animals, and microorganisms that live in soil and water. Nitrogen takes on a variety of chemical forms throughout the nitrogen cycle, including nitrous oxide (N_2O) and nitrogen oxides (NOx).

Nitrogen gas (N_2) Gaseous form of nitrogen, N, also called atmospheric nitrogen. It comprises approximately 79% of atmospheric gas.

Nitrogen oxides (NOx) Gases consisting of one molecule of nitrogen and varying numbers of oxygen molecules. Nitrogen oxides are produced in the emissions of vehicle exhausts and from power stations. In the atmosphere, nitrogen oxides can contribute to formation of photochemical ozone (smog), can impair visibility, and have health consequences; they are thus considered pollutants.

Nitrogenous biochemical oxygen demand (NBOD) The quantitative measure of the amount of oxygen required for the biological oxidation of nitrogenous material, such as ammonia-nitrogen and organic nitrogen (Measured by subtracting carbonaceous BOD from total BOD values).

Nitrosomonas A genus of nitrifying bacteria that oxidize ammonia to nitrite.

Nitrous oxide (N_2O) A powerful greenhouse gas with a global warming potential of 298 times that of carbon dioxide (CO_2). Major sources of nitrous oxide include soil cultivation practices, especially the use of commercial and organic fertilizers, fossil fuel combustion, nitric acid production, and biomass burning. The GWP is from the IPCC's Fourth Assessment Report (AR4). Natural emissions of N_2O are mainly from bacteria breaking down nitrogen in soils and the oceans. Nitrous oxide is mainly removed from the atmosphere through destruction in the stratosphere by ultraviolet radiation and associated chemical reactions, but it can also be consumed by certain types of bacteria in soils.

NOAA Weather Radio A radio with a special receiver to receive information from the network of radio stations that broadcast continuous weather information from the National Weather Service.

Nocardia It is a group of irregularly bent, short branching filamentous organisms that cause dense dark foam in aeration basins. Associated with high fats, oils, and greases. Easily identified under a microscope.

Nonionic polymer A neutrally charged colloidal big organic substance used as a coagulant to aid in separating (settling or floating) solids in water or wastewater

Nonmethane volatile organic compounds (NMVOCs) Organic compounds, other than methane, that participate in atmospheric photochemical reactions.

Nonstructural elements Any portion of the building or grounds not connected to the main structure (e.g., bookshelves, file cabinets, furnishings).

NPDES permit National Pollutant Discharge Elimination System (NPDES) permit is the US regulatory agency document issued by either a federal or state agency which is designed to control all discharges of pollutants from all point sources and storm water runoff into US waterways. A treatment plant that discharges to a surface water will have a NPDES permit.

Nutrient ratio The ratio of carbon, nitrogen, phosphorus and sometimes other trace elements that is essential to optimum growth of the biomass in activated sludge.

Nutrient A substance that is taken in by organisms and promotes growth. Carbon, nitrogen, and phosphorus are essential nutrients to most aquatic organisms.

Ocean acidification Increased concentrations of carbon dioxide in sea water causing a measurable increase in acidity (i.e., a reduction in ocean pH). This may lead to reduced calcification rates of calcifying organisms such as corals, mollusks, algae and crustaceans.

Organic A volatile, combustible, or biodegradable compound containing carbon bound with other elements.

Organic nitrogen It is an organic nutrient containing nitrogen which is chemically bound in organic molecules such as proteins, amines, and amino acids. It is calculated by subtracting the Ammonia Nitrogen analysis results from Total Kjeldahl Nitrogen analysis results.

Organic phosphate (a) It is a molecule of phosphate combined with an organic compound. Total phosphorus includes orthophosphate, polyphosphate and organic phosphate. (b) It is an organic phosphorus originating from organic sources, such as body and food waste, and sometimes industrial sources. Some can biologically decompose into inorganic orthophosphate; other types are non-biodegradable and will pass through the wastewater system without treatment.

Organic solvent A carbon-based solvent that is capable of dissolving other substances.

Ortho Imagery It is the aerial photograph that has been rectified to produce an accurate image of the Earth by removing tilt and relief displacements, which occurred when the photo was taken.

Orthophosphate It is an inorganic nutrient containing phosphorus which is required for plant and animal growth. An inorganic, soluble form of phosphorus that is readily available to plants and animals without further breakdown, and accounts for 70–90% of total phosphorus in wastewater. The easiest form of phosphorus that can be treated chemically. Total phosphorus includes orthophosphate, polyphosphate and organic phosphate.

OSHA It stands for the Occupational Safety and Health Administration of the United States Department of Labor, formed by the Occupational Safety and Health Act of 1970.

Output rating It refers to the percent of solid material production which may be expected from the dredge under normal operating conditions. Not to be confused with such phrases as output capacity, which usually refers to theoretical pumping output which could be more than twice as much under certain conditions.

Oxic It is an environmental condition in which an aquatic environment contains limited dissolved oxygen only in the range of 1.5 to 2.0 mg/L.

Oxidation ditch It is a ring-shaped channel suspended-growth biological process system similar to extended aeration process for mainly carbonaceous oxidation and nitrification. It was developed to minimize waste-activated sludge production through endogenous decay of the sludge mass. It is equipped with mechanical aeration devices, such as, mechanical brush aerators surface aerators and jet aerator devices to aerate and pump the wastewater. Screened wastewater enters the ditch is aerated, and circulates at about 0.8 to 1.2 ft./sec. The economics of oxidation ditches appear most favorable when the solids retention time (SRT) is long, particularly where nitrification is required. When both nitrification and denitrification are required, certain portion of the ditch must be operated under anoxic condition (instead of normal aerobic oxidation condition), the process should be renamed as "oxidation-denitrification ditch."

Oxidation reduction potential (ORP) The potential required to transfer electrons from the oxidant to the reductant; the quantitative measure, in mV, of the state of oxidation in wastewater treatment.

Oxidation The addition of oxygen, removal of hydrogen, or removal of electrons from a compound (For instance, organic matter may be oxidized to a more stable compound).

Oxidize To chemically transform a substance by combining it with oxygen.

Ozonation It is an oxidation process or a disinfection process, or both, involving the use of ozone.

Ozone (O_3) Ozone is a toxic and corrosive oxidant, or disinfectant. Ozone gas cannot be stored and must be produced on site. Ozone is produced by passing air or oxygen between oppositely charged plates or through tubes in which a core and the tube walls serve as the oppositely charged surfaces. Ozone is an excellent disinfectant or oxidant, but is harmful lung irritant. Therefore, un-reacted ozone must be vented through a destruction unit (usually thermal destruction).

Ozone-depleting substance (ODSs) A family of man-made compounds that includes, but are not limited to, chlorofluorocarbons (CFCs), bromofluorocarbons (halons), methyl chloroform, carbon tetrachloride, methyl bromide, and hydrochlorofluorocarbons (HCFCs). These compounds have been shown to deplete stratospheric ozone, and therefore are typically referred to as ODSs.

Ozone layer The layer of ozone that begins approximately 15 km above Earth and thins to an almost negligible amount at about 50 km, shields the Earth from harmful ultraviolet radiation from the sun. The highest natural concentration of ozone (approximately 10 parts per million by volume) occurs in the stratosphere at approximately 25 km above Earth. The stratospheric ozone concentration changes throughout the year as stratospheric circulation changes with the seasons. Natural events such as volcanoes and solar flares can produce changes in ozone concentration, but man-made changes are of the greatest concern.

Ozone precursors Chemical compounds, such as carbon monoxide, methane, nonmethane hydrocarbons, and nitrogen oxides, which in the presence of solar radiation react with other chemical compounds to form ozone, mainly in the troposphere.

Ozone Ozone, the triatomic form of oxygen (O_3), is a gaseous atmospheric constituent. In the troposphere, it is created by photochemical reactions involving gases resulting both from natural sources and from human activities (photochemical smog). In high concentrations, tropospheric ozone can be harmful to a wide range of living organisms. Tropospheric ozone acts as a greenhouse gas. In the stratosphere, ozone is created by the interaction between solar ultraviolet radiation and molecular oxygen (O_2). Stratospheric ozone plays a decisive role in the stratospheric radiative balance. Depletion of stratospheric ozone, due to chemical reactions that may be enhanced by climate change, results in an increased ground-level flux of ultraviolet (UV) B radiation.

Package treatment plant A prefabricated water or wastewater treatment plant consisting of two or more unit processes.

Packing Also called fill, this is a product placed within the tower to promote uniform water flow distribution and to enhance water-air contacts. (Note: Again, note the similarity to wet scrubbing applications.)

Parasite It is an organism that lives in or on another organism (its host) and benefits by deriving nutrients at the host's expense.

Particulate matter (PM) Very small pieces of solid or liquid matter such as particles of soot, dust, fumes, mists, or aerosols. The physical characteristics of particles, and how they combine with other particles, are part of the feedback mechanisms of the atmosphere.

Particulate Solids suspended in water, wastewater, or air that can vary widely in shape, size, density, and charge.

Parts per billion (ppb) Number of parts of a chemical found in one billion parts of a particular gas, liquid, or solid mixture.

Parts per million by volume (ppmv) Number of parts of a chemical found in one million parts of a particular gas, liquid, or solid.

Parts per trillion (ppt) Number of parts of a chemical found in one trillion parts of a particular gas, liquid, or solid.

Pathogen It is a bacterium, virus, or other microorganism that can cause disease.

Pathogenic organisms Bacteria, viruses, or cysts, which can cause disease (typhoid, cholera, dysentery) in a host such as a human. Also called pathogens.

Percolation The movement or flow of water through soil or rocks.

Perennial crops They are crops developed to reduce inputs necessary to produce food by greatly reducing the need to replant crops from year to year.

Perfluorocarbons (PFCs) A group of chemicals composed of carbon and fluorine only. These chemicals (predominantly CF4 and C2F6) were introduced as alternatives, along with hydrofluorocarbons, to the ozone-depleting substances. In addition, PFCs are emitted as by-products of industrial processes and are also used in manufacturing. PFCs do not harm the stratospheric ozone layer, but they are powerful greenhouse gases: CF4 has a global warming potential (GWP) of 7390 and C2F6 has a GWP of 12,200. The GWP is from the IPCC's Fourth Assessment Report (AR4). These chemicals are predominantly human-made, though there is a small natural source of CF4.

Performance Normally expressed as the amount of cooling for a given quantity of water at a given wet bulb temperature.

Permafrost Perennially (continually) frozen ground that occurs where the temperature remains below 0 ° C for several years.

Pest A destructive insect or other animal that attacks crops, food, livestock, human, or structure, etc. For instance, the tomato plant attracts a pest called whitefly, and the wood frame house attracts a pest called termite.

Pesticide It is a substance, usually a chemical, used for destroying insects or other organisms harmful to cultivated plants or to animals

pH (a) The negative logarithm of the hydrogen ion concentration ($-\log10[H^+]$) where H^+ is the hydrogen-ion concentration in moles per liter. Neutral water has a pH value of 7. (b) An expression of both acidity and alkalinity on a scale of 0 to 14, with 7 representing neutrality; numbers less than 7 indicate increasing acidity and numbers greater than 7 indicate increasing alkalinity. Acid rain can increase the pH level of the water in a lake, thereby killing all life.

pH adjustment A means of maintaining the optimum pH through the use of chemical additives.

Phenology The timing of natural events, such as flower blooms and animal migration, which is influenced by changes in climate. Phenology is the study of such important seasonal events. Phenological events are influenced by a combination of climate factors, including light, temperature, rainfall, and humidity.

Pheromones They are chemical substances secreted from bee glands and used as a means of communication; honey bees secrete many different pheromones.

Phosphate PO_4^{3-} ion

Phosphorus (P) P is an essential element and nutrient for all life forms. Occurs as orthophosphate, polyphosphate, and organic phosphates, the sum of which is total phosphorus.

Phosphorus-accumulating organisms (PAOs) Microorganisms (bacteria) that are capable of uptaking and storing orthophosphate in excess of their biological requirements.

Photosynthesis (a) The process by which plants take CO_2 from the air (or bicarbonate in water) to build carbohydrates, releasing O_2 in the process. There are several pathways of photosynthesis with different responses to atmospheric CO_2 concentrations. (b) The manufacture by plants of carbohydrates and oxygen from carbon dioxide and water in the presence of chlorophyll, using sunlight as an energy source. Plants thereby absorb carbon from the atmosphere with the process, and store it in their forms.

Physical recovery Identification of possible relocation areas for operations as well as plans to restore services, equipment, materials, and buildings and grounds after an incident.

Phytoplankton Usually microscopic aquatic plants, sometimes consisting of only one cell.

Pit latrine (Also Simple pit latrine) It is the cheapest and most basic form of improved sanitation, which consists of a pit dug into the ground with some kind of sturdy cover with a hole in it through which excreta goes into the pit.

Plankton Those organisms that are unable to maintain their position or distribution independent of the movement of water or air masses.

Plugging It is a term used to indicate when the material within a slurry that is being hydrotransported slows down, then settles, and plugs or clogs the discharge pipeline.

Point, data It is a single x,y coordinate that represents a geographic feature too small to be displayed as a line or area at that scale.

Pollination It is a process involving the transfer of pollen from the anthers of a flower to the stigma of the same flower or of another flower; pollination is a prerequisite for fertilization, and fertilization allows the flower to develop seeds.

Pollinator It is the biotic agent (vector) that moves pollen from the male anthers of a flower to the female stigma of a flower to accomplish fertilization or "syngamy" of the female gametes in the ovule of the flower by the male gametes from the pollen grain.

Pollution The contamination of a natural ecosystem, especially with reference to the activity of humans.

Pollution control A wide range of techniques, solutions, practices, and services for treating wastewater and waste.

Polygon It is a multisided figure that represents area on a map. Polygons have attributes that describe the geographic feature they represent.

Polyhydroxyalkanoates (PHAs) Energy-rich carbon polymers inside a bacterial cell, which are converted from readily available organic molecules, such as volatile fatty acids (VFAs), in the wastewater. PHAs are the intracellular energy storage of the phosphorus-accumulating organisms (PAOs). The PAOs utilize PHA as an energy source to uptake phosphorus from the wastewater in the aerobic zone of the biological phosphorus removal process.

Polymer Used with other chemical coagulants to aid in binding small suspended particles to larger chemical flocs for their removal from water.

Polyphosphate (Poly-P) (a) Inorganic phosphorus derived from synthetic detergents. May be hydrolyzed into orthophosphates. (b) It is a large molecule containing many individual molecules of orthophosphate. Total phosphorus includes orthophosphate, polyphosphate, and organic phosphate.

Ponding A condition occurring on trickling filters when the hollow spaces (voids) become plugged to the extent that water passage through the filter is inadequate. Ponding may be the result of excessive slime growths, trash, or media breakdown.

Postaeration It is one of the tertiary treatment processes. Because of the more stringent water quality standards being adopted by various regulatory agencies, the practice of postaeration has increased substantially in recent years. The introduction of water-quality-based effluent standards and permits that include high dissolved oxygen levels has made it necessary for many wastewater treatment plants to postaerate the plant effluent before discharge.

Pour-flush latrine It is a latrine where a small quantity of water is poured in to flush excreta through a water seal into a pit.

Powdered activated carbon adsorption Powdered activated carbon (PAC) is used in both water and wastewater treatment. For water treatment, PAC systems generally involve dosing powdered activated carbons in slurry form to rapid mixers and flocculators (for mixing and flocculation, and adsorption) before reaching clarifiers for PAC-floc separation and clarifier effluent discharge. For wastewater treatment, powdered activated carbon in slurry form is dosed to either mixing/flocculation chambers (if they exist), or to an aeration basin directly to be mixed with mixed liquor for adsorption reactions. The spent PAC again is recovered or removed by a subsequent clarifier (either sedimentation clarifier or flotation clarifier). When PAC is used in conjunction with a flotation clarifier, it is also called "adsorption flotation."

Preaeration Preaeration is one of the preliminary treatment processes. It promotes a more uniform distribution of suspended and floating solids. Aerating wastewater prior to primary clarification (sedimentation or flotation) can also improve its treatability, provide grease separation, odor control, grit removal and flocculation, and increase BOD removals. It is now common to combine grit removal with preaeration as one unit process. Since dissolved air flotation (DAF) will provide aeration for grease separation, odor control, and oxidation, preaeration is not needed if DAF is used for primary flotation clarification (instead of sedimentation clarification).

Precession The wobble over thousands of years of the tilt of the Earth's axis with respect to the plane of the solar system.

Precipitate An insoluble, finely divided substance, which is a product of a chemical reaction, or a natural hydrological reaction.

Precipitation Any and all forms of water, whether liquid or solid, that fall from the atmosphere and reach the Earth's surface. A day with measurable precipitation is a day when the water equivalent of the precipitation is equal to or greater than 0.2 mm.

Precision The degree to which a set of observations or measurements of the same property, usually obtained under similar conditions, conform to themselves; a data quality indicator. Precision is usually expressed as standard deviation, variance, or range, in either absolute or relative terms.

Preliminary effluent The effluent from a preliminary treatment system (i.e., bar screen, comminutor, and grit chamber) by which most of large objects, such as rocks, logs and cans, grit, etc., in raw wastewater have been removed.

Preliminary treatment It is the first treatment step, or preliminary step, of either a conventional wastewater treatment system or an independent physicochemical treatment system. Preliminary treatment consists of bar screen, comminutor, and grit chamber mainly for removing large objects, such as rocks, logs and cans, grit, etc., from raw wastewater. Comminutor is an option depending on the nature and characteristics of raw wastewater.

Prevention Actions taken to avoid an incident or to intervene to stop an incident from occurring.

Primary effluent The effluent from a primary treatment system (either primary sedimentation clarification or primary flotation clarification) by which most of total suspended solids in wastewater have been removed.

Primary flotation clarification A dissolved air flotation clarification process, which is used for primary clarification of wastewater with a main purpose of total suspended solids removal.

Primary flotation clarification It is a unit process or unit operation for removal of mainly total suspended solids (settleable solids and floatable solids) from screened wastewater using a primary flotation clarifier.

Primary flotation clarifier A dissolved air flotation (DAF) reactor is used to float total suspended solids (TSSs) from screened wastewater by decreasing their apparent density. DAF consists of saturating a portion or all of the wastewater feed, or a portion of recycled effluent with air at a pressure of 25 to 90 lb./square inch (gage). The pressurized wastewater is held at this pressure for 0.5 to 3 minutes in a retention tank and then released to atmospheric pressure to the flotation chamber. The sudden reduction in pressure results in the release of microscopic air bubbles, which attach themselves to TSS and oil particles in the wastewater in the flotation chamber. This results in agglomeration, which, due to the entrained air, have greatly increased vertical rise rates of about 0.5 to 2 ft./min. The floated materials rise to the surface to form a froth layer (float). Specially designed scrapers or other skimming devices continuously remove the froth (or float). The retention time in the flotation chambers is usually about 20--60 minutes for rectangular flotation clarifier, and about 3–15 minutes for circular flotation clarifier using zero-horizontal velocity design. The effectiveness of dissolved air flotation depends upon the attachment of bubbles to the suspended solids and/or oil, which are to be removed from the waste stream. The attraction between the air bubble and particle is primarily a result of the particle surface charge and bubble-size distribution. The more uniform the distribution of water and microbubbles, the shallower the flotation clarifier can be. Generally, the depth of effective flotation units is between 3 and 9 feet.

Primary sedimentation clarification It is a unit process or unit operation for removal of mainly total suspended solids (settleable solids and floatable solids) from screened wastewater using a primary sedimentation clarifier.

Primary sedimentation clarifier It is a tank used to settle mainly total suspended solids (TSSs) from screened raw wastewater by gravity. The main objectives of a primary sedimentation clarifier are removal of settleable solids by settling them to the clarifier bottom, and removal of floatable solids by skimming them from the clarifier's wastewater surface. In a rectangular sedimentation clarifier, the wastewater flows from one end to the other and the settled sludge is moved to a hopper at on end, either by scrapers called "flights" set on parallel chains, or by a single bottom scraper set on a traveling bridge. Floating materials, such as grease and oil, are collected by a surface skimmer and then removed from the rectangular sedimentation clarifier. In a circular sedimentation clarifier, the wastewater usually enters in the middle and flows toward the outside edge. Settled sludge is pushed to the hopper that is in the middle of the circular clarifier's tank bottom. Floating material is removed by a surface skimmer connected to the sludge collector.

Primary swarm It is the first swarm to leave the parent colony, usually with the old queen (see secondary swarm).

Primary treatment It is an important wastewater treatment step (either primary sedimentation clarification or primary flotation clarification mainly for removing total suspended solids from preliminary treatment effluent) after the preliminary treatment (i.e., bar screen, comminutor, and grit chamber mainly for removing large objects from raw wastewater), but before secondary treatment (either biological treatment or physicochemical treatment mainly for removing dissolved organic/inorganic pollutants from primary effluent).

Procedure A set of systematic instructions for performing an operation.

Propolis (a) It is called bee glue, which is sap or resinous material that is collected from trees or plants by bees and used to strengthen the comb and to seal cracks. (b) They are sap or resinous materials collected from trees or plants by bees and used to strengthen the comb and to seal cracks; also called bee glue

Protozoan Small, one-celled animals such as amoebae, ciliates, and flagellates.

Queen bee It is a female bee with a fully developed reproductive organ, and larger and longer than a worker bee. The queen is the center of the hive. She produces all of the eggs, the majority of which develop into worker bees, some into male drones, and a few occasionally become young queens, only when necessary.

Radiation Energy transfer in the form of electromagnetic waves or particles that release energy when absorbed by an object.

Radiative forcing A measure of the influence of a particular factor (e.g., greenhouse gas (GHG), aerosol, or land use change) on the net change in the Earth's energy balance.

Ramming It is a term applied to the excessive shock to which the dredge pump is subjected to as a result of suddenly losing and regaining suction.

Rapid mixing A water treatment unit process of quickly mixing a chemical solution uniformly through the process water.

Raw wastewater Plant influent or wastewater BEFORE any treatment.

Reactivity Any substance exhibits the characteristic of reactivity if it has any of the following properties: (a) it is normally unstable and readily undergoes violent change without detonation; (b) it reacts violently with water; (c) it forms potentially explosive mixtures with water; (d) when mixed with water, it generates toxic gases, vapors, or fumes in a quantity sufficient to present a danger to human health or the environment; (e) it is a cyanide- or sulfide-bearing substance, which, when exposed to pH conditions between 2 and 12.5, can generate toxic gases, vapors, or fumes in a quantity sufficient to present a danger to human health or the environment; (f) it is capable of detonation or explosive reaction if subjected to a strong initiating source or if heated under confinement; (g) it is readily capable of detonation or explosive decomposition or reaction at standard temperature and pressure; (h) it is forbidden explosive as defined in 49 CFR 173.51, or a Class A explosive as defined in 49 CFR 173.53 or Class B explosive as defined in 49 CFR 173.88.

Reagent A chemical substance used to cause a reaction for the purpose of chemical analysis.

Reagent blank A sample consisting of reagent(s), without the target analyte or sample matrix, introduced into the analytical procedure at the appropriate point and carried through all subsequent steps to determine the contribution of the reagents and of the involved analytical steps to error in the observed value.

Reagent grade The second highest purity designation for reagents, which conform to the current specifications of the American Chemical Society Committee on Analytical Reagents.

Reagent solution A distilled water containing a chemical substance used to cause a reaction for the purpose of chemical analysis.

Receiving water A stream, river, lake, ocean, or other surface or groundwater into which treated or untreated wastewater is discharged.

Recirculation The return of part of the effluent from a treatment process to the incoming flow.

Recovery plant (a) It is a plant designed to recover minerals such as ore. Dredges can pump a slurry via the discharge pipeline directly to a recovery plant where water and material are separated. (b) It is a plant designed to recover wastewater for water reuse.

Recycling Collecting and reprocessing a resource so that it can be used again. An example is collecting aluminum cans, melting them down, and using the aluminum to make new cans or other aluminum products.

Reduction A chemical reaction involving the addition of electrons to a chemical entity.

Refinery A refinery is a production facility composed of a group of chemical engineering unit processes and unit operations refining certain materials or converting raw material into products of value.

Refining waste A refining waste or refinery waste is the waste or wastewater from a refinery.

Reflectivity The ability of a surface material to reflect sunlight including the visible, infrared, and ultraviolet wavelengths.

Reforestation Planting of forests on lands that have previously contained forests but that have been converted to some other use.

Relative humidity The ratio of the partial pressure of water vapor in the air to the vapor pressure of water at the temperature of the air, which contains the water vapor.

Relative sea level rise The increase in ocean water levels at a specific location, taking into account both global sea level rise and local factors, such as local subsidence and uplift. Relative sea level rise is measured with respect to a specified vertical datum relative to the land, which may also be changing elevation over time.

Renewable energy Energy resources that are naturally replenishing such as biomass, hydro, geothermal, solar, wind, ocean thermal, wave action, and tidal action.

Reservoir dredging It is common for cities and municipalities to source the community's water from open-air reservoirs. Sometimes these reservoirs will silt-in or become contaminated with material that drains from the surrounding area. Dredging is an efficient and effective way to continually keep these reservoirs clean.

Residence time The average time spent in a reservoir by an individual atom or molecule. With respect to greenhouse gases, residence time refers to how long on average a particular molecule remains in the atmosphere. For most gases other than methane and carbon dioxide, the residence time is approximately equal to the *atmospheric lifetime*.

Resilience A capability to anticipate, prepare for, respond to, and recover from significant multihazard threats with minimum damage to social well-being, the economy, and the environment.

Respiration The process whereby living organisms convert organic matter to CO_2, releasing energy and consuming O_2.

Retention time The time water, sludge, or solids are retained or held in a clarifier or sedimentation tank.

Return activated sludge (RAS) (a) Settled activated sludge, returned from the bottom of final sedimentation clarifiers, to mix with incoming raw or primary effluent; (b) floated activated sludge, returned from the top of flotation clarifiers, to mix with incoming raw or primary effluent.

Reunification A process to ensure that children are safely reunited with parents or legal guardians when an emergency occurs.

Riparian It is an environment of, relating to, or situated on the banks of a river.

Rising sludge Rising sludge occurs in the secondary clarifiers of activated sludge plants when the sludge settles to the bottom of the clarifier, is compacted, and then starts to rise to the surface, usually as a result of denitrification.

Risk management In business, or science, or engineering, the forecasting and evaluation of public health, environmental, financial risks together with the identification of procedures to avoid or minimize their impact.

River dredging The most common use for cutter suction dredge equipment. River projects vary in purpose. Some river dredging projects work to maintain adequate channel depths, while others are aimed at flood control and mitigation.

Robbing It is an action of stealing nectar or honey by bees from other colonies, which occurs more often during a nectar dearth.

Rotary drum thickening It is sludge preliminary concentration process for removal of water from dilute sludge using a rotary screen drum. In operation, polymer is mixed with dilute sludge, and the conditioned sludge is then passed through rotating screen drums, which separate the flocculated solids from water. Thickened sludge rolls out the end of the drums, while separated water decants through the screens.

Rotating biological contactor (RBC) In an RBC system, a series of closely spaced circular polystyrene or polyvinyl chloride disks is partially submerged in wastewater. Biological growth (biomass) attached to the surface of the disks. The biomass is kept in aerobic conditions (if carbonaceous oxidation and nitrification are intended), or anoxic conditions (if denitrification is intended) by rotation of the disks. Aerobic RBC is opened, so biomass may utilize the oxygen in the air, while anoxic RBC for denitrification is normally closed to block off the air. The RBC effluent discharges to a clarifier (either a sedimentation clarifier or a flotation clarifier) for solid-water separation, returning sludge, wasting sludge, and clarifier effluent discharge.

Salmon (a) A large edible fish that is a popular game fish, much prized for its pink flesh. Salmon mature in the sea but migrate to freshwater streams to spawn. (b) Salmon is the common name for several species of ray-finned fish in the family Salmonidae. Other fish in the same family include trout, char, grayling, and whitefish. Salmon are native to tributaries of the North Atlantic and Pacific Ocean. (c) Representatives of salmon are sockeye salmon, Atlantic salmon, Chinook salmon, Coho salmon, and pink salmon.

Salt water intrusion Displacement of fresh or ground water by the advance of salt water due to its greater density, usually in coastal and estuarine areas.

Sample A part of a larger whole or a single item of a group; a finite part or subset of a statistical population. A sample serves to provide data or information concerning the properties of the whole group or population.

Sampling The process of obtaining a representative portion of the material of concern.

Sand and gravel dredging It is an extremely common industry worldwide. Sand and gravel dredged from rivers or landlocked mines provide aggregate materials for building, construction, and infrastructure projects.

Sandfloat It is combined circular dissolved air flotation and filtration (DAFF) clarifier designed by the Lenox Institute of Water Technology (LIWT) and manufactured by Krofta Engineering Corporation (KEC).

Sanitary landfill Burying sludge or solid wastes to minimize nuisance conditions or public health problem is called sanitary landfill. Sludge and solid wastes are buried in layers, which are covered with fill material excavated at the site. Landfills must be located where nearby wells and groundwater supplies will not be contaminated by leachate from the landfill operation. Sludge landfills are generally separate from refuse landfills.

Sanitation The development and application of measures for the sake of cleanliness and protecting health, especially the disposal of excreta and solid waste.

Scale, map It is the ratio or relationship between a distance or area on a map and the corresponding distance or area on the ground.

Scenarios A plausible and often simplified description of how the future may develop based on a coherent and internally consistent set of assumptions about driving forces and key relationships.

Screen A device used to retain or remove suspended or floating objects in wastewater. The screen has openings that are generally uniform in size. It retains or removes objects larger than the openings. A screen may consist of bars, rods, wires, gratings, wire mesh, or perforated plates.

Sea surface temperature The temperature in the top several feet of the ocean, measured by ships, buoys, and drifters.

Secondary effluent The effluent from a secondary treatment step, which may be either (a) biological treatment, such as activated sludge aeration or equivalent plus secondary clarification, or (b) physicochemical treatment, such as chemical precipitation/coagulation plus secondary clarification. Secondary treatment step removes most of dissolved organic/inorganic pollutants from primary effluent. Since secondary clarification (either secondary sedimentation clarification or secondary flotation clarification) is the final step of secondary treatment, the secondary effluent is also the secondary clarification effluent.

Secondary flotation clarification It is a unit process or unit operation for removal of the bio-oxidation process generated activated sludge, and/or the chemical precipitation/coagulation process generated chemical sludge using a secondary flotation clarifier.

Secondary flotation clarifier A dissolved air flotation (DAF) reactor is used to float biological sludge (activated sludge) from biologically oxidized wastewater (such as aeration tank effluent) and/or chemical sludge from chemically coagulated /flocculated wastewater (such as flocculator effluent) by decreasing the sludge's apparent density. DAF consists of saturating a portion or all of the wastewater feed, or a portion of recycled effluent with air at a pressure of 25 to 90 lb./square inch (gage). The pressurized wastewater is held at this pressure for 0.5 to 3 minutes in a retention tank and then released to atmospheric pressure to the flotation chamber. The sudden reduction in pressure results in the release of microscopic air bubbles, which attach themselves to suspended solids and other particles in the wastewater in the flotation chamber. This results in agglomeration, which, due to the entrained air, has greatly increased vertical rise rates of about 0.5 to 2 ft./min. The floated materials rise to the surface to form a froth layer (float). Specially designed scrapers or other skimming devices continuously remove the froth (or float). The retention time in the flotation chambers is usually about 20–60 minutes for rectangular flotation clarifier, and about 3–15 minutes for circular flotation clarifier using zero-horizontal velocity design. The effectiveness of dissolved air flotation depends upon the attachment of bubbles to the suspended solids, which are to be removed from the waste stream. The attraction between the air bubble and particle is primarily a result of the particle surface charge and bubble-size distribution. The more uniform the distribution of water and microbubbles, the shallower the flotation clarifier can be. Generally, the depth of effective flotation units is between 3 and 9 feet.

Secondary membrane clarification (or secondary membrane separation, or secondary membrane filtration) A water-solid separation/clarification process uses membrane device instead of conventional sedimentation clarification in the secondary wastewater treatment step.

Secondary sedimentation clarification It is a unit process or unit operation for removal of the bio-oxidation process generated activated sludge, and/or the chemical precipitation/coagulation process generated chemical sludge using a secondary sedimentation clarifier.

Secondary sedimentation clarifier It is a tank used to settle the chemical precipitation/coagulation process generated chemical sludge, and/or the bio-oxidation process generated activated sludge for removing dissolved organic/inorganic substances from wastewater. The main objectives of a secondary sedimentation clarifier are removal of settleable chemical and/or biological sludge solids by settling them to the clarifier bottom. In a rectangular sedimentation clarifier, the wastewater flows from one end to the other and the settled sludge is moved to a hopper at on end, either by scrapers called "flights" set on parallel chains, or by a single bottom scraper set on a traveling bridge. In a circular sedimentation clarifier, the wastewater usually enters in the middle and flows toward the outside edge. Settled sludge is pushed to the hopper that is in the middle of the circular clarifier's tank bottom.

Secondary treatment It is a wastewater treatment step after primary treatment (either primary sedimentation clarification or primary flotation clarification). Secondary treatment may be either biological treatment (such as activated sludge aeration plus secondary clarification) or physicochemical treatment (such as chemical precipitation/coagulation plus secondary clarification) mainly for removing dissolved organic/inorganic pollutants from primary effluent.

Sediments They are the materials that settle on the bottom of a waterway that has been transported from the surrounding area over time as a result of weathering and erosion.

Selector hydraulic retention time (HRT) The given time it takes wastewater, including any return flows, to pass through a tank (selector).

Selector A zone in a biological treatment process with specific environmental conditions that allow for the growth or lack of growth of certain microorganisms (such as an anoxic or anaerobic zone).

Seminar Session designed to orient participants to new or updated plans, policies, or procedures through informal discussion.

Sensitivity The degree to which a system is affected, either adversely or beneficially, by climate variability or change. The effect may be direct (e.g., a change in crop yield in response to a change in the mean, range, or variability of temperature) or indirect (e.g., damages caused by an increase in the frequency of coastal flooding due to sea level rise).

Septic A condition produced by anaerobic bacteria. If severe, the wastewater produces hydrogen sulfide, turns black, gives off foul odors, contains little or no dissolved oxygen, and creates a high oxygen demand.

Sequencing batch exchanger (SBE) A sequencing batch process involving the use of separate batch process steps in sequence of filling water/wastewater, reacting with ion exchanger, settling insoluble spent ion exchanger, discharging treated effluent, and recycling the settled spent ion exchanger for regeneration, resulting in removing ionic pollutants from water or wastewater. (US Patent 5,354,458).

Sequencing batch flotation (SBF) A sequencing batch process (either biological or physicochemical) involving the use of separate batch process steps in sequence of filling water/wastewater, reacting with chemicals or biomass, floating light weight-insoluble flocs and/or biomass, discharging treated effluent, and wasting floated insoluble scums, for water or wastewater treatment is also called sequencing batch flotation (SBF) process. (US Patent 5,354,458).

Sequencing batch reactor (SBR), biological A biological SBR system performs all the necessary functions of carbonaceous and nutrient removals in a single tank with variable water levels and timed aerobic aeration or anoxic mixing. This SBR system requires a minimum of three tanks and advanced automation equipment to control the cycle times and phases. The SBR control systems allow the operation to be configured to operate as almost any other suspended growth biological reactor by adjusting the cycle phases between fill phase, react phase (either aerobic aeration or anoxic mixing), separation phase (either sedimentation or flotation), and decant phase. The biological SBR system involving the use of flotation in the separation phase is invented by Dr. Lawrence K. Wang,, Dr. Lubomyr Kurylko, and Dr. Mu-Hao Sung Wang in 1994 (US Patent 5,354,458).

Sequencing batch reactor (SBR), physicochemical A physicochemical SBR system performs all the necessary functions of organic and inorganic pollutants removals in a single tank with variable water levels and timed mixing, coagulation/flocculation, or clarification (either sedimentation or flotation). This SBR system requires a minimum of three tanks and advanced automation equipment to control the cycle times and phases. The SBR control system allows the operation to be configured to operate as almost any other suspended floc physicochemical reactor by adjusting the cycle phases among fill phase, react phase (chemical precipitation/coagulation, flocculation), separation phase (either sedimentation or flotation clarification), and decant phase. The physicochemical SBR system involving the use of either sedimentation or flotation in the separation phase is invented by Lawrence K. Wang,, Lubomyr Kurylko, and Mu-Hao Sung Wang in 1994 (US Patent 5,354,458).

Sewage The used water and water-carried solids from homes that flow in sewers to a wastewater treatment plant. The preferred term is "wastewater."

Shelter-in-place A procedure to keep you safe by remaining inside. It involves selecting an interior room or area in which to take refuge. Where you shelter depends on the type of hazard or threat.

Short-circuiting Shortening of hydraulic detention time, such as when plant flow exceeds design flow.

Short ton Common measurement for a ton in the United States. A short ton is equal to 2000 lbs. or 0.907 metric tons.

Short-circuiting A condition that occurs in tanks or basins when some of the water travels faster than the rest of the flowing water. This is usually undesirable, since it may result in shorter contact, reaction, or settling times in comparison with the theoretical (calculated) or presumed detention times.

Shredding It is a unit operation for reducing the particle size of objects or debris in the influent wastewater, also termed grinding or comminuting. These devices may be installed with a screen directly in the wastewater flow or separately out of the wastewater flow, with the shredded particles returned to the flow downstream of the screen. Only those shredding and grinding devices that are installed directly in the influent channel are termed comminuting devices.

Sidestreams Sources of inflow from within the wastewater treatment facility, such as supernatant return from digesters, centrate from centrifuge thickening, filtrate from filter presses, etc. These sidestreams may be high in solids, BOD, or nutrients and may add to the plant organic or hydraulic loading.

Single-sludge activated sludge system It is a biological activated sludge process system involving the use of one suspended activated sludge solids going through aerobic and anoxic zones for carbonaceous oxidation, nitrification, denitrification, and phosphorus removal. The single-sludge activated sludge system is commonly a continuous flow system using sedimentation clarification. Lenox Institute of Water Technology (LIWT) has developed three innovative single-sludge activated sludge systems all for carbonaceous oxidation, nitrification, denitrification, and phosphorus removal: (a) continuous flow single sludge system using flotation clarification; (b) sequencing batch reactors (SBR) using sedimentation for carbonaceous removal and nutrient removal; and (c) SBR using flotation for carbonaceous removal and nutrient removal.

Sink Any process, activity, or mechanism, which removes a greenhouse gas, an aerosol or a precursor of a greenhouse gas or aerosol from the atmosphere.

Sintering Using heat or pressure to form a solid mass of material, without melting the material to the point of liquefying.

Sludge (a) The settleable solids separated from liquids during processing. (b) The deposits of foreign material on the bottoms of streams or other bodies of water.

Sludge age (SA) SA is the length of time a particle of activated sludge stays in the biological treatment plant, measured in days. In a biological phosphorus removal plant, sludge age is the amount (pound or kilogram) of mixed liquor suspended solids in all the biological reactors divided by the suspended solids withdrawn from the system per day (pound per day of waste activated sludge, or kilogram per day of waste activated sludge).

Sludge dewatering Sludge dewatering is the second step of sludge handling, after sludge thickening. Sludge dewatering is required for all facilities that do not dispose of their sludge in liquid form. It is a physical (mechanical) process to reduce the moisture content of sludge. The purpose is to reduce sludge volume, in turn, to reduce subsequent sludge digestion costs and/or final sludge disposal (land disposal or incineration) costs. Dewatering is usually required prior to sludge incineration. There are a variety of dewatering processes available: (a) vacuum filtration; (b) centrifuge; (c) belt filter press; (d) plate and frame press; (e) sludge drying beds or infrared sludge dryer; (f) sludge lagoons.

Sludge digestion The process of changing organic matter in sludge into a gas or liquid or a more stable solid form. These changes take place as microorganisms feed on sludge in anaerobic (more common) or aerobic digesters.

Sludge dryer, infrared It is a sludge dewatering process using regular direct radiant energy from infrared heating elements. Dewatered sludge is transferred from the dewatering equipment to the dryer by a conveyor belt. The sludge then drops into the augers, where the auger blades agitate the sludge, while infrared heating elements heat the sludge. This process then repeats once the sludge is dropped into the secondary drying zone.

Sludge dryer, microwave It is a new sludge dewatering process developed by Lawrence K. Wang of the Lenox Institute of Water Technology involving the use of microwave energy from a microwave dryer. Dewatered sludge is transferred from the dewatering equipment to the dryer by a conveyor belt. The sludge then drops into the augers, where the auger blades agitate the sludge, while microwaves heat and dewater the sludge. This process then repeats once the sludge is dropped into the secondary drying zone.

Sludge drying beds It is a sludge dewatering process using solar energy to evaporate water from the sludge slurries.

Sludge handling It includes every required processes or management regarding waste sludge, such as sludge thickening, sludge dewatering, sludge digestion, sludge storage, sludge transportation, sludge disposal, etc.

Sludge lagoons It is an earthen basin used for holding and dewatering sludge slurries by solar energy.

Sludge thickening and dewatering combined, belt filter press Conditioned sludge is first placed on a gravity drainage section where it is allowed to thicken. In this section, the bulk of the free water is removed from the sludge by gravity. On some units, this section is equipped with a vacuum assist, which enhances drainage and may help to reduce odors. Following gravity drainage, pressure is applied in a low-pressure section, where sludge is squeezed between opposing porous cloth belts. On some units, the low-pressure section is followed by a high-pressure section where the sludge is subjected to shearing forces that induce the release of additional quantities of water from the sludge. The final dewatered sludge cake is removed from the belts by scrapper blades.

Sludge thickening/dewatering centrifuge, basket centrifuge They are suitable for small plants. It is a semibatch type operation. Sludge cake is collected on the sides of the spinning bowl, while the centrate (dilute effluent stream from the centrifuge) overflows the bowl rim. Once solids have built up to a maximum thickness, the feed sludge is stopped and scrapper blades peel the sludge from the walls. The process is then resumed.

Sludge thickening/dewatering centrifuge, solids bowl centrifuge Sludge is fed at a constant flowrate into the rotating bowl where it is separated into a dense cake and a dilute stream called centrate. The units can be used with no prior chemical conditioning, but solids caputer and centrate quality are improved considerably when the sludge is conditioned with polymer.

Sludge thickening Sludge thickening is the first step of sludge handling. It is employed prior to subsequent sludge dewatering processes to increase the efficiency of the sludge dewatering equipment. There are at least five types of sludge thickening processes: gravity thickening, flotation thickening, centrifuge thickening, gravity belt thickening, and rotary drum thickening.

Sludge volume index (SVI) SVI is the ratio of the volume (in mL) of sludge settled from a 1000 mL sample in 30 minutes to the concentration of mixed liquor (in mg/L) multiplied by 1000. It indicates settling capabilities, compaction, and it is an indication of filamentous organism overgrowth and several other indicators of process problems.

Slurry It is a term used to describe the mixture of solids with a liquid (typically water). In dredging, these solids can include silt, sand, gravel, clay, or coal.

Snowpack A seasonal accumulation of slow-melting snow.

Sodium hypochlorite It is a liquid sodium hypochlorite solution to be used as a disinfectant for water or wastewater disinfection.

Soil carbon A major component of the terrestrial biosphere pool in the carbon cycle. The amount of carbon in the soil is a function of the historical vegetative cover and productivity, which in turn is dependent in part upon climatic variables.

Solar radiation Radiation emitted by the Sun. It is also referred to as short-wave radiation. Solar radiation has a distinctive range of wavelengths (spectrum) determined by the temperature of the Sun.

Solid wastes A solid waste is defined by the US Environmental Protection Agency (USEPA) as the following: Any garbage, refuse, sludge from a waste treatment plant, water supply treatment plant or air pollution control facility, and other discarded materials including solid, liquid, semisolid, or contained gaseous material resulting from industrial, commercial, mining, and agricultural operations and from community activities Sect.6903(27). The regulatory definition is found at 40CFR261.2(a)(1). A solid waste is defined by the US regulation as any "discarded material" not expressly excluded by the US regulations. A material is considered "discarded" if it is "abandoned," "recycled," or "considered inherently waste-like." Accordingly, the discarded, abandoned, recycled, or waste-like materials generated from a laboratory facility are all considered to be solid wastes.

Solidification/stabilization (S/S) The process utilizes chemically reactive formulations that, together with the water and other components in sludges and other aqueous hazardous wastes, form stable solids.

Solids retention time (SRT) The theoretical length of time, usually in days, that solids are retained in an aeration basin, clarifier, or other structure. SRT is used to calculate wasting rates.

Soluble BOD (SBOD) Soluble BOD is a sample that has been filtered through a 0.45μm filter. Soluble BOD includes volatile fatty acids (VFAs) and organic material that will readily ferment to create more VFAs in an anaerobic selector.

Soluble Dissolved in a solution and more readily available for biochemical or chemical reactions. For instance, soluble BOD or soluble COD is readily available as food to microorganisms in a biochemical reaction. Soluble BOD includes volatile fatty acids (VFAs) and organic material that are readily available to microorganisms in a biological process.

Solution (a) A liquid (solvent) that contains a dissolved substance (solute). (b) A liquid mixture of dissolved substances. In a solution, it is impossible to see all the separated parts.

Solvent A liquid used to dissolve another substance.

Spatial Analysis It is the process of modeling, examining, and interpreting model results. Spatial analysis is useful for evaluating suitability and capability, for estimating and predicting, and for interpreting and understanding.

SPDES permit **S**tate **P**ollutant **D**ischarge **E**limination **S**ystem (SPDES) permit is each State's regulatory agency document issued by either a state agency, which is designed to control all discharges of pollutants from all point sources and storm water runoff into US waterways. A State's treatment plant that discharges to a State's surface water will have an SPDES permit.

Spectrophotometer (a) Meter that uses a specific wavelength of light to measure light absorption in a sample; used for phosphorus, ammonia, nitrite, and nitrate analyses. (b) A spectrophotometer is an instrument designed for physical sample analysis via full spectrum color measurement. By providing wavelength-by-wavelength spectral analysis of a sample's reflectance, absorbance, or transmittance properties, it produces precise data beyond that observable by the human eye. Spectrophotometers offer a higher level of flexibility and versatility than colorimeters due in part to the fact that they offer multiple illuminant/observer combinations and are capable of measuring metamerism, identifying colorant strength, analyzing a comprehensive range of sample types, and giving users a choice between including or excluding specular reflectance to account for geometric attributes. Full spectrum analysis also provides for greater specificity, potentially identifying color differences missed by colorimeters.

Spoils, dredging It is the name applied to both beneficial and nonbeneficial discharged dredged material—also referred to as discharge, fill, or deposit.

Spuds, dredge They are large steel cylinders located at the end of the dredge ranging from 20′ (6 m) to 100′ (30 m) in height. Spuds hold the dredge in place during operation and are used to advance the dredge by "walking" it forward.

Standard solution A solution containing a known concentration of analytes, prepared and verified by a prescribed method or procedure and used routinely in an analytical method.

Storm sewer A separate pipe, conduit, or open channel (sewer) that carries runoff from storms, surface drainage, and street wash, but does not include domestic and industrial wastes.

Storm surge An abnormal rise in sea level accompanying a hurricane or other intense storm, whose height is the difference between the observed level of the sea surface and the level that would have occurred in the absence of the cyclone.

Stratosphere Region of the atmosphere between the troposphere and mesosphere, having a lower boundary of approximately 8 km at the poles to 15 km at the equator and an upper boundary of approximately 50 km. Depending upon latitude and season, the temperature in the lower stratosphere can increase, be isothermal, or even decrease with altitude, but the temperature in the upper stratosphere generally increases with height due to absorption of solar radiation by ozone.

Streamflow The volume of water that moves over a designated point over a fixed period of time. It is often expressed as cubic feet per second (ft3/sec).

Structural elements Any component of the building whose primary function is to support the dead load (e.g., building, roof).

Structured Query Language (SQL) It is a syntax for defining and manipulating data from a relational database. Developed by IBM in the 1970s, it has become an industry standard for query languages in most relational database management systems.

Struvite It is magnesium ammonium phosphate ($MgNH_4PO_4(H_2O)_6$). Struvite forms hard, very insoluble, white, yellowish-white, or brownish-white crystals.

Subnatant The water or liquid remaining below a floated sludge (float) after flotation.

Subsiding/subsidence The downward settling of the Earth's crust relative to its surroundings.

Substrate The food or chemical substance on which organisms depend for growth. The organic matter in wastewater (as measured by the BOD_5 test) is a substrate for the microorganisms in a biological process. For instance, in biological phosphorus removal systems, volatile fatty acids are a readily available substrate used by phosphorus-accumulating organisms (PAOs).

Sulfate aerosols Particulate matter that consists of compounds of sulfur formed by the interaction of sulfur dioxide and sulfur trioxide with other compounds in the atmosphere. Sulfate aerosols are injected into the atmosphere from the combustion of fossil fuels and the eruption of volcanoes like Mt. Pinatubo. Sulfate aerosols can lower the Earth's temperature by reflecting away solar radiation (negative radiative forcing). General Circulation Models, which incorporate the effects of sulfate aerosols more accurately, predict global temperature variations.

Sulfur hexafluoride (SF_6) A colorless gas soluble in alcohol and ether, slightly soluble in water. A very powerful greenhouse gas used primarily in electrical transmission and distribution systems and as a dielectric in electronics. The global warming potential of SF6 is 22,800. This GWP is from the IPCC's Fourth Assessment Report (AR4).

Super It is a part of commercial or other managed beehive that is used to collect surplus honey. Normally, it is placed over or above the brood chamber.

Supernatant Liquid removed from settling sludge. Supernatant commonly refers to the liquid between the sludge on the bottom of an anaerobic digester and the scum on the surface. The liquid is usually returned to the influent wet well or to the primary clarifier.

Supracell It is a circular dissolved air flotation (DAF) clarifier designed by the Lenox Institute of Water Technology (LIWT) and manufactured by Krofta Engineering Corporation (KEC).

Surface water Surface water is water on the surface of continents such as in a river, lake, or wetland. It can be contrasted with groundwater and atmospheric water. Nonsaline surface water used is replenished by precipitation and by recruitment from ground-water.

Surfactant A surface active compound that lowers the surface tension (or interfacial tension) between two liquids, between a gas and a liquid, or between a liquid and a solid. Surfactants may act as detergents, wetting agents, emulsifiers, foaming agents, etc.

Surplus honey It is the honey removed from the hive, which exceeds what is needed by bees for their own use.

Suspended solid Solids that either float on the surface or are suspended in water, wastewater, or other liquids, and which are largely removable by laboratory filtering.

Suspended-growth biological processes In a suspended-growth biological process system, biomass or microorganisms are suspended in the biological reactors for aerobic and/or anoxic biochemical reactions. They include all kinds of activated sludge processes as well as oxidation ditch process, such as (a) conventional activated sludge process using diffused aeration; (b) conventional activated sludge process using mechanical aeration; (c) high rate activated sludge process using diffused aeration; (d) pure oxygen activated sludge process using covered reactor; (e) pure oxygen activated sludge process using uncovered reactor; (f) extended aeration activated sludge process with nitrification; (g) activated sludge process separate reactors for carbonaceous oxidation, nitrification, and denitrification; (h) single sludge activated sludge process for carbonaceous oxidation, nitrification, and denitrification; (i) sequencing batch reactor (SBR); (j) membrane bioreactor (MBR); (k) oxidation ditch process; (l) aerated lagoon, etc. The effluent of the suspended-growth bioreactor (aeration tank, or oxygenation tank, or mixing tank) discharges to a solid-water separator (sedimentation clarifier, flotation clarifier, or membrane filter) for additional processing.

Swarm It represents a large number of worker bees and drones, and usually the old queen leaves the parent colony to establish a new colony.

Swarming It is the natural process of honey bee reproduction at colony level.

Tabletop exercise A discussion-based exercise intended to stimulate discussion of various issues regarding a hypothetical situation. Tabletop exercises can be used to assess plans, policies, and procedures or to assess types of systems needed to guide the prevention of, protection against, response to, or recovery from a defined incident. Tabletop exercises are typically aimed at facilitating understanding of concepts, identifying strengths and shortfalls, and/or achieving a change in attitude. Participants are encouraged to discuss issues in depth and develop decisions through slow-paced problem-solving rather than the rapid, spontaneous decision-making that occurs under actual or simulated emergency conditions. Tabletop exercises can be breakout (i.e., groups split into functional areas) or plenary (i.e., one large group).

Tailings It is a maintenance operation for reclaiming or eliminating mine tailings through dredging that supports efficient mine operations. Typical sectors include iron ore, gold, and oil sands where reclamation of tailings can have a positive return on investment.

Technology It is professional field involving the application of known knowledge as a "tool" that extends human capability (e.g., from hammers to the internet).

Teragram 1 trillion (1012) grams = one million (106) metric tons.

Terminal elevation, dredge It is the vertical height above the centerline of the dredge pump and the final point of discharge.

Tertiary flotation clarification A dissolved air flotation clarification process, which is used for advanced or tertiary treatment with a main purpose of nutrient removal or wastewater renovation.

Tertiary membrane clarification (or tertiary membrane separation, or tertiary membrane filtration) A water-solid separation/clarification process uses membrane device instead of conventional sedimentation clarification in the tertiary wastewater treatment step, such as nitrification and denitrification.

Tertiary treatment It is a wastewater treatment step after secondary treatment, meaning tertiary treatment is for treating the secondary effluent in that most of organic substances have been removed. Normally, tertiary treatment includes nitrogen removal processes, phosphorus removal processes, granular activated carbon, granular media filtration, membrane filtration, and postaeration.

Theme, ArcView An ArcView theme stores map features as primary features (such as arcs, nodes, polygons, and points) and secondary features such as tics, map extent, links, and annotation. A theme usually represents a single geographic layer, such as soils, roads, or land use.

Thermal discharge An industrial, agricultural, or commercial discharge of heated liquid (such as wastewater, or gas, or steam from a factory) into natural waters or air at a temperature harmful to the environment.

Thermal expansion The increase in volume (and decrease in density) that results from warming water. A warming of the ocean leads to an expansion of the ocean volume, which leads to an increase in sea level.

Thermal stabilization A heating process for metal incorporation into certain stable products.

Thermohaline circulation Large-scale density-driven circulation in the ocean, caused by differences in temperature and salinity. In the North Atlantic, the thermohaline circulation consists of warm surface water flowing northward and cold deep water flowing southward, resulting in a net poleward transport of heat. The surface water sinks in highly restricted sinking regions located in high latitudes.

Thickener, flotation A tank or basin in which liquid sludge is held for a period of time during which most of the lighter air-sludge flocs float to the liquid surface as the concentrated-thickened sludge and very small amount of the very heavier solids will settle to the tank bottom.

Thickener, gravity A tank or basin in which liquid sludge is held for a period of time during which most of the heavier solids settle to the tank bottom as the concentrated-thickened sludge and very small amount of the lighter solids will float to the water surface as the scum.

Thickening, sludge Sludge thickening is employed prior to subsequent sludge dewatering processes to increase the efficiency of the sludge dewatering equipment. There are at least five types of sludge thickening processes: gravity thickening, flotation thickening, centrifuge thickening, gravity belt thickening, and rotary drum thickening.

Threat Natural, technological, or human-caused occurrence, individual, entity, or action that has or indicates the potential to harm life, information, operations, the environment, and/or property.

Thunderstorm A storm with thunder and lightning, often accompanied by rain or hail.

Tidal rivers (a) A *tidal river* is a *river*, or more typically a stretch of a *river*, whose flow and level is influenced by *tides*; (b) a river that enters a sea or estuary and is thus affected by tides.

Tornado A violent, rotating column of air that extends between, and is in contact with, the ground and a cloud.

Total Kjeldahl nitrogen (TKN) The combined amount of organic and ammonia nitrogen determined by a laboratory analysis.

Total nitrogen (TN) All forms of nitrogen including ammonia nitrogen, nitrate nitrogen, nitrite nitrogen, and organic nitrogen.

Total phosphorus (TP) All forms of phosphorus, including orthophosphate, polyphosphates, and organic phosphorus.

Toxic A substance which is poisonous to a living organism.

Toxicity The relative degree of being poisonous or toxic. A condition which may exist in wastes and will inhibit or destroy the growth or function of certain organisms.

Trace gas Any one of the less common gases found in the Earth's atmosphere. Nitrogen, oxygen, and argon make up more than 99 percent of the Earth's atmosphere. Other gases, such as carbon dioxide, water vapor, methane, oxides of nitrogen, ozone, and ammonia, are considered trace gases. Although relatively unimportant in terms of their absolute volume, they have significant effects on the Earth's weather and climate.

Transpiration Similar to Evapotranspiration.

Tropical storm A severe storm that develops over tropical seas with winds from 39 to 73 miles per hour.

Troposphere The lowest part of the atmosphere from the surface to about 10 km in altitude in mid-latitudes (ranging from 9 km in high latitudes to 16 km in the tropics on average) where clouds and "weather" phenomena occur. In the troposphere, temperatures generally decrease with height.

Tsunami Enormous waves caused by underwater disturbances such as earthquakes.

Tundra A treeless, level, or gently undulating plain characteristic of the Arctic and sub-Arctic regions characterized by low temperatures and short growing seasons.

Turbid Having a cloudy or muddy appearance.Two-stage DAFA series of two consecutive dissolved air flotation treatment using different chemicals or operational conditions.

Two-stage treatment A series of two consecutive water or wastewater treatment using different chemicals, operational conditions, or even different unit processes.

Ultraviolet radiation (UV) (a) Most UV disinfection units are constructed with an array of VU lamps installed in a water, wastewater, or air channel, for water disinfection, wastewater disinfection, or air disinfection, respectively. A proper dosage of UV radiation has shown to be an effective bactericide and virucide while not contributing to the formation to toxic disinfection by-products. However, certain compounds may be altered by UV radiation. Because the effective distance of UV light is very limited, most UV disinfection occurs when the light penetration depth can be minimized, termed the thin film approach. The term "thin film" refers to the separation distance between the UV tubes. (b) The energy range just beyond the violet end of the visible spectrum. Although ultraviolet radiation constitutes only about 5 percent of the total energy emitted from the sun, it is the major energy source for the stratosphere and mesosphere, playing a dominant role in both energy balance and chemical composition. Most ultraviolet radiation is blocked by Earth's atmosphere, but some solar ultraviolet penetrates and aids in plant photosynthesis and helps produce vitamin D in humans. Too much ultraviolet radiation can burn the skin, cause skin cancer and cataracts, and damage vegetation.

Urban runoff Storm water from city streets and adjacent domestic or commercial properties that may carry pollutants of various kinds into the sewer systems and from there to rivers, lakes, or oceans.

Vacuum filtration This sludge dewatering process consists of a cylindrical drum which rotates partially submerged in a vat of sludge. The filter drum is divided into compartments by partitions or seal strips. A vacuum is applied between the drum deck and filter medium, causing filtrate to be removed and filter cake to be retained on the medium during the pickup and cake drying cycle. The filter medium may be a cloth made of natural or synthetic fibers, stainless steel wire mesh, or coil springs. Dewatered sludge is ordinarily removed by a fixed scraper blade.

Vector (a) In environmental science, an organism, typically a biting insect or tick, that transmits a disease or parasite from one animal or plant to another. (b) A quantity having direction as well as magnitude, especially as determining the position of one point in space relative to another. (c) In mathematics and physics, a vector is an element of a vector space. For many specific vector spaces, the vectors have received specific names, which are listed below. Historically, vectors were introduced in geometry and physics before the formalization of the concept of vector space.

Vegetable oil An oil derived from plants, e.g., canola oil, olive oil, sunflower oil.

Ventilated improved pit latrine (VIP) It is a pit latrine with a black vent pipe going from the pit to above the superstructure with a fly screen on top. VIP latrines have the advantage over simple pit latrines of removing flies and odors if designed, built, and used properly.

Volatile fatty acids (VFAs) Fatty acids containing fewer carbon atoms, and are dissolvable in water. When organic material undergoes fermentation reactions, smaller and more readily available organic molecules are formed known as volatile fatty acids (VFAs). VFAs typically found in wastewater are acetic acid (acetate) and propionic acid. VFAs provide the food for PAOs.

Volatile solids Those solids in water, wastewater, or other liquids that are lost on ignition of the dry solids at 550 ° C.

Volcano A vent in the Earth that, when pressure builds and it erupts, releases dangerous molten rock and gases.

Vulnerability The degree to which a system is susceptible to, or unable to cope with, adverse effects of climate change, including climate variability and extremes. Vulnerability is a function of the character, magnitude, and rate of climate variation to which a system is exposed; its sensitivity; and its adaptive capacity.

Warning A notice issued when a hazardous weather event is occurring, is imminent, or has a very high probability of occurring. A warning is used for conditions posing a threat to life or property.

WASH It is an acronym that means Water, Sanitation, and Hygiene. Access to WASH usually includes safe water, appropriate sanitation, and hygiene education. Adequate access to WASH can greatly reduce water-borne diseases in a community.

Waste activated sludge (WAS) Solids removed/wasted from a biological process, such as activated sludge process, to prevent an excessive buildup in the system.

Waste disposal Removing and destroying or storing damaged, used or other unwanted domestic, agricultural or industrial products and substances for final disposal by burning, burial at landfill sites or at sea, and recycling.

Waste management Waste management includes the activities and actions required to manage waste from its inception to its final disposal. This includes the collection, transport, treatment, and disposal of waste, together with monitoring and regulation of the waste management process.

Wastewater renovation Treatment of wastewater to a degree that the wastewater becomes a useful water resource.

Wastewater (a) Water that has been used and contains dissolved or suspended waste materials. (b) The used water and solids from a community that flow to a treatment plant. Storm water, surface water, and groundwater infiltration also may be included in the wastewater that enters a wastewater treatment plant. The term "sewage" usually refers to household wastes, but this word is being replaced by the term "wastewater."

Watch A notice used when the risk of a hazardous weather event has increased significantly, but its occurrence, location, and/or timing is still uncertain. It is intended to provide enough lead time so that those who need to set their plans in motion can do so.

Water quality management The water-quality-related activities include the systematic collection of physical, chemical, and biological information, and the analysis, interpretation, and reporting of those measurements, according to a preplanned design and structure.

Water quality It refers to the chemical, physical, biological, and radiological characteristics of **water**. It is a measure of the condition of water relative to the requirements of one or more biotic species, or to any human need or purpose.

Water reclamation It is a process consisting of proper wastewater treatment to a degree that the treated water can be directly or indirectly reused/recycled.

Water resources Water in various forms, such as rain, snow, ice, clouds, surface water, groundwater, reclaimed water, or reused water that is potentially useful for domestic, agricultural, industrial, commercial, or recreational applications.

Water vapor The most abundant greenhouse gas, it is the water present in the atmosphere in gaseous form. Water vapor is an important part of the natural greenhouse effect. While humans are not significantly increasing its concentration through direct emissions, it contributes to the enhanced greenhouse effect, because the warming influence of greenhouse gases leads to a positive water vapor feedback. In addition to its role as a natural greenhouse gas, water vapor also affects the temperature of the planet because clouds form when excess water vapor in the atmosphere condenses to form ice and water droplets and precipitation.

Weather Atmospheric condition at any given time or place. It is measured in terms of such things as wind, temperature, humidity, atmospheric pressure, cloudiness, and precipitation. In most places, weather can change from hour to hour, day to day, and season to season. Climate in a narrow sense is usually defined as the "average weather," or more rigorously, as the statistical description in terms of the mean and variability of relevant quantities over a period of time ranging from months to thousands or millions of years. The classical period is 30 years, as defined by the World Meteorological Organization (WMO). These quantities are most often surface variables such as temperature, precipitation, and wind. Climate in a wider sense is the state, including a statistical description, of the climate system. A simple way of remembering the difference is that climate is what you expect (e.g. cold winters) and "weather" is what you get (e.g., a blizzard).

Weir A wall or plate placed in an open channel and used to measure the flow of water. The depth of the flow over the weir can be used to calculate the flow rate, or a chart or conversion table may be used.

Wet bulb temperature The temperature measured by an ordinary thermometer as a thin film of water on the thermometer is evaporated into a surrounding air stream.

Wet infrastructure It is the spectrum of *water*-related infrastructures relating to potable *water* supply, wastewater treatment, resources recovery, water storage, *water* resource management, flood management, coastal restoration, hydropower, and renewable energy facilities.

Wet oxidation A method of treating or conditioning sludge before the water is removed. Compressed air is blown into the sludge; the air and sludge mixture is fed into a pressure vessel where the organic material is stabilized.

Wet well A compartment or tank in which wastewater is collected. The suction pipe of a pump may be connected to the wet well or a submersible pump may be located in the wet well.

Winter storm A weather event resulting from low temperatures that can include extreme low temperatures, strong winds, and precipitation like snow, sleet, and ice.

Worker bee It is a female bee with undeveloped reproductive organ; the majority of the honey bees in a colony are worker bees, and they do all the work in the colony except laying fertile eggs. The worker honey bees visit flowers to obtain nectar and pollen as their food, and construct the waxy combs made from secretions of the wax glands. The worker bees tend the eggs and larvae as well as feed the queen and carry away the eggs produced by the queen.

Workshop A type of training focused on increased participant interaction and achieving or building a product (e.g., plans, policies). A workshop is typically used to test new ideas, processes, or procedures; train groups in coordinated activities; and obtain consensus. Workshops often use breakout sessions to explore parts of an issue with smaller groups.

XRD analysis A technique can provide both qualitative and quantitative information on the phases in a tested sample.

Zoogleal mass Jelly-like masses of bacteria found in both the trickling filter and activated sludge processes.

References

1. Wang LK (2021) Humanitarian engineering education of the Lenox Institute of Water Technology and its new potable water flotation processes. In: Wang LK, Wang MHS, Shammas NK (eds) Environmental flotation engineering, Springer Nature Switzerland, pp. 1–72
2. Wang LK, Wang MHS, Fahey EM (2021) An innovative lee, Massachusetts USA dissolved air flotation potable water filtration plant. In: Wang LK, Wang MHS, Shammas NK (eds) Environmental flotation engineering, Springer Nature Switzerland, pp. 73–94
3. Shammas NK, Wang LK (2016) Water engineering: hydraulics, distribution and treatment. Wiley, New York, USA, p 806
4. Holtorff MS (1999) Municipal waste treatment by innovative physical-chemical and biological processes. Master Engineering thesis, Lenox Institute of Water Technology, Lenox, MA, USA
5. Wang LK (2010) Conversion factors for environmental engineers. In: Wang LK, Shammas NK, Selke WA, Aulenbach DB (eds) Flotation technology. Humana Press, Totowa, NJ, USA, pp 618–666
6. Wang MHS, Wang LK (2014) Glossary ad conversion factors for water resources engineers. In: Wang LK, Yang CT (eds) Modern water resources engineering. Humana Press, Totowa, NJ, USA, pp 759–851
7. Wang MHS, Wang LK (2016) Glossary of land and energy resources engineering. In: Wang LK, MHS W, Hung YT, Shammas NK (eds) Natural resources and control resources. Springer, New York, USA, pp 493–623
8. Wang MHS, Wang LK (2015) Environmental water engineering glossary. In: Yang CT, Wang LK (eds) Advances in water resources engineering. Springer, New York, pp 471–556
9. Wang LK, Wang MHS (2015) Environmental engineering glossary, 2nd edn. Lenox Institute Press, Newtonville, NY, USA
10. Shammas NK, Wang LK, Wang MHS (2021) Biological processes for water resources protection and water reuse. In: Wang LK, Wang MHS, Hung YT, Shammas NK (eds) Environmental and natural resources engineering. Springer Nature Switzerland, pp 73–168
11. Li C (2021) Understanding, conservation and protection of precious natural resources – bees. In: Wang LK, Wang MHS, Hung YT, Shammas NK (eds) Environmental and natural resources engineering. Springer Nature Switzerland, pp 1–52
12. Su M, Shi K, Chen D (2021) Advances in Cadmium detoxification by sintering with ceramic matrices. In: Wang LK, Wang MHS, Hung YT, Shammas NK (eds) Environmental and natural resources engineering. Springer Nature Switzerland, pp 299–324
13. Bungay HR (2021) Environmental engineering education in modern age. In: Wang LK, Wang MHS, Hung YT, Shammas NK (eds) Environmental and natural resources engineering. Springer Nature Switzerland, pp 377–386
14. Wong JM, Hess RJ, Wang LK (2021) Operation and performance of the AquaDAF® process system for water purification. In: Wang LK, Wang MHS, Shammas NK (eds) Environmental flotation engineering, Springer Nature Switzerland, pp 301–338
15. Wang LK, Wang MHS, Hung YT , Shammas NK (eds) (2021). Integrated natural resources research. Springer Nature Switzerland
16. Wang LK, Wang MHS, Hung YT, Shammas NK (eds) (2021) Environmental and resources engineering. Springer Nature Switzerland
17. Wang LK, Wang MHS, Shammas NK, Aulenbach DB (eds) (2021) Environmental flotation engineering. Springer Nature Switzerland
18. Wang LK, Wang MHS, Hung YT, Shammas NK (eds) (2021) Integrated natural resources management. Springer Nature Switzerland

Index

A
Abrupt climate change, 423
Absconding swarm, 44, 423
Accelo-filter, 123
Acetate metabolism, 81
Acetyl-CoA, 81
Acidity, 423
Acinetobacter, 424
Activated biofilters (ABF), 114
Activated carbon, 184, 424
Activated sludge, 424
  aeration requirements, 95, 97
  biological solids, 94
  biological system, 87
  biomass, 94
  conventional (*see* Conventional activated sludge process)
  definition, 78, 79
  effluent substrate concentration *vs*. substrate utilization rate, 94
  F/M, 93
  sludge growth/substrate removal
    biological flocs, 87
    biological growth, 88
    biological reactor, 88
    complete mix model, 89, 90
    dilution rate, 88
    Michaelis-Menten enzymatic kinetics, 88
    no recycle model, 89, 91
    plug flow model, 90, 91
    sludge recycle model, 90, 91
  sludge production *vs*. retention time, 96
  sludge retention time, 92–94
  substrate removal, 95
  substrate utilization rate, 93
  substrate utilization *vs*. retention time, 95
  temperature effect, 97, 98
  variables, 87
Activated sludge process, 414, 415, 424, 425
Acute bee paralysis virus (ABPV), 19
Adaptation, 425
Adaptive capacity, 425
Adsorbed organic matter, 79
Adsorption flotation, 425
Advanced waste treatment, 425
Advisory, 425
Aerated-contact beds, 113
Aerated static pile system, 439
Aeration, 95, 97, 426
Aerobes, 426
Aerobic, 426
Aerobic bacteria, 115, 426
Aerobic denitrification, 109
Aerobic digestion, 426
Aerobic treatment, 344
Aero-filter, 123
Aerosols, 426
Afforestation, 426
Africanized honey bees (AHBs), 3
After-action findings, 426
Air, 95
Air lift, 426
Air pollution, 379
Albedo, 426
Algae, 115, 426
Aliquot, 426
Alkali bee (*Nomia melanderi*), 426

L. K. Wang, M. -H. S. Wang, Y. -T. Hung, N. K. Shammas (eds.),
Environmental and Natural Resources Engineering, Handbook of Environmental Engineering 19, https://doi.org/10.1007/978-3-030-54626-7

Alkalinity, 427
Alkylphenol ethoxylate (APE) surfactants, 182
Alkylphenols (APs), 182
Allochthonous sources, 280
Alternative energy, 427
Ammonia nitrogen, 427
Ammonia stripping, 427
Ammonification, 427
Anaerobic, 427
Anaerobic-aerobic treatment method, 350
Anaerobic bacteria, 427
Anaerobic degradation route, 348
Anaerobic digestion, 427
Anaerobic filter, 114
Anaerobic processes, 346
 acetogenesis, 348
 acidogenesis, 348
 foaming/bulking, 349
 limitations, 349
 methanogenesis, 348
 solubilization/hydrolyis, 347
 temperature, 348
 toxic materials, 349
Analysis (chemical), 428
Analytical reagent (AR), 428
Anchor booms, 428
Androgens, 172
Anionic polyelectrolyte, 428
Anionic polymer, 428
Anionic surfactant, 428
Annual plant, 44, 428
Anoxic, 428
Anthropogenic, 428
Apiary, 44, 428
Apiculture, 44, 428
Apis Iridescent virus (AIV), 19
Apis mellifera, 12, 28
Approach, 428
AquaDAF, 428
Aquatic plants, 402, 403
Aquiclude, 266
ArcView, 488
Arrhenius equation, 98
Arthropods, 397–399
Artificial wastewater treatment, 112
Atmosphere, 428
Atmospheric lifetime, 429
Attached-growth biological processes, 429
Attached-growth systems
 ABF, 114
 aerated-contact beds, 113
 anaerobic biological reaction, 111
 anaerobic filter, 114
 application, 114
 artificial wastewater treatment, 112
 biological filtration, 113
 biological slime film, 111, 112
 $C_5H_7O_2N$, 111
 $C_aH_bO_CN_dS_e$, 111
 contact beds, 112
 diffused oxygen, 112
 dissolved organic pollutants, 110
 effluent, 112
 fixed nozzle distribution system, 113
 microorganisms, 112
 $NaNO_3$, 112
 NH_3, 111
 organic pollutants, 110, 111
 plastic packings development, 113
 primary flotation clarification, 138
 rotary distributor, 113, 114
 rotating disk systems, 113
 secondary flotation clarification, 138
 trickling filters, 112
 wastewater flows, 110
Attribute, 429
Autotrophic organisms, 429
Auxiliary engine, 429

B

Bacillus, 115
Bacteria, 429
Bacterial infection, 19
Baffle, 429
Bar screens, 429
Base, alkalinity, 429
Base layer, 429
Basic plan, 430
Basin, 430
Basket centrifuge, 484
Beekeepers
 africanized bees, 33, 34
 management practices, 32
 recommendations, 31
 Russian bees, 34, 35
 VSH, 32, 33
Bees, 423, 426, 430, 432, 456, 475, 488
 agricultural economy, 3
 beekeepers, 4
 CCD, 4
 economic impact, 4
 food resource, 3
 habitat diversity, 4
 honey bees, 3, 4
 pesticides, 5

pheromones, 3
plant pollination, 3
plants, 3
queen bees, 3
selective breeding programs, 4
wild (native) bees, 3
Belt filter press, 430, 484
Best available technology (BAT), 175
Biochemical oxidation demand (BOD), 458
Biochemical oxygen demand (BOD), 62, 125, 336, 430, 431
Biochemical pesticides, 395
Biofilm model, 119, 131, 132
Biofilter, 123
Biofuels, 430
Biogas production, 349
Biogeochemical cycle, 430
Biological filtration, 113
Biological flocs, 78
Biological growth, 88
Biological magnification, 407, 408
Biological nutrient removal (BNR), 430, 447
Biological oxidation
acetate metabolism, 82
activated sludge microorganisms, 79, 80
activated sludge system, 79
energy flow, 84–86
hydrogen transfer, 82, 83
intracellular enzymatic actions, 81
physical adsorption, 79
respiration, 86, 87
substrate intake, 80, 81
synthesis, 86, 87
tricarboxylic acid cycle, 81, 82
wastewater, 77, 79
Biological phosphorus removal (BPR), 430
Biological processes, 426, 428–431, 433, 438, 442, 446, 458, 461, 464, 469, 486, 487, 492
Biological reactor, 88
Biological SBR system, 139
Biological slime film, 116
Biological solids, 86, 94
Biological waste treatment processes
biological SBR system, 139
DAF, 136
denitrification
flotation clarification, 137, 138
membrane clarification, 139
DGF, 136
primary flotation clarification, 136
SBF, 139
secondary flotation clarification, 136
single-sludge activated sludge system, 137
Biomass, 430
Biomass separation, 351
Biopesticides, 393, 395
Biosolids, 431
Biosphere, 431
Biotower, 431
Bisphenol A, 182, 183
Black carbon (BC), 431
Blank, 431
Blank sample, 431
Bleaching earth, 358
Blow-down, 431
Blowoff, 431
Boat plant, 431
Booster pump, 431
Borehole, 431
Breeding, 44, 431
Briefing, 432
Brood, 44, 432
Buffer, 432
Buffer solution, 432
Bulking, 432
Bumble bee (*Bombus occidentalis*), 432
Business recovery, 432

C
Cadmium (Cd), 432
crystalline product formation, detoxification, 313, 314
definition, 301
food chains, 301
metal stabilization effect, 314, 316
methods
incorporation, 310
materials/chemicals, 310
qualitative/quantitative XRD analysis, 310
stabilixation effect evaluation, 311
pollution/control technologies, 302–304
XRD-based qualitative/quantitative analysis, 308
XRD technique, 316
Cadmium oxide (CdO), 301
Calibrate, 432
Calibration, 432
Calibration curve, 432
Calibration standard, 432
Carbonaceous biochemical oxygen demand (CBOD), 433
Carbon capture and sequestration (CCS), 432, 433

Carbon cycle, 433
Carbon dioxide, 433
Carbon dioxide equivalent (CDE), 433
Carbon dioxide fertilization, 433
Carbon footprint, 433
Carbon monoxide (CO), 357
Carbon sequestration, 433
Carpenter bee (*Xylocopa micans*), 433
Cationic polymer, 434
Cationic surfactant, 434
Caustic refining, 331
Cavitation, 434
Centrate, 434
Centrifugal separators, 340
Centrifuge, 434
Centrifuge thickening, 434
Ceramic matrices, 434
  CdO, 306, 307
Ceramic sintering processes, 301
Characteristic wastes, 434
Chemical analysis, 435
Chemical changes, wastewater microorganisms, 83
Chemical control program, 403
Chemical engineering, 378, 379, 434
Chemical flocculation, 54
Chemical oxygen demand (COD), 62, 435, 458
Chemical pesticides
  biopesticides, 395
  control of targeted pests, 393
  in developed nations, 394
  in environmental control, 388
  federal regulations, 394
  FIFRA, 393
  as fumigant, 396
  inorganic, 395
  IPM, 389, 390
  organic, 394, 405
  pharmaceuticals, 405
  physical removal, aquatic weeds, 392
  pollution and human health risks
    biological magnification, 407, 408
    fate and transport, 407
    toxic effects, 408
    zone of influence, 406, 407
  registered products, 393
  resistance monitoring, 405
Chemical process control, 379
Chemical sludge/lime sludge, 354
Childproofing, 435
Chlorination, 435
Chlorine demand, 435
Chlorine requirement, 435
Chlorine sponge, 435
Chlorofluorocarbons (CFCs), 435, 451, 457
Chloroform, 435
Citizen science, 406
Civil engineering, 380, 435, 436
Civilization, 379
Clari-DAF, 436
Clarifier, 436
Clean sample, 436
Clean Water Act (CWA), 465
Climate, 436
Climate change, 425, 436
Climate feedback, 436
Climate lag, 436
Climate model, 436
Climate sensitivity, 436
Climate system, 437
Coagulant, 437
Coagulation, 437
  and filtration processes, 186
  and flocculation, 343, 344
Coalbed methane, 437
Coal-fired power plant, 225
Coal mine methane, 437
Cobenefit, 437
Coliform, 437
Collector, 437
Colony, 44, 437
Colony Collapse Disorder (CCD), 4, 19–21
Colorimetric measurement, 437
Comminuting, 437
Comminutor, 438
Community Emergency Response Team (CERT), 438
Complete mix activated sludge process, 438
Complete mix process, 101
Complex, 438
Compliance Safety and Health Officer (CSHO), 441
Composite, 438
Compost, 438
Composting, 438
Composting latrine, 438
Comprehensive Environmental Response, Compensation and Liability Act (CERCLA), 290
Comprehensive Preparedness Guide (CPG), 439, 446
Computer-aided design (CAD), 439
Computer programs, 381, 382
Concentration, 439
Concept of operations (CONOPS), 439
Confined space, 439

Constant-pH leaching test (CPLT), 309
Contact beds, 114
Contact stabilization process, 102, 103, 439, 440
Contamination, 440
Conventional activated sludge process
  applications, 100
  biological sludge, 99
  complete mix process, 101
  contact stabilization process, 102, 103
  design criteria, 104, 106, 107
  design parameters, 104, 105
  domestic wastewater, 99
  extended aeration process, 101, 102
  high-rate aeration, 108
  Kraus process, 104
  membrane bioreactor, 109, 110
  MLSS, 99
  modified aeration process, 105, 106
  nitrogen removal
    denitrification, 109
    nitrification, 109
  oxidation ditch, 108
  quasi-plug flow mode, 99
  sludge concentration *vs*. removable BOD, 104
  step aeration process, 100, 101
  step feed, 108
  tapered aeration, 105
Conventional biological wastewater treatment system, 440
Conventional physicochemical wastewater treatment system, 440
Cooling, 436, 441, 458, 466, 471
Cooling ponds
  costs, 199
  design
    balance of accuracy *vs*. facility, 207
    basic design equations, 201
    completely unmixed cooling pond, 209
    dikes and baffles, 210
    economics, 211
    evaporation of water, 211
    fogging, 211
    geographic locations, 210
    infiltration, 211
    land formations, 200
    local weather patterns, 200
    longitudinal mixing, 206
    net plant temperature rise, completely mixed *vs*. unmixed pond, 209
    once-through cooling ponds, 208
    pond exit temperature, 206, 208
    pond performance, 210
    recirculating cooling pond, 201–204
    recreation, 211
    seasonal weather variations, 210
  design chart, 205
  design rule, 200
  exit temperature, 204, 205
  mechanism of heat dissipation, 199, 200
  recirculating/once-through type, 198
  slow process, 199
  temperature, 198
Cooling range, 441
Cooling towers
  advantages and disadvantages, 217
  causes, poor performance
    clogged spray nozzles, 222
    correct tower, chosen, 223
    632,000 KW coal-fired electric power plant, 223
    LD (*see* Legionnaire's disease (LD))
    poor airflow, 222
    poor pump performance, 222
    scale formation, 222
  convection, 214
  counter-flow natural draft, 214
  cross-flow natural draft, 215
  current research, 225
  design requirements, 224
  dry cooling towers, 214
  forced draft, 217, 221
  hybrid-induced draft towers, 220
  hybrid types, draft cooling towers, 220
  induced draft (mechanical) cross-flow, 218
  induced draft towers, 214
  induced draft type, 220, 221
  industrial process, 211
  mechanism of heat dissipation, 212
  natural draft atmospheric, 214–216
  natural draft towers, 214
  natural draft, wet hyperbolic, 218, 219
  as natural/mechanical draft, 199
  ozone treatment, 218
  size, cooling system, 212
  in southern California, 220
Coordinate, 441
Coral bleaching, 441
Corrosivity, 441
Cross connection, 441
Crude vegetable oil, 328
Cryosphere, 441
Cuckoo bee (*Epeolus compactus*), 441
Cutter, 441
Cutter suction dredge (CSD), 461
Cytochrome, 83

D
DAF boat plane, restoration, 293
Data, 441
Database, 441
Debris flow, 441
Dechlorination, 441, 442
Deforestation, 442
Deformed wing virus (DWV), 19
Degumming, 330
Denitrification, 109, 443
Denitrification filter, 442
Deodorization, 332
Deodorizer distillate, 354
Derived-from wastes, 443
Desertification, 443
Design parameters, 104
Desulfovibrio, 115
Detention time, 443
Detoxification, 443
Detritus, 443
Di-basic acid injection system, 223
Di-(2-ethylhexyl) phthalate (DEHP), 183, 184, 189
Dichlorodiphenyltrichloroethane (DDT), 176, 177
Diethyl phthalate (DEP), 183, 189
Diethylstilbestrol (DES), 175
Diffused air, 95
Diffused-air aeration, 443
Diffused oxygen, 112
Diffuser, 443
Digester, 443
Digestive enzymes, 80, 81
Digger bee (*Ptilothrix bombiformis*), 443
Digging depth, dredging, 444
Digital Elevation Model (DEM), 444
Digitize, 444
Dioxin (2,3,7,8-tetrachlorodibenzo-p-dioxin TCDD), 181, 190
Disinfection, 397, 444
Dispersion, 444
Dissolved air flotation (DAF), 54, 68, 136, 339, 341, 441, 444
Dissolved air flotation and filtration (DAFF), 68, 441
Dissolved gas flotation (DGF), 68, 136, 444
Dissolved organic pollutants, 110
Dissolved oxygen (DO), 444
Distilled water, 444
Domestic wastewater, 360
Dredge, 445
Dredge mining, 444
Dredge pump, 445
Dredger, 445
Dredging, 445
Drift, 445
Drift eliminators, 445
Drill, 445
Drone, 44
Drone bee, 445
Dry bulb temperature, 445
Dry cooling towers, 214
Dryland farming, 445
Dual Wheel Excavator (DWE), 445

E
Earthquake, 445
Earth system, 437
Eccentricity, 445
Eckenfelder models, 129, 130
Ecology, 445
Ecosystem, 445
Educational videos, 384
Effluent, 445
Electric power plants, 198
Electrode potential, 85
Electroflotation methods, 340
Electron activity, 85
Electron transport systems, 82, 84
Elevated process water, 198
Elevated water temperature, 198
Elutriation, 445
Emergency action plan, 446
Emergency kit, 446
Emergency operations plan, 446
Emissions, 446
Emissions factor, 446
Emotional and psychological recovery, 446
Endocrine-disrupting compounds (EDCs), 446
  adverse effects, 171
  APs and APEs, 182
  chemicals, endocrine disruption, 179
  definition, 171
  fetal development and early childhood, 172
  highly chlorinated compounds
    dioxin, 181
    furan, 181
    PCBs, 180, 181
  isotherm constants, 189
  NPDWRs, 188
  pesticides
    DDT, 176, 177
    as endocrine disruptors, 176
    endosulfan, 177, 178
    MCLs, 176

methoxychlor, 178
toxic compounds, 176
plastic additives
bisphenol A, 182, 183
DEHP, 183, 184
DEP, 183
potential EDCs, 175
reproductive problems, 172
treatment processes, 188
APs and APEs, 190
DDT removal, 189
DEHP removal, 189
DEP removal, 189
dioxin removal, 190
endosulfan, 188
methoxychlor removal, 188
PCB-1221 and PCB-1232, 190
water treatments (*see* Water treatments, EDC removal)
Endocrine disruptors, 446
abnormalities, 174
blocker chemicals, 175
chemicals, 173
DES, 173
description, 171, 174
EDCs (*see* Endocrine-disrupting compounds (EDCs))
endocrine disrupting chemical, 175
environmental pollutants, 173
estrogen mimics, 172
imbalance, 174
industrial chemicals and pesticides, 173
mimic chemicals, 175
potential EDCs, 173
triggers chemicals, 175
Endocrine Disruptor Screening Program (EDSP), 172
Endocrine systems, 171, 446
biological processes, 171
hormones, 174
pituitary gland, 174
Endogenous nitrate respiration, 111
Endogenous oxygen respiration, 112
Endosulfan, 177, 178, 188
Energy, 447
Energy consumption, 335
Energy-controlled kinetic rate process, 98
Energy efficiency, 447
Energy flow, 84–86
Energy star, 447
Engineering, 447
Engineering education
environmental awareness, 378
with internet, 378
pollution, 378
Enhanced biological phosphorus removal (EBPR), 447
Enteric fermentation, 447
Environmental conservation
activated sludge effluent, 62, 63
applications, 55
features, 55
flotation/filtration cell, 55, 56
ground and fresh surface waters, 54
lagoons effluent, 67
multiple processes, 54
primary municipal effluent, 59, 60
processing, 54
raw municipal wastewater, 60, 62
RBC effluent, 63–66
recycling, 55
reuse, 55
risk-based wastewater, 54
technical feasibility, 68
trickling filter effluent, 66, 67
two-stage flotation system, 57, 58
water clarification package plant, 54
water quality, 54
water resources, 54
Environmental control
attractants, 392
chemical pesticide approaches, 388
cold, heat, dehydration and UV radiation, 393
competition, 392
construction and packaging, 391
elimination, breeding places, 391
hormonal pesticides, 392
IPM, 390
maintenance, healthy plants, 391
measures, 391
multiple crops on agricultural land, 392
natural predators, 392
nonchemical, 390
pest management, 390
physical removal, aquatic weeds, 392
preventative programs, 408
sanitation, 391
sterile mating, 392
Environmental curriculum, 380, 384
Environmental endocrine disruptors, 176
Environmental engineering, 378, 447
approval and financial recourses, 379
courses, 380
expensive equipment, 378
information retrieval, 382

Environmental engineering education
  chemical process control, 379
  freshman chemistry, 379
  laboratory courses, 380
  mass and energy balances, 379
  microbiology, 380
Environmental pollutants, 173
Environmental protection, 448
Environmental Protection Agency, 79
Environmental remediation, 447
Environmental resource management, 447
Environmental risk management, 448
Environmental science, 448
Environmental studies, 448
Environmental technology, 448
Environment impacts
  exhaust gas, 357
  solid waste, 353, 355
  VOC, 356
  wastewater sources/quantities, 335–337
  wastewater treatment, 338
  water/energy consumption, 334, 335
Epilimnion, 276
Equalization basin, 448
Equilibrium temperature, 200, 201, 203, 210
Estrogens, 172
Estuary, 448
EU Conservation Reserve Program (CRP), 37
Eutrophication, 448
Evacuation, 448
Evaporation, 198, 212, 245, 247–249, 449
Evapotranspiration, 249, 250, 449
Excessive heat, 449
Exclusionary methods, 391
Excreta, 449
Exercise, 449
Exoenzymes, 80
Extended aeration activated sludge process, 449
Extended aeration process, 101, 102

F
Factory emissions, 359
Factory wastewaters, 360
Fallow, 44, 449
Farmer
  agricultural economy, 36
  agricultural industry, 35
  agrochemical practices
    habitat conservation, 39
    land-use practices, 39
    managing pesticide drift, 39, 40
    pesticide usage, 39
    timing of application, 39
  annual cropping systems, 36
  fallow areas, 37
  farm practices, 38
  field margins, 38
  honey bees, 35
  landscapes, 36
  noncropped lands, 37, 38
  perennial cropping systems, 37
  plant selection, 38
  rable farmland, 36
  wild pollinators, 35
Fat-oil-grease (FOG), 340
F-cell, 68, 449
Federal Emergency Management Agency (FEMA), 446, 449
Federal Food, Drug and Cosmetic Act (FFDCA), 394
Federal Insecticide, Fungicide and Rodenticide Act (FIFRA), 393
Federal Water Pollution Control Act, 79
Feedback mechanisms, 449
Fermentation, 449
Fick's Law, 118
Field margins, 44, 449
Filamentous organism, 449
Filter flies, 116
Filter media, 134
  characteristics, 134, 135
Filter washing, 360
Filtration, 450
Fire-resistant material, 450
First responder, 450
Fixed-film aerobic processes, 346
Fixed nozzle distribution system, 113
Flavine adenine dinucleotide (FAD), 82
Floc, 450
Flocculation, 450
Flocculator, 450
Flood, 450
Flood routing, 257
Flotation cell, 68, 450
Flotation clarification, 138, 442
Flotation/filtration cell (FF-cell), 54, 68, 449, 451
Flotation processes, 340
  coagulation and flocculation, 343, 344
  DAF, 341
  electroflotation, 342
  IAF, 341
  vacuum, 341
Flotation tank, 436
Flotation thickening, 451
Fluidized bed incinerator, 451
Fluidized bed system, 458
Fluorinated gases, 451
Fluorocarbons, 451

Food/microorganisms (F/M), 345
Food Quality Protection Act, 173
Food-to-biomass ratio (F/M), 93
Forage, 44, 451
Force, 451
Forced draft cooling towers, 217, 221
Forcing mechanism, 451
Forests, 41
Fossil fuel, 451
Free fatty acids (FFA), 328
Freeboard, 451
Freshman chemistry, 379
Freundlich equation, 185
Fuel switching, 451
Full-scale exercise, 452
Functional annex, 452
Functional exercise, 452
Fungal-based mycopesticide, 395
Fungi, 115
Fungicides, 22, 23, 39, 44, 452
Furan, 181

G

Galler and Gotaas model, 130, 131
General circulation model (GCM), 452
Geocode, 452
Geographic Information System (GIS), 452
Geosphere, 452
Gibbs free energy, 85
Glacier, 452
Global average temperature, 452
Global Positioning System (GPS), 452
Global warming (GW), 453, 457, 465, 468, 487
Global warming potential (GWP), 433, 452
Glucose, 84
Glycogen, 453
Glycogen-accumulating organisms (GAOs), 453
Glyphosate, 23
Good laboratory practices (GLPs), 453
Grab sample, 453
Granular activated carbon (GAC), 184–186, 424, 453
Grasslands, 42
Gravity belt thickening, 453
Gravity filters, 450
Gravity grease traps, 340
Gravity thickening, 453
Greasy cooling water, 336
Greenhouse effect, 447, 453
Greenhouse gas (GHG), 454
Green solvent, 453
Grinding, 454
Grit, 454
Grit chamber, 454
Grit characterization, 414
Grit removal, 454
Grit sampling, 413
  and characterization methods, 414
Ground water, 454
  acquifer, 267
  advantage, 266
  aquiclude, 266
  monitoring, 269
  piezometric surface, 270
  Theis method, 273, 274
  water level, 268

H

Habitat fragmentation, 454
Habitat loss, 24, 25
Halocarbons, 454
Hardening, 332
Hazard, 455
Hazardous wastes, 455
Hazard-specific annex, 455
Headloss, 455
Headworks, 455
Health hazards, 397
Health risks
  assessment, 396
  and pollution (*see* Chemical pesticides)
Heat dissipation mechanism
  in cooling ponds, 199, 200
  in cooling towers, 212
Heat island, 455
Heat load, 455
Heat waves, 455
Heavy metal, 455
Herbicides, 23, 24, 45, 455
Heterotrophic, 455
High-rate aeration process, 108
High-rate filters, 120, 123
Hive, 45, 456
Homo sapiens, 232
Honey bees (*Apis mellifera*), 456
  beekeeping and agriculture industries, 6
  communication
    abscond, 14
    Africanized, 12
    aggressive defense behavior, 12
    characteristics, 12, 13
    dances, 10, 11
    European, 11, 12

Honey bees (*Apis mellifera*) (*cont.*)
heightened defensiveness, 12, 14
nesting locations, 14
pheromones, 9, 10
survival strategy, 14
swarming, 14
reproduction
individual, 8
swarming, 8, 9
social structure, 6–8
Honey crop, 45, 456
Hormonal pesticides, 392
Hormones, 456
biological function, 174
blockers, 175
chemical messengers, 171
definition, 173
endocrine system, 172
regulations, 174
Horsepower, 456
Howland models, 128, 129
Hudson River, 59
Human diseases, 398
Humanitarian engineering, 456
Humanitarian technology, 456
Hurricane, 456
Hybrid draft cooling towers, 220
Hydraulic dredging, 456
Hydraulic loadings, 117, 120, 123
Hydraulic retention time (HRT), 90, 91, 110, 345, 456, 481
Hydrochlorofluorocarbons (HCFCs), 451, 457
Hydrofluorocarbons (HFCs), 451, 457
Hydrogen ions, 83
Hydrogen sulfide gas (H_2S), 457
Hydrogen transfer, 82, 83
Hydrolases, 81
Hydrologic cycle, 457
Hydrology, 236, 238, 240, 243, 245, 447, 457
Hydrophilic groups, 81
Hydrosphere, 457
Hyperbolic cooling towers, 216
Hypochlorination, 457
Hypolimnion, 277
Hyponeuston, 235

I
Ice core, 457
Ignitability, 457
Impeller, pump, 458
Incident Command System (ICS), 458
Incineration, 458
Incinerator, 458
Independent physicochemical wastewater treatment system (IPCWWTS), 458
Indirect emissions, 459
Induced air floatation (IAF), 341
Industrial agriculture systems, 24
Industrial ecology (IE), 459
Industrial revolution, 459
Industrial wastewaters, 360
Industry, 459
Infiltration/percolation, 251
Inflow, 459
Influent, 459
Infrared radiation, 459
Infrared sludge dryer, 459
Innovative biological wastewater treatment system, 459
Innovative physicochemical flotation wastewater treatment system, 460
Inorganic chemical pesticides, 395
Insecticide, 45, 460
Integrated Pest Management (IPM)
definition, 390
environmental and chemical approaches, 389
Integrated systems approach, 417
Interactive web pages, 381
Intergovernmental Panel on Climate Change (IPCC), 436, 460
Internet, 378, 381, 382, 384, 461
Intracellular enzymatic actions, 81
Inundation, 461
In-vessel system, 439
Ion, 461
Iron (Fe), 307
isohyetal method, 243
Israeli acute paralysis virus (IAPV), 19

K
Kashmir bee virus (KBV), 19
Kraus process, 104, 461
Krofta Advanced Municipal Effluent Treatment (KAMET), 461
Krofta Engineering Corporation (KEC), 479, 487

L
Laboratory courses, 380
Ladder, 461
Lagoon, 461, 462
Lake, 462

Lake dredging, 462
Landfill, 463
Land reclamation, 463
Landslide, 463
Latitude, 463
Latrine, 463
Laundromat, 463
Layer, 463
Leaching test, 309
Leaf-cutting bee (*Megachile pascoensis*), 463
Least developed country, 463
Lecithin, 354
Legionnaire's disease (LD)
  airborne, 224
  biocide and bronopol use, 225
  causes, 224
  control measures, 224
  in environment, 225
  legionella DNA from cooling towers, 225
  outbreak, 225
  pneumonia, 224
Lenox Institute of Water Technology (LIWT), 137, 293, 479, 482, 487
Lime-softening (LS) process, 187
Line data, 463
Liquid, 9
Listed wastes, 463
Long-wave radiation, 463
Low-rate filter, 120, 122

M
Make-up water, 464
Marina dredging, 464
Masking agents, 464
Mason bee (*Osmia cobaltina*), 464
Mathematical modeling, 464
Maximum contaminant level (MCL), 176
Mechanical aeration, 464
Mechanical dredging, 464
Megacities, 464
Melliferous, 45, 464
Membrane bioreactor (MBR), 109, 110, 464
Membrane clarification, 442
Membrane filtration
  biomass separation, 351
  inorganic membrane, 351
  NF, 350
  RO, 350
  UF, 350
Membrane fouling, 352
Membrane systems/membrane bioreactors (MBR), 351
Metadata, 464
Metazoan, 79, 464
Methane (CH_4), 465
Methoxychlor, 178, 188
Metric ton, 465
Michaelis-Menten enzymatic kinetics, 88
Microbial slime systems, 115, 116
Microbiology, 380, 385
Microfiltration (MF), 350
Microorganisms, 115, 465
Microsoft PowerPoint, 382
Microsoft Word, 382
Microzone denitrification, 109
Mimic chemicals, 175
Miticide, 45, 465
Mitigation, 465
Mixed liquor (ML), 425, 465
Mixed liquor suspended solids (MLSSs), 99, 345, 425, 465
Mixed liquor volatile suspended solids (MLVSS), 465, 466
Mixed wastes, 465
Modified aeration process, 105, 106
Molecular oxygen, 82, 84
Monocultures, 24
Multiple hearth furnace incinerator, 466
Multi-stage trickling filter plant, 124, 125
Municipal solid waste (MSW), 303, 466

N
Nanofiltration (NF), 350
National Pollutant Discharge Elimination System (NPDES), 468
National Primary Drinking Water Regulations (NPDWRs), 187
National Research Council (NRC), 113, 126, 127
National Weather Service, 468
National Wildlife Refuge (NWR) System, 40
Natural biopesticides, 395
Natural draft atmospheric cooling towers, 215, 216
Natural-draft wet hyperbolic cooling towers, 218, 219
Natural gas, 466
Natural predators, 392
Natural resources, 448, 466
Natural variability, 466
Nectar flow, 45, 466
Neonicotinoids, 20
Neuston, 235
Neutralization process, 330, 331

Nicotinic acetylcholine receptors, 20
Nitrate (NO_3), 466
Nitrification, 78, 109, 110, 466, 467
Nitrifier, 467
Nitrifying bacteria, 467
Nitrite (NO_2), 467
Nitrite Lock, 467
Nitrobacter, 467
Nitrogen (N), 79, 104, 467
Nitrogen cycle, 467
Nitrogen gas (N_2), 467
Nitrogen oxides (NOx), 467
Nitrogen removal, 109
Nitrogenous biochemical oxygen demand (NBOD), 467
Nitrosomonas, 467
Nitrous oxide (N_2O), 468
NOAA Weather Radio, 468
Nocardia, 468
Noncompressible fluid, 235
Nongovernmental Organization (NGO), 466
Nonionic polymer, 468
Nonmethane volatile organic compounds (NMVOCs), 468
Nonstructural elements, 468
Nonylphenol (NP), 182
Nonylphenol ethoxylates (NPEs), 182
No recycle model, 90
Nosema infection, 22
Nuisance birds, 389
Nuisance plants and pesticide control
  aquatic plants, 402, 403
  terrestrial plants, 401, 402
Nuisances, 388, 389
Nutrient, 468
Nutrient ratio, 468
Nutrient removal, 415, 416

O
Occupational Safety and Health Administration (OSHA), 469
Ocean acidification, 468
Octylphenol, 182
Oil traps, 340
Oleophilic (oil-attractive) plates, 340
Open Courseware, 382, 383
Organic, 468
  chemical pesticides, 394
  loading, 120
  nitrogen, 468
  nutrients, 81
  phosphate, 468
  pollutants, 110
  products, 39
  solvent, 469
Ortho Imagery, 469
Orthophosphate, 469
Output rating, 469
Oxic, 469
Oxidation, 469
Oxidation ditch, 108, 469
Oxidation process, 425
Oxidation-reduction potential (ORP), 84, 85, 109, 469
Oxidize, 469
Ozonation, 469
Ozone (O_3), 470
Ozone-depleting substance (ODSs), 470
Ozone layer, 470
Ozone precursors, 470

P
Package treatment plant, 68, 470
Packing, 470
Parasites, 25, 26, 45, 470
Particulate, 471
Particulate matter (PM), 358, 431, 470
Parts per billion (ppb), 471
Parts per million by volume (ppmv), 471
Parts per trillion (ppt), 471
Pathogen, 471
Pathogenic organisms, 395, 471
  arthropods, 397–399
  biopesticides, 395
  control of insects, 400
  microbes and worms, 397
  vertebrates, 399, 400
Percolation, 471
Perennial crops, 45, 471
Perfluorocarbons (PFCs), 451, 471
Performance, 471
Perlite, 358
Permafrost, 471
Personalized books, 382
Personalized web page, 383
Pesticides, 19, 45, 393, 471
  organic, 396
  pentachlorophenol, 402
  vector arthropods, 399
  See also Chemical pesticides
Pests, 471
  animal, 404
  definition, 389
  environmental control, 390

IPM, 389, 390, 409
management, 390
microorganisms, 395
organic pesticides, 396
pest control, 388, 394, 405
pesticide-based control, 408
pesticides, 393
resistance, 405
use of poisons, 404
and vectors, 389
See also Environmental control
pH, 471
pH adjustment, 471
Phenology, 471
Pheromones, 45, 472
Phosphates, 79, 458, 472
Phosphorus (P), 472
Phosphorus-accumulating organisms (PAOs), 424, 472, 486
Photosynthesis, 472
Physical adsorption, 79
Physical recovery, 472
Physicochemical (PC) process technology, 458
Phytoplankton, 472
Pieris rapae, 28
Pit latrine, 472
Plankton, 472
Plastic packings, 113
Plastic super-rate filters, 124
Pleuston, 235
Plug flow model, 90, 92, 99
Plugging, 472
PO_4^{3-} ion, 472
Point, data, 472
Point-of-entry (POE) treatment, 187
Points-of- use (POU) treatment, 187
Pollination, 45, 472
Pollinator, 45, 472
Pollinator Health Task Force, 41
Pollinator Research Action Plan, 41
Pollinators
agriculture and human health, 15, 42
animal-assisted pollination, 5, 15
animal-pollinated crops, 17
causes, 18
CCD, 19–21
climate change, 27, 28, 43
crops, 15
disease, 26
drug companies, 15
economy, 42
effects
climate change, 28
crop improvements, 28
direct crop yield reduction, 29, 30
economic impacts, 30
human health impacts, 31
food products, 44
fruits, 16
fungicides, 22, 23
governments
agriculture industry, 40
arid and semiarid western shrub lands, 41
beekeeper, 40
forests, 41
grasslands, 42
legislation, 40
outcomes, 41
regulations, 40
riparian areas, 42
roadsides, 41
wildlife management, 40
wildlife openings, 42
habitat loss, 24, 25
herbicides, 23, 24
history, 17, 18
insecticides, 20, 22
insects, 5
management practices, 43, 44
micro-habitats, 43
nutritional value, 16
nuts, 16
parasites, 25, 26
pesticide industry, 43
quality, 16
quantitative and qualitative assessments, 15
recommendations, 43
reproductive cycle, 5
seeds/fruit, 5
statistics, 17, 18
vitamins, 16
wild bees, 6
Pollution, 378, 472
and human health risks (*see* Chemical pesticides)
prevention technologies, 406
Pollution control, 472
Polychlorinated biphenyls (PCBs), 180, 181
Polygon, 472
Polyhydroxyalkanoates (PHAs), 473
Polymer, 473
Polyphosphate (Poly-P), 473
Polyvinylidene fluoride membrane, 351
Ponding, 473
Postaeration, 473

Pour-flush latrine, 473
Powdered activated carbon (PAC), 186, 189, 190, 424, 425, 473
PowerPoint presentation, 382
Preaeration, 473
Precession, 474
Precipitate, 474
Precipitation, 474
Precision, 474
Pre-emergents, 401
Prefabricated package plants, 102
Preliminary effluent, 474
Preliminary treatment, 474
Presentation skills, 380, 381
Prevention, 474
Primary effluent, 474
Primary flotation clarification, 68, 136–138, 474
Primary flotation clarifier, 474
Primary sedimentation clarification, 475
Primary sedimentation clarifier, 475
Primary swarm, 45, 475
Primary treatment, 475
Procedure, 475
Propolis, 45, 475
Protozoa, 79, 116
Protozoan, 475
Putative endocrine, 172

Q
Quantitative X-ray diffraction (QXRD) analysis, 308
Quasi-plug flow mode, 99
Queen, 45
Queen bee, 475

R
Radiation, 476
Radiative forcing, 476
Ramming, 476
Rapid mixing, 476
Raw wastewater, 476
Reactivity, 476
Reagent, 476
Reagent blank, 476
Reagent grade, 476
Reagent solution, 476
Receiving water, 476
Recirculation, 476
Reclaimed water (reuse), 224, 225
Recovery plant, 476
Recycling, 477
Redox potential, 85
Reduction, 477
Redwood-medium trickling filter, 114
Reference intensity ratio (RIR) method, 308
Refinery, 477
Refinery soap-stock processing, 364
Refinery wash water processing, 364
Refining waste, 477
Reflectivity, 477
Reforestation, 477
Relative humidity, 477
Relative sea level rise, 477
Renewable energy, 477
Reproductive abnormalities, 172
Reservoir dredging, 477
Residence time, 477
Resilience, 477
Resource Conservation and Recovery Act (RCRA), 290
Respiration, 477
Respiration oxygen demand, 96
Retention time, 477
Return activated sludge (RAS), 425, 477, 478
Reunification, 478
Reuse, thermal discharges, 224, 225
Reverse osmosis (RO), 350
Rietveld method, 308
Riparian, 45, 478
Rippl method, 264
Rising sludge, 478
Risk management, 478
River dredging, 478
Robbing, 45, 478
Rock media filters, 124
Rotary drum thickening, 478
Rotating biological contactors (RBCs), 63–66, 478
Rotating disk systems, 113
Roughing filters, 121, 124

S
Safe Drinking Water Act (SDWA), 173, 290
Salmon, 478
Salt water intrusion, 478
Sample, 478
Sampling, 479
Sand and gravel dredging, 479
Sandfloat, 479
Sanitary landfill, 479
Sanitation, 391, 479
Saving America's Pollinators Act of 2013, 40

Scale, map, 479
Scenarios, 479
Screen, 479
Sea surface temperature, 479
Secondary effluent, 479
Secondary flotation clarification, 68, 136–138, 479
Secondary flotation clarifier, 479, 480
Secondary sedimentation clarification, 480
Secondary sedimentation clarifier, 480
Secondary treatment, 480
Second-order nonlinear ordinary differential equation, 119
Sedimentation basin, 436
Sedimentation clarification, 442
Sediments, 481
Selector, 481
Seminar, 481
Sensitivity, 481
Septic, 481
Sequencing batch exchanger (SBE), 481
Sequencing batch flotation (SBF), 139, 481
Sequencing batch reactor (SBR), 137, 139, 481
Settling tank, 436
Sewage, 482
Sewage treatment, 379
Sewage treatment plant (STP), 60
Shelter-in-place, 482
Short-circuiting, 482
Short ton, 482
Shredding, 482
Sidestreams, 482
Silicon (Si), 307
Single-sludge activated sludge system, 137, 482
Single-stage trickling filter plant, 124, 125
Sink, 483
Sintering, 483
Sintering process, 301, 304, 305
Slime layer and liquid film
  oxygen transfer, 116, 117
  substrate transfer, 117–119
Sloughing, 112
Sludge, 359, 483
Sludge age (SA), 86, 483
Sludge dewatering, 483
Sludge digestion, 483
Sludge dryer, 483
Sludge drying, 483
Sludge handling, 483
Sludge lagoons, 483
Sludge production, 102
Sludge retention time, 91–94, 96
Sludge thickening/dewatering, 434, 484, 489
Sludge volume index (SVI), 484
Slurry, 484
Snowpack, 484
Soap-stock, 354
Sodium hypochlorite, 484
Soil carbon, 484
Solar radiation, 484
Solid wastes, 333, 353, 358, 484
  bleaching earth, 358
  perlite, 358
  sludge, 359
Solidification/stabilization (S/S), 303, 485
Solids bowl centrifuge, 484
Solids retention time (SRT), 469, 485
Soluble, 485
Soluble BOD (SBOD), 485
Solution, 485
Solvent, 485
Solvent extraction process, 327
Spatial analysis, 485
Spectrophotometer, 485
Spent bleaching earths (SBE), 353
Spent nickel catalyst, 354
Spoils, 486
Spuds, 486
Standard solution, 486
State Pollutant Discharge Elimination System (SPDES) permit, 485
Steady-state equation, 119
Step aeration process, 100, 101
Step feed process, 108
Sterile mating, 392
Storm sewer, 486
Storm surge, 486
Stratosphere, 486
Streamflow, 486
Structural elements, 486
Structured Query Language (SQL), 486
Struvite, 486
Students as teachers, 379, 380
Study guide, 417
Submerged-contact aerators, 113
Subnatant, 486
Subsiding/subsidence, 486
Substrate, 486
Substrate intake, 80, 81
Substrate removal, 87, 88
Sulfate aerosols, 487
Sulfur hexafluoride (SF_6), 487
Super, 487
Supernatant, 487
Super-rate filters, 120, 124
Supracell, 487

Surface heat transfer, 201, 202, 204, 207
Surface water, 487
Surfactant, 487
Surplus honey, 45, 487
Suspended-growth activated sludge system, 137
Suspended-growth biological processes, 487
Suspended solids (SS), 340, 487
Swarm, 45, 488
Swarming, 45, 488
Synthesis, 86

T
Tabletop exercise, 488
Tailings, 488
Tank car washings, 337
Tapered aeration, 105
Technical presentation, 380
Technology, 488
Temperature effect, 97, 98
Teragram, 488
Terminal elevation, 488
Tertiary flotation clarification, 69, 136, 488
Tertiary membrane clarification, 488
Tertiary treatment, 425, 488
Testosterone
  androgen, 172
Theis method, 270
Thermal discharge, 489
Thermal expansion, 489
Thermal pollution, 198, 224, 225
Thermal stabilization, 304–306, 489
Thermohaline circulation, 489
Thiamethoxam, 20
Thickener, 489
Threat, 489
Thunderstorm, 489
Tidal rivers, 489
Tornado, 489
Total Kjeldahl nitrogen (TKN), 489
Total nitrogen (TN), 489
Total phosphorus (TP), 489
Total suspended solids (TSSs), 62, 458, 475
Toxic, 489
Toxicity, 490
Toxicity characteristic leaching procedure (TCLP), 309
Trace gas, 490
Transpiration, 249, 490
Trickling filters
  accelo-filter, 123
  aero-filter, 123
  attached-growth systems (*see* Attached-growth systems)
  construction consideration, 133, 134
  description, 119
  design, 122, 133, 134
  ecology, 115, 116
  high-rate, 120, 123
  low-rate, 120, 122
  microbiology, 115, 116
  models
    biofilm, 131, 132
    Eckenfelder, 129, 130
    Galler and Gotaas, 130, 131
    Howland, 128, 129
    NRC, 126, 127
    Upper Mississippi River-Great Lake Board, 128
    US Army Design Formulas, 132
    US EPA, 133
    Velz, 127, 128
  multi-stage plant, 124, 125
  plastic super-rate, 124
  rock media, 124
  single-stage plant, 124, 125
  super-rate, 120, 124
  variations, 121
Trickling filter system, 347
Trigger chemicals, 175
Trihalomethanes, 288
Tropical storm, 490
Troposphere, 490
Tsunami, 490
Tundra, 490
Turbid, 490
Two-stage DAF, 69, 490
Two-stage flotation system, 57, 58
Two-stage treatment, 69, 490

U
Ultrafiltration (UF), 350
Ultra-/micro-filtration membrane, 109
Ultraviolet radiation (UV), 490
Unit hydrograph, 255
United States Fish and Wildlife Service, 40
Upper Mississippi River-Great Lake Board model, 128
Urban runoff, 491
US Army Corps of Engineers (US ACE), 132
US customary units, 417
US Environmental Protection Agency (USEPA), 40, 290, 434, 484
  general EPA model, 133

V
Vacuum drying, 331
Vacuum filtration, 491
Vacuum flotation, 341
van't Hoff-Arrhenius equation, 97
Vapor, 9
Varroa destructor, 25
Varroa-sensitive hygiene (VSH) trait, 4, 32, 33
Vectors, 491
  definition, 389
  environmental control, 408
  houseflies, 399
  mosquito, 399
  pathogenic worms, 389
  pesticides, 389, 393
  and pests, 389
  rat, 399
Vegetable oil refining
  bleaching, 331, 332
  chemical, 328
  degumming, 330
  deodorization, 332
  environment (*see* Environment impacts)
  filtration, 331, 332
  flow diagram, 329
  hardening, 332
  neutralization, 330, 331
  physical, 328, 329
  processing steps, 328
  wastewater, 333
Vegetable oils, 327, 491
  process chart, 327
Velz model, 127, 128
Ventilated improved pit latrine (VIP), 491
Ventilation, filter, 134
Vitrification, 304
Volatile fatty acids (VFAs), 473, 491
Volatile organic compounds (VOCs), 327, 355
Volatile solids, 491
Volatile suspended solids (VSS), 96, 339
Volcano, 491
Vulnerability, 491

W
Warning, 491
Waste activated sludge (WAS), 425, 492
Waste disposal, 492
Waste management, 492
Waste treatment process, 378
Wastewater, 492
Wastewater biology, 412
Wastewater generation, 333
Wastewater prevention, 338
Wastewater renovation, 69, 492
Wastewater sources, 333
Wastewater treatment, 413, 415
  analysis, 360, 362, 363
  biological treatment, 344, 366
    aerobic, 344–346
    anaerobic, 346–349
    coagulation/flocculation, 352
    membrane filtration, 350, 351
    membrane fouling, 352
  caustic neutralization, 363, 364
  chemical treatment, 339, 365
  coagulation and flocculation, 343, 344
  DAF, 341
  dewatering process, 366
  electroflotation, 342
  flotation, 340
  IAF, 341
  oil traps, 340
  physical treatment, 363, 364
  physicochemical, 339
  primary, 339
  secondary, 339
  sludge removal, 366
  vacuum flotation, 341
Wastewater treatment plants (WWTPs), 416
Watch, 492
Water Environment Federation (WEF), 415–417, 419
Water pollution, 232
Water pollution control, 289–292
Water quality, 54, 492
Water quality management, 492
Water reclamation, 492
Water resource recovery facilities (WRRFs), 412, 416, 417
Water resources, 54, 492
  aquatic environment, thermal impacts, 283–285, 287
  lakes, pollution, 275–279, 281, 282
  life on earth, 234, 235
  protection, 492
  runoff
    calculations, 259, 261
    flood, 262
    flood routing, 257
    freshwater, 251
    ground water, 251
    rational method, 254, 255
    ripple method, 264
    storm flow, 256
    time factor, 253

Water resources (*cont.*)
unit hydrograph, 255
toxics, 287, 288
Water reuse, 476
Water, Sanitation and Hygiene (WASH), 491
Water treatments, EDC removal
coagulation/filtration, 186
feasibility, 184
GAC treatment, 184–186
LS process, 187
PAC treatment, 186
physicochemical processes, 184
treatment processes, 184
Water vapor, 492
Weather, 492
Web page assignment, 381
Web pages, 380–384
Weir, 493
Wet bulb temperature, 493
Wet infrastructure, 493
Wet oxidation, 493
Wet weather design, 417
Wet well, 493
Wild bees, 6
Windrow system, 439
Winterization, 358
Winter storm, 493
Worker bee, 45, 493
Workshop, 493
World Health Organization (WHO), 302
World Meteorological Organization (WMO), 436

X
X-ray diffraction (XRD), 308, 493

Z
Zoogleal mass, 493
Zoogleal slime film, 110